PRENTICE-HALL SERIES IN THE BIOLOGICAL SCIENCES
William D. McElroy and Carl P. Swanson, *editors*

2nd edition

MECHANISMS OF BODY FUNCTIONS

DEXTER M. EASTON

Florida State University
Tallahasse

Prentice-Hall, Inc., Englewood Cliffs, New Jersey

Library of Congress Cataloging in Publication Data

Easton, Dexter M
Mechanisms of body functions.

Prentice-Hall series in the biological sciences)
1. Physiology. I. Title. [DNLM: 1. Physiology.
QT104 E13m 1974]
QP34.5.E37 1974 612 73-12960
ISBN 0-13-572388-4

10 9 8 7 6 5 4 3 2 1

Printed in the United States of America

Prentice-Hall International, Inc., *London*
Prentice-Hall of Australia, Pty. Ltd., *Sydney*
Prentice-Hall of Canada, Ltd., *Toronto*
Prentice-Hall of India Private Limited, *New Delhi*
Prentice-Hall of Japan, Inc., *Tokyo*

Contents

Preface

The successive editions of a general textbook perhaps reflect the author's maturing judgment and developing insights as much as they show advances in the field. In this second edition of *Mechanisms of Body Functions*, the reader will find a greatly expanded coverage of elementary biophysical concepts, especially in the early chapters. Each section thereafter has been extensively rewritten, for clarity and completeness, with emphasis on a relatively small number of fundamental topics.

The general organization of the first edition of the book has been found to be satisfactory, and has been retained in this second edition. Therefore the following paragraphs from the original preface remain appropriate:

This book is intended as a general introduction to human physiology for the student who has had little or no training in the biological and physical sciences at the college level. Since physiology is founded on physics, chemistry, and anatomy, the elementary concepts of these sciences are discussed when they are necessary to clarify the physiological principles. Since physiology is partly a quantitative science, a few simple examples of physiological measurements are provided. In most instances, the anatomical descriptions are functional in approach and serve as a background for the discussions of physiological mechanisms. If there is a special emphasis in this book, it is on the role of the nervous system in regulating the diverse functions of the body machinery.

A large part of the text is necessarily concerned with simple word descriptions of functional aspects of human biology. The student will probably encounter some unfamiliar words, as well as familiar words used in unfamiliar ways. He should not hesitate to use a dictionary when reading these pages.

The sequence in which the physiological topics are treated is a logical one, preferred by many teachers. The general field and methods of physiological thought, particularly in terms of cells, are mentioned in the introductory chapter. Thereafter, the topics are arranged in an organ system sequence: skeletal, muscle, and nervous systems occupy the first half of the book. The autonomic nervous system provides a transition to the discussion of adjustments controlled by the hormones carried in the circulation. Then follow particular contributions made by muscle, nerve, hormones, and circulatory system to the major body functions of respiration, digestion, metabolism, excretion, and reproduction. The circle is completed in the final chapter with a discussion of the physiology of heredity, which brings us back to the cell.

The illustrations have been drawn specifically to illustrate the text material. Careful attention should be paid to them and to their legends. In some instances, these legends serve as summaries of portions of the text. The figures have been synthesized from many sources, original, published, and unpublished. The final rendering of the figures was performed by Mr. Felix Cooper. I appreciate his efforts and those of the staff of Prentice-Hall, especially Mr. Ken Cashman and Mr. Edward Lugenbeel, in bringing my thoughts into print.

Dexter M. Easton
Tallahassee, Florida

MECHANISMS
OF
BODY FUNCTIONS

1

PHYSIOLOGY, ANATOMY AND LIFE

1.1 The Meaning of Physiology and Science

What are physiology and anatomy all about?

Can you describe the form and structure of a living thing? Then you are an *anatomist*. Do you understand how it functions? Then you are a *physiologist*. Anatomy and physiology are complementary sciences. The anatomy without the function is dead. The function without the anatomy seems a misty abstraction. Physiology is a way of thinking about the world of life—in terms of functional relationships of the structures of living things.

Physiology comes from the same root as *physics*. Both sciences deal with the activities and functions of objects in the real world. Compared to physics, the subjects of concern to physiology have an extra dimension called *life*, which we can usually recognize but find hard to define. Physiologists seek to understand and describe the activities and functions of living things in relation to and in terms of the physical and chemical principles that govern the rest of the physical world.

Anatomy is the form, physiology the function

Anatomy is conveniently studied in dead, preserved specimens in which physiological processes have ceased. Anatomical description ranges from the large and easily visible down to the ultramicroscopic and molecular level. Physiological processes may be described in physical and chemical terms at each level along the way: at one level in terms of the operation of levers, the flow of liquids through tubes, and the physical laws governing the effects of stress on materials; at another level in terms of electrical charges moving in a salt solution and of chemical reactions occurring between molecules.

A knowledge of physics and chemistry may not, however, provide an immediate understanding of physiology. There are, in fact, many biological situations that present-day physics and chemistry cannot adequately explain. In some instances, the background in the physical sciences may be poorly developed; for these sciences, like the body of biological knowledge, are still evolving as knowledge accumulates and understanding grows. In other instances, we simply do not know enough yet about the structural details of the complex biological system to allow us to say how the physicochemical events may occur in the biological context. Often, however, we can unravel a morphological situation and understand its functions at least partly in terms of simple physical and chemical principles. Some of these instances are included in this book.

All the overt behavior, the movements that seem to be the obvious mark of life in every one of us, result directly from forces exerted by muscles controlled by nerves whose commands are programmed in the central nervous system (CNS). All our perceptions of the world inside and outside our bodies depend on sense organs and their nerves. Circumstances, internal and external, that derange these functions may distort or eliminate the perceptions and disorganize or paralyze the movements. If we can understand the conditions regulating reflex movements, heart rate, respiration, and other vital signs, we have the power to control life. That is an awesome responsibility, and physiology is an awesome subject.

Human physiology is part of the study of life

Life on earth depends on the sun. The green chlorophyll of plants captures the energy of sunlight and locks it into chemical compounds that provide fuel to keep the plants alive. Animals that eat the plants use the same fuel to keep themselves functioning. The explanations of how these events take place in living things—from absorption of sunlight by plants to the motions of animals—all lie within the science of physiology. Like any other animal, the human being is a kind of machine, for a machine is a " . . . device . . . which may serve to transmit and modify force or motion so as to do some desired kind of work."* Human physiology is the study of how our own body machinery functions.

How can we account for the difference between an active, vibrant, living human being and the same individual stilled in death? How can we restore vitality to limbs made motionless by illness, accident, or age? How can we prevent such deterioration? Not miracles, but physiological principles are involved in these questions.

* *Webster's New Collegiate Dictionary* (Springfield, Mass.: G. & C. Merriam Co., 1973). A dictionary is an indispensable aid to the beginner in physiology.

If a person knows only a little about physiology, he may think of it as a branch of medicine, and it is true that a consideration of physiological mechanisms underlies most of the study of the healing arts. But physiology is a *science* in its own right and in its larger meaning refers to the functional aspects of *all* living things. Indeed, the study, by experimental methods, of the physiological mechanisms in other living things provides a foundation for, and is a continuing concern of, experimental medicine.

A mechanistic description of the human body may help us to learn to use and control that body so that it will function properly. We may even gain insight about how the quality and richness of our intellectual and emotional life depend on the integrity of the body. Physiology is not merely a division of science or an intellectual discipline. To know human physiology is to understand human life.

What is science all about and where is physiology among the sciences?

All sciences are really attempts to make, in our heads, models of the real world. A model is expected to behave like the real world if it is a good model. If it is perfect, then it is a complete description, exactly like the thing itself.

The accuracy of our models of the world grows according to the level of technology that gives us tools to extend the range of sensitivity of our sense organs, through which we get the information needed to construct those models. What we really would like to do in science is to describe each particular, complicated phenomenon in terms of something simpler that we understand better. Taking this point of view, we see all sciences as one, compartmentalized for convenience of study. The *one* science can encompass the entire universe and everything in it. You may ask how a scientific view of the universe differs from any other. Scientific workers have shown the interrelatedness of events; they have often shown how each effect must have its cause, as well as the manner of the cause-and-effect relationship; and they have explained apparent exceptions to this relationship. Such activity is the essence of science.

When a system to be analyzed is simple enough, the word description of it can be translated into a mathematical formula that we can use to predict what will happen to one aspect of the phenomenon when we change another aspect. This is the goal of the *experimental method of science*.

When we try to understand an event, we imagine a cause that might explain it; that is, we construct an hypothesis. To see if our explanation is correct (i.e., to test the hypothesis), we make a change in or a change related to the particular cause and note the effect, while trying to keep all other possible causes constant. Some phenomena are as yet inaccessible to our control, like the stars in their courses, and it is then sufficient to show a consistency of the observation with an adequate model based on other events in our experience. *Verifiability*, *consistency*, *reasonableness*, these are characteristics of scientific information.

Verifiability means that a scientific "truth" or relationship, once discovered, can then be observed by anyone who will take the trouble to obtain the necessary background to understand it. But any truth in science is only an approximation appropriate to the general understanding in the field at the time when that truth is determined. *But for that moment in history it may be the best approximation possible.* Herein does scientific truth differ from some other kinds: it is not authoritarian but changes as new information becomes available. In fact, that information, together with the general statements that may be made about it and because of it, *is* science.

In a consistent world, actions have predictable consequences, and, as a matter of fact, only in such a world can we behave intelligently. The inconsistency and unpredictability we encounter are usually a result of our lack of complete information, the inadequate view, the imperfect model we have in our heads of what the world is "really like."

What is *reasonable* at one time in history may be unreasonable at another, and so certainly some things that are presented here as a reasonable picture of human physiology may, in the near future, turn out to be unreasonable. This book is intended to help you construct in your head verifiable, consistent, and reasonable models of the mechanisms of body functions. Such models will be inadequate because, first, the available physiological knowledge, although impressive, is quite meager in terms of what we may understand a few hundred years from now. Second, our communication is imperfect. What we try to write is not necessarily what you think you read. We hope, however, that the models you construct will be flexible enough to be readjusted later and made more complete as your knowledge and experience grow.

Measurement

A scale of measurement allows us to give precision to our statements and, by providing standards for comparison, can help us to understand relationships. You should be acquainted with the scientific scale for measurement of time and distance, mass and volume, shown in Table 1-1. Names are given to certain commonly used multiples of the fundamental units. With practice, you will be able to convert easily

Table 1-1 Standard dimensions (D)*

$\frac{D_{st}}{D}$	Time	Length (D)	Mass	Volume	$\frac{D}{D_{st}}$
10^{-3}		kilometer (km)			10^{3}
1	second (s)	meter (m)	gram (g)	liter (l)	1
10^{3}	millisecond (ms)	millimeter (mm)	milligram (mg)	milliliter (ml)	10^{-3}
10^{6}		micrometer (μm)			10^{-6}
10^{9}		nanometer (nm)			10^{-9}
10^{10}	–	Ångström (Å)	–	–	10^{-10}

* The reference standard dimension (D_{st}) is in each instance underlined. Column D_{st}/D shows the numbers of units of each standard contained in one unit of the reference standard. Column D/D_{st} shows the size of each standard compared to the reference standard. The multiples are consistently named, except Å is a special unit convenient for measuring atomic and molecular differences. The reader may wish to fill in the blanks.

from one expression to another. Note the convenient notation in terms of powers of 10. A large number of other dimensions may be derived by combinations of these fundamental units.

Relations of the sciences

There are many ways to describe the relationships of the sciences. A science might be considered in terms of the extent to which a complete analysis of its subject matter depends on other sciences. Then the sciences could be placed along a scale according to that dependency. In the series shown in Fig. 1-1, *physics* is most

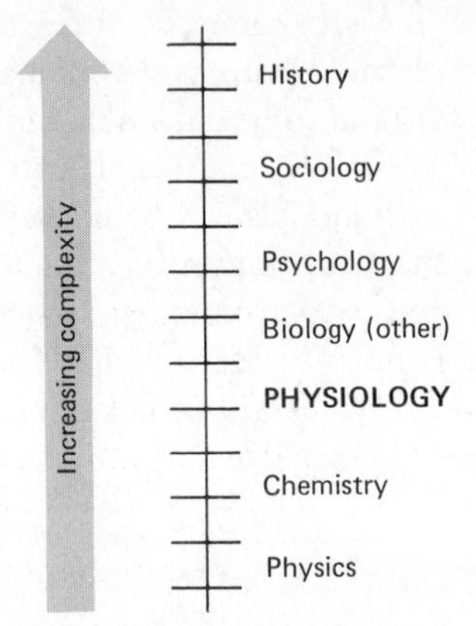

Fig. 1-1. Levels of complexity in various sciences.

independent. *Chemistry* uses principles of physics, and *physiology* makes use of both. *Biology* includes physiological ideas and provides the substantive reality in which the events of *psychology* occur. *Sociology* treats of the behavior of groups and thus includes psychology, while *history*, the interpretation of events in time, is most inclusive of all.

The arrangement of the sciences along a scale implies some sort of measurement. We are not actually in a very good position to put numbers on this scale. If you will consider how you might set up, for this list, a scale of "dependence on other sciences" or of "complexity," you will have some idea of how difficult it may sometimes be to make a measurement. Often we must be satisfied with *word* descriptions about our world.

All the sciences begin as pictures and word descriptions about the nature of things. As relationships become more clearly understood, the models become more complete, and more complete mathematical statements can be made about them. The more fundamental sciences—that is, those bodies of knowledge that deal with the simpler relationships—can be most completely described in mathematical language. The mathematical models are far less complete in the more complex sciences.

A hierarchy of complexity is perhaps clear if we place, along a scale, classes of structures that come into consideration in physiological thinking (Fig. 1–2). The

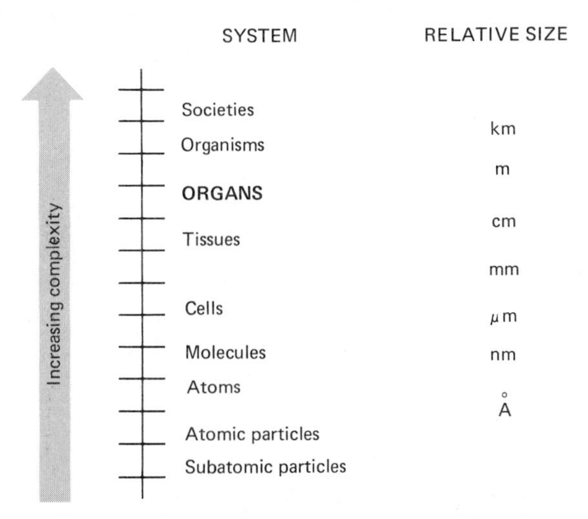

Fig. 1-2. Levels of complexity involved in physiology.

structures at each level in this list include the classes of items falling lower in the list. In a complete description of a unitary and deterministic science, the events at each level would depend on and be predicted by the events at the next lower level.

In practice, we may construct models that describe the behavior of items at each level independently of the nature of the simpler particles of which they are composed. Thus much of chemistry does not have to take into account the particles of the atomic nucleus, and the behavior of groups of individuals in societies may be described without reference to the condition of the various body tissues of each person.

This book is concerned largely with the functions of organs, but it will range up and down the scale of complexity in order to visualize organ function within the context of science in general.

1.2 Life and the Structure of Cells

The meaning of living

A description of life is a list of physical attributes. In considering something that we think of as alive, we may ask, for example, "Does it consume oxygen and produce carbon dioxide?" However, many things that are alive utilize not oxygen but some other molecule in an oxidative process. Most plants produce oxygen, although they consume it in the dark. The physician uses the heartbeat as a "vital sign" in the human. A plant has no heartbeat, although there are other rhythms of life. An iron bar, which we would not consider to be alive consumes oxygen when it rusts, and a water pump rhythmically moves its load through the plumbing. What, then, is the difference between living and not living? Between life and nonlife?

The many kinds of living things on our planet appear superficially different from one another, but they are alike in many ways. Of the more than one hundred chemical elements in the universe, less than twenty—and these the smaller atoms—make up the main substance of living things. Every living thing is remarkably similar to every other in the proportions of these elements.

Four small atoms, hydrogen (H), carbon (C), oxygen (O), and nitrogen (N), provide the structure of about 95 percent of any animal or plant. These atoms are attached together to form overwhelmingly complex molecules that are remarkable because of their organization into arrangements that can duplicate themselves. This characteristic is so outstanding that it provides a definition of living things: *Living things are chemical machines that grow, replace their own worn-out parts, and reproduce themselves.*

We would expect machines with self-replacement and reproductive abilities to be complicated, and, indeed, a multitude of chemical changes—collectively described as *metabolism*—go on in living things.

During *digestion* the food materials are broken down into their constituent molecules and are thus made available to be absorbed into the circulation, assimilated by the cells, and used by the cells in all the pathways of *metabolism* (Fig. 1-3). In *secretion* and *excretion*, we see chemical end products of metabolism. The chemical reactions of metabolism provide the material substance for *growth* and *reproduction*, plus the energy that maintains *excitability* and muscle *contractility*.

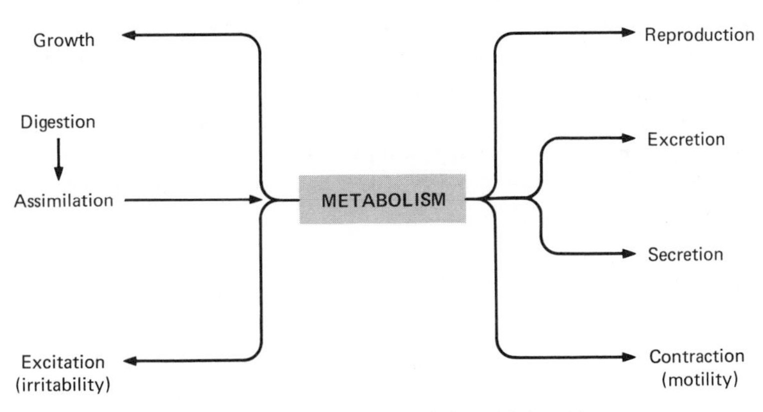

Fig. 1-3. Functional characteristics of living things.

The unit of life is the cell

The chemical machinery of any living thing is arranged in small packets called *cells*. In biology, the name *cell* was originally given to the microscopic structures seen in a slice of cork. These chambers are, however, only the dead cellulose walls that have been manufactured by the living structure of the bark of the cork tree. The term is used now to mean the smallest part into which a living thing may be divided and still show the properties of life. Some animals and plants consist of only one cell. This cell can assimilate and metabolize food; it responds to its environment and is capable of reproduction, usually in conjuction with another cell.

Although cells are generally considered the most elementary forms of life, viruses (which are smaller) are sometimes described as the simplest living things. Actually, viruses are incomplete forms of life, parasitic in cells. The virus particle isolated from a cell appears as lifeless as a crystal of table salt. A virus must be inside a cell to function. In the presence of a virus specific to it, part of the metabolic machinery of the cell is diverted to manufacture virus, rather than normal cell substance. Virus particles outside the kind of cell in which they are parasitic cannot reproduce themselves.

Cells vary in size and shape. A fertilized egg, such as that from which each of us began, is a sphere about 100 μm (0.1 mm or about 0.004 in.) in diameter, barely visible to the unaided eye. The nerve cells that signal a touch of the fingertips are fibers longer than the arm and thinner than the finest hair (0.01 mm or less in diameter).

The material and the structure of cells

PROTOPLASM AND THE MEMBRANE. The material that is the substance of the living cell is called *protoplasm*. The protoplasm is mostly water. It may be considered a very special kind of solution of organic and inorganic material. A spoonful of gelatin in hot water seems to be a very ordinary kind of solution. Allowed to cool, it becomes a semisolid mass. The organic matrix is then a continuous, loose meshwork, in the interstices of which the water molecules are held. Protoplasm is somewhat similar in its consistency but entirely different in composition. Unlike the gelatin, the protoplasm of a cell is not homogeneous but instead is organized into several different structures in each cell (Fig. 1-4).

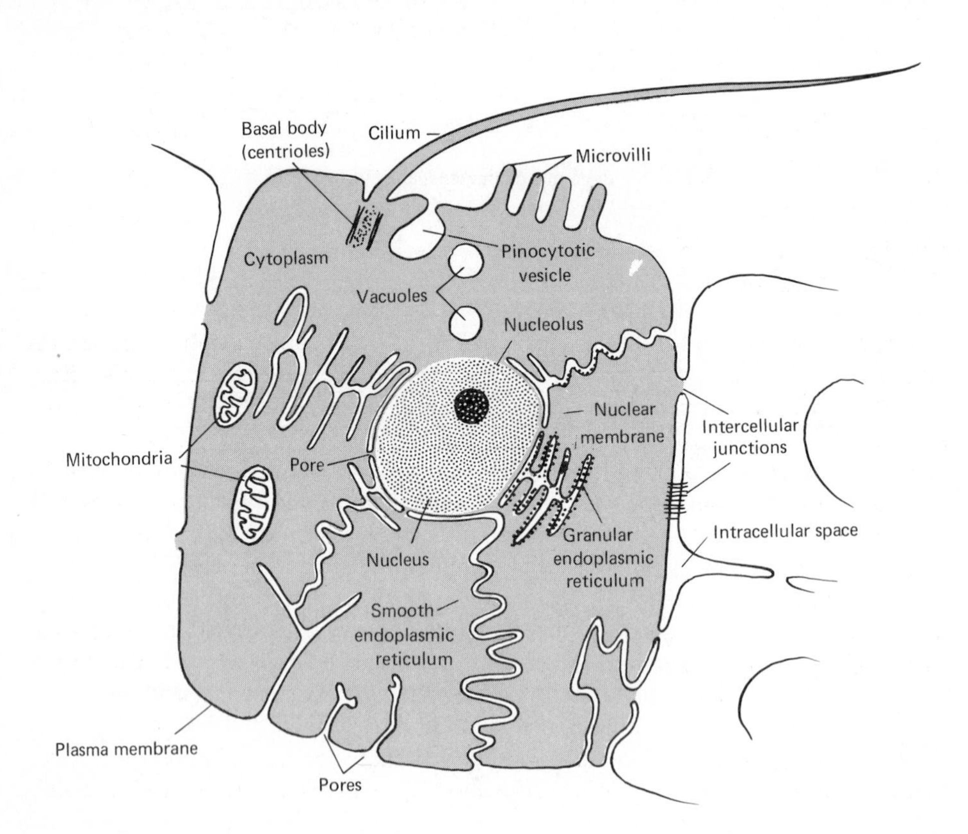

The outer boundary of a cell is called the *cell membrane*. It provides a barrier between the cell substance and the watery environment. The membrane itself is part of the living substance, but it may be surrounded by nonliving walls, such as the cellulose walls around plant cells or the material of bone that encloses bone cells.

The cells may manufacture molecules that become new cells, or carry electrical charges, or produce movement. Still other molecules manufactured by the cells reach the outside of the cell membrane as *secretions*. Some secretions are carried away in the surrounding fluid, but others remain attached to the cell membrane and form part of a matrix within which the cell lives.

The chemical reactions that go on in the cells use up some materials and produce other substances that may be waste products or special secretion products. Both the raw materials and the final products must move through the cell membrane. Almost all the different kinds of molecules in the solution normally bathing the cells penetrate into the cells to a greater or lesser extent, and only some are permitted to remain inside. This special ability to regulate the intracellular composition depends on the integrity of the cell membrane.

This thin boundary, only 200 Å thick, is a network mosaic of protein and lipid. It is a remarkable fabric, organized in such a way that some molecules smaller than a certain size can dissolve through regions of lipid, whereas other kinds will move mainly through the watery zones of the membrane. We say that the cell membrane is *differentially* or *selectively permeable* because it allows only particular substances to pass through it. Differential permeability of the cell membrane is maintained as long as a cell is alive.

CYTOPLASM AND NUCLEUS. The cytoplasm is the fluid, organic material and structure within the boundary of the cell membrane but outside the *nucleus*, which is the largest separate structure within the cell and the most important for

---

Fig. 1-4. *Organelles of cells.* This generalized cell illustrates the kinds of organelles that may be found in various cells. Any particular cell is unlikely to have all these structures at any one time or in the form shown. The external *plasma membrane* is invaginated into *pores* that are apparently continuous with the *endoplasmic reticulum*. Large particles may enter into the tubular system via the pores, or may appear in *pinocytotic vesicles* by which materials may enter or leave the cell. Some substances dissolve through the plasma membrane and enter directly into the *cytoplasm*. *Microvilli* are evaginations of the plasma membrane that increase the surface area for absorption. The membrane system encloses the *nucleoplasm* which has access to the cytoplasm directly via pores in the *nuclear membrane*. On the cytoplasmic side of the tubular system, ribosomes originating in the nucleus, give the *granular* appearance to the *endoplasmic reticulum*. *A cilium* is an enormous outpocketing of the plasma membrane, supported internally by tubular structures that arise from the *basal body* (which alternatively functions as a *centriole* in cell division). Fibrillar structures are enormously hypertrophied in muscle cells. In some cells (sensory epithelia) cilia become modified as mechano-, chemo-, and photoreceptors. The membrane system of the *mitochondria*, sites of respiratory enzymes, appears to arise separately from the membrane system of the rest of the cell. *Intercellular junctions* are of special importance in epithelia and in the synapses between nerve cells.

continued life of the cell. Within the nucleus is the *nucleoplasm*, separated from the surrounding *cytoplasm* by the nuclear membrane. The nucleus renews the chemical machinery of the cytoplasm, which can function only for a limited time if the nucleus is destroyed. If the nucleus is experimentally removed from a cell, the cell will generally live only a few days. Yet in our own bodies the red blood cells normally lose their nuclei before being released into the circulation. These cells live as long as 120 days.

The nucleus is also the part of the cell most clearly capable of self-reproduction. By manufacturing replicas of themselves, the chromosomes of the nucleus are responsible for the multiplication of cells. As cell division proceeds, differences that are controlled by the nucleus develop among the cells. The nucleus is responsible for reproduction of the species. Differences among individuals and among species of animals depend ultimately on small differences in the structure of the material of the nucleus, which consists mainly of *deoxyribonucleic acid* (DNA) and *protein*. At special times in the life of a cell, the material of the nucleus is arranged into visibly filamentous structures, the *chromosomes*, which are parceled out in a precise fashion when cell division occurs.

In the cytoplasm, a large surface area is made available by the *endoplasmic reticulum*, which in many cells seems to be a tremendously folded and crumpled membrane that may be a continuous infolding of the cell membrane. Associated with the reticulum are tiny spherical particles called *ribosomes*, which are about 100 nm in diameter. The *mitochondria* are variously shaped particles in the cytoplasm outside of the reticulum. They are 1 to 5 μm in greatest dimension. Within each mitochondrion, a much-folded internal structure provides a large surface area.

It is speculated that the intake of food materials into the cell may be along channels from the outside of the cell into the endoplasmic reticulum. Cellular chemical changes that occur in appropriate sequence for the work of the cell apparently are related to the ribosomes and mitochondria. The ribosomes are small chemical factories where molecules of protein are synthesized. A mitochondrion is another kind of chemical factory where, among other things, utilization of oxygen by the cells occurs.

The origin of cells and of ourselves

ONTOGENY AND PHYLOGENY. The secret of anything lies in its history, and the existence of any living thing makes sense only in terms of its origin and development. Consider your own beginning. You remember when you were a little child, but you cannot recall your still-earlier career as an embryo and then as a fetus absorbing nutrients from your mother's blood. You have been told, but hardly believe, that you were once a single cell, the fusion of an egg and a sperm joined by an act of union like that in virtually all living creatures. This must surely seem to be, but indeed was not, your beginning.

Each human being, in growing from a fertilized egg to a newborn child, passes through a sequence of changes that reflects the consequences of billions of years of evolutionary process. From one cell came a multicellular organism in which some cells could, in turn, become a similar multicellular organism. *Similar*, but not exactly the same. There are changes in each generation. Other forms of life are our distant relatives, descendants of ancient forebears, some of whose offspring trod a different route to the present. Follow your ancestors and theirs back to the beginning and come to a single cell or to several similar single cells.

ORIGIN OF LIFE. We find it not too difficult to trace by inference and imagination every multicellular organism back to a single cell in its life and in the history of living things. It is more difficult to describe the evolution of a living cell from nonliving molecules, but the outlines of even that story are becoming clear. In the waters and gases of the young earth, heat, pressure, ultraviolet radiation, and the forces of atomic interaction drove the elements into combinations of molecules of enormous complexity that had an identity separate from the rich salt broth that spawned them.

It seems reasonable that the hundred-odd elements of the universe have evolved from the simplest element, hydrogen. It seems equally reasonable that in the properties of some of these elements lie the potentialities for organization into the complex molecules of living things.

Survival of cells

EXTERNAL ENVIRONMENT OF CELLS. Life as we know it depends on certain special circumstances. Most important, the materials that the cellular chemical machinery uses must ultimately be available from the environment of the cells, and the conditions of the environment must permit the cellular chemical reactions to occur. The list in Fig. 1-5 shows some environmental limitations on life. Water is the dominant constituent of protoplasm, and cells must have continuous access to it. Various salts are essential to cells for several reasons: to maintain normal size, to keep the cell membrane functioning properly, and to allow the cellular enzymes to work. Salts of Na^+, K^+, Ca^{++}, and Mg^{++} are present in body fluids as chlorides, phosphates, and sulfates.

Temperatures much below the freezing point or above the boiling point of water, a range of about 100°C, are inimitable to the active continuation of life processes. In ice, at 0°C or less, molecular events are too restrained and in steam, 100°C or more, too rapid and dispersed. The limits for survival of cells of a warm-blooded animal, such as man, are even narrower. The cells of warm-blooded animals survive best at their normal temperature, around 37°C. Although some kinds of cells are able to

1. **Water.**
2. **Various mineral salts** somewhat similar to sea water in composition.
3. **Energy source.** Variable according to the type of cell. Each animal has food requirements according to its particular set of enzymes. All need C, O, H, N, P for synthesis of organic materials. Oxygen, as O_2, is generally needed by many organisms.
4. **Temperature.** The range for liquid water (about 0° to 100°C).

Fig. 1-5. Some environmental requirements for cell survival.

survive extreme pH, most cells do better in neutral or slightly alkaline solutions. Except for some special cells that utilize other molecules, most cells require molecular oxygen.

The environmental requirements necessary for cells to show many elementary functional processes are thus rather simple, but without a source of energy the calls are like machines that consume their own substance to keep themselves going. In a simple salt solution, numerous fundamental physiological processes, such as nerve conduction, muscle contraction, and secretion, may be studied in tissues isolated from the body. Solutions that more precisely duplicate the normal tissue fluids and are maintained free of bacteria and other foreign organisms may permit growth and reproduction of the cells.

ORGANIC REQUIREMENTS OF CELLS. For continued survival, cells require organic materials made of carbon, hydrogen, oxygen, and nitrogen.

Certain bacteria can manufacture their protoplasm from simple compounds, such carbon dioxide (CO_2) and ammonia (NH_3). Cells of higher organisms have lost the ability to use the simplest organic materials and instead must be supplied with more complex molecules. For example, amino acids are necessary to build protein, required for structural and enzymatic functions in the cells. The cellular biochemical machinery can transform and rearrange some organic molecules into particular forms for its needs. Molecules that cannot thus be obtained must be secured from the substance of an animal or plant. The vitamins are in this category. Finally, carbohydrates and fats must be available as sources of energy and for structural uses.*

The survival of tissues removed from the body

The ground-up residues of cells provide information on how the pieces of protoplasm might function, but properties of living cells are most suitably observed in whole cells. Knowledge of the mechanisms of cell and organ functions in the human body has been obtained only partly from direct study of the human. Most basic knowledge in physiology has first been obtained by observation and experimentation on tissues and organs of lower animals, such as monkeys, cats, dogs, and frogs, whose body functions are basically the same as man's.

Fortunately for the physiologist, an animal need not be intact for its organs and tissues to function. After the death of the whole animal, tissues and organs may survive for long periods of time if they are removed and kept in a suitable salt solution. This solution, basically similar to blood, may be quite simple. Seawater suitably diluted with distilled water will provide a fairly good environment for the temporary survival of tissues removed from an animal body. Incidentally, this observation lends support to the idea that living matter originated in ancient seas.

* The metabolic roles of carbohydrates and lipids, as well as of protein, are discussed in Chap 14.2.

The internal environment of the body

In our own bodies, the living cells are almost never in direct contact with the *external environment.* Everywhere there is a layer of dead keratinized cells or a layer of mucus serving as a protective barrier for the living cells underneath.

Our relatively impermeable skin protects us from the stresses of the environment, preventing us from shrinking in dry air or in very salt water, and from swelling if we are immersed in ordinary water.

Under this surface, all the cells (and we have about 26 billion cells at birth) are continually bathed in a salt solution that remains quite constant in composition. This solution is rather similar to but somewhat more dilute than ordinary seawater.

For a free-living cell, the environment is the entire watery world in which it lives. The cells of the body also have access to a fluid environment, but they have other cells as close neighbors. When these cells are of like kind, the entire assemblage of similar cells is called a *tissue.* The fluid in direct contact with the cells fills the *intercellular* (or *tissue*) *spaces* among the cells and is called *tissue fluid.* The volume occupied by fluid is small compared with the dimensions of the cells. The fluid, moving slowly through the intercellular spaces, forms a continuous aqueous environment for all the cells. This is the *internal environment* of the body.

The composition of the tissue fluid depends on what the cells secrete into it and what it receives from the blood. Although the blood flows in closed vessels and is thus separated from the fluid in the tissue spaces, some substances can move through the permeable walls of certain of these vessels. The internal environment, in that sense, is continuous with the bloodstream, which flows in the tiny capillaries near all the body cells. Substances in the aqueous medium move continually from blood to tissue fluid and from tissue fluid to blood. By means of this exchange, the internal environment is kept uniform and constant in its properties. Constancy of temperature, of salt concentration, of food materials, and of oxygen are all necessary if the cells are to function as they should.

Homeostasis

The condition of stability of the internal environment is called *homeostasis.* Homeostasis is maintained by a multitude of small adjustments in such variables as secretory activity, muscle contraction, and blood flow. These adjustments are made in response to any influence on the body that tends to upset the balance of body functions.

The numerous homeostatic mechanisms of the body resemble certain man-made devices. By means of a thermostat, for example, a heating or cooling device compensates for loss or gain of heat by the building in which it is installed and thus keeps the temperature of the building constant. The temperature of the body must also be kept constant in the face of great fluctuations in the outside world. In the thermostatic machinery of the body or of the building, a small departure from a prearranged level serves as a signal to return to that level.

Section 1: *Questions, Problems, and Projects*

1. Cite examples of properties possessed by both inanimate objects and by living things.
2. What organisms get along continually without O_2?
3. Name some "vital signs" of a human.
4. What is the "scientific method"?
5. Does the level of technology affect the accomplishments of scientists? Give examples.
6. How is science to be distinguished from technology?
7. Do you think there is any way Science is like Art, or scientists are like artists?
8. List distinctly "vital" characteristics of a living thing, such as (a) a bacterium, (b) a tree, (c) a cat.
9. What is the volume of a human egg cell in cubic millimeters? In cubic micrometers?
10. What useful purposes does a cell membrane serve?
11. Name some important cellular organelles and suggest their functions.
12. What is cell membrane permeability?
13. What effects do low and high temperatures have on cells (a) in a polar bear? (b) In an alga that grows in hot springs?
14. (a) An electron micrograph is described as showing an enlargement of 420,000 diameters. From that statement write a simple equation to show the relation between the length L of an image in the picture and the length l of the corresponding object on the original specimen.
    (b) If the original object is 10 Å long, what would be its length in the picture? Show the answer also in exponential notation.
    (c) There are 10^4 Å in one micrometer. How many angstroms are there in one millimeter?
    (d) Express answer of part (b) in micrometers and millimeters.
    (e) If an image in the picture is 12.6 mm long, what is the length in the original object?

2

MULTICELLULAR COMPLEXITY

2.1 Cell Division and the Development of Tissues

The rationale of multicellularity and the continuity of life

Although the functions of life can be expressed in individual cells, some organisms consist of many different groups of cells that have separate, special functions. In order to understand this situation, we must first assume a tendency of cells to stick together; then we must realize that in a mass of cells nutrients cannot be supplied nor waste products adequately removed because many cells are not directly in contact with the external environment (see Chap. 3.1). Special biological plumbing has developed to restore this contact.

The origin of an individual is usually described as resulting from the combination of an egg and a sperm. But these cells came from two separate parent individuals who had similiar origins; and the process goes back through billions of years to mammals, reptiles, amphibians, fishes, and even farther back to ancestors who were unicellular organisms. Through all of these various groups runs a thread of life that can be followed backward in time, from any one individual now alive. That thread is in the chromosomes of the nuclei of cells (see Chap. 3.2). That same thread runs in similar fashion in time, from every cell in your body back to the egg and sperm that made you an indivudual. Whatever the final differentiated form of the various specialized body cells, all the cells originated through the division of preexisting ones. The capacity for this division lies in the cell nucleus.

The nucleus of a cell contains a most remarkable machine—the deoxyribose nucleic acid (DNA) and protein—which manufactures more of itself and can direct the manufacture of more nuclear cell substance. The substance of a cell that is going to divide increases in amount until at a critical time it is parceled out so that two cells exist where one existed earlier. In order to carry out this distribution into two cells, the nucleus undergoes a particular series of changes called *mitosis*.

Multiplication of cells

CELL DIVISION AND CHROMOSOMES. When a cell divides, it is replaced by two *daughter cells*, superficially identical to one another. The process of cell division has two aspects: first, the material of the nucleus is separated into two equal portions, and second, the cytoplasm is separated into two portions—one for each mass of nuclear material.

During cell division the material of the nucleus can be observed to change from a more or less homogeneous mass to a number of separate bodies that become deeply colored with certain dyes and are therefore called *chromosomes*.

The number of chromosomes is constant for most cells of any particular species of animal or plant. In the egg and sperm (the *germ cells* or *gametes*) there are only half this characeristic number in each instance. The chromosome number in other body cells (*somatic cells*) is said to be *diploid*; in germ cells, *haploid*. In man, the diploid number (2N) is 46, the haploid (N), 23, The chromosomes in the haploid cell are generally all slightly different from one another in form, whereas in the diploid cells there are two chromosomes of each form. The members of any one pair are normally indistinguishable from one another in appearance, with the exception of *sex chromosomes*. A pair of chromosomes are *homologous chromosomes*.

During division in somatic cells, each chromosome is revealed as a double structure of two intertwined coils. The two parts separate as daughter chromosomes, and one from each of the original chromosomes is allocated to each daughter cell. This process of parcelling the chromosomes into two groups during somatic cell division is *mitosis*, a name that refers to the threadlike appearance of the chromosomes during the division process. From one cell division to the next, there is no significant decrease in the amount of nuclear material. The nucleus has the remarkable property of controlling the synthesis of more of itself. The process of chromosomal self-duplication is called *replication*. *Mitosis*, involving the allocation of chromosomes to separate new cells, is visible evidence of the replication, but the doubling in amount of chromosome material occurs in the interphase periods, between successive mitoses.

STAGES OF MITOSIS IN SOMATIC CELL DIVISION. A cell showing the earliest evidence of mitosis is said to be in *prophase*. At an early stage of prophase, each chromosome appears as two long, thin, slightly coiled threads, attached parallel to one another at one point, the *centromere* of the chromosome. This point is at a characteristic place along each particular chromosome; the length of the arms on each end helps to identify the chromosomes (Fig. 2-1). The chromosomes have doubled to become *chromatids* prior to prophase. Mitosis is a means of separating the chromatids.

During *prophase* the double chromosomes grow shorter and thicker as the coils of each chromatid pull tighter. In the cytoplasm, the *centriole*, a microscopic point of activity, appears double. In most animals, including man, the two separate centrioles move to opposite poles of the cell. Tiny filaments arise and extend in all directions as *astral rays* around each centriole. Some of the filaments pass from one centriole to the other, forming the *spindle*. Other filaments run from each centromere toward each centriole. The fibrils of the spindle and asters can be isolated after appropriate treatment to dissolve away the surrounding cytoplasm. The chromosomes are in continuous motion, seen as a twisting and writhing in time-lapse motion pictures. At the end

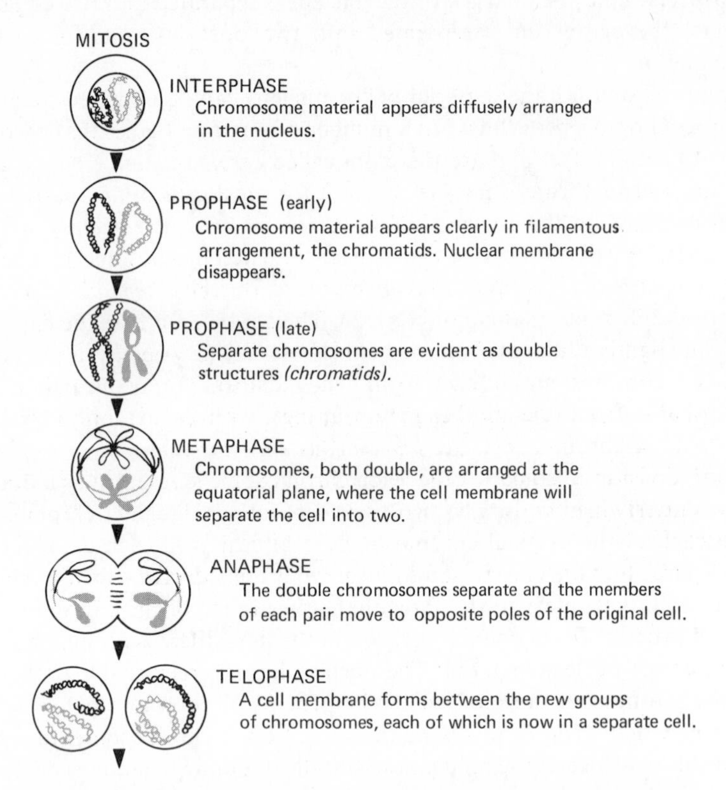

Fig. 2-1. Sequence of changes during division (mitosis) of ordinary cells. Sequential changes in the cell nucleus are shown during division of one diploid cell into two diploid cells. Prophase, metaphase, anaphase, telephase are indicated at appropriate places. Only one pair of homologous chromosomes is illustrated. Between prophase and anaphase, coils of the chromonemata are omitted. Behavior of each of the other pairs of homologues in the nucleus would be similar to that of the pair illustrated.

of prophase, the nuclear membrane has disappeared, and the chromosomes occupy the entire central region of the cell.

During *metaphase* the spindle is well organized, and the chromosomes, assembled at the midregion between the centrioles, form the *equatorial plate*. By this time fibers are seen between centromeres and centrioles. The chromosomes are radially arranged at the equatorial plate, as if repelling one another. At the end of metaphase, the centromere of each chromosome has divided into two, and the two chromatids, now called *daughter chromosomes*, move in opposite directions toward the centrioles at opposite poles of the cell.

The stage of nuclear division during which the chromosomes move away from the equator and from each other is called *anaphase*. Separation starts at the newly split centromeres, and as the daughter chromosomes move apart, the ray between centromere and the centriole shortens. Whether these rays actually pull the chromosome apart is not entirely clear. The chromosomes separate at their ends last of all as they move toward the centrioles.

When the chromosomes reach the poles of the spindle, anaphase has ended; the dividing cell is said to be in the *telophase* stage. During telophase a nuclear membrane is established around each set of daughter chromosomes, and each chromosome can be seen to consist of two coiled filaments or *chromatids*. By the end of telophase, the chromosomes are no longer visible as such, and two new nuclei have become established. At the end of telophase, the cell membrane has constricted around the fibers of the spindle that were pulled out by the separate nuclei, and the original cell is pinched into two separate cells. Having completed telophase, the nucleus of each daughter cell now enters *interphase*—the stage between divisions, when preparations are made for further mitosis; for example, synthesis of chromosomal RNA will be carried out, preparatory to another cell division cycle.

Early development

The egg from which an adult human develops is a sphere with a diameter somewhat less than the dot on the letter "i" (less than 100 μm). A sperm is less than $\frac{1}{100}$ as large, completely beyond the range of the unaided eye. Its addition to the egg produces imperceptible change in the volume of the egg. In spite of the discrepancy in size, each cell makes an equal contribution of hereditary material during fertilization. Most of the volume of the egg is food, providing the energy and raw materials for cell division.

The sperm serves two useful functions: it stimulates cell division of the egg, and it contributes a packet of hereditary material. Both the sperm and the egg have only half the number of chromosomes found in other body cells. Eggs and sperms are produced during a modified mitosis called *meiosis*, which is described in Chap. 16.2. During fertilization the normal somatic (diploid) number of chromosomes (46) is restored, and ordinary mitosis then follows.

Cell division of an egg can be stimulated to occur without sperm. Fatherless sea urchins, and even fatherless rabbits, can be produced if an egg is stimulated to divide by contact with a suitable salt solution or the point of a pin. There is no reason to suppose that the same thing could not happen to a human egg appropriately treated. In some animals, especially insects, development from unfertilized eggs (*parthenogenesis*) is a normal event. An individual developing parthenogenetically inherits all its characteristics from the mother. This does not mean, incidentally, that such an individual would be identical to its mother. Instead, it would carry a new combination of hereditary factors from the mother's parents, as will be explained in Chap. 16.2.

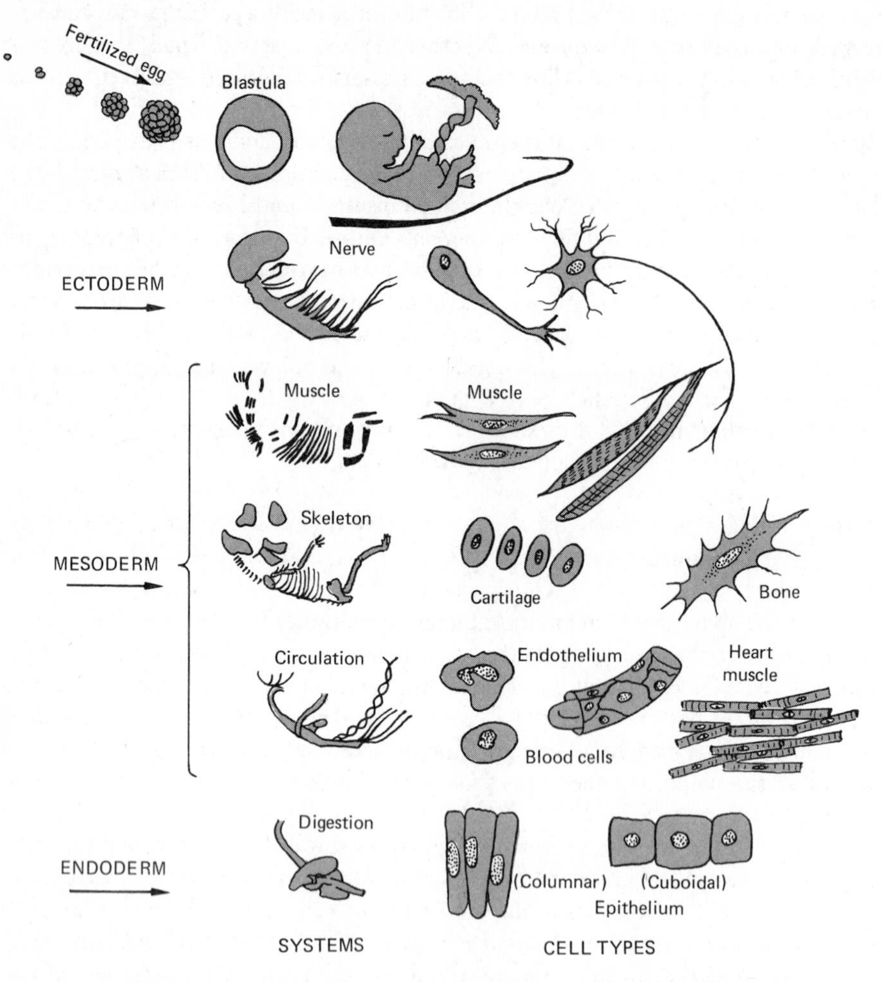

Fig. 2-2. Cells differentiated from a fertilized egg. Only some of the organ systems and cell types are shown. The digestive and respiratory systems are nonfunctional in the embryo.

The first cell division in a human egg yields two cells; division of these two cells produces four, and so on. Thus, by cell division, the number of cells multiplies (see Fig. 2-2). At first the cells become smaller during successive divisions, and the many cells form a ball no larger than the original egg. The cells become smaller as they divide, but the relative amount of cytoplasm in each cell decreases more than does the relative amount of nuclear material. Evidently the cells manufacture the substance of the nuclei out of the food materials stored in the cytoplasm. Each nucleus is apparently a replica of the original one formed when the sperm nucleus combined with the egg nucleus. In spite of this identity of the nuclei, differences that result from the successive cell divisions begin to occur in the cells.

The ball of cells is called a *blastula*, the individual cells are *blastomeres*, and the hollow center of the ball is the *blastocoele*. The blastula sinks into the endometrium, the lining of the uterus. One end of the blastula becomes the embryo; the other end develops the chorion, a membranous structure that becomes the embryonic contribution to the placenta, where the growing embryo receives nourishment from the mother.

Several other extraembryonic tissues also develop. The *yolk sac*, only rudimentary in humans, is important in reptiles and birds because it holds the large food supply on which such embryos depend for their development inside the eggshell. Similarly the *allantois* is more important in these other animals, serving as a storage place for excretory products. In mammals like ourselves, the waste products are removed by the mother's circulation. In reptiles, birds, and mammals, the *amnion* is a sac entirely enclosing the embryo, which floats in a bath of amniotic fluid.

The primary germ layers

At a rather early age in the embryogeny of any vertebrate, including humans, three fundamental cell layers can be distinguished. As cell division continues, these layers become quite extensive; they fold in upon one another, and out of them emerges the form of the developing embryo (Fig. 2-2).

Each of the tissues in the adult body can be traced back in its origin to one or another of these primary cell layers, or *germ layers*, as they are called. The *ectoderm* is the outside germ layer. From it come in succession the entire central nervous system, the peripheral nerves, sense organs, and the outer skin with all its derivatives, such as hair, teeth, and nails. The ectoderm surrounds the primitive digestive tract formed of a second germ layer, the *endoderm*. This layer is continuous with the ectoderm at each end of the digestive tract. From it arises the lining of the digestive tract, and the liver and pancreas. The lungs, although different from the digestive tract in function, develop from the endoderm. Finally, between ectoderm and endoderm, is formed the *mesoderm*, from which come all bone and all muscle, both smooth and striated. In addition, the entire circulatory system (including the vessels and the blood that flows within them), the kidneys, and the reproductive organs are of mesodermal origin. Tissues from all the primary germ layers cooperate to provide the substance of the organs of the differentiated individual.

Cellular differentiation

In single-celled animals, all the functions of life are adequately carried out in one cell. In multicellular organisms, groups of cells become specialized in their functions; that is, certain cells carry on one or more of the functions of life more effectively than others. For example, nerve cells carrying impulses specialize in irritability, but muscle cells exert force, thus providing motility. Such specialized cells are said to be *differentiated;* that, is they have become different from other cells. The fertilized egg from which we begin is *un*differentiated. During the course of cell division, groups of cells appear that are unlike their neighbors (Fig. 2-2). In general, the cells that are present early in the development of an individual are relatively undifferentiated; similarly, young cells of an individual are relatively undifferentiated compared with older cells of the same individual.

The mechanisms of differentiation are little understood. The nucleus, which controls the life of a cell, is generally assumed to be the same from one generation of cells to the next. The constituents of the cytoplasm may become altered, however. Changes in the course of cytoplasmic chemical processes that follow such alterations may give rise to what we see as functional and anatomical differentiation of the cells.

Specialization of function by cells is normally accompanied by relative loss of other functions. For example, following differentiation, many cells lose the ability to reproduce. There is no multiplication of nerve cells or muscle cells after a relatively young age, although these cells continue to grow by increasing in size.

Some types of cells continue to multiply even in the older individual. This is true of certain cells of the skin, of the reproductive system, and of cells that line the digestive tract. In continually growing tissues of this sort, the differentiated cells do not, however, reproduce. It may not, therefore, always be possible to demonstrate in any particular cell all the functions that living things are supposed to show. It may be necessary to look into the history of the cell to find a time when some of these features were present.

In many cells the ability to reproduce may appear to be lost but may actually only be latent, as in the restoration of injured bone and the regeneration of liver. Cancers are pathological examples in which the cells attain the ability to reproduce, but they do so in an unorganized and uncontrolled fashion.

Varieties of tissues

Most cells of the body remain relatively fixed in position. Tissues made up of flat sheets of cells cover the surface of the body to form the skin; they also line the inner and outer surface of such structures as the lungs and digestive tract. Such tissues are called *epithelia*. The cells of an epithelium are arranged side by side and are held together by an *intercellular cement*. They have access to the fluid environment at their free sides. Forming continuous sheets in this fashion, epithelia may control passage of materials in and out of the larger tissue spaces.

Protection at the skin surface is afforded by the rapidly proliferating layers of the *stratified squamous epithelium*, which consists of overlapping layers of flat cells (Fig. 2-2). Passage of oxygen into the blood, and of carbon dioxide out, is permitted by the thin, single-layered, *simple squamous epithelium* that lines the final subdivisions of the lungs. A similar thin layer, lining the blood vessels, permits selective passage of materials from the bloodstream into the tissue spaces, thus allowing the cells to receive nutrients. The layer of cells, in this instance being *within* the body, is called an *endothelium*, and the capillary walls, where the exchange occurs, consist exclusively of this single thin layer.

At surfaces where there is active secretion from one side to the other, tall, column-like cells form a *columnar epithelium*. Such tissue is found, for example, along the excretory tubules of the kidney, and along the digestive tract. A special *ciliated columnar epithelium* lines parts of the lungs. The whiplike *cilia* in these cells assist in moving undesirable particles out of the lungs.

Connective tissues of various kinds help hold all the other tissues together, as the name *connective* implies. There is very little extracellular material between the cells of an epithelium, but the cells of many types of connective tissue are surrounded by an obvious matrix of material secreted by the cells.

Adipose connective tissue cells store fats in their cytoplasm, but *fibrous connective tissue* cells secrete tough materials to provide tendons and ligaments. *Cartilage* cells secrete a tough, waxy material that in many instances is replaced by *bone*, a special, hard connective tissue that provides the frame for the entire body. Blood is also a connective tissue. It differs from other connective tissues in that the cells are not fixed in position but are carried about in the fluid part of the blood.

Nervous tissue is composed of *nerve cells* having long extensions that, distributed among all the tissues of the body, carry messages throughout the body. *Neuroglial cells* surround and provide them with protection, nutrition, and mechanical support.

The most obvious *muscle tissue* is *skeletal*, or striated, *muscle*, but *smooth muscle* can be discovered in all blood vessels, in the digestive tract, in the excretory, and in the reproductive system. Muscle of heart is a third type of muscle that is different from both skeletal and smooth muscle.

2.2 Organs, Systems, and the Body as a Whole

Organs and systems

In complicated animals like ourselves, *organs* are formed of various combinations and proportions of several tissues such as *muscle* for movement, *nerve* for impulse conduction, and *connective* for holding the structure together (Fig. 2-3). The brain, stomach, and heart are organs. The brain is mostly nerve tissue, but there is muscle tissue in its blood vessels. Stomach and heart are mostly muscle, supplied with a rich network of nerves. The organs, in turn, are assembled into *body systems*—each of which is especially suited to a particular function. Body systems, organs, and tissues concerned with physiological functions are listed in the accompanying figure.

The organ systems of the body are not entirely separate structures; each is functionally dependent on, and often anatomically intertwined or continuous with, other systems. The "transmission lines" of the nervous system ramify throughout all the other organ systems. The "pipelines" of the circulatory system have an equally wide distribution. In each organ system there are some large and obvious organs: the brain and spinal cord of the nervous system and the heart of the circulatory system, the bones of the skeletal system, the stomach, intestine, liver, and pancreas of the digestive system, the lungs of the respiratory system, the kidney of the excretory system, the external sex organs and the gonads of the reproduction system, the adrenal glands, the pituitary gland and the gonads of the endocrine system.

Anatomical System	Typical Organs	Physiological Function	Tissues and Cells
Skeletal	Bone	Support (cells secrete tough matrix)	Bone cells
Muscular	Muscle	Movement; apply force (contraction)	Muscle cells
Nervous	Brain Spinal cord Eye Ear	Detection and transfer of information (irritability and conductivity)	Nerve cells Neuroglia
Circulatory	Heart Blood vessels Capillaries	Movement of fluid environment	Endothelium Heart muscle Smooth muscle Connective tissue
Respiratory	Lungs	Gas exchange	Epithelium Smooth muscle
Digestive	Stomach Liver Intestine	Digestion of food (absorption, secretion)	Epithelium Smooth muscle
Excretory	Kidneys	Removal of wastes, secretion, filtration	Epithelium
Endocrine	Pituitary Thyroid Adrenal	Chemical integration of body functions Secretion	Epithelium Connective tissue
Reproductive	Testis Ovary	Reproduction Cell division	Epithelium

Fig. 2-3. Anatomical and physiological relationships.

The *skin*, the *respiratory system*, the *excretory system*, and the *digestive system* control the passage of certain substances into and out of the body. The *skeletal system* provides support; the *muscle system* moves the parts of the body. The *nervous system* carries information to and from the brain by means of nerve impulses. The *circulatory system* also carries messages in the form of chemical substances that have particular effects on the cells with which they come in contact. The *endocrine*, or *ductless gland, system* is a special source of such chemical messengers.

The separate organs of most systems of the body are easy to find. In the endocrine system each endocrine gland is a separate organ. In the circulatory system the heart is an obvious organ and is easily distinguished from the blood vessels. In the digestive system the stomach is separate from the intestine. The numerous glands and sense organs of the skin are microscopic.

The skin is an example of a system, the *integumentary* system, for it contains several different organs that contribute to the proper function of the entire layer (Fig. 2-4). At the surface of the skin is an epithelial layer, the *epidermis*. This rests on the *dermis*, a connective tissue layer through which blood vessels run. The blood vessels are lined by an *endothelium* surrounded by smooth muscle and connective

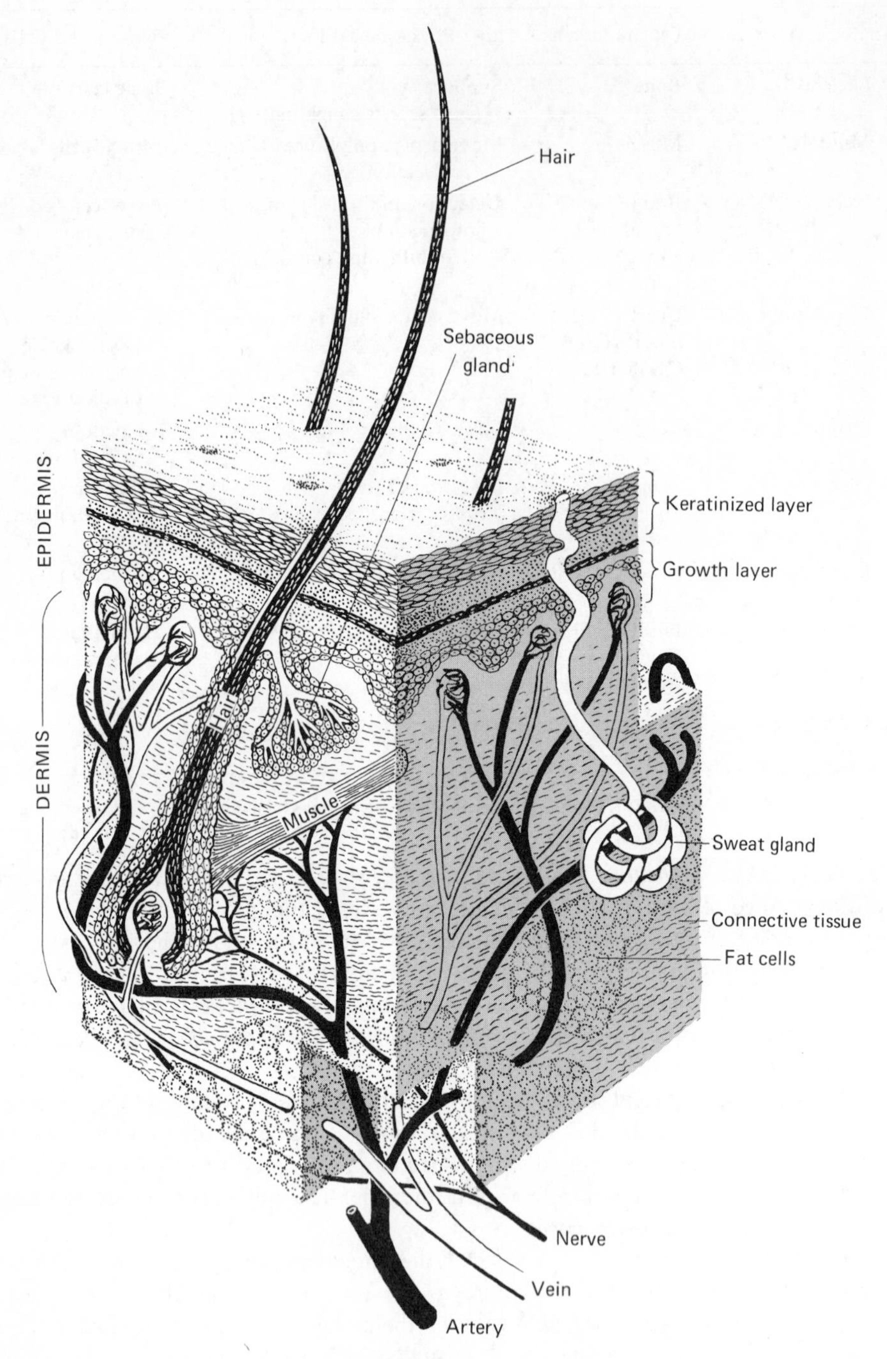

Fig. 2-4. The skin as a system.

tissue fibers. Nerve fibers run liberally throughout the skin. The receptors in the skin that permit us to sense pain, pressure, and changes in temperature are called *sensory end organs.* A sensory nerve ending enfolds the base of each hair in the skin. Other organs of the skin are the *sweat glands* that moisten and the *sebaceous* glands that lubricate the skin. The skin is a complex cover system that serves several functions. It provides a barrier to the external environment; it permits the organism to be aware of changes in that environment; and it aids in temperature regulation and in the excretion of wastes.

Finding our way in body structure

In describing the structure and functions of large animals such as ourselves, we need directional terms to help us find our way around. The locations of structures in the body are described by means of terms such as those shown in Figs. 2-5 and 2-6. The terminology used with reference to the four-legged stance of most animals may

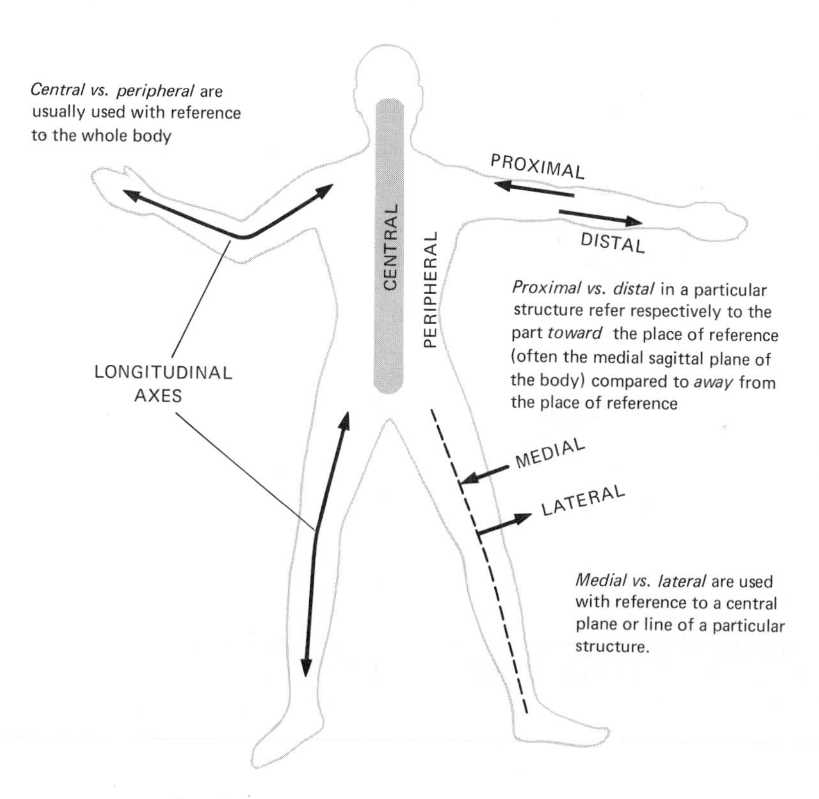

Fig. 2-5. Directional terms in anatomical descriptions.

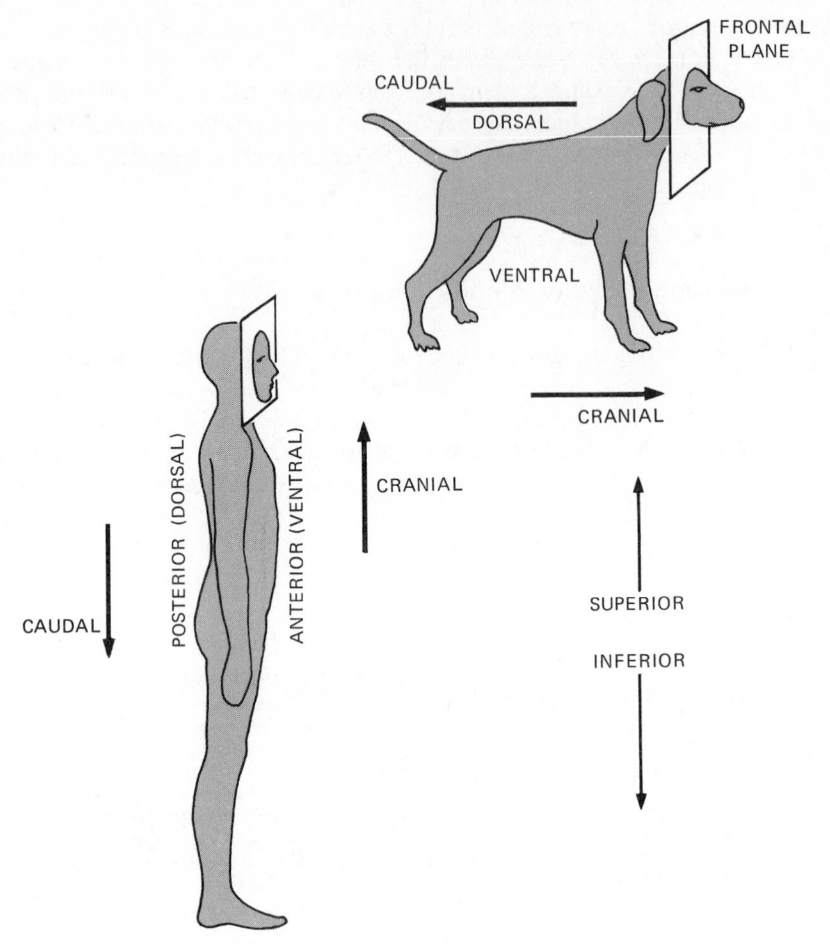

Fig. 2-6. Directional terms in bipedal and quadrupedal situations—**dorsal** ("back"), **ventral** ("stomach"), **cranial** (head end) **rostral** (snout end), **caudal** (tail end), **frontal**—have reference to particular parts of the body. **Horizontal, vertical, transverse, superior** (above) and **inferior** (below) are more general terms of orientation and position.

not always appear to be consistent with that used in human anatomy. The main problem concerns the relationship of the head to the body. The head is carried snout forward in both situations, but the long axis of the body is horizontal in the one instance and vertical in the other. The terminological consequences of these orientations are illustrated in Figs. 2.5 and 2.6 Note, for example, that a frontal plane is a vertical section in the human and a horizontal section in the quadruped.

Section 2: *Questions, Problems, and Projects*

1. What useful purpose is served by mitosis? By meiosis?
2. Predict what would happen if functional eggs and sperm were produced only by cell division, and meiosis did not occur.
3. Name the tissues you would expect to find in an organ such as a muscle, or in the stomach, or in the brain.
4. What factors limit the size of a multicellular organism?
5. What is the sequence of nuclear events in cell division called?
6. What would happen if half the liver were removed?
7. What kind of effects can one type of cell have on another type? Explain these effects.
8. What is the usual relation of numbers of one kind of cell to another kind in the body?
9. What is a cancer?
10. When does mitosis occur?
11. Does mitosis occur in all cells?
12. What determines that mitosis shall not occur?
13. What is the relation of rate of mitosis to age?
14. Name some tissues in which mitosis continues into the adult. Explain the circumstances that make this reasonable.
15. Name some tissues in which mitosis does not normally occur in the adult. Explain.
16. Does the duplication of genetic material occur before or during mitosis?
17. What is replication at nucleic acid and at chromosomal level?
18. Describe the sequence of chromosomal events that occurs during mitosis.
19. What happens to the chromosomes during each of these stages: prophase, metaphase, anaphase, telophase?
20. When are chromatids evident? When is the centriole doubled? What is a daughter chromosome? Where is the mother?
21. What causes the chromosomes to move apart during mitosis?
22. What useful purpose is served by mitosis?
23. What is fertilization?
24. Define and explain differentiation. What may be the relative role of nucleus and cytoplasm?

25. There are 6×10^{13} cells in your body. How many million cells is that?
26. In general, what sort of cells make collagen and elastic fibers?
27. Why is a red cell red?
28. What is an epithelium?
29. Name two places where you can find an epithelium.
30. A certain virus is a sphere with a diameter of 0.15 μm. What is its volume in cubic micrometers? A bacterium parasitized by that virus is the shape of a rod roughly 1 μm in diameter and 10 μm long. What is the volume of the bacterium? How many of the bacteria could be present in the volume 1 mm $\times$ 5 mm $\times$ 0.1 mm under your fingernail?

3

CHEMICAL ANATOMY OF CELLS AND THEIR ENVIRONMENT

3.1 The Watery Matrix of Cells

There is no more important compound than water, H_2O, for the survival and growth of cells. About three-fourths of the body weight is H_2O, distributed in cells and in the fluid environment surrounding the cells.

Most of the water in the body is inside of cells, but one-fifth or more is extracellular—that is, outside the cells—ultimately all continuous, but to a considerable extent compartmentalized as various separate body fluids. Each compartment has a combination of constituents not quite the same as any other; each has a different range over which its constituents may change in the course of normal physiological processes; each has a different special role in the biological economy of the body.

Some of the body fluids must be kept rather constant in composition because they provide the *internal milieu* or internal environment of the body in which the cells are bathed—plasma within the blood vessels, lymph derived from intercellular spaces, cerebrospinal fluid in the nervous system are all in this category. Other fluids are delivered to the outside of the body and may vary greatly in composition, depending on the activity of different parts of the body. These fluids include sweat, saliva, gastrointestinal juices, and urine, which is temporarily stored, then excreted from the body.

The body fluids consist mainly of water in which varying amounts of inorganic salts, as well as organic substances, are dissolved. The inorganic salts give rise to various ions, such as Na^+, Ca^{++}, Mg^{++}, together with their negatively charged counterparts, such as Cl^-, $SO_4^=$, $PO_4^{\equiv}$. The organic substances are sugars, amino acids, protein, vitamins, and so on. All the body fluids arise from filtration of fluid through cellular layers or are secretory in origin, in the sense that, to a greater or lesser degree, they are the result of metabolic processes of cells. We will consider here chiefly such features as are common to all the fluids. Special functions and properties will be considered at a later time in relation to particular system when the specific fluids are of special interest. Here we will be concerned with some properties of solutions that are of general interest.

Movement of particles in solution: Diffusion

The particles of which any substance is composed continually and randomly move with respect to one another. The distances traversed in a solid are small even by molecular measurements; and each individual molecule strikes another so soon when it moves that it tends to remain more or less in its own place. The molecules of a gas move farthest and fastest before they collide. Particles in solution also move about randomly. Each particle follows a straight line until it hits another particle.

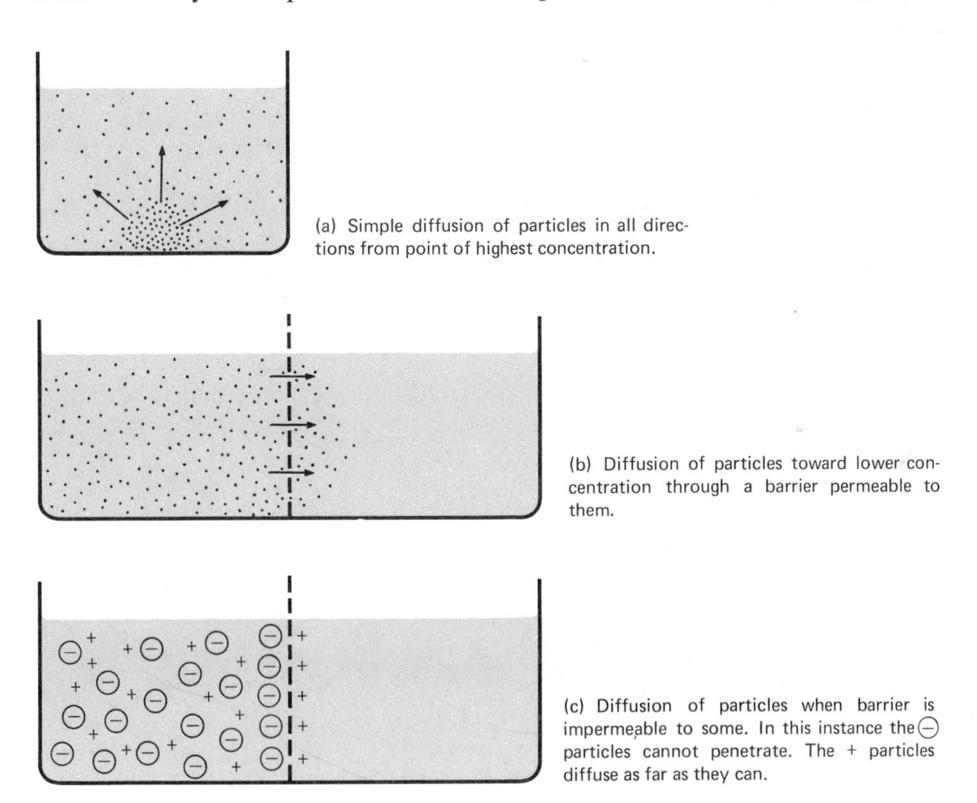

(a) Simple diffusion of particles in all directions from point of highest concentration.

(b) Diffusion of particles toward lower concentration through a barrier permeable to them.

(c) Diffusion of particles when barrier is impermeable to some. In this instance the ⊖ particles cannot penetrate. The + particles diffuse as far as they can.

Fig. 3-1. Diffusion.

Upon collision, both particles bounce away in different directions, only to hit other particles and rebound again. In this way, you can see that a particle at one side of a vessel might eventually reach the other side. As the molecules of a lump of sugar go into solution in a beaker of water, they will hit and be hit by other sugar molecules, and especially by other water molecules. They take a new course each time until they have moved throughout the volume of solvent available to them. This process is called *diffusion* (Fig. 3-1). It is mainly by diffusion that dissolved substances—gases, salts, food substances—move into cells and waste products of metabolism move out. Diffusion is in the direction *from* the region of higher concentration of the substance *toward* the region of lower concentration. Restricted diffusion through membranes underlies osmotic pressure and controls cell volume. When the particles have an electric charge, the restricted diffusion may generate an electric potential across the membrane.

Cell membrane permeability and osmotic pressure

DIFFUSION AND MEMBRANE PERMEABILITY. Diffusion can build up pressure if it occurs in relation to a differentially permeable membrane. *Permeability* refers to the extent to which a membrane allows particles to pass through it. A cell membrane allows some substances to diffuse into the cell and allows others to diffuse out. The cell membrane is *permeable* to those substances that pass through it. Suppose that a cell membrane were effectively *impermeable* to all the normal constituents of its environment *except* water molecules. Such a membrane, mostly imaginary, is called a *semipermeable membrane* in the physicochemical sense. Only water, and no other substances, will pass through its holes. Although all kinds of substances *can* pass through most cell membranes, the membranes behave to a limited extent *as if* they were nearly semipermeable.

The principal constituents of the *extracellular fluid* are sodium (Na^+) and chloride (Cl^-), and both ions are excluded from the interior of cells. The seeming impermeability of most cells to Na^+ ions is a result of the "sodium pump," metabolic machinery that transports the Na^+ ions outward after they have entered the cell. Some substances not excluded by a pump mechanism may enter so slowly that the cell appears to be impermeable to them over the short term.

ORIGIN OF OSMOTIC PRESSURE. We may effectively prevent the diffusion of the *solute* (the material in solution) through the membrane and allow *water*, the *solvent*, to move. Like the solute, the water tends to diffuse in the direction determined by its concentration difference. The average concentration of water molecules in pure water is higher than the average concentration of water in a solution diluted with a solute. Therefore there will tend to be a net movement of the water *through* the semipermeable membrane *from* the region where the concentration of *water is high to* the region where the concentration of *water is low. Where the water concentration is high is a dilute solution so far as the solute is concerned. Where the water concentration is low is a concentrated solution as far as the solute is concerned.* Thus the water moves from the dilute solution to the more concentrated solution. Such a movement of water through a semipermeable membrane is called *osmosis.* If the concentration of solute is the same on both sides of the membrane, we expect no net movement. The driving force for the osmotic flow of the water is *diffusion*, acting through the concentration difference, across the membrane, and it is called *osmotic pressure*. We can think of the water being pulled toward the higher concentration of solute (the solution of higher concentration), which is therefore said to have the higher osmotic ("pulling") pressure. If any cells of the body are placed in ordinary water, they will swell by taking up water. The contents of the cells thus appear to have a higher osmotic pressure than the water, and the cell membrane seems to act like a semipermeable membrane in the osmotic sense. Actually, cell membranes allow the passage of many materials in addition to water, but metabolism, acting as a pump, effectively keeps some substances on one side or the other of the cell membrane.

OSMOTICITY AND TONICITY. Figure 3-2 shows what happens to a cell placed in a solution that is more concentrated (a) or more dilute (c) than the solution in which the cell is of normal volume (b). The changes in volume would be expected if

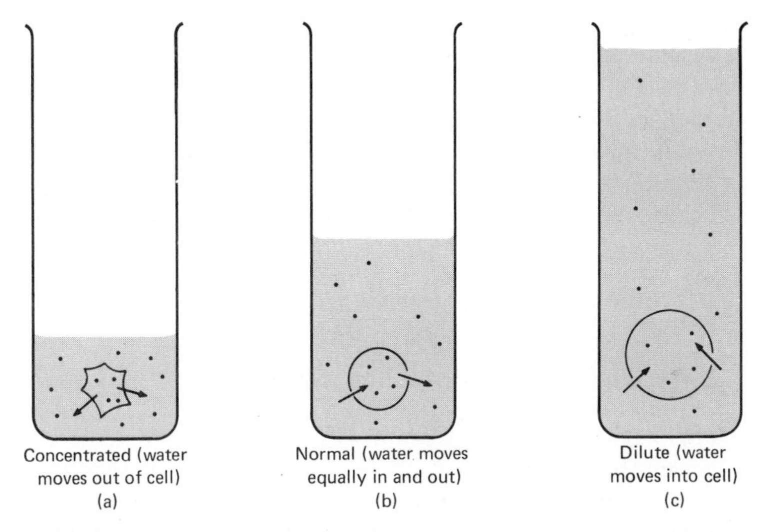

Fig. 3-2. Cell volume affected by the osmotic pressure of the surrounding solution. (a) Water has been removed from the plasma which has therefore become a more concentrated solution of salts, higher in osmotic pressure than is normal blood. Water moves out of the cell. (b) The cell is in normal blood plasma. The osmotic pressure of the external and internal solutions are equal; water moves equally into and out of the cell, and the cell volume remains constant. (c) The plasma has been diluted. The cell interior now has an osmotic pressure that is relatively higher than that of the plasma, and water moves inward, causing the cell to swell.

the cell membrane is impermeable to the solute. If the cell has a semipermeable membrane, it will keep its normal volume in an *isosmotic* (normal) solution; it will swell in a *hypo-osmotic* (diluted) solution; and it will shrink in a *hyperosmotic* (concentrated) solution.

If the cell membrane is permeable to a particular solute molecule, the cell will behave as if that molecule were water. For example, most mammalian cells will remain unchanged in volume if immersed in 0.154 M NaCl but will swell in the 0.12 M KCl, even though these two solutions are isosmotic when considered in terms of an ideal semipermeable membrane. (M means moles per liter; see pp. 43 and 49.)

In order to account for these differences, which are so important biologically, the term *tonicity* is used to describe the apparent "biological osmotic pressure." A solution in which the cell keeps its normal volume is *isotonic* to the normal solution and to the internal cell contents. A solution in which the cell swells must be relatively dilute and is said to be *hypotonic*. One in which it shrinks is a *hypertonic* solution.

The NaCl and KCl solutions mentioned above are isosmotic in the physicochemical sense, but they are not isotonic. The KCl solution is hypotonic to the cells because the KCl easily penetrates the cell membrane, along with the water. Taking advantage of these ideas, we may determine whether a particular substance in solution penetrates a cell membrane by noting whether the cell swells when the substance in question is substituted for normal constituents of the extracellular fluid. If a substance penetrates a cell membrane, it exerts no osmotic pressure so far as the cell is concerned. Common salt (NaCl) and sucrose sugar behave as if they do *not* penetrate the membrane, and they *do* exert osmotic pressure, directly proportional to their concentration. However, if equal concentrations of both substances are used, the salt will appear to have an osmotic pressure twice that of the sugar. This interesting result is due to the fact that each "molecule" of salt NaCl acts like two particles in solution, the negative and positive ions, Na^+ and Cl^-. Therefore, for equal molar concentrations of sugar and salt, the total concentration of particles in the salt is twice as great.

Electrical properties of water solutions

POLAR SOLUTIONS. We have emphasized several times that the cell membrane is a semipermeable barrier. Suppose that it selectively obstructs the passage of particles according to the electric charges on the particles. If it allows only positive electric charges to go through, then these charges will tend to accumulate on one side of the membrane, leaving the negative electric charges on the other side (Fig. 3-1).

Electric forces are of fundamental and general importance for our understanding of physiological processes, as well as of the underlying chemical processes. If we consider these forces at several levels, we shall better understand the physiological events in nerves and in the brain, in muscles and in the heart, in the stomach and the kidney, and we will more easily comprehend the metabolic processes involving molecular breakdown and rearrangement in all cells.

One remarkable feature about water is that an enormous number of substances will dissolve in it. We can explain this solubilizing action by the property of *electric polarity* that is possessed by both water and its solutes. The ions of NaCl, for example, bear opposite *electric charges*, and if we look at the *ion pair*, we see that one end of it is a positively charged *pole* and the other end or pole is negatively charged. Water is also polar, as the drawing of the structure of a water molecule shows (Fig. 3-3). The

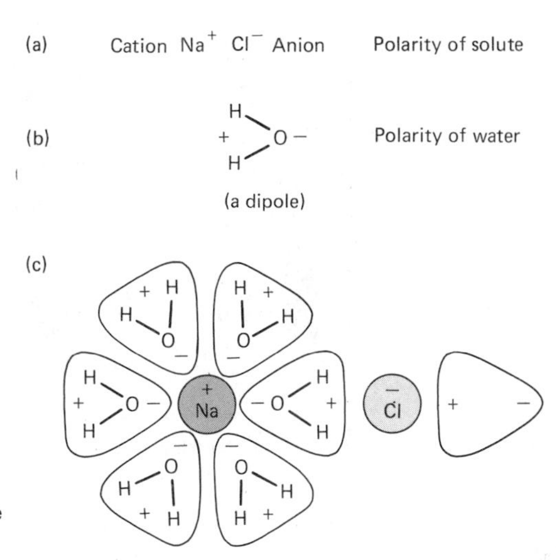

Fig. 3-3. How water molecules become oriented around solute ions.

oxygen (O) is relatively negative; the hydrogen (H) is relatively positive. The water molecules tend to interact with the salt ions, and they mix easily with the ions because the charges of the *solute* tend to orient toward the opposite charges of the *solvent*.

The water molecule is unlike the NaCl in that the H_2O tends to hang together as a unit, a condition that is described by saying that the water molecule is *dissociated* or pulled apart only to a small extent. The NaCl, on the other hand, when present in water solution is very much dissociated.

ELECTRIC CURRENT: A DIRECTED MOVEMENT OF CHARGES. The electrically charged particles in a polar solution such as NaCl in water will move if an *electric field* is applied to the solution. By means of an electrical cell (several together in series are called a *battery*), negative charges may be held separated from the positive charges (Fig. 3-4).

The anode ("an" = plus) is the positively charged pole of the electrical cell or battery; the cathode ("cath" = down) is the negatively charged pole. Each pole will attract charges of the sign opposite to it. The ⊕ ions of the solution will tend to move toward the cathode and are therefore called *cat*ions, whereas the ⊖ ions move toward the anode and are called *an*ions. Na^+ is a CAT*ion*; Cl^- is an AN*ion*. H_2O is a polar substance, but since its constituents tend to hang together more firmly than the NaCl, more work is required to pull them apart. The H^+ will tend to move toward the cathode and the O^- will tend to move toward the anode. If sufficient electric work

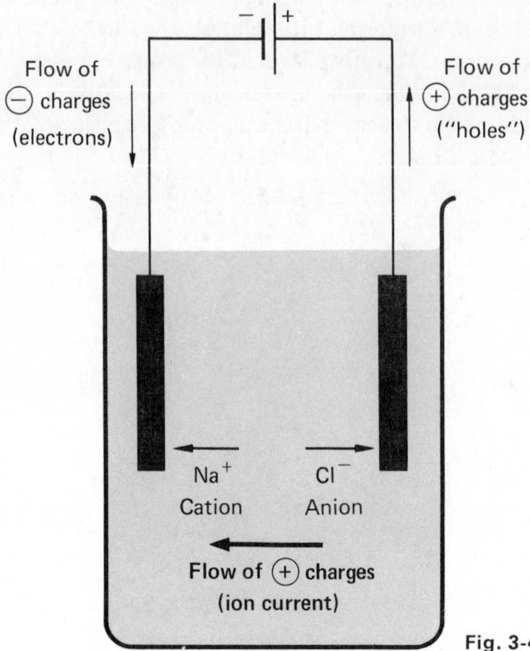

Fig. 3-4. Action of an electric cell.

can be done—that is, if the voltage between the poles of the electric cell is large enough—then water molecules may be pulled apart. There will then occur an immediate transfer of electrons from the O and to the H, so that molecular oxygen and hydrogen gas are produced ($2O^- - 2e \longrightarrow O_2$ and $2H^+ + 2e \longrightarrow H_2$).

That is why a water salt solution is called an *electrolyte*, a word that means to split or cleave by electricity. As we saw, an electric potential can split the water into its component parts.

The movement of the ions through the solution is called an *electric current*. The ions continue to move only as long as there is a driving force—that is, there is an electric charge that is not "satisfied" or balanced by an opposite charge. We may think of the electric cell as having a surplus of positive ⊕ charges at the anode and a surplus of negative ⊖ charges at the cathode. These charges can be delivered to the solution by way of a wire, the *electrode*, in contact with the solution. The whole system is in balance, and a ⊖ charge can be made available at the cathode only if an opposite ⊕ charge is able to be freed at the anode. The charged particles that flow in the wire are the electrons that carry a ⊖ charge (Fig. 3-4).

At the cathode, the electrons flow *from the electric cell to the solution*, but at the anode, the electrons flow *from the solution to the electric cell.* Just as the cations tend to neutralize or cancel out or pair up with the negative charges at the cathode, so we may regard the anions as neutralizing, or pairing up with a corresponding number of *holes* from which electrons are absent, at the anode. The *holes* are ⊕ charged entities.

There is, in effect, a "flow of holes" along the anode from the electric cell to the solution but in the opposite direction compared to the flow of electrons. The direction of "flow of holes" or ⊕ charges through the wire and of movement of cations through the solution is the same. For convenience, this direction is conventionally defined as the direction of flow of electric current.

For convenience and consistency in analysis, *the direction of flow of electric current is arbitrarily considered to be the direction in which the ⊕ charges move.*

The reader may well question the relevance to physiology of a discussion of electricity, which may be more suitably related to the automobile ignition system or to house lights. We can justify our concern when we realize that electrical (or, more accurately, electromagnetic) forces, along with gravity, run the universe, and living things are no exception. In respect to the fluid compartments of the body, the electric charges help determine the direction of movement of constituents of the compartments.

Nonelectrolyte solutions

POLAR NONELECTROLYTES. A substance such as glucose (sugar) is as easily dissolved in water as is table salt, but it is not an electrolyte. The structure of the molecule shows why it dissolves easily in water, nonetheless (Fig. 3-5). Notice that it

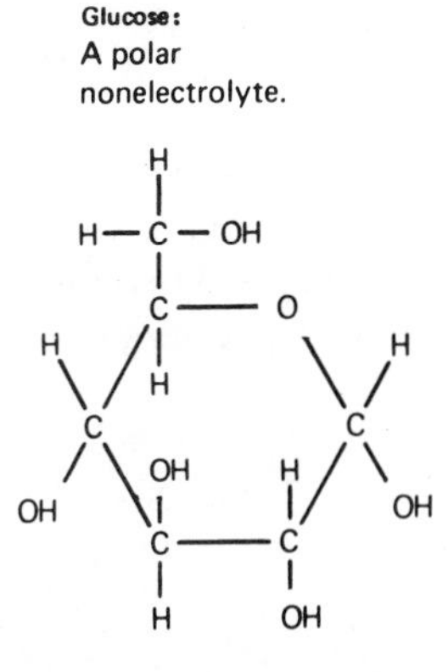

Glucose:
A polar
nonelectrolyte.

Fig. 3-5. Nonpolar compounds—relatively insoluble in water.

fairly bristles with O atoms (specifically, OH structures). These atoms, like the O in H_2O, tend to be relatively negative regions. The relatively positive H atoms tend to be attracted to the relatively negative O ends of water molecules. Many water molecules can, therefore, closely approach a glucose molecule, which thus becomes hydrated, or *solvated*, and goes into solution. However, the glucose molecule does not become ionized and dissociated into oppositely charged portions, so it does not move in an electric field in the manner of Na^+ and Cl^-. It is a *nonelectrolyte*.

NONPOLAR COMPOUNDS IN WATER. The *nonpolar* compounds constitute another large class of substances. Such compounds do not show a polar structure; they cannot effectively interact with water; and they are therefore rather insoluble in water. Many oily and fatty substances are of this category. Straight chain and ring hydrocarbons, such as those in gasoline and similar products, are particularly nonpolar and insoluble in water (Fig. 3-6). Similar compounds having OH rather than

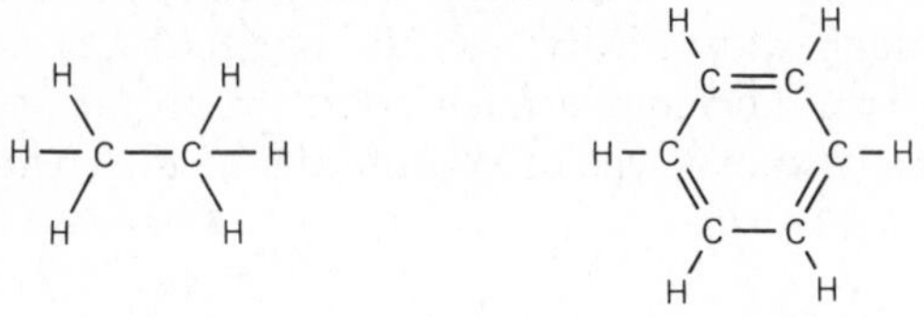

Ethane

Benzene

Figure 3-6. Straight chain and ring hydrocarbons.

H are polar compounds and soluble in water (Fig. 3-7). The difference that results in water solubility is the relative tendency of the molecule to have an electrically polar structure. In the ethane molecule, the hydrogen atoms are held closely to the carbon chain; whereas in the ethanol molecule, the OH sticks out at one end with its H hanging loose, ready to be pulled toward the relatively negative oxygens of water. When such a molecule goes into solution, it is somewhat under protest, for the hydrocarbon end does not "feel comfortable" in the water. As you might expect, such substances will tend to accumulate at the surface, with the polar end sticking in the water, the nonpolar end away from the water. On the other hand, the nonpolar end will tend to mix completely (dissolve) in nonpolar solvents. The tendency of the nonpolar molecules to keep to themselves is reflected in the fact that ethane is a gas at ordinary temperature, whereas ethanol is a liquid. In ethanol the polar ends tend to interact with one another and form aggregates that are held together and do not so easily fly off separately after the manner of particles of a gas.

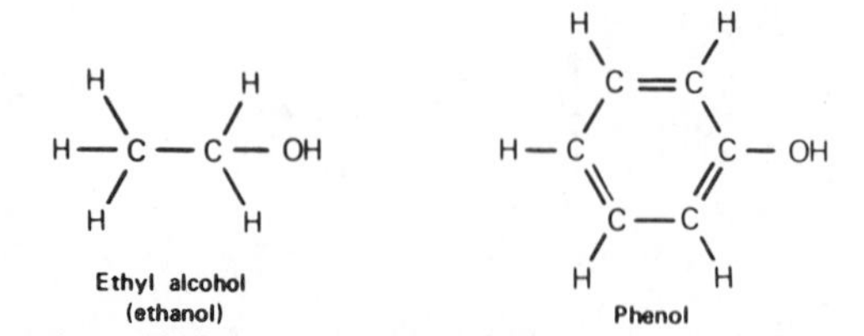

Fig. 3-7. Water-soluble polar derivatives of molecules shown in Fig. 3-6.

The proton: The H^+ ion

STRONG AND WEAK ACIDS. It should be clear from this discussion that hydrogen seems to play a special role in polar solutions—the tendency for it to be pulled off or held to a molecule tells something about the solution. In a solution under the conditions we are talking about, the hydrogen nuclei, or *protons*, are the particles of interest. The relatively free proton is eager to hook on to a particle of opposite sign (—), and a solution that easily sets free its protons—that is, the protons are greatly dissociated in solution—is said to be a *strong acid*. One that holds tenaciously to its protons is a *weak acid*—that is, it is much less dissociated in solution. When a molecule is weakly dissociated, it is relatively nonpolar; when it is strongly dissociated, it is relatively polar. The abundance of protons in the vicinity also determines the probability that a molecule will be associated or dissociated.

We speak of the abundance of H^+ as the hydrogen ion *activity* of H^+, which is expressed in gram atom equivalents per liter, or, loosely, as moles per liter, or M/l.* You might expect that this means the same as the *concentration* of H^+, but that is not quite true. Suppose that you have a solution of a weak acid and you try to soak up the H^+. Each time you take out a H^+, a molecule of the weak acid dissociates and provides another to take its place. After you have soaked up practically all the "free" H^+ ions, there are still almost as many left.

Here is what happens: if the weak acid is HA, then $HA \rightleftharpoons H^+ + A^-$. Every time a H^+ ion is removed from its condition shown on the right-hand side of this equation, another takes its place by dissociation of HA.

Evidently the "total available" number of protons is much greater than the number that seem to be free at any moment. Therefore we cannot titrate the "free" acid because every proton neutralized is immediately replaced by another, and the concentration of "free" proton tends to remain constant. From this situation comes the need to speak of hydrogen ion activity (H^+), but it also gives rise to the very useful H^+ *buffer action* of a weak acid.

*p*H, A MEASURE OF ACIDITY. Because the amount is very small, a special notation is used to describe how much H^+ there appears to be in water and salt solutions. The concentrations of (H^+) in body fluids are generally very low indeed. Pure water has the equivalent of only 1/10,000,000 g (H^+) per liter of water. 1/10,000,000 = $1/10^7$, which is a rather awkward number. In order to provide a more convenient scale of measurement, the power of 10 (or the logarithm) in the fraction, in this case 7, is used to describe the (H^+) activity in terms of a *p*H scale. By definition, $1/10^7 = 10^{-7}$. Now, $\log 10^{-7} = -7$, and $-\log 10^{-7} = 7$; or, in general, $-\log (H^+) = p\text{H}$. In this scale, one unit increase in *p*H means a 10 times decrease in (H^+), and vice versa. In a fraction like $1/10^7$, the larger the power of 10, the smaller the fraction. Therefore the larger the *p*H, the smaller the (H^+). For the body fluids generally, the *p*H is usually kept around 7.4, which is somewhat on the alkaline side of neutrality. To find what (H^+) *p*H 7.4 corresponds to, you need only, find in a table of logarithms the number of which the logarithm is 7.4. The relation of (H^+) and *p*H is shown in Fig. 3-8(a).

* A mole is the molecular weight of a compound, expressed in grams. Equal numbers of moles of different compounds have equal numbers of molecules.

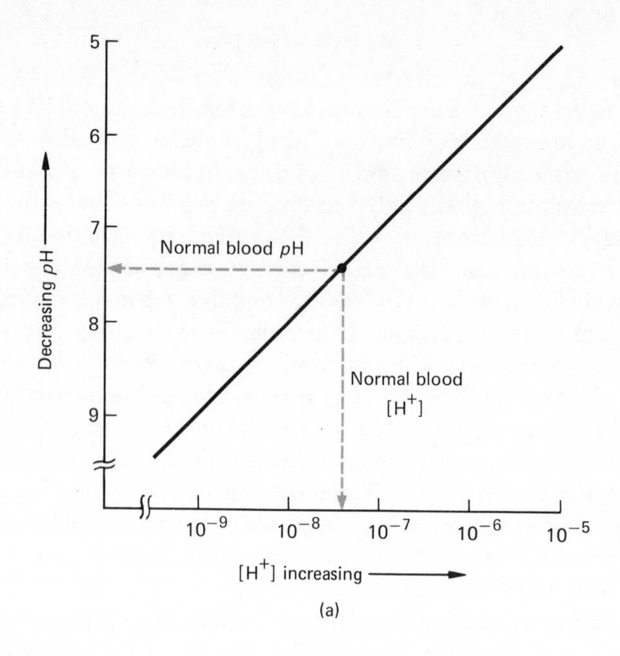

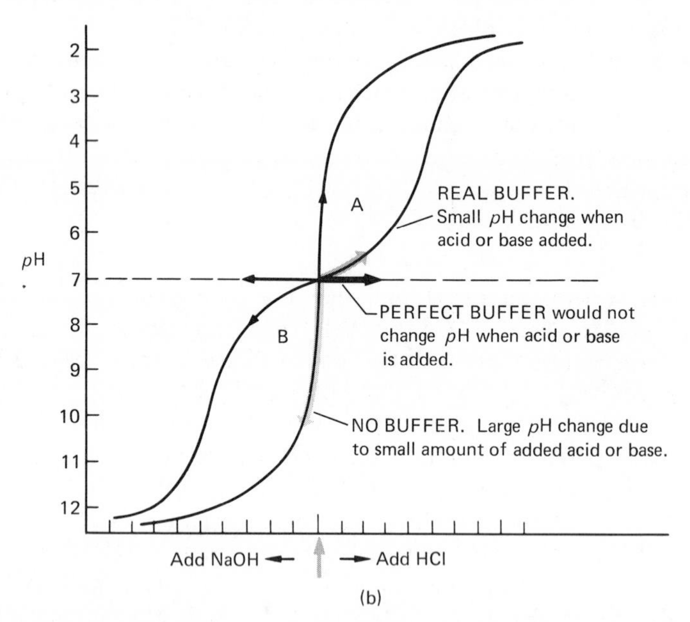

Fig. 3-8. Relation of (H) and *p*H—*p*H titration. (a) **Abscissa** (H) shown on a scale in which equal increments are multiples of 10 (a logarithmic scale). **Ordinate**, corresponding values of *p*H, shown on an arithmetic scale of the exponents of the numbers on the abscissa. (b) Starting at the dot (neutral *p*H in this instance), we see a large change in *p*H when a small amount of base (NaOH) is added to a solution containing no *p*H buffer, and we see a smaller change if the solution does contain a buffer. A similar change, but in the opposite direction, is observed when acid is added.

*p*H Buffer. If we know the concentration of acid in a solution like hydrochloric acid, HCl, a strong acid, we can calculate the *p*H, as described above. Suppose, however, that a small quantity of HCl is added to a solution, and the *p*H does not change as much as we would predict. We say that the solution has acted as a *buffer* against the acid.

In order to illustrate *p*H buffer actions, let us compare a solution having a buffer action at about *p*H 7 and another solution at *p*H 7 but having no buffer capacity. To each we will add a strong acid, such as HCl, in 0.01-ml increments. Let the acid be one molar (M) in concentration, and let the volume of solution to be titrated be 1 liter. 0.01 ml of 1 M HCl in 1 liter of the unbuffered solution gives a *p*H of about 2. 0.1 ml gives a *p*H of about 1. But if 0.01 ml of 1 M HCl is added to a liter of *buffered* solution the *p*H may decrease only one or two *p*H units or less, depending on the kind of buffer and its concentration. Figure 3-8 illustrates this idea.

As you may have suspected from the discussion so far, and as will become more obvious later, the H^+ activity of the body fluids may be important in determining the molecular structure of substances dissolved in those fluids. Consequently, the buffering power of the body fluids is important in order to keep the *p*H rather constant. Specific examples of buffer action will be described in the following pages.

*p*H by Electrical Measurement. In solutions of certain dyes, color depends on *p*H which determines the proportion of the dye that is in the associated and dissociated form. A better method is an electrical one. To understand the electrical measurement of (H^+), we must discuss the measurement of electric potentials in general. The ideas we will consider will also help us understand electric potentials found in relation to cell membranes.

Electric potentials at a boundary

We have already mentioned the existence of opposite electric charges: plus ⊕ and minus ⊖. Take into account the fact of diffusion and the existence of differentially permeable boundaries. Diffusion, you will remember, refers to the movement of the particles of a substance to occupy the total volume available to them. In general, if the concentration of the substance at A is greater than at B, then there is a tendency for diffusion of the substance to occur from A to B (Fig. 3-1). Now if there is a boundary impermeable to the substance, diffusion will occur up to the boundary and no further. Suppose that a potassium salt, K^+A^-, is enclosed by a boundary permeable to the K^+ but not to the A^-. You will expect the K^+ to diffuse across the boundary, but how far will it go?

Each K^+ is paired up with an A^-, and this balance of charges cannot readily be violated. We cannot usually have positive or negative charges in large numbers by themselves. Each must, in general, be balanced by an equal charge of the opposite sign. Each K^+ ion, as it tends to diffuse across the boundary, is held back by the A^-, which cannot permeate the boundary. But the diffusion from the region of high concentration of K^+ to the region of low concentration of K^+ tends to leave behind the negatively charged ions A^-. The compartment containing the high K^+ is therefore left with a negative charge of a magnitude determined by the extent to which the K^+ has managed to diffuse away. Conversely, the region into which the K^+ diffuses acquires a relatively positive charge (see Fig. 3-1).

Now let us return to *p*H. We imagine a boundary permeable only to H^+. Then the electric potential difference across the boundary is a logarithmic function of the (H^+) on the two sides. Therefore, the description of (H^+) as a logarithmic function is not entirely arbitrary, and *p*H is a sensible description of (H^+) seen in electrical terms.

3.2 The Organic Matrix of Cells

Structural chemistry of cells

One way to understand a machine is to take it apart and find out how the pieces fit together. This method can be as useful in working with organisms as in working with physical machines. In their efforts to understand the functioning of living things, physiologists receive assistance from other biologists who take organisms apart in various ways: the *anatomist* who demonstrates the relations of the parts of an animal, usually by dissecting dead specimens; the *histologist*, who describes the structure of the tissues; the *cytologist* who describes the appearance of cell contents; and the *biochemist*, who analyzes the chemical composition of the materials of protoplasm.

The structure of a cell cannot be seen without a microscope. The cell usually must be killed and stained with dyes that possess different affinities for the various cellular components. Under special lighting conditions, however, structures like the nucleus and the mitochondria can be made visible in living cells. Details of ultramicroscopic structure in dead cells are revealed by means of the electron microscope.

Chemical analysis usually requires that the cells first be broken apart. Many substances in various proportions may be found in the protoplasmic extract obtained by grinding up the cells. When cell material is centrifuged at high speed, the substances having the highest specific gravity will sink to the bottom most rapidly. By this method, separate masses of mitochondria or pieces of the endoplasmic reticulum as *microsomes* may be obtained for study. The biochemist can show that some materials produced by the actions of living cells can also be produced by these isolated particles that remain after the organization of the cells has been disrupted. Insofar as he does this, he contributes to physiology and to the understanding of function in the living cell.

Chemical composition of protoplasm

If we make a chemical analysis of the average human body, we find it to be mostly water, as Fig. 3-9 shows. The 40 percent of the body that is not water is organic matter and mineral salts. Of that part of the body that is not water, 56 percent is protein, while the rest is fat, carbohydrate, and minerals.

Organ or Tissue	Percent
Brain	84
Muscle	77
Liver	73
Cartilage	67
Bone	40
Fat	15
Body average	60%

Fig. 3-9. Percent of weight that is water in various organs and tissues of the body.

Another way of looking at body constituents is in terms of the elementary composition. The elements that form the organic matter and water are present in the proportions by weight shown in Fig. 3-10.

Element	Percent
O Oxygen	65
C Carbon	18
H Hydrogen	10
N Nitrogen	3
P Phosphorus	2
S Sulfur	0.25
	98.25%

Fig. 3-10. Percentage composition of body—some elements, by weight. The remaining 1.75% of the body is comprised mainly of mineral salts. These salts will be discussed later in the book.

The comparison by weight is actually quite misleading. The chart seems to suggest that the most common atom in the body is oxygen. This is not at all true. The most abundant element is hydrogen. How this comes about from the chart can be shown by calculating the relative numbers of atoms of each kind. To do so, we must take into account the atomic weights of the elements. According to Fig. 3-10, 65 g of every 100 g of body material consists of oxygen atoms. We know from elementary chemistry that different atoms are of different weights (Fig. 3-11). The oxygen atom, for example, is about 16 times as heavy as the H atom.

Element	Atomic Weight
Hydrogen	1
Carbon	12
Nitrogen	14
Oxygen	16
Phosphorus	31
Sulfur	32

Fig. 3-11. Approximate atomic weights of the most common elements of the body.

The atomic weights show the relative mass of individual atoms, and when weights of these elements are taken in proportions indicated by the atomic weights, there will be equal numbers of atoms of each kind in each sample. That is, if you wanted to get equal numbers of H and O atoms, you would select, say, 1 g H and 16 g O. These weights, in fact, may be called the *gram equivalent atomic weights.* If we compare *molecules*, we speak of *gram equivalent molecular weights*, or *moles*. How can we determine relative numbers of atoms of each kind in the body? According to Fig. 3-10, 65 g out of every 100 g of total body material are oxygen. Therefore there are $\frac{65}{16}$ or about 4 gram equivalent atomic weights of oxygen in each 100 g of material.

There are 10 g of H in every 100 g of body weight. That amounts to 10 gram atoms of H. Calculate the corresponding values for the other atoms. Add the numbers to obtain the total gram atom weight in 100 g of material. Then you can calculate the percent of atoms of each kind in the body. You should get figures somewhat like those shown in Fig. 3-12.

Element	Percent
Oxygen	26
Carbon	10
Hydrogen	60
Nitrogen	1

Fig. 3-12. Approximate percentage composition of body—for some atoms.

Concentrations of particular atoms or molecules are often expressed on a simple weight basis. But *meaningful relations of atoms and molecules generally lie in their relative numbers.* Chemical reactions, for example, which are the foundation of the life process, depend on the relative numbers of participating atoms or molecules. There will be other opportunities to consider these ideas elsewhere in this book.

Consider now the remaining mineral salts, which constitute less than 2 percent of the total body weight (Fig. 3-13). The elements that are present in small amounts are not necessarily less important. NaCl, for example, present in only 0.3 percent of the total weight, is the main constituent dissolved in the blood and in fluids surrounding the cells generally. Many of the other elements are important because they work together with certain of the organic molecules.

Element	Percent
Ca^{++} calcium	2
K^{+} potassium	0.35
Na^{+} sodium	0.15
Mg^{++} magnesium	0.05
Fe^{+++} iron	0.004
Mn^{++} manganese	0.0003
Cu^{++} copper	0.00015
Cl^{-} chloride	0.15
I^{-} iodide	0.00004
plus trace elements	

Fig. 3-13. Percentage composition of body—some mineral salts.

Organic constituents of protoplasm

Several chemical compounds do not occur naturally in large amounts except as part of, or as products of, living organisms. These are the *organic compounds* —mainly combinations of carbon, hydrogen, oxygen, and nitrogen. Some of these compounds can be synthesized from the elements or from simple compounds in the chemical laboratory. Some of the simple organic compounds can be produced under conditions of high temperature that provide some insight into how living matter may have originated on earth a billion or more years ago. A few groups of organic compounds are of outstanding importance in protoplasm because of the relatively large amounts present and the important roles they play. They may be separated from ground-up cell material by simple extraction procedures.

PROTEIN. Protein is the most prevalent organic constituent, comprising somewhat less than 20 percent of the body material. The only compound that forms a larger part is water, which provides about three-quarters of the total fat-free body weight. If the water is removed, then more than half of the remaining dry body weight is protein.

The approximate average percentage of atoms in proteins is shown in Fig. 3-14. Again it is evident that H is the most common element.

Element	Percent
Hydrogen	50
Carbon	30
Oxygen	9
Nitrogen	8
Sulfur	2
Phosphorus	2

Fig. 3-14. Percent of total number of atoms in average protein.

Proteins are constructed of small molecules, *amino acids*, each of which consists of a particular arrangement of about a half-dozen carbon atoms, a dozen or so hydrogen atoms, two or three atoms of oxygen, and one or two atoms of nitrogen.

Sequences of amino acids, attached to one another in a particular, regular, sequential manner, are protein molecules. Figure 3-15 shows the chemical formulas of several amino acids. They are drawn to show the features that make them similar. The amino and carboxylic acid structures give the class name to the amino acids:

```
      H
     /
  —N
     \
      H

      OH
     /
  —C
     \\
      O
```

The "backbone" of the amino acid molecule is a series of carbon atoms, like a string of beads. The amino acids of most biological interest have the amino group attached to the carbon atom situated next to the carboxylic acid. That carbon atom is the α (alpha) or first one in the sequence, for the carbon atoms are identified by the sequence of letters in the Greek alphabet. The molecules we are describing are therefore all α amino acids. There are about 22 different, biologically important amino acids. All have, at one end of the molecule, the structure shaded in the figure. They differ in the arrangement and number of atoms in the rest of the molecule. Some, but not all, of these amino acids will be represented in the composition of any protein molecule. Generally several hundred of each of the various types will be present. The vast number of possible combinations accounts for the large number and large size of specifically

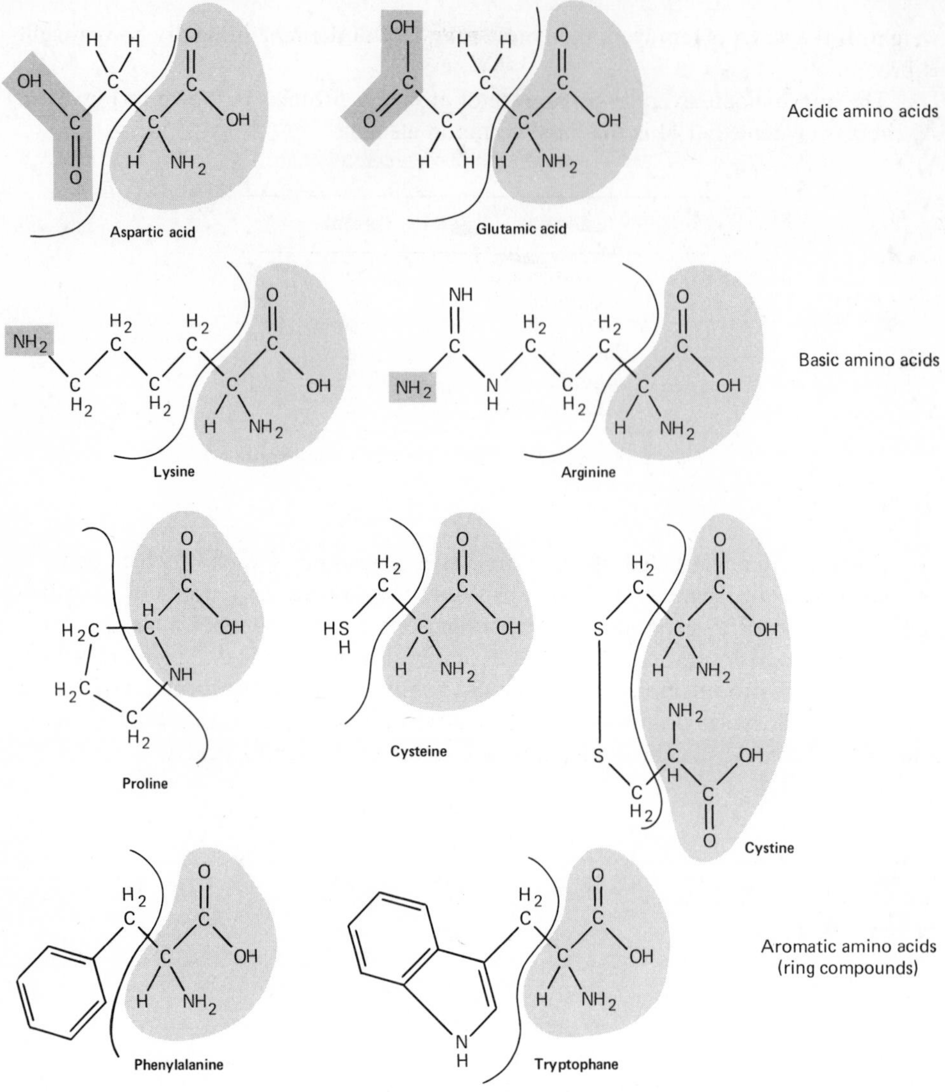

Fig. 3-15. Classes of amino acids.

different protein molecules that are made in the body by combining the α amino acids with one another in particular ways.

Amino acids are not very large molecules, but, joined together, they form protein molecules that are enormous from a molecular standpoint. The structure of a protein molecule is not a simple string, but instead the molecule is coiled, doubled back upon itself, and twisted in a way that is determined by the particular sequence of amino acids that constitute the protein structure. The simplest sequence forms a *peptide chain.*

Carbon has four bonds available, and if a separate atom is attached to each of these bonds, the arrangement of the structure is tetrahedral. This tetrahedral arrangement is a fundamental feature determining the morphology of the protein molecule and ultimately, therefore, the morphology of cells and their constituents.

THE ANATOMY OF PROTEIN MOLECULES. Consider the consequences of attaching one carbon to another in sequence at corresponding corners of the tetrahedra. The sequence will be a three-dimensional zigzag [Fig. 3-16 (a)]. Now imagine joining the carboxyl of one amino acid molecule to the amino of the next, with the loss of water, as shown in Fig. 3-16 (b). To make each connection, rotate each amino acid molecule one after the other, so as to get the amino N of one molecule at a suitable angle next to the carboxyl C of another. Notice that the structure becomes coiled upon itself (Fig. 3-17). Note, also, that there is a terminal amino group at one end and a terminal carboxyl group at the other. There are three peptide bonds (C—N linkage) in each turn. Imagine the same process repeated. The same thing can happen again and you will get another turn. Continue turn upon turn in this fashion and you have a helix. The arrangement of the amino acids is like the threads of a screw. This is the *α helix* arrangement found in many proteins. The existence of the α helix depends on the constituent parts of the molecule being rotated in just the right way with respect to one another.

In this helix, shown in Fig. 3-16, arrow (2) of the turn shown would come to lie above the point in the second turn of the helix that would correspond to arrow (1). Between these two points, a mutual attraction can occur via the H atom, as shown in the figure. This is the *hydrogen bond* that tends to stabilize or hold firm the shape of this large molecule.

Imagine that the double bond between the C and the O splits so that the H can occupy one of these bonds. But the H is held by the N. In effect, the H becomes associated with both the O of the one amino acid and the N of the other. The structure tends to be held stabilized in this arrangement. You can imagine that this stable configuration might be disarranged and not be restored. This is one kind of *denaturation* of the protein; that is, the stabilized helix might be transformed into a random coil structure. The biological significance of this change is that the denatured protein would probably not have the enzymatic properties of the *native* compound. It would be of no use to the metabolism of the cell in which it might be found.

The specific shape of a protein molecule will depend partly on the *R* (for *R*esidue) groups that stick out from the helical backbone. In the example, each constituent amino acid has a different *R* group. Several properties of protein depend on the characteristics of the constituent amino acids. Some acids have an additional acidic or basic group (Fig. 3-17). The *R* groups that are on the surface of the protein molecule will be acidic in the first instance and basic in the second.

Proline is a very special amino acid—a structure that would result if a three-carbon chain *R* group twisted back and attached to the amino N. This makes a bulky structure that does not fit well in the α helix, which must have a bend in the chain at the point where a proline molecule exists.

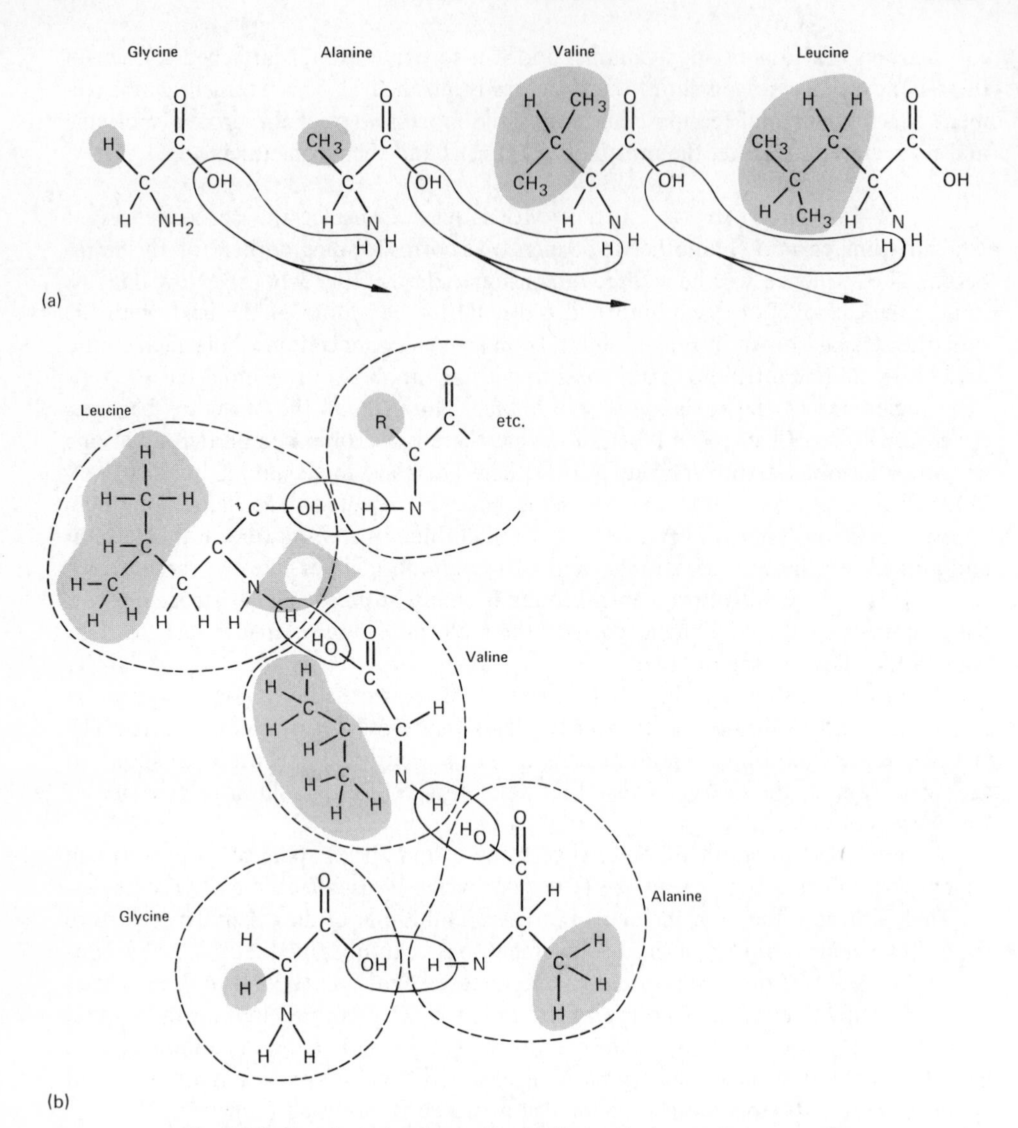

Fig. 3-16. Coupling of amino acids. (a) Removal of water between amino acids to form peptides. (b) The same amino acids have been rotated to form the α helix configuration, which is seen more completely in Fig. 3-17. The part that is different from one amino acid to another, and is known as the residue group, is shaded. The parts that are the same from one amino acid to the next form the peptide backbone of the protein molecule.

LEVELS OF STRUCTURAL COMPLEXITY IN PROTEIN. The morphology of the protein molecule is fundamental to all other anatomical and physiological levels of discussion of human anatomy and physiology. We certainly cannot know all the different configurations that protein molecules can assume. Only consider for a moment the different shapes that can result from the assembling of a few amino acids having

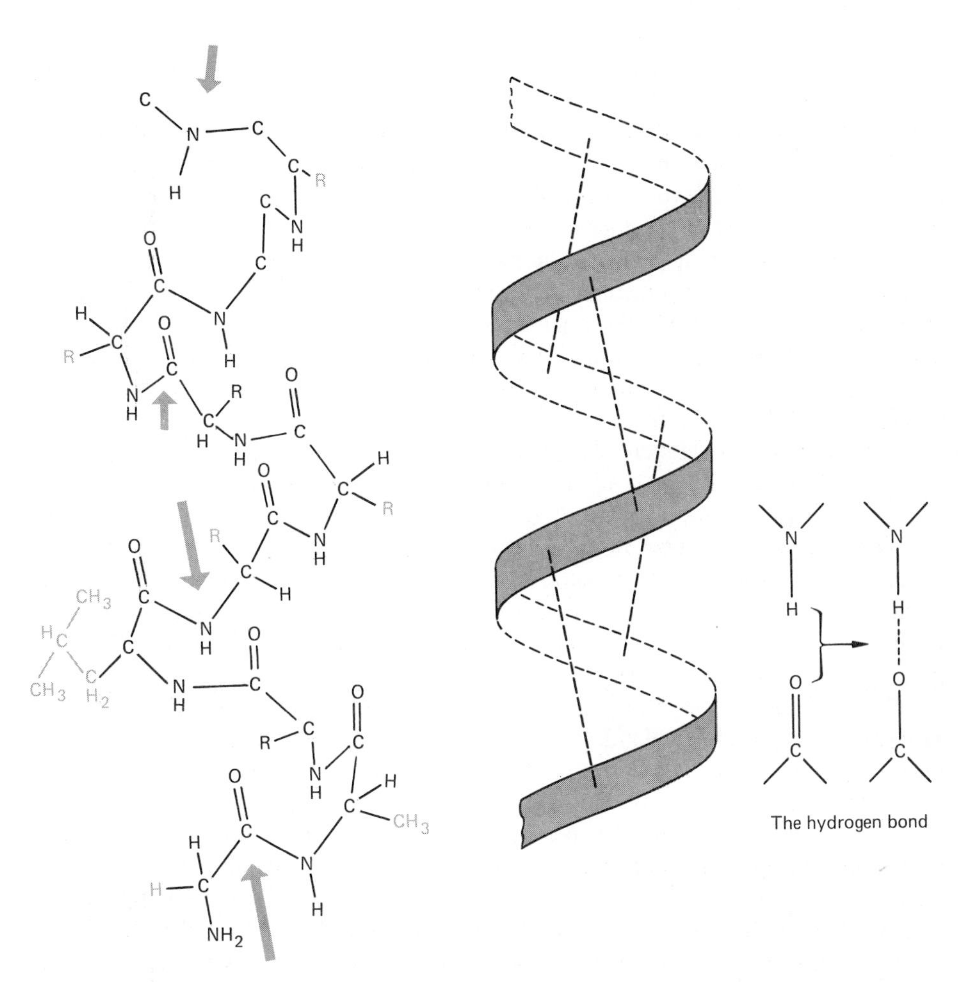

Fig. 3-17. Construction of protein alpha helix. Water has been taken out to allow formation of the peptide bond between the molecules shown in Fig. 3-16; several amino acids have been added to make a total of 10 in the chain. Note the assembly assumes the form of a right-hand screw, the α helix. The structure is stabilized by the hydrogen bond. Location of two sites of H-bond formation are shown by arrows on the molecular structure, and the location of four such sites are shown on the helical equivalent. A protein molecule is constructed of hundreds of amino acids that are linked in this manner.

different *R* groups. The particular sequence of amino acids in the protein molecule determines its *primary structure*. The helix (generally right handed), or other form that the sequence assumes, stabilized by H bonds, is the *secondary structure*. The relation of one section of helix (or other arrangement) with respect to another may be fixed by linkages made at particular places. This relative orientation of the segments of the molecule provides the *tertiary structure*.

Consider the coil to be like a very long spring, bending back upon itself. In such a situation, contact between *R* groups of different amino acids may occur. Chemical interaction between *R* groups of acidic and basic amino acids and H bonding may occur. The sulfur (S) in cysteine has a special role to play in these interactions, for cysteines in two different places in the molecule generate a disulfide bond, S—S, and stabilize the folded arrangement of the helix bent back upon itself. This level of organization is tertiary structure. For a particular sequence of amino acids, there is only one most likely tertiary structure that the molecule will acquire because of the interaction of its side groups. Therefore a particular molecular anatomy follows from the sequence in which the amino acids are arranged.

The entire molecule may now interact with its environment in ways that depend on what side groups are exposed. The acidic and basic side groups are very polar and water soluble (see Chap. 3.1). Methyl groups (CH_3) and benzene ring compounds are nonpolar and more soluble in fats and oils (organic solvents). There is, therefore, still another level of protein structure—*quaternary*—in which we consider the arrangement of one chain with relation to another chain. This arrangement, too, is determined by the chemical properties of the side groups. Polar groups will tend to aggregate with polar groups, for example, and nonpolar with nonpolar.

The arrangement of side groups that appear on the face of a protein molecule provides a surface with which any particular molecule will interact to an extent, depending on its conformity to that surface. As there are many possible surface configurations, so there are also many possible interactions. The special importance of the different shapes arises from the likelihood that the stresses imposed on a molecule thus momentarily pulled and pushed in relation to the surface on which it sits may help or hinder certain chemical reactions to occur. This is the basis for enzymatic action. In nonbiological chemistry, a substance that increases the rapidity of a chemical reaction, but is not itself significantly consumed, is called a *catalyst. An enzyme is a biological catalyst*. Many of the different protein configurations are associated with different specific enzyme reactions. Some protein molecules provide structural features inside or outside of cells and may have no enzymatic activity.

To a considerable extent, the differences among individuals and species of animals can be described in terms of differences in structure of protein molecules. Immunity reactions, for example, as seen in blood types, rest on subtle differences in protein molecules that are similar but not exactly alike.

The protein framework of a cell is not a static one but is continually being broken down and resynthesized from the same and other amino acids in the neighborhood. Nevertheless, protein molecules remain rather constant in composition and structure. If they did not, they could not carry out their enzymatic activities. But their stability is like that of a river, which, ever changing, fundamentally remains the same. The proteins of a body cell, and, indeed, probably of all the body constituents, exist, therefore, in a *steady state* or *dynamic equilibrium* with their environment. New molecules and parts of molecules move in from the bloodstream in exchange for similar parts lost from the cell.

The protean multiplicity of protein structure is written in the structure of *nucleic acids*, another class of large organic molecules. Nucleic acid has the remarkable ability of determining the synthesis of more of itself, and the self-reproducing portions of the chromosomes are nucleic acids. This remarkable property of self-duplication is characteristic of the *deoxyribonucleic acid* that is found almost exclusively in the nucleus*. *Ribonucleic acid* is found in the ribosomes and confers on these structures their ability to govern the synthesis of protein by the cell.

Nucleic acids and the chemical basis of replication

The increase in chromosome material that occurs during multiplication of cells takes place just before cell division during the *interphase*, between divisions. The elaborate motions of mitosis merely provide the means of parceling this material evenly to the daughter cells (see Fig. 2-1).

The exactness of mitosis implies that the chromosomes have an important role in the life of the cell; yet apparently almost the entire cycle of biochemical reactions that make energy available in the cell can proceed for a while without the nucleus. However, in the absence of the nucleus, the synthesis of protein by the cell comes to a stop, and the cell has a limited life span.

The synthesis of protein in the cytoplasm occurs in relation to the ribosomes of the endoplasmic reticulum. The ribosomes are so named because they contain, as part of their structure, the five-carbon sugar, *ribose*. Several hundred ribose molecules are attached to one another through bonds of phosphate to form the backbone of *ribonucleic acid*, RNA, the characteristic type of large molecule in the ribosome (Fig. 3-18c). Attached to each ribose is another ring structure, called a *base*, generally one of two *purines* (*adenine* or *guanine*) or of two *pyrimidines* (*cytosine* or *uracil*), which, arranged in a particular sequence, provide the patterns for organizing the amino acids in the specific protein structures. The purine or pyrimidine, plus ribose and phosphoric acid, constitute a *nucleotide*. The building blocks of the nucleic acids and the way in which they are attached to one another are shown in Fig. 3-18. Figure 3-18 (a) shows the pyrimidine or purine base linked to ribose to form nucleoside, and a phosphate group attached to the nucleo*s*ide to form a nucleo*t*ide. A nucleic acid molecule is a chain of nucleotides. The phosphate of one (at the left), *a*, is coupled to the ribose of another (at the right), *b*, at the points indicated by the arrows in the figure, to produce the structure shown in Fig. 3-18(b) and (c).

The nucleus is the source of the ribonucleic acid. The nucleus is, therefore, not

* The role of deoxyribonucleic acid in cell reproduction is discussed in more detail in Chap. 16.2.

only responsible for replication, but it also gives rise to the enzymatic versatility of the cytoplasm. The agent responsible for these remarkable capacities is the *deoxyribonucleic acid*, DNA, which, together with protein, forms the structure of the chromosomes.

DNA is fundamentally similar to its offspring, RNA. It differs in that the five-carbon sugar molecule of DNA has one less oxygen atom, hence *de-oxy*. The nucleotides are alike, except for *thymine* in DNA, which is slightly different from the *uracil* of the RNA. Finally, the total number of the four kinds of nucleotides is far greater in DNA and, consequently, the length of the molecule is greater. A long molecule of DNA may have a molecular weight of several million compared with a few thousand for the RNA molecule.

Within the chromosomes, pairs of DNA molecules are intertwined like double springs (Fig. 3-19). The pyrimidines and purines of the pairs are matched up in a special way. Each coil is held to its companion by means of weak chemical attraction (*hydrogen bonds*). Note that the large purine with two H bonds (adenine) matches with the small pyrimidine having two H bonds (thymine). Similarly, the large purine with three H bonds (guanine) matches with the small pyrimidine having three H bonds (cytosine). These are the only allowable combinations. Any others will be too short or too long to span the distance between the edges of the helices or will not have matched H bonds. Figure 3-19(b) shows, at a turn of the helix, the two combinations in both arrangements between the helices.

The precise way in which the molecules of DNA are organized in a chromosome, hundreds of times larger, is not known. Nevertheless, the double-stranded, coiled

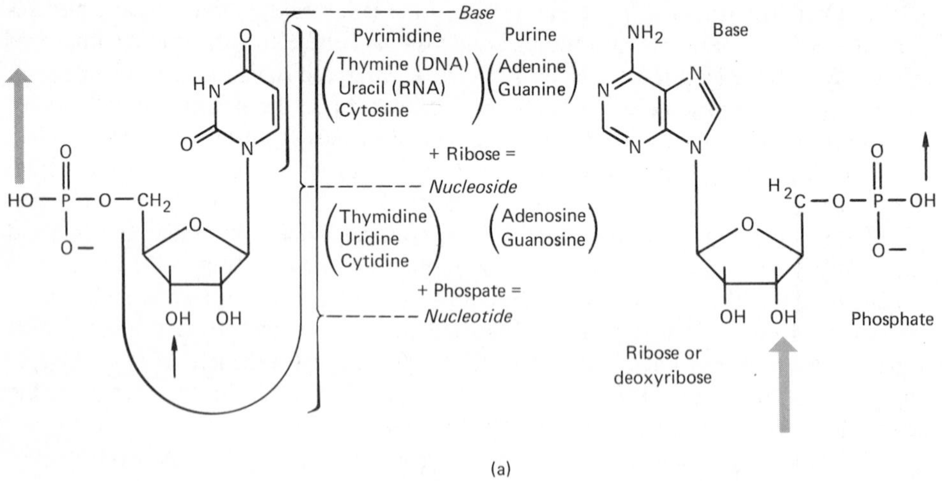

Base
Pyrimidine (Thymine (DNA), Uracil (RNA), Cytosine) Purine (Adenine, Guanine)
+ Ribose =
Nucleoside
(Thymidine, Uridine, Cytidine) (Adenosine, Guanosine)
+ Phospate =
Nucleotide
Base
Ribose or deoxyribose
Phosphate

(a)

Figure 3-18(a)

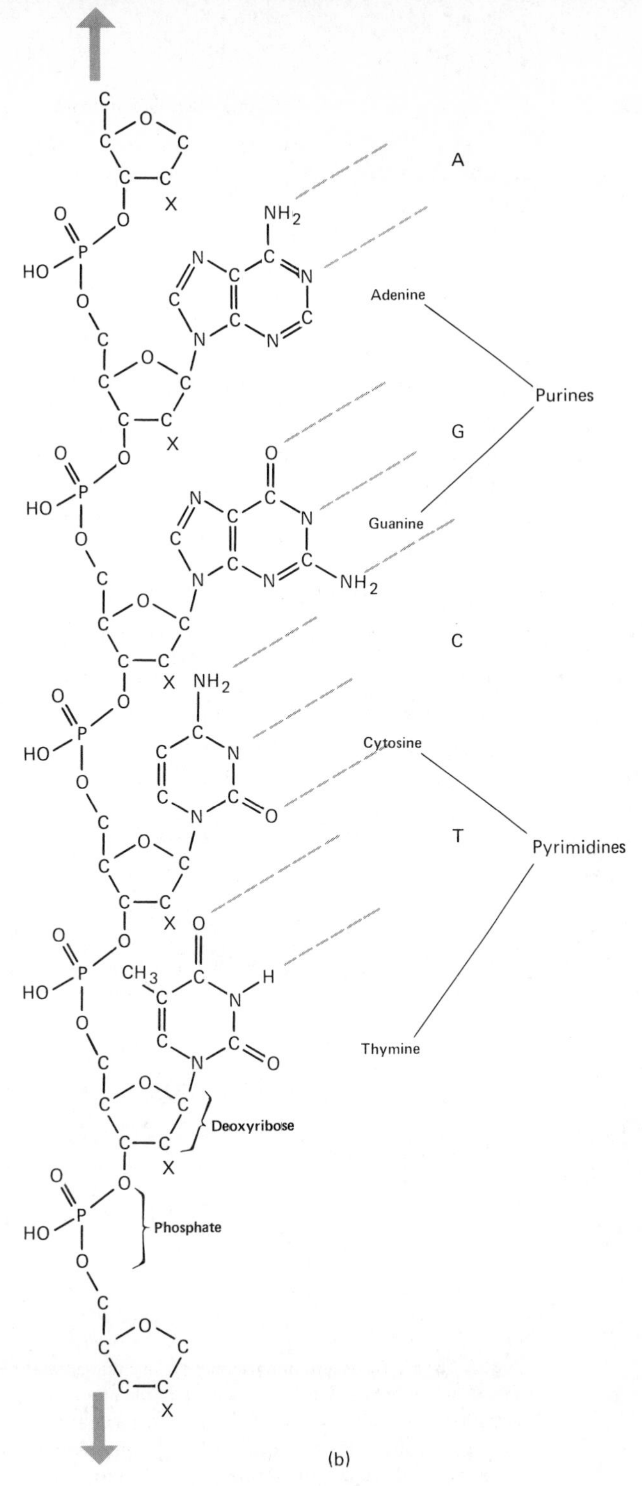

Fig. 3-18(b). A single strand of deoxyribonucleic acid (DNA). x shows the point where oxygen is present in RNA. DNA also differs from RNA in the presence of thymine instead of uracil. CH_3 makes this difference.

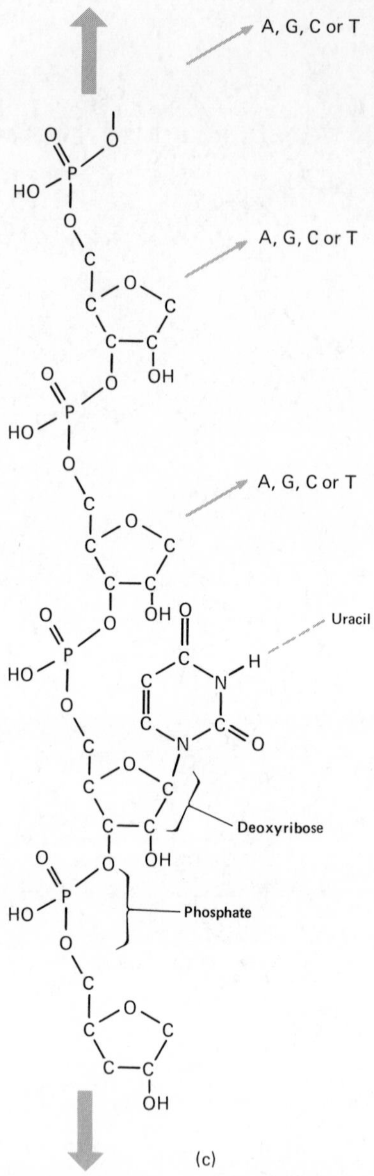

A, G, C or T

A, G, C or T

A, G, C or T

Uracil

Deoxyribose

Phosphate

(c)

Fig. 3-18(c). Structure of a fragment of a single strand of a ribonucleic acid molecule. The ribose-phosphate chain, rendered in light lines, extends in both directions, as indicated by the arrows. To this backbone of several hundred ribose-phosphate units, the four bases are attached in sequences that determine the synthesis of protein specific for the organism in which the particular ribose nucleic acid is found. For simplicity, H atoms (except for OH and NH_2) have been omitted from the structure.

nature of the molecules of DNA is also revealed in the structure of a chromosome in which DNA (about 20 percent) and protein (about 80 percent) are intertwined.

The doubling of chromosome material that is seen during prophase of mitosis occurs before a double strand is visible through the light microscope. Imagine the members of the double strand separating. The nucleotides along the chain then pair up, by means of hydrogen bonds, with other purines and pyrimidines that happen to be available in the immediate vicinity. Each single strand will match with the nucleotides in such sequence that a replica of the lost partner will be regained, after the adjacent phosphate and ribose moieties have been coupled together by loss of water between them.

Figure 3-19(b) shows that the space between the strands is just fitted by a purine (adenine or guanine) plus a pyrimidine (cytosine or thymine). Any other combination would be too long or too short for the distance between the edges of the helices. The molecules with two available H bonds will match, and those with three will match. Only those combinations shown in the figure are allowable. This picture makes it easy to understand how replication of chromosomes can be so precise from one generation of cells to the next. When the strands are separate, available nucleotides line up to produce the complementary strand, and twice as much DNA exists as was present

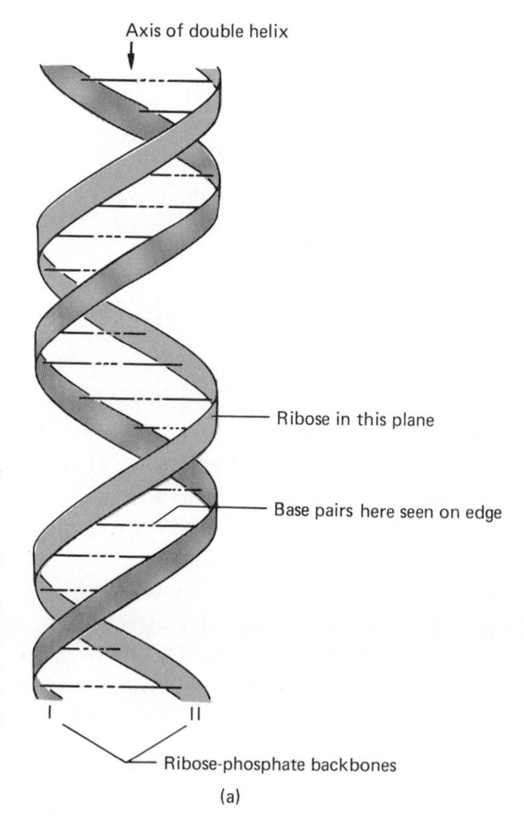

Fig. 3-19(a). Two separate helices of DNA are shown entwined. In the deoxyribose-phosphate backbone, the deoxyribose molecules are flat in the plane of the ribbon. The base pairs are seen on edge. In a view along the long axis of the helix, the base pairs would be seen in full face view. (See Fig. 3-19b.) They are directed inward toward the axis and are perpendicular to the surface of the "ribbon backbone." The pairing of the bases occurs at the points where the horizontal lines are interrupted.

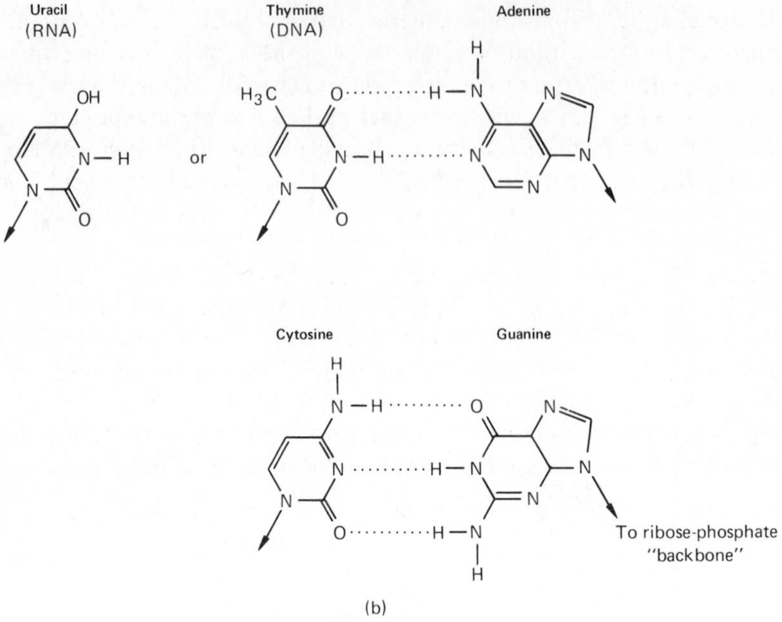

(b)

Figure 3-19(b)

at first. During mitosis the original and the replicated material are parceled out in equal portions.

The role of DNA in differentiation

It is not entirely clear whether differentiation of cells during development follows from loss of ability of the nucleus to provide the appropriate substances to the cytoplasm or whether it is a result of cytoplasmic changes controlled in some other way by the nucleus. Whether the nuclear material remains unaltered in molecular structure in successive generations of cells or whether it undergoes changes during cell differentiation, the DNA is still responsible for the form in which the life of the cell expresses itself.

The arrangement of the nucleotide pairs along the DNA strand is characteristic for any particular species, but it is different for different species of animals. Variations among animals appear to depend finally on differences in the sequences of these pairs in the nucleic acids. How can the sequence of the arrangement of nucleotides in the nucleic acid molecules determine the multitude of morphological and functional differences among individuals and among species?

To put the question another way, how can the nucleic acids control the specific forms of protein that are synthesized in an individual? In a superficial way, the question is easy to answer. The nucleotide sequence of the RNA molecules in the cytoplasm is determined by the nucleotide sequence in the DNA molecules of the nucleus, and the arrangement of amino acids in the protein is in turn fixed by the sequence of nucleotides in the RNA.

Protein synthesis

Researchers in molecular biology are justly proud of their achievements in working out the steps by which the nucleic acids, together with certain enzymes, bring about the synthesis of protein in cells. They have found that the *ribosomes* attached to "rough" endoplasmic reticulum have *ribonucleic acid* molecules as part of their structures. The AGCU base combinations of these RNA molecules (see Figs. 3-18 and 3-19) are determined by the base sequence of the DNA molecules in the nucleus in the same manner that the base sequence of one strand of DNA is determined by the base sequence of the other strand. These RNA molecules, complementary to DNA in their base sequences, are messengers that carry the DNA *code* out into the cytoplasm. Thus they are called *messenger RNA*. In the ribosomes, these single-stranded messenger RNA molecules serve, in their turn, as templates along which the AGCU bases of *transfer RNA* may arrange themselves in a complementary fashion. Like messenger RNA, transfer RNA also arises from the nucleus, by the joining together of AGCU bases that assemble in a complementary fashion along single strands of DNA.

The transfer RNA molecules are much smaller than the messenger RNA molecules, and each molecule is twisted about itself by pairing of complementary bases, like a twisted hairpin. At the loop in this structure is a sequence of three bases, a *triplet*. Consider how many combinations of these three bases there might be, when each can be taken as many times as necessary, in any combination with any two of the other three, from the four bases available. Of those possible, about 20 sets of combinations are relevant to the task of the transfer RNA. That task is to carry a particular amino acid and to line it up along the messenger RNA. There is one set of triplets for each amino acid. For example, a transfer RNA molecule with the sequence AGG or UGG at the loop will carry glycine; CCG or UCG will carry alanine. Other combinations determine the other 18 common amino acids. The amino acid is carried at the free ends of the transfer RNA molecule, whereas the triplets are at the loop. The triplets at the loop become arranged alongside the messenger RNA according to the complementarity between messenger and transfer RNA base sequence. The amino acids at the opposite ends now lie in a sequence determined by the base sequence along the messenger RNA (and ultimately by the base sequence in a particular segment of DNA in the nucleus). The amino acids then combine with one another in the sequence in which they are arranged, and a specific sequence of amino acids constituting a specific protein is formed.

This is a very broad outline of some of the events in protein synthesis. Inasmuch as the synthesis of specific protein determines the molecular character of one kind of cell as distinct from another, this is the sequence of processes that underlie differentiation and the determination of specific kinds of cells.

The physical characteristics of the mature individual arise from the sequence of arrangement of nucleotide pairs, as a complicated machine may arise from a construction manual if there are means of carrying out the instructions. The sequence of arrangement of nucleotide pairs is in effect a code—an arrangement of symbols that can convey information. Ultimately it ought to be possible to describe hereditary characteristics in terms of this code.

Section 3: *Questions, Problems, and Projects*

1. Explain how a pH buffer functions.
2. Using a table of logarithms, calculate (a) the pH when $(H^+) = 2, 5, 7.5, 8, 10$. (b) The (H^+) when $p\text{H} = 1, 3.5, 9$.
3. Describe how you would make a 0.2 M solution of NaCl; a 0.2 M solution of glucose. Compare the two solutions as to the osmotic pressure they would exert. How many moles of NaCl in a 0.3–percent solution? If that is the concentration of NaCl in your body, how many ions of Na^+ do you contain? (There are about 6.06×10^{23} molecules in one mole.)
4. Explain how diffusion may give rise to an electric potential difference.
5. The potential difference across a membrane permeable to ion C^+, in the presence of an ion species to which the membrane is not permeable, may be calculated from the equation

$$E = 60 \log \frac{(C^+)_1}{(C^+)_2} \text{ mV}.$$

   Here 1 and 2 represent the two sides of the membrane. Calculate E if (a) $C_1 = 100C_2$. (b) If $C_2 = 100C_1$. (c) If $C_1 = C_2$.
6. Would you expect a potential difference to arise from the presence of a different concentration of glucose on each side of a membrane? Why?
7. Explain how an amino acid may act as an acid or as a base. Describe what happens when you add base and when you add acid to a solution of amino acid.
8. In a person weighing about 150 lb, about 40 liters is fluid, and of this amount 25 liters is intracellular fluid. Assume that 1 Kg $\approx$ 2.2 lb $\approx$ 1 liter, and work the following problems.
   (a) What is the person's approximate weight in kilograms?
   (b) What percent of his weight is fluid?

(c) What percent of the fluid volume is extracellular?
(d) What percent of the total body volume is extracellular fluid?
(e) What is the probable extracellular fluid volume of a person weighing 50 kg?

9. If a is the amount of a substance and V is the volume the solution containing it occupies, write an equation describing the concentration c of the substance in the volume. You have a sample that has volume V_1, containing an amount a_1, and a smaller volume V_2 containing an amount a_2. If $c_1 = c_2$, write the equation expressing the equality of concentration, but using V_1, a_1, V_2, and a_2. Write the equation to solve for a_2.
10. Two solutions are separated by a membrane, through the apertures of which molecules in solution can diffuse.
    (a) List the factor(s) that the rate of diffusion is directly proportional to.
    (b) List the factor(s) that the rate of diffusion is inversely proportional to.
    (c) In what direction will diffusion through the membrane occur?
    (d) In what direction does the Na^+ pump move Na^+ across the membrane?
11. A salt solution is placed on side A of a freely permeable membrane and pure water on side B.
    (a) In which direction will there be a net transfer (diffusion) of water?
    (b) Of salt?
    (c) What will be the condition at equilibrium?
    (d) If the membrane is semipermeable, in which direction will water move?
    (e) Osmotic pressure is directly proportional to what factors? What effect does solute particle size have on osmotic pressure? Why?
12. (a) How many moles of H_2O are there in 1 liter of water?
    (b) How many moles sucrose ($C_{12}H_{22}O_{11}$) are in 2 liters of a 0.1 M solution of sucrose?
13. A spherical cell 6 μm in diameter is placed in a hypertonic solution and increases 10 percent in diameter. Calculate the volume change in mm^3.
14. (a) Explain why cells will swell in urea solution or in KCl solution isosmotic to blood.
    (b) What would be the molar concentrations of the following solutions, if made isosmotic to mammalian blood?
        (1) Glucose
        (2) KCl
        (3) NaCl

4

FRAMEWORK AND MOVEMENT

4.1 The Skeletal System

A *skeleton* is a framework. Nearly all animals have some means of stiffening their body structure and of maintaining a more or less constant body form. The skeleton is an external structure in an insect and in other arthropods, but it is internal in the vertebrates and consists of *bones* attached to one another by *ligaments*. The skeleton provides a means of attachment for the muscles, which are a large part of the body of any higher animal. The skeleton also protects softer structures, such as the brain, heart, and lungs, and it provides a place for the manufacture of blood cells, in the bone marrow.

Structure of bone

Bone consists partly of materials nearly identical to certain minerals found in rock. In bone, however, the crystals of this material are formed by the action of bone cells that take out of the bloodstream the necessary materials, which are then precipitated around the bone cells. The minerals can be dissolved out of a bone by means of dilute acid (vinegar, for example). After such treatment, the form of the bone remains as a dense, resilient meshwork of collagen, a tough protein matrix (Fig. 4-1). This organic matrix is also secreted by the bone cells. If a bone is burned, the organic material will be removed, and the inorganic salts will remain as a brittle model of the original bone.

At the microscopic level, collagen and minerals form a homogeneous material that is arranged in an open meshwork of plates and bars (*spongy bone*) or in several relatively uniform concentric layers around each of the smallest blood vessels of the bone (*compact bone*). In either instance, the material has been formed by the bone cells (*osteocytes*) that lie embedded within it. The *osteon* (more commonly called the *Haversian* system in reference to its discoverer) is the unit of structure in compact bone. It is a set of concentric layers of bony material around a central canal, in which there may be one or more small blood vessels (Fig. 4-2). The bone cells lie in spaces (*lacunae*) in the bony matrix and are connected to one another by protoplasmic strands in little canals (*canaliculi*) that pass from each lacuna to those nearby. Similar strands pass to the central canal from the cells of the innermost layer. The exchange of waste products and food materials between the cells and blood takes place through these connections.

In spite of its solid appearance, bone is not unchanging. The osteons are continually undergoing rearrangement and reorganization, as the stresses on the bone

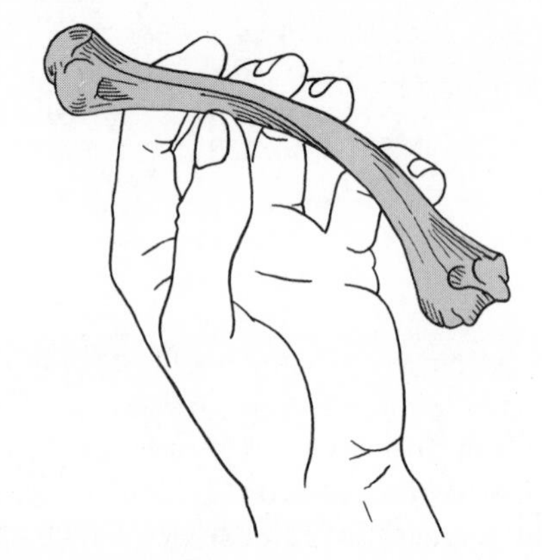

Fig. 4-1. A bone is easily bent if the mineral salts are removed.

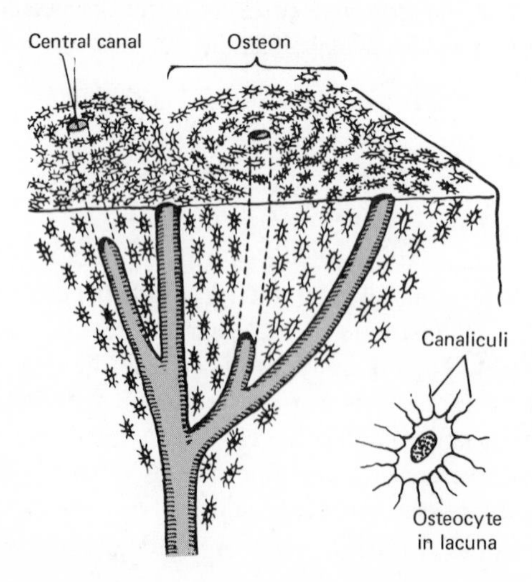

Fig. 4-2. Structure of compact bone. Osteocytes deposit bone material in concentric layers around blood vessels. Four osteons are shown overlapping one another. Osteocyte lies in a lacuna and cellular extensions lie in canaliculi radiating out from the central chamber of the lacuna.

change during the life of the individual. When muscles contract, they pull on the bones to which they are attached. The forces thus applied help, to a certain extent, to mould the architecture of each bone, although the form of each embryonic bone is roughly appropriate to the kind of stresses to which the bone will be subject in the adult.

Origin and growth of bone

The bones of the skeleton are described from their shapes as *long*, *short*, *flat*, and *irregular*. Microscopically, they are very much alike, but they do not all originate in the same way. The flat bones of the *skull* are formed by deposition of bone material by young bone cells (*osteoblasts*) arranged in a flat connective tissue layer or membrane. This method of bone growth is said to be direct and is called *intramembranous bone formation*. The *frontal*, *parietal*, and parts of the *temporal* and *occipital* bones of the skull originate in this manner (Fig. 4-5). These bones can be traced back in evolution to the great bony plates under the skin of primitive fish.

The long bones of the limb, on the contrary, do not arise directly as bone but begin as small structures of *cartilage* in the embryo. Because the cartilage structure serves as a pattern for the true bone, it is called a *model* of the bone. Cartilage, a waxy, organic material, is softer than bone. It is secreted by cartilage cells. During development, cartilage is replaced by bone. The indirect formation of bone from a cartilaginous model is called *intracartilaginous bone formation*. Several steps are involved in this transformation (Fig. 4-3). Early in embryonic development, bone is formed directly in the connective tissue on the surface of the shaft of the cartilage model. This surface layer becomes the *periosteum*. At first this is a loose structure through which the growing blood vessels can penetrate. At about the same time bone is being formed by the periosteum, calcium salts are being deposited inside the shaft around the cartilage cells. As *calcification* proceeds, the cartilage cells swell and degenerate, finally leaving only empty capsules in linear array. Blood vessels penetrating through the periosteum grow into these empty capsules and bring bone-building cells, or *osteoblasts*, that line the walls of these capsules, lay down bone, and become *osteocytes* (Fig. 4-3). Thus the interior of the bone becomes a set of interconnecting bars and plates (*spongy bone*) built on the cartilage between the capsules. Tissue that has thus become bone is said to be *ossified*. The cavity penetrated by blood vessels at the center of the bone is the *marrow cavity*.

As the bone continues to grow in thickness at the periosteum, and in length by growth of cartilage that is replaced by bone, the marrow cavity within must increase in size. The removal of bone material necessary for this enlargement is apparently the responsibility of the *osteoclasts*. Bone material at the inside is dissolved away by osteoclasts, while osteoblasts deposit bone material outside the cavity. In a sense, the growth of a long bone is like the enlargement of a house in which new bricks are placed on the outside while old bricks are removed from inside.

Although much intracartilaginously formed *spongy bone* is subsequently trans-

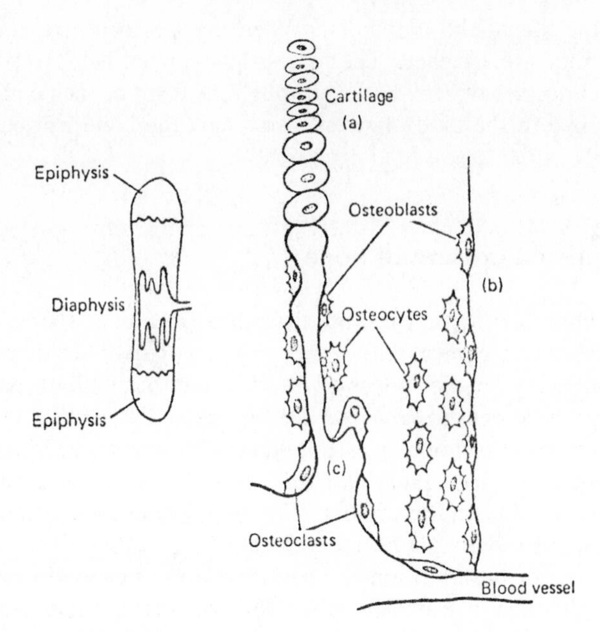

Fig. 4-3. Intracartilaginous bone formation. Bone formation is shown for an enlarged section of the diaphysis of the growing bone. (a) Inside the bone, growth results from the multiplication of cartilage cells. Older cartilage cells degenerate as the cartilage becomes calcified. Osteoblasts move into the spaces left by the degenerated cartilage cells. (b) Intramembranous bone formation at surface where bone is deposited by osteoblasts. (c) Osteoclasts remove bone material, thus enlarging the marrow cavity.

formed into *compact bone* characterized by closely packed osteons, some spongy bone does remain at the center of most bones and forms the marrow cavity. Early in the life of the individual, blood cells multiply, mature, and move from these cavities into the circulation. With increasing age, some of this *red* marrow becomes *yellow* as fat replaces the blood-forming cells.

In the early cartilage model of a long bone, the central shaft is the *diaphysis*. At each end, separated from the diaphysis, is an *epiphysis*. These terms apply to the corresponding parts of a long bone at any age. Lengthening of the original model occurs by means of division of the cartilage cells (in uncalcified cartilage) at each end. Centers of ossification finally arise in the epiphysis; only a thin plate of cartilage, the *epiphyseal plate*, remains at each end between diaphysis and epiphysis. This plate of cartilage continues to be the growth center mainly responsible for bone elongation. The line that marks it can even be seen in bones from relatively old individuals, although it ossifies generally by about the twentieth year.

When ossification is complete, bone is still alive. Its living quality is clearly revealed if the bone is broken. Unlike a rock, bone heals itself. During healing, as during development, destruction of bone by osteoclasts and construction by osteoblasts go on simultaneously. Cells involved in healing come from the periosteum where they form a connective tissue framework into which bone material is deposited subsequent to injury. In the mature bone these cells do not secrete bone material unless they are stimulated by injury.

Changes in the bones provide the most striking evidence of body growth. The long limb bones provide the extra height of tall people. In some abnormal instances, growth of bones may be resumed late in life. The bones then grow where cartilage still remains in the projecting parts of the skeleton—the hands, the feet, the jaw; the condition is therefore called *acromegaly*, a term meaning enlargement of terminal parts. Secretion of growth hormone from the pituitary gland is responsible for excessive skeletal growth.

If excessive pituitary secretion starts early in life, the individual may become a *giant* whose body is normally proportioned but large. If a deficiency begins early in life, the individual will remain a *dwarf* or midget, normally proportioned but small. Still another error in bone growth is *achondroplasia*. In this condition, intracartilaginous bone formation in the long bones is defective, while periosteal bone formation is unimpaired. Consequently, the legs and arms are shorter than usual, although the rest of the body may be normally proportioned.

Regulation of bone growth is accomplished by hormonal control of protein retention by the cells. The deposition of the mineral salts, especially calcium, is secondary and depends on several factors, including dietary calcium, blood phosphate levels, secretions from the *parathyroid gland*, the *pituitary gland*, and the *gonads*.

Some specific bones of the skeleton

There are more than 200 separate bones in the young adult human. In older people some of the bones may fuse; thus the total number may be slightly decreased.

The bones of the skeleton may be conveniently divided into two major groups (Fig. 4-4). The *skull* and *vertebral column*, together with the *ribs* and the *sternum*, constitute the *axial skeleton*. The rest of the skeleton—*pectoral* and *pelvic girdles*, and the *appendages* (limbs) attached thereto—is the *appendicular skeleton*.

Axial skeleton: The skull

The bones of the *skull* join or *articulate* with one another at irregular, serrated edges or *sutures* (Fig. 4-5). These joints are completely immovable in the adult. Such immovable joints are called *synarthroses* (fused joints). In the newborn child, the skull bones can move relative to one another, facilitating passage of the head through the birth canal of the mother. The regions of ossification are separated from one another by thin membranes. These soft regions of the skull of the young child are called *fontanels*, of which the occipital and frontal are most prominent. The latter remains a soft spot until the end of the second year. In several of the skull

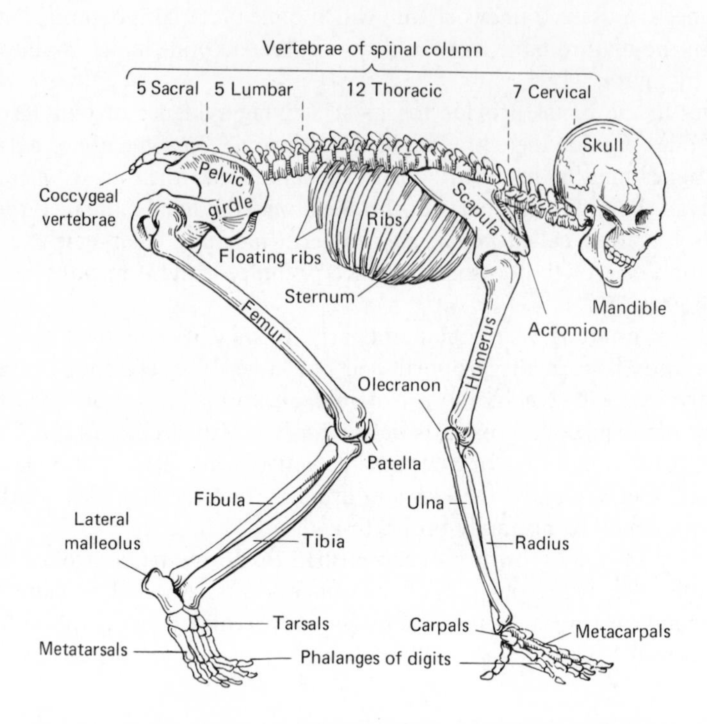

Fig. 4-4. Human skeleton. The skeleton is here oriented so that the corresponding parts of the limbs may be easily compared.

bones there are large cavities continuous with the nasal cavity. These cavities serve as a resonating chamber for the voice. They are generally larger in men than in women. The *frontal sinuses* in the *frontal bones* are a common site of infection brought in via the nasal cavity. On the inside of the cranium there are prominent smooth grooves where the *venous sinuses* run. These sinuses are between the inside wall of the skull and a tough layer, the *dura mater*, that covers the brain. They are large veins draining the blood supply of the brain.

The *temporal bone* is especially important because it contains the *inner ear apparatus* that is sensitive to sound, to movement, and to the position of the head. The *middle ear bones* that transmit sound to the inner ear are contained within the *middle ear chamber* of the temporal bone. This chamber is covered externally by the tympanic membrane, forming the blind end of the *external auditory meatus*. Like the temporal bone, the *occipital* bone has a double origin—partially from intramembranous and partially from intracartilaginous bone formation. The *occipital condyles* of this bone are smooth round structures that constitute a portion of the joint between the vertebral column and the base of the skull. They are on each side of the *foramen magnum*, a large aperture through which the brainstem reaches the spinal cord. The base of the brain rests on the *sphenoid* bone, whose irregular borders touch several other

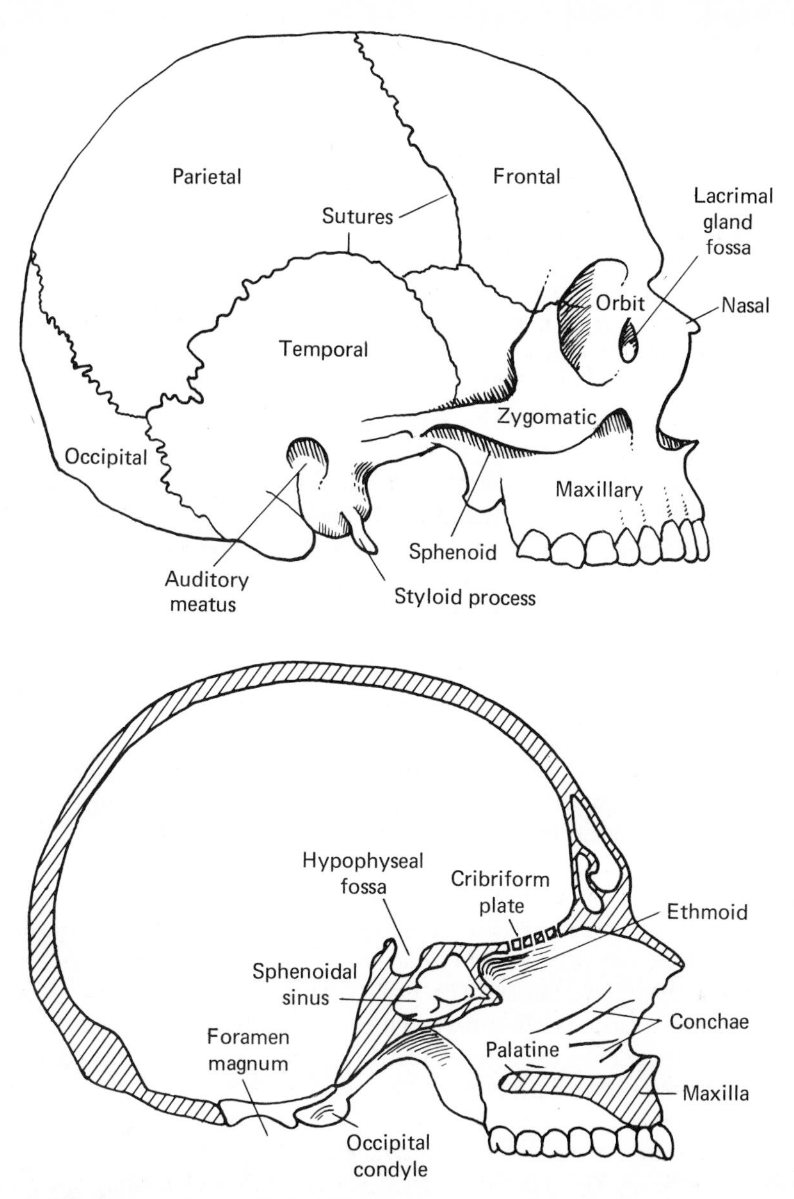

Fig. 4-5. The skull. Large flat "membrane" bones enclose the brain. Sinuses are air spaces in the bones.

bones of the skull. Rostrally (at the nose end) the sphenoid bone sends branches up into the *orbit* or eye cavity. This cavity is formed by contributions from several bones, including the temporal. The *pituitary* (or *hypophysis*), an important endocrine gland, rests in the *hypophyseal fossa*, a saddle-shaped pocket in the sphenoid bone.

The upper teeth are imbedded in the *maxillary* bone; the lower teeth are all held by the single bone of the lower jaw, the *mandible*. The nasal cavity is separated from the oral cavity by the *palatine* bone, which, together with a contribution from the maxillary bone, forms the hard palate. Olfactory nerve fibers travel to the brain by way of foramina in the cribriform plate of the ethmoid bone.

The vertebral column

The vertebrae are joined to one another by means of strong *ligaments*. The vertebral column (or spine) formed by the several vertebrae (about 24) is sufficiently stiff to prevent excessive twisting of the spinal cord, yet considerable flexibility is permitted by the movement of the vertebrae with respect to one another. The limited movement possible between vertebrae justifies the name *amphiarthrosis* for this type of joint. In older people the functional flexibility of the vertebral column decreases, due to lack of exercise of the appropriate muscles and deposition of calcium salts. The vertebrae move on one another at the *articulation surfaces*, but the bodies of the vertebrae are separated from one another by the *intervertebral disks*. These are tough fibrous pads with an inner core of pulpy material that can be squeezed as the vertebral column is bent. Occasionally the material of the disk may be displaced, an occurrence that may cause pressure on nearby nerve roots and consequent severe pain.

Each vertebra has a main body, or *centrum*, that derives from the *notochord*, which is a continuous, dorsal stiffening rod in very primitive vertebrates—and of which evidence remains in embryos of higher vertebrates. The *neural arch* encloses the spinal cord dorsally and is surmounted by the *spinous process*. The spinous and the *transverse* processes provide places for attachment of vertebral muscles. Between the successive vertebrae, spinal nerves pass from the spinal cord to the rest of the body. There are five groups of vertebrae (Fig. 4-4). The seven *cervical vertebrae* are the bones of the neck. Vertebra number one, the *atlas*, holds the skull and rotates on the *axis*, the second vertebra. Rotational movements of the head occur at this articulation between first and second vertebrae. Mammals generally (even the long-necked giraffe) possess seven cervical vertebrae. The twelve *thoracic vertebrae* have characteristic lateral articulation surfaces where they connect to the ribs. The five *lumbar vertebrae* in the small of the back, like the cervical vertebrae, articulate only with one another but have short, blunt, transverse processes. In contrast to those of the neck, however, they have a large centrum and prominent transverse processes. The five *sacral vertebrae* are fused into a solid structure, the *sacrum*. The variable number of *coccygeal vertebrae* are generally fused as the *coccyx* (the base of the tail in many other animals).

The ribs

The *ribs* provide protection for the heart and lungs. Together with the diaphragm muscles that form the floor of the chest cavity (*thorax*), they make possible the efficient breathing of mammals. Intercostal muscles move the ribs, between which they are attached, upward and outward, increasing the size of the thorax. Ventrally, the first seven ribs articulate via cartilages, directly with the *sternum*. The cartilages of the next three ribs do not articulate directly on the sternum but on one another. Consequently, they are called *false* ribs. The last two ribs on each side are free at their ventral ends and are therefore called *floating* ribs. The bars of cartilage give resiliency to the thoracic cage but may become ossified in elderly people.

Appendicular skeleton: The girdles

The *appendicular skeleton* consists of the limbs and the girdles to which they are attached (Figs. 4-4 and 4-6). The *pectoral girdle articulates* (forms a joint)

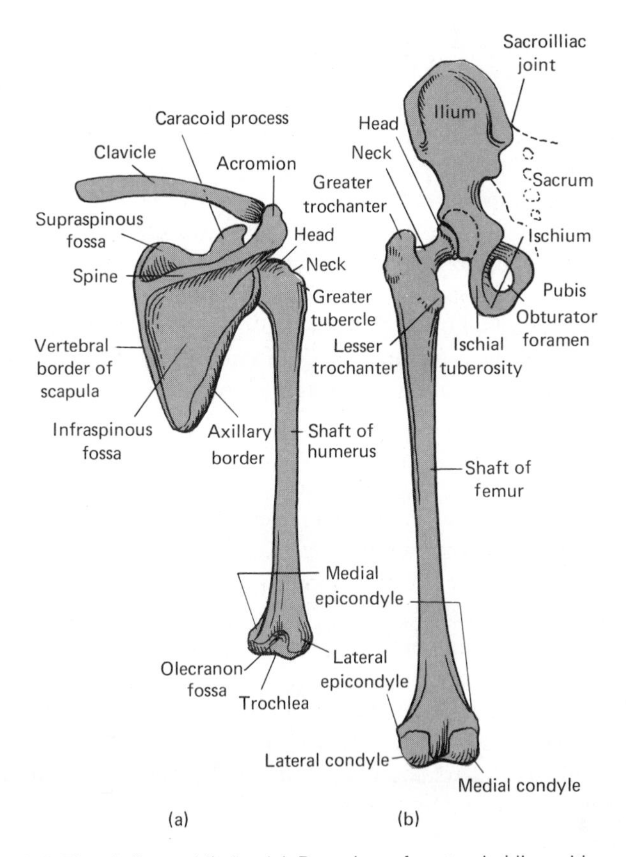

Fig. 4-6. The girdles and limbs. (a) Rear view of pectoral gidle and humerus (right side). The pectoral girdle articulates ventrally on the manubrium of the sternum. (b) Rear view of the pelvic girdle and femur (left side). The pelvic girdle articulates dorsally on the sacrum, while ventrally the pubes join at the pubic symphysis.

with the axial skeleton only on the ventral side, where the proximal end of the collar bone (*clavicle*) joins the rostral end of the sternum. It is also anchored to the ribs by strong ligaments as well as by the muscles that attach to it (Fig. 4-6). Laterally, the shoulder blade or *scapula* articulates with the clavicle at the *acromion of the scapula* (the point of the shoulder). The scapular *spine*, a sort of keel, divides the large, flat, dorsal surface of the scapula into two equal parts, the *supraspinous fossa* and the *infraspinous fossa*. Large masses of muscle are attached on both these surfaces. In its distal part, the scapula spine is prolonged into the acromion. The *vertebral border* of the scapula faces the vertebral column; the *axillary* border forms the dorso-medial edge of the *axilla* or arm pit. You can easily find some of these structures on your own body. The *glenoid* ("like a valley") *fossa* is part of a ball-and-socket joint formed where the *head* of the humerus articulates. The *coracoid* ("like a crow's beak") *process* projects ventrally from the scapula and forms an attachment point for certain arm muscles.

The *pelvic girdle* is rather firmly attached at the *sacrum*. In the adult, each half of the pelvic girdle is one bone, the hipbone, which is formed from the fusion of three separate embryonic bones corresponding to three named parts in the adult. Each of these three bones makes an equal contribution to the *acetabulum*, the socket of the ball-and-socket joint of the hip.

The *pubis* of one side meets with the corresponding one of the other side at the *pubic symphysis*, which may be discovered just above the external sex organs. This symphysis undergoes considerable loosening in the female during pregnancy, so that the birth canal becomes sufficiently wide for childbirth.

A second part, the *ilium*, can be detected as a bony protuberance at the waist level on each side. These protuberances are the tops of the wide wings of the ilia. The articulation of the pelvic girdle on the vertebral column is at the *sacroiliac joint*. The third division of the hipbone is the *ischium*. The ischial *tuberosity* bears the weight of the body in the sitting position. These structures have prominent external pads, the *ischial callosities*, in some of the anthropoids other than man. In addition to participating in the formation of the acetabulum, the pubis and ischium join at their ends, leaving a large aperture or foramen between these bones. This space becomes almost completely obturated or covered by a tough membrane. The aperture is called the *obturator foramen* for this reason.

The pelvic girdles are somewhat different in the two sexes. The width of the pelvic cavity constitutes a larger proportion of the overall width of the pelvic girdle in the female than it does in the male. In addition, the overall width of the pelvis compared to the shoulders is greater in the female than it is in the male. This difference in bone development is related to the fact that women bear children. Space must be available for growth of the foetus and for childbirth.

The limbs

The joints of the limb bones join two separate, easily movable parts. A joint of this kind is called a *diarthrosis*. The forelimb (anterior limb) and hind limb (posterior limb) are similar to one another in construction. Their corresponding parts are shown in Figs. 4-4 and 4-6. Both limbs articulate on their respective *girdles* by way of a ball-and-socket joint: the *head* of the *humerus* of the upper arm in the glenoid fossa of the scapula, and the *head* of the femur of the thigh in the acetabulum. The *neck* of the femur is set at a considerable angle to the shaft, but the corresponding two parts in the humerus pass imperceptibly into one another.

Distally, the *humerus* forms a hinge joint with the *ulna*, which projects beyond the hinge as the *olecranon*. When the arm is straightened by the large muscles at the back of the arm, the olecranon is stopped in the *olecranon fossa* of the humerus. The femur articulates distally with the tibia, with which it forms the hinge joint of the knee. Muscles on the ventral side of the thigh straighten the leg and are attached to the tibia by way of the *patella* or kneecap, a flat bone (called *sesamoid*) imbedded in the tendon of these muscles.

The *radius* of the arm and the *fibula* of the leg are long, thin bones associated, respectively, with the *ulna* and the *tibia*. The hand may be moved palm upward or downward when the radius is rotated at the elbow, the ulna remaining fixed in position. The *carpals* of the wrist and the *tarsals* of the ankle occupy similar positions in the limbs and serve comparable functions. They are, therefore, corresponding bones. The *metacarpals* of the palm of the hand correspond to the *metatarsals* of the foot. The separate bones of the digits (toes and fingers) distal to the metatarsals and metacarpals are called *phalanges*.

Elevations and projections on the limb bones have various names according to location and function. On the head of the humerus, the *greater tubercle* is lateral to the *lesser tubercle*. The *greater trochanter* is a large projection on the lateral side of the femur where the neck of the femur joins the shaft. The *lesser trochanter* is on the medial side. Certain round, smooth, articular eminences are designated *condyles*. *Medial and lateral condyles* are at the distal end of the femur. The *medial* and *lateral epicondyles* are projections at the sides of articular prominences on the distal end of the humerus. The ankle is on the medial side formed by a protuberance of the tibia, the *medial malleolus*, and on the lateral side by the *lateral malleolus*, part of the fibula.

Several parts of the skeleton never become bone but remain as cartilage. The outer ear, the end of the nose, and the cartilages between ribs and sternum rarely become ossified even in old age.

There is no bone at the contacting surfaces between bones or at movable joints where the parts of the joint are subject to continual abrasion. Instead there is a layer of cartilage that wears away at the surface but is continually replaced by cells below the surface. Each movable joint is enclosed in a capsule. The *joint capsule* is a thin, tough, connective tissue layer formed partly by the periosteum, which continues as a cover over the joint from one bone to the next. On the inner surface of the capsule is the *synovial membrane*, a layer of cells that secrete the *synovial fluid* to provide lubrication for the moving parts of the joint.

4.2 Skeletal Muscles

A muscle is an organ that can exert force. It consists of muscle tissue with connective tissue and vascular tissue interwoven among the muscle cells.

The muscle cells that provide the force are long and thin and are called *muscle fibers* because of the fibrous texture they impart to a muscle (Fig. 4-7). The simplest muscles are found in the viscera. For example, *smooth muscle* of the digestive tract and of blood vessels consists of tiny, spindle-shaped cells, each with a centrally placed nucleus. Muscles that move the bones are *skeletal muscles*. They are also called *voluntary muscles* because we have conscious control over their action. *Heart*, or *cardiac muscle*, is still a third type, in which the fibers are interconnected and have central nuclei but otherwise are similar in appearance to skeletal muscle fibers.

Movement in a muscle cell is carried out by bundles of tiny protein *myofibrils* packed lengthwise inside each cell. The fibrils lie in *sarcoplasm*, a noncontractile fluid cytoplasm. Each muscle fiber of a voluntary muscle is actually a compound structure, and the numerous peripherally placed nuclei in each fiber remain as an indication of its multiple origin. The fibrils in a skeletal muscle cell are arranged in such a way that they present a banded appearance under appropriate lighting conditions or after suitable staining. The bands are perpendicular to the long axis of the fibers and are called *striations*. Voluntrary muscle is therefore said to be *striated*. Small striated muscles consist of a few hundred muscle fibers, while the larger muscles are made up of many

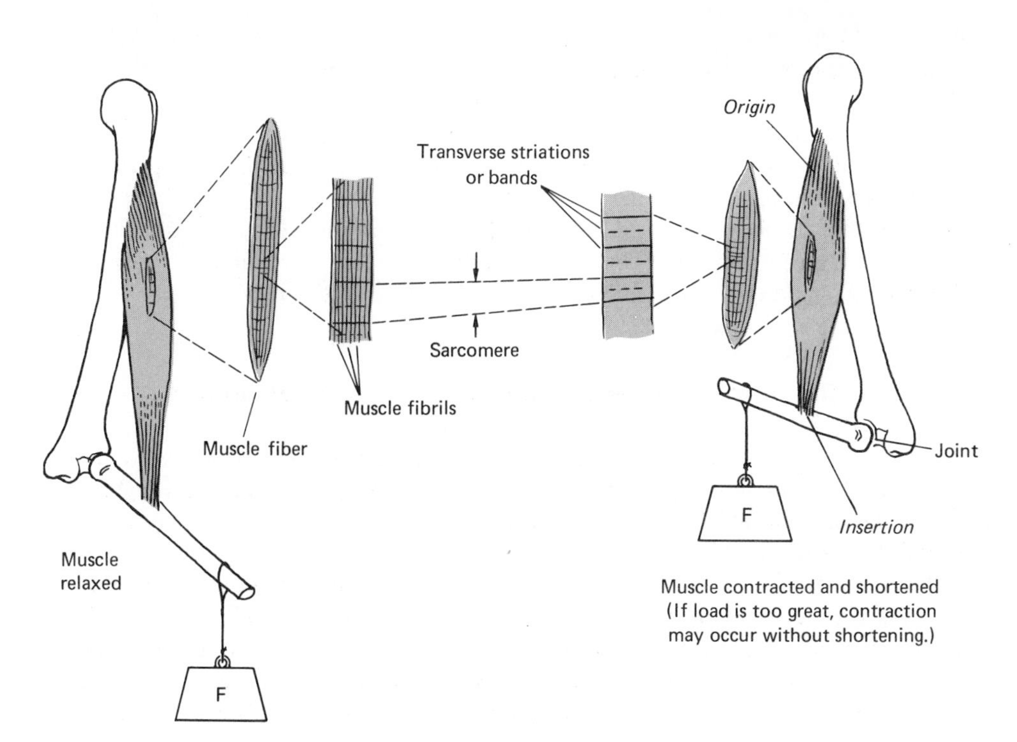

Fig. 4-7. Muscle contraction involves the production of a force to pull between the attachments of the muscle. During contraction the end of the muscle, which is the insertion, may move toward the origin which remains fixed in position. The opposing force may be too great for the muscle force to overcome, in which case contraction may occur without shortening by the muscle.

thousand fibers. The fibers or cells of both *smooth muscle* and *skeletal muscle* are generally arranged more or less parallel to one another, often overlapping at their ends. Fibers of *heart* or *cardiac muscle* are striated, and the interconnected cells appear to form a *syncytium*, but there are zones (*intercalated disks*) between them that mark the boundaries of the cells.

The molecules of contractile material exist in groups of long filaments or *muscle fibrils*. In a striated muscle fiber, they are arranged in repeating sequences of bands within the fibril (see Fig. 4-7 and Chap. 5.2). Each set of bands is called a *sarcomere*. The boundary of the sarcomere is marked by a line (the *Z* line) which appears to be attached to the cell membrane, the *sarcolemma*. Numerous tiny filaments consisting of molecules of the protein *actin* extend from both sides of this boundary. The middle of the sarcomere is occupied by another set of filaments, the larger, less-numerous *myosin* filaments. The ends of the myosin filaments interdigitate among the ends of the actin filaments. It is this arrangement of material that produces the striated or banded appearance of the muscle cell. During contraction the widths of the bands change, as the bands of actin filaments at the end of each sarcomere are brought closer to the actin filaments at the opposite end of the same sarcomere. This process occurs as actin and myosin filaments slide along one another and become hooked together successively farther along as they do so. Force is thus developed, and if the attachments of the muscle are allowed to move, the sarcomere shortens. If the attachments remain fixed in position, the contraction occurs without shortening of the muscle. The molecular basis of muscle contraction is described in Chap. 5.2.

The form and function of striated muscles as organs

In a skeletal muscle the connective tissue that binds the fibers together surrounds them as a tough *fascia* that is drawn out at the ends of the fibers to form inelastic *tendons* that attach the muscle, usually to definite points on the skeleton. The fibers of skeletal muscles are arranged in various ways with respect to one another and to the tendons. Most of the muscles of the limbs, especially those attached to the bones of the feet and hands, are long, thin, and pointed at both ends; that is, they are cigar shaped, or *fusiform*. In fusiform muscles the fibers overlap one another in such a way that most of the fibers are in the thick middle section, and fewer are at each end. Tapered ends of the separate fibers help to establish the fusiform shape, but generally the fibers do not run the entire distance between the ends of a muscle. In some instances, the fibers may fan out to connect a small point with a large area, or they may meet at a central tendon as the barbs of a feather meet at the central shaft.

When movement occurs during contraction of a skeletal muscle, the central portion or *belly* of the muscle becomes thicker, while the two ends of the muscle are brought closer together. During shortening of the muscle, one end tends to remain fixed in position, while the other end moves. The end that remains fixed is called the *origin;* the end that moves is called the *insertion* of the muscle (Fig. 4-7).

The motions produced by muscle contraction are described with reference to the *standard position* of the body—that is, the body standing erect with arms straight down and palms directed forward. In Fig. 4-8 this position has been modified to show an individual in motion, in order to illustrate some of the terms describing motion and positions of the limbs. See, also, Figs. 2-5 and 2-6 for anatomical terms of position.

A lengthwise line down the center of the body is the *long axis* of the body. The *midsagittal* plane may be imagined to divide the body into symmetrical right and left halves along this line. Limb movements and the resultant positions can be described with reference to the midsagittal plane of the body in the standard position. A folding together at a joint, in a direction parallel to the midsagittal plane, is *flexion*. An unfolding in the opposite direction is *extension. Abduction* refers to a movement *away* from the midsagittal plane, but *adduction* means a movement *toward* the midsagittal plane of the body. Usually the body is *not* in standard position, and in order to describe the position of any part of the body, it is useful to imagine the body first restored to

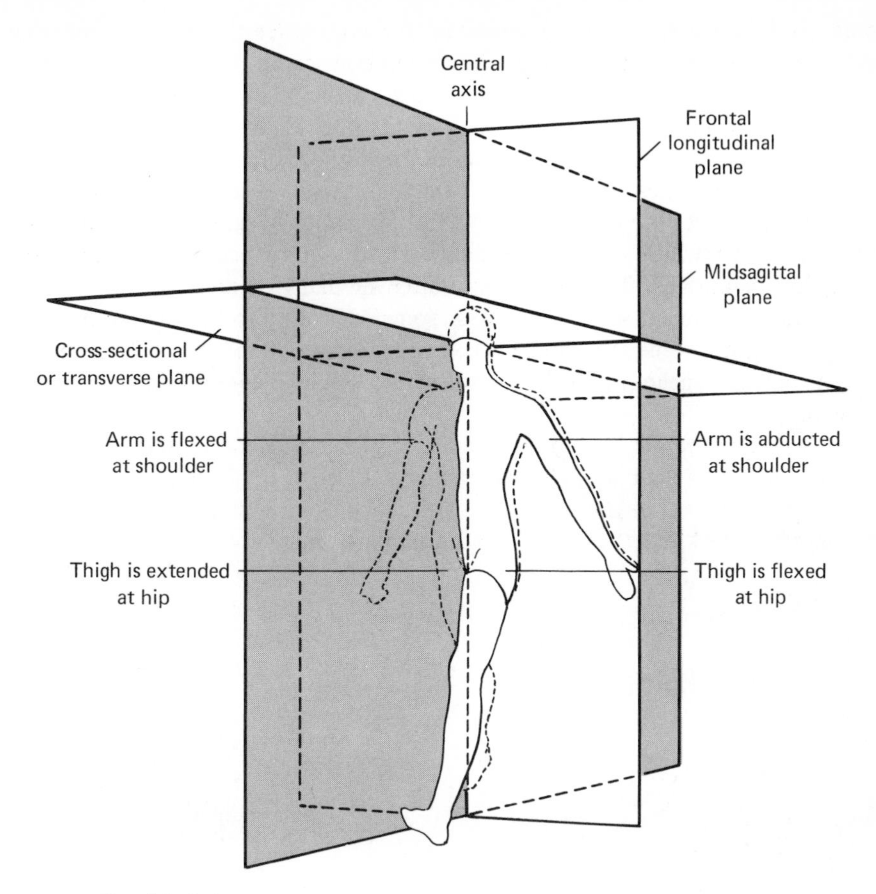

Fig. 4-8. Reference planes for describing movement and position of parts of the body.

the standard position. Then carry out in sequence, beginning proximally, near the trunk, the motions that would be necessary to bring the joint into the position in question.

Muscles that produce movement in directions opposite to one another are said to be *antagonists*. Muscles that work together to accomplish a particular movement are *synergists*. Two muscles that bend the arm at the elbow are synergistic with one another, but both are antagonistic to a muscle that straightens the arm.

It is obvious that, to produce a particular movement, the muscles opposing a motion must relax or they may prevent the accomplishment of the movement. The manner in which the nervous system coordinates these muscle actions will be described in a later chapter.

In several instances, muscles are attached across two joints. Generally, however, one principal or *prime movement* can be assigned to each such muscle. In the discussion that follows, the prime movement is emphasized in most instances, but the reader should note the secondary movements that the muscles can carry out. Only flexion and extension are possible in simple hinge joints, such as the elbow and knee, but several joints are capable of motion in more than one plane. *Rotation* can occur at the shoulder and hip, for example.

In general, the more complicated motions will not be considered in the description of specific muscles that follows. Many of these other motions require additional muscles; other movements can be brought about by suitable combinations of the muscles listed, which are mainly the antagonistic pairs at each joint. Only the larger muscles are mentioned. The reader may discover many of these on his own body.

Muscles are named in various ways: some according to their location, or their origin or insertion; some according to their appearance or their action. A few moments spent in analyzing meanings of the muscle names will help considerably in remembering them.

Because the body is bilaterally symmetrical, each skeletal muscle has a twin on the opposite side of the body. Corresponding muscles of opposite limbs are independent of one another, but the twinned muscles associated with structures of the axial skeleton often work together to bring about a particular movement.

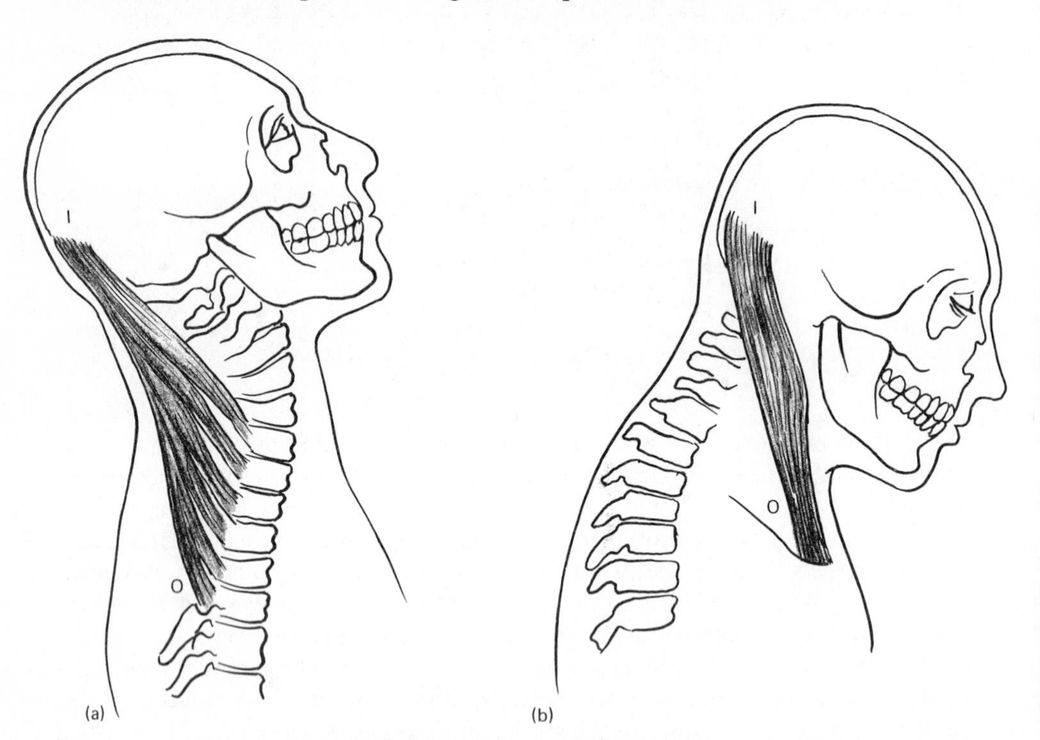

Fig. 4-9. Movements of the head on the neck. (a) Extension of the head: **semispinalis capitis**—*origin* along cervical and thoracic vertebrae spines; *insertion* on occipital region of cranium. (b) Flexion of head: **sternomastoid**—*origin* on sternum and clavicle; *insertion* on mastoid region of temporal bone.

Movements of the axial skeleton

The head is flexed at the joint between atlas and occipital condyles, when the *sternocleidomastoid* muscles of both sides contract (Fig. 4-9). The origin on the sternum and clavicle and insertion on the mastoid provide the name sternocleidomastoid.

Contraction of the *semispinalis capitis* muscles of both sides will bring about extension of the head if the sternocleidomastoids are sufficiently relaxed. Semispinalis refers to the origin along half of the vertebral column, while capitis refers to the insertion on the head.

Simultaneous contraction of sternocleidomastoid and semispinalis capitis of one side abducts the head with respect to that side. In this abducted position, contraction of the corresponding pair of muscles on the other side will adduct the head to the midline, and abduct it beyond to the other side.

Contraction of any one of these muscles alone would tend to tilt the head. With a suitable combination of these and other muscles associated with the neck, rotation at the joint between atlas and axis will occur.

Flexion of the trunk depends on the *quadratus lumborum* muscles that lie deep at the back of the abdominal cavity (Fig. 4-10). This name refers to the rectangular shape of the muscle and its location in the lumbar region of the back. The abdominal muscles that hold the abdominal contents firmly in place and assist in breathing may also help in flexion of the trunk.

The trunk is moved backward—that is, extended—by powerful muscles that lie dorsally along both sides of the vertebral column. One important trunk extensor is the *sacrospinalis,* which originates partly on the sacrum and inserts partly on the upper vertebrae.

When the sacrospinalis and quadratus lumborum of one side contract, the trunk is abducted on that side. When the body is in this position, contraction of the corresponding muscles on the opposite side will result in adduction of the trunk.

Movements of the pectoral girdle

The scapula has a large number of muscles originating or inserting on it, and it can be moved in many directions. A large muscle covering the upper back and shoulders—the *trapezius*—pulls the scapula toward the vertebral column [Fig. 4-11 (a)]. The trapezius is thus an adductor of the scapula. Because it inserts on the clavicle as well as on the spine of the scapula, it also tends to move the clavicle upward and backward.

Abduction of the scapula is brought about especially by the *serratus anterior,* which inserts on the vertebral edge of the ventral side of the scapula [Fig. 4-11 (b)]. The name is descriptive of the sawtooth or serrated appearance of the origins of the muscle from several ribs. *Anterior* merely distinguishes this muscle from another of similar appearance that has a more posterior location.

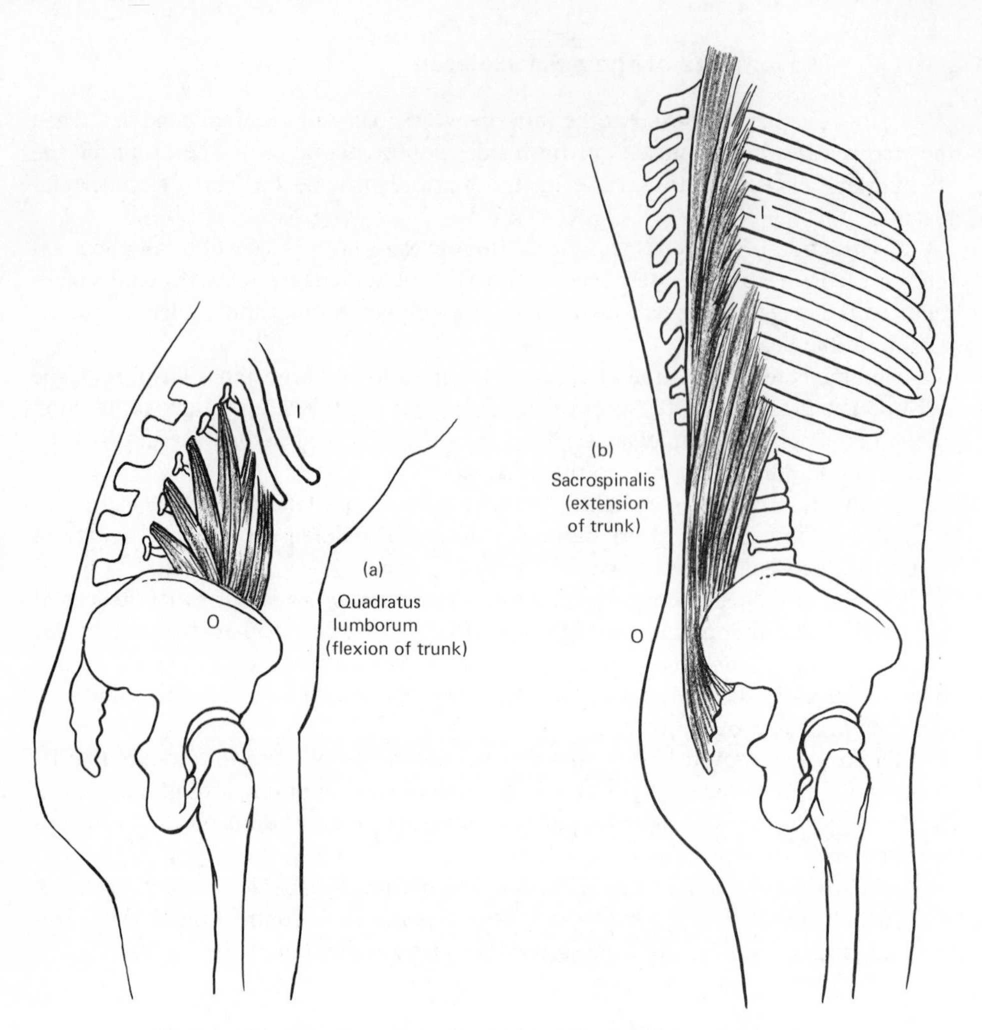

Fig. 4-10. Movements on the trunk. (a) Flex trunk: **quadratus lumborum**—*origin* on iliac crest; *insertion* on last rib and on transverse processes of upper lumber vertebrae. Extend trunk: **sacrospinalis**—origin on sacrum and adjacent iliac crest, insertion on ribs, spines, and transverse processes of vertebrae.

Movements of the anterior limb

The flexor action of the *coracobrachialis* muscle can be easily predicted from the origin of this muscle on the coracoid process of the scapula and its insertion on the humerus [Fig. 4-12 (a)]. *Brachium* (referring to arm) is a root used in many words. *Brachiation*, for example, means to travel by means of the arms, as monkeys do from tree to tree. The coracobrachialis would be very important for this manner

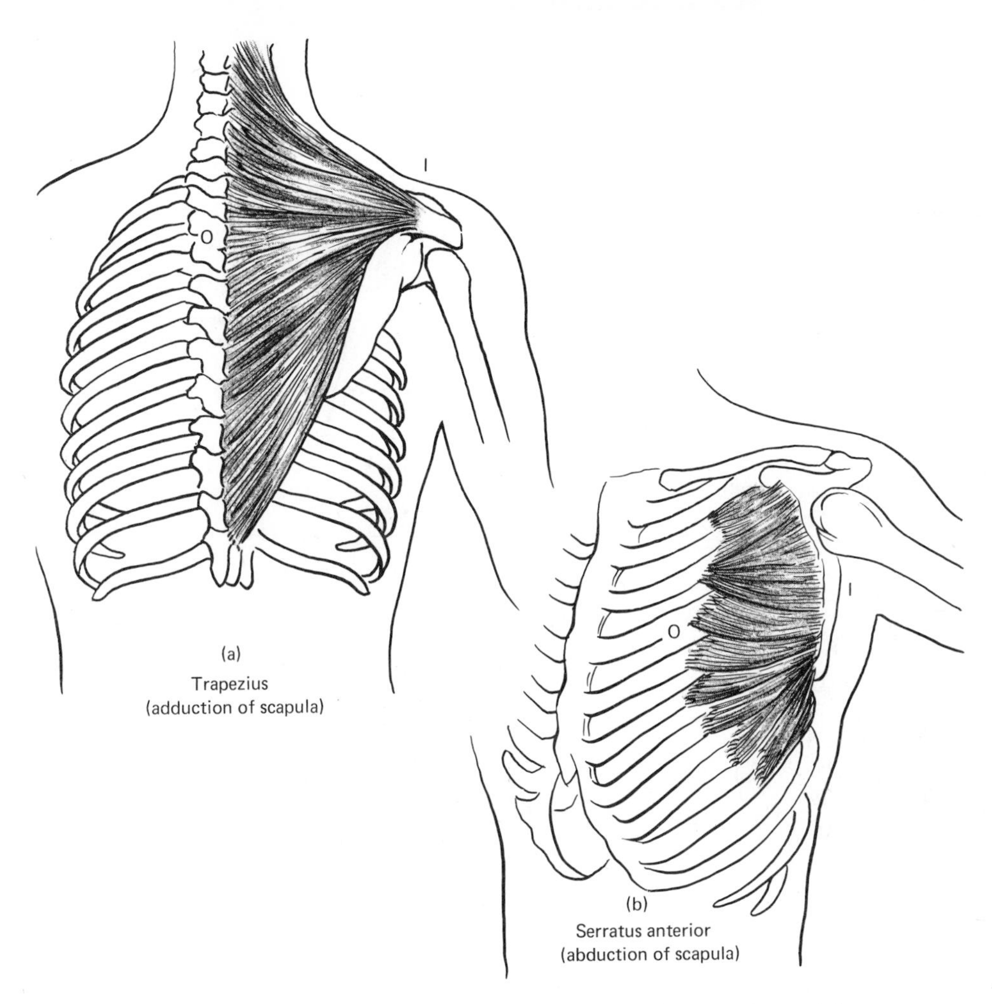

Fig. 4-11. Movements of the scapula. (a) Adduct scapula: **trapezius**—*origin* from occipital protuberance down to the last thoracic vertebrae; *insertion* on lateral one-third of clavicle and on spine of scapula. (b) Abduct scapula: **serratus anterior**—origin on upper nine ribs, insertion along vertebral border of scapula.

of locomotion. The flexor action of the coracobrachialis is opposed by the *teres major*, an extensor of the arm at the shoulder [Fig. 4-12(b)]. "Major" distinguishes this muscle from a smaller, "minor" one. Teres refers to the rounded shape of the muscle in cross section.

Abduction of the humerus at the shoulder results when the *deltoid* muscle contracts [Fig. 4-12(d)]. From a narrow insertion laterally on the head of the humerus, this muscle fans out to a wide origin on the scapula and clavicle. Thus it has the shape of the Greek letter delta. "oid" is an ending that means "similar to."

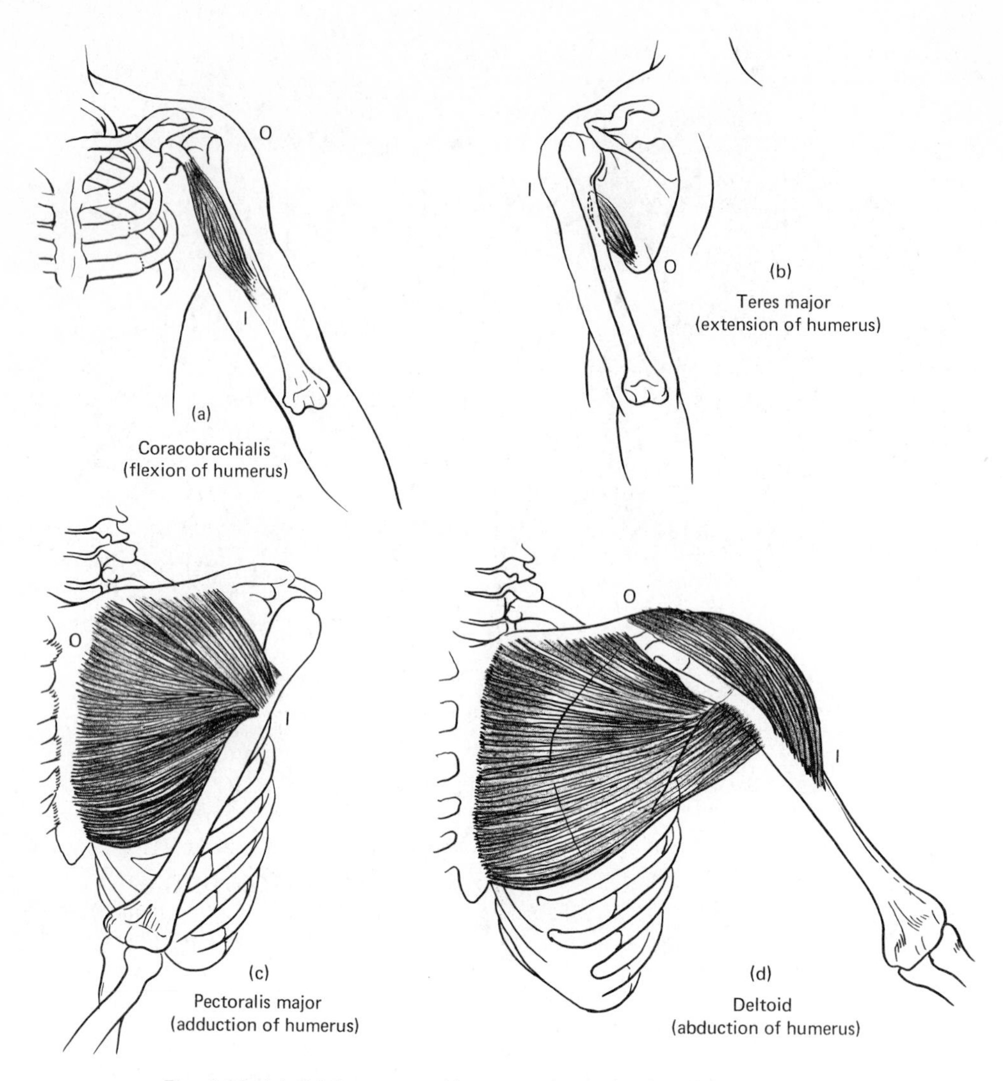

Fig. 4-12. (a) *Origin* on coracoid process of scapula; *insertion* along middle of medial surface of humerus. The coracobrachialis of the left arm shown from the front, brings the arm forward. (b) Extend humerus: **teres major**—*origin* on lower axillary border of scapula; insertion on intertuberlar groove of humerus. The teres major of the left arm, seen from in back, moving the arm backward. (c) Adduct humerus: **pectoralis major**—*origin* on sternum and medial part of the clavicle; insertion on greater tubercle of humerus. (d) Origin on acromion and lateral one-third of clavicle; insertion on deltoid tubercle of humerus.

The upper arm may be adducted as far as the vertical position, where it is stopped by the ribs. Further adduction can be brought about if the limb is pulled somewhat ventrally—that is, if it is flexed. The *pectoralis major* is important in this adduction.

Pectoralis major means "large muscle of the chest" [Fig. 4-12(c)]. Adduction of the arm in extension can be accomplished with the help of certain muscles on the dorsal side.

The main movement of the arm at the elbow is a hinge action of the ulna on the humerus. Flexion of the arm at the elbow results when the *brachialis* muscle contracts. This muscle originates on the surface of the shaft of the humerus and inserts on the head of the ulna [Fig. 4-13(a)].

The *biceps brachii* muscle inserts on the radius in such a way that it rotates the radius, in addition to aiding the brachialis in flexing the arm at the elbow. Brachialis and biceps brachii are synergists, in spite of their different insertions. Biceps means two-headed, in reference to the two origins of this muscle. The long head of the biceps brachii arises from the coracoid process; the short head from just above the glenoid fossa. The outward rotation of the radius, brought about by the biceps brachii, brings the hand into the palm-up or *supine* position. Small muscles between radius and ulna assist this action, but others, especially the *pronator quadratus*, reverse the direction of this motion. Palm down is the *prone* position of the hand, and a muscle that moves the hand into this position is a pronator.

Extension of the arm at the elbow is the responsibility of one muscle, the *triceps brachii*. As the name implies, this muscle has three origins. Lateral and medial heads arise from the humerus, while the long head originates just below the glenoid fossa. All three heads insert together via a common tendon on the olecranon process.

Movements of the wrist and fingers depend on several muscles of the forearm. The *flexor carpi ulnaris* and the *flexor carpi radialis* both originate on the medial epicondyle of the humerus and can aid flexion at the elbow. The flexor carpi ulnaris also flexes the hand (carpus) by means of its insertion on metacarpals at the ulnar side of the hand. The flexor carpi radialis has a similar action through its insertion on metacarpals at the radial side.

The *extensor carpi ulnaris* and the *extensor carpi radialis* originate on the lateral epicondyle of the humerus. They may be synergistic with the flexor carpi muscles in their action across the elbow joint. More imporant, they extend the hand at the wrist through their insertions on the dorsal side of metacarpals at the ulnar and radial sides of the hand.

Muscles in the forearm also control flexion and extension of the fingers. The proximal joints of the fingers are flexed by the superficial *flexor digitorum* muscles that arise from the proximal head of the ulna; the terminal phalanges are flexed by a deeper-lying *flexor digitorum* muscle that originates more distally on the ulna.

The flexors of hand and fingers have long tendons, which pass over the wrist to the points of insertion [Fig. 4-13(b)]. These tendons are held close to the wrist and palm of the hand by a strong cuff of connective tissue, the *flexor retinaculum*.

Extension of the fingers is carried out for the most part by the *extensor digitorum communis*, which, with the extensor muscles of the hand, originates on the lateral epicondyle of the humerus [Fig. 4-13(c)].

Like the flexor tendons, the extensor tendons are held close to the wrist and palm by means of a tough connective tissue cuff, the *extensor retinaculum*.

The fingers can be fanned out (abducted) or moved together (adducted) by

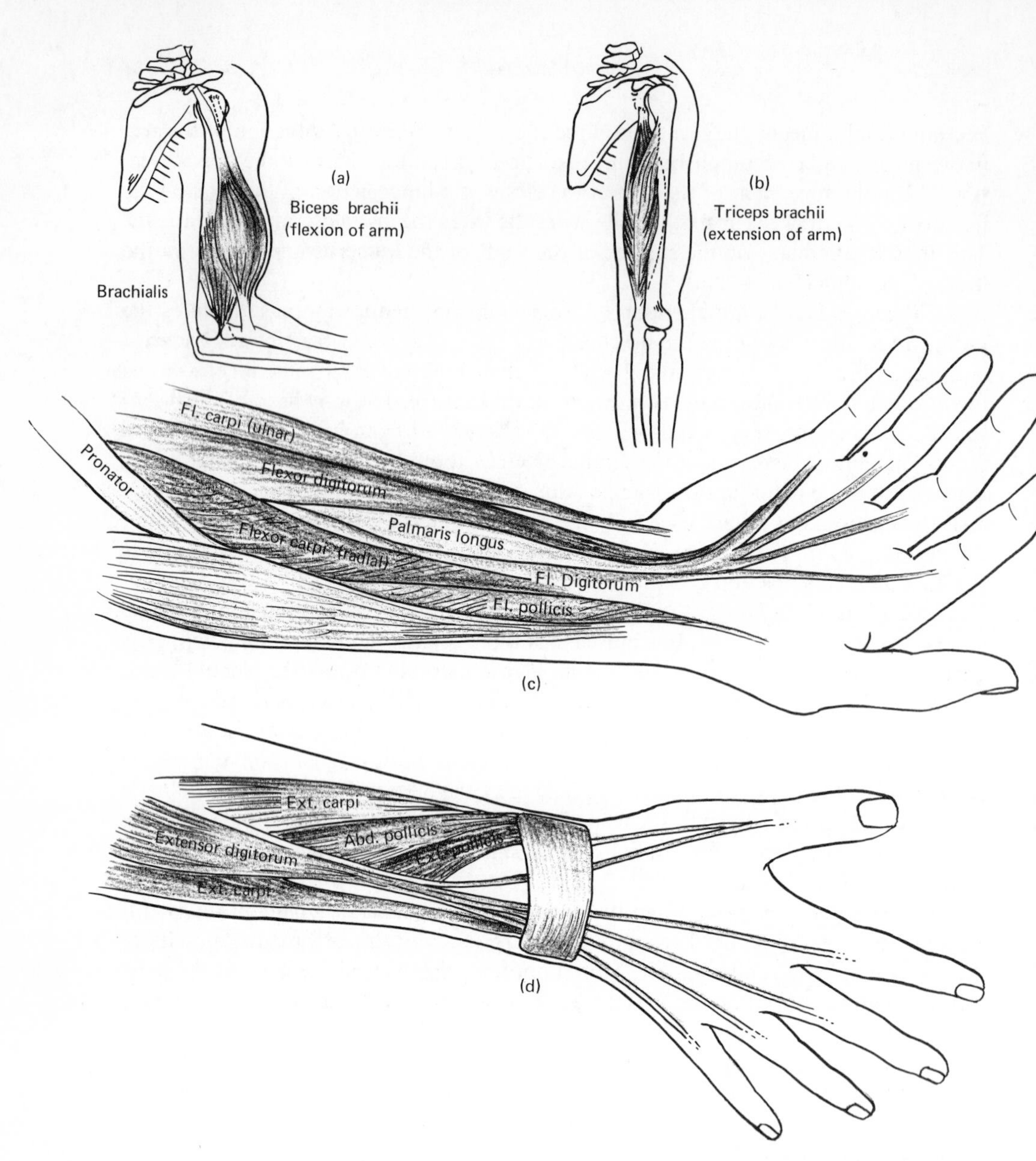

Fig. 4-13. Movements on the lower arm at the elbow. (a) Flex radius and ulna: **brachialis**—*origin* on middle of the humerus; *insertion* on coronoid process of ulna. **Biceps brachii**—*origin* of short head on coracoid process of scapula, of long head at supraglenoid tubercle; *insertion* of both heads on tubercle of radius. The biceps brachii and the brachialis of the right arm are seen from the right side. (b) Extend radius and ulna: **triceps brachii**—origin of the three heads are separately on the infraglenoid tuberosity, the dorsal distal part of the humerus, and on the lateral and proximal part of the humerus; insertion of all three on olecranon process of ulna via a common tendon. The triceps brachii of the right arm, is seen from the right side. (c) Extensions of fingers and wrist.

means of small muscles in the palm of the hand. These muscles are called *interosseus* muscles, because they are between the carpals of the palm.

Movements of the lower limbs

Flexion of the thigh during walking is the responsibility mainly of the *iliopsoas* muscle. This muscle has two origins: one, on the inner surface of the ilium; the other, along the lateral spines of the lumbar vertebrae [Fig. 4-14(a,b)]. The two parts have a common insertion on the greater trochanter of the femur. While lying on your back, you can flex the trunk at the hips if your legs are held down. In this instance, the conventional origin of the iliopsoas muscle moves and thereby becomes a functional insertion, while the conventional insertion on the femur remains stationary and thereby becomes a functional origin.

Action of the iliopsoas is opposed by the *gluteus maximus*, the largest muscle of the rump. Another gluteal muscle, the *gluteus medius*, brings about abduction of the thigh at the hip. Originating from the lateral surface of the ilium, its fibers insert on the greater trochanter of the femur. Its action is opposed by the massive adductor muscles of the thigh [Fig. 4-14(c)].

The thigh adductors are in three parts: The *adductor brevis* stretches from the lower part of the pubis to the upper part of the shaft of the femur. Parallel to the adductor brevis, but traveling a longer route, is the *adductor longus*, arising near the pubic symphysis and inserting along the middle of the shaft of the femur. The *adductor magnus* is rather different from the other two. It originates on the ischium and inserts by a long tendon, medially, just above the medial condyle of the femur. Its attachments also allow it to assist in extension of the hip at the thigh.

Several muscles of the leg act across two joints. This is the situation in the *hamstring muscles*, named from the stringlike appearance of their exposed tendons. The largest of the hamstrings is the *biceps femoris* [Fig. 4-15(a)]. The long head of this muscle originates on the ischial tuberosity and thus helps in extension of the thigh. The short head originates on the surface of the femur. Both heads insert via a common tendon into the proximal part of the fibula.

The *semimembranosus* and the *semitendinosus* muscles, named from their appearance, also arise from the ischial tuberosity, but they insert on the head of the tibia. Their action is, in effect, synergistic with that of the long head of the biceps femoris.

Two large muscles that flex the leg at the knee arise on the ventral side of the pelvic girdle. The *gracilis* passes from the pubis to the head of the tibia, while the *sartorius*, inserting close to the gracilis, originates on the anterior superior iliac spine.

Quadriceps femoris is the powerful extensor muscle of the leg [Fig. 4-15(b)]. It must keep the knees straight in spite of their tendency to bend from the weight of the entire body. As its name implies, it is a four-headed muscle. All four heads insert via a common tendon into the head of the tibia. In this tendon the kneecap or *patella* serves as a protective cover over the large knee joint. Each head of the quadriceps

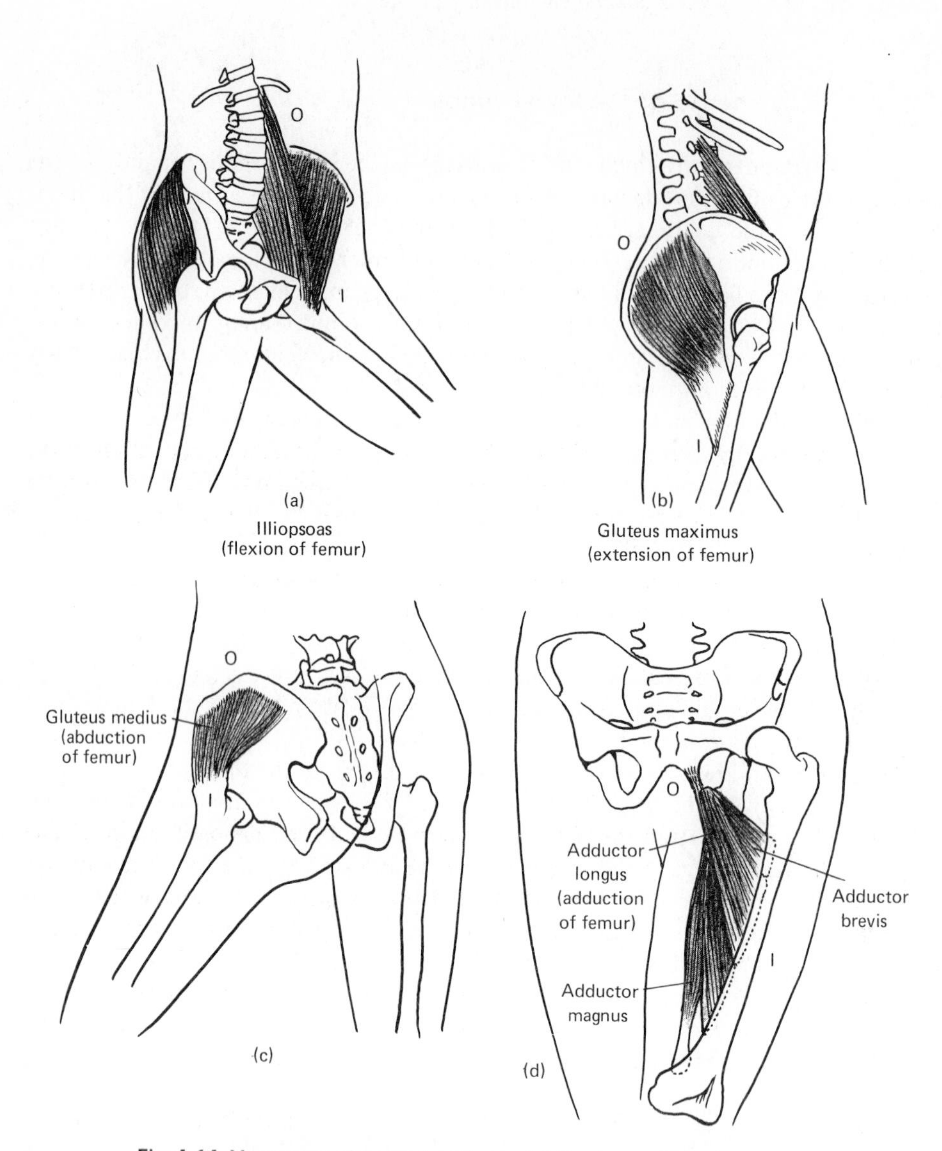

Fig. 4-14. Movements of the thigh at the hip. (a) Flex femur: **iliopsoas**—origin on lumbar vertebrae and iliac fossa; insertion on lesser trochanter of femur. (b) Extend femur: **gluteus maximus**—origin on coccyx and posterior surface of ilium; insertion on gluteal tuberosity of femur. (c) Abduct femur: **gluteus medius**—origin on the outer surface of the ilium; insertion on the lateral surface of the ilium; insertion on the lateral surface of the greater trochanter. (d) Adduct femur: **adductor brevis, longus, magnus**—origin on pubus and ischium; insertion on linea aspersa of femur.

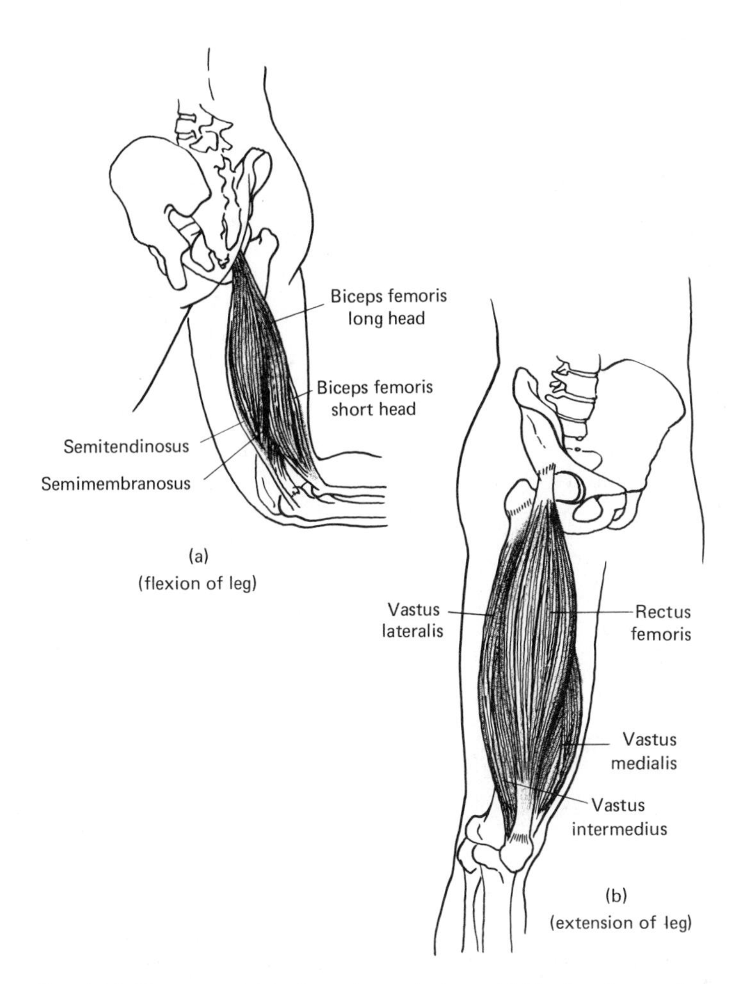

Fig. 4-15. Movements of the lower leg at the knee. (a) Flex leg: **hamstring**: **semitendinosus** and **semimembranosus**—*origin* on ischial tuberosity; *insertion* on proximal part of tibia. **Biceps femoris**—origin of the long head from ischial tuberosity, of the short head from linea aspersa of femur; *insertion* of both heads on proximal part of fibula. Not shown: **gracilis** and **sartorius.** (b) Extend leg: **quadriceps femoris**: **rectus femoris**—origin on anterior inferior iliac spine, **vastus lateralis** and **medialis**—origin on linea aspersa of femur, and **vastus intermedius**—origin on anterior surface of femur. All these parts of the quadricepts femoris insert on the tibial tuberosity via the patella.

femoris is given a separate name. The *rectus femoris* is straight along the femur from its origin on the anterior inferior iliac spine to its insertion via the patella. It assists in flexion of the thigh at the hip as well as extension of the leg at the knee. The remaining three heads arise exclusively on the femur. The large *vastus lateralis* is on the lateral

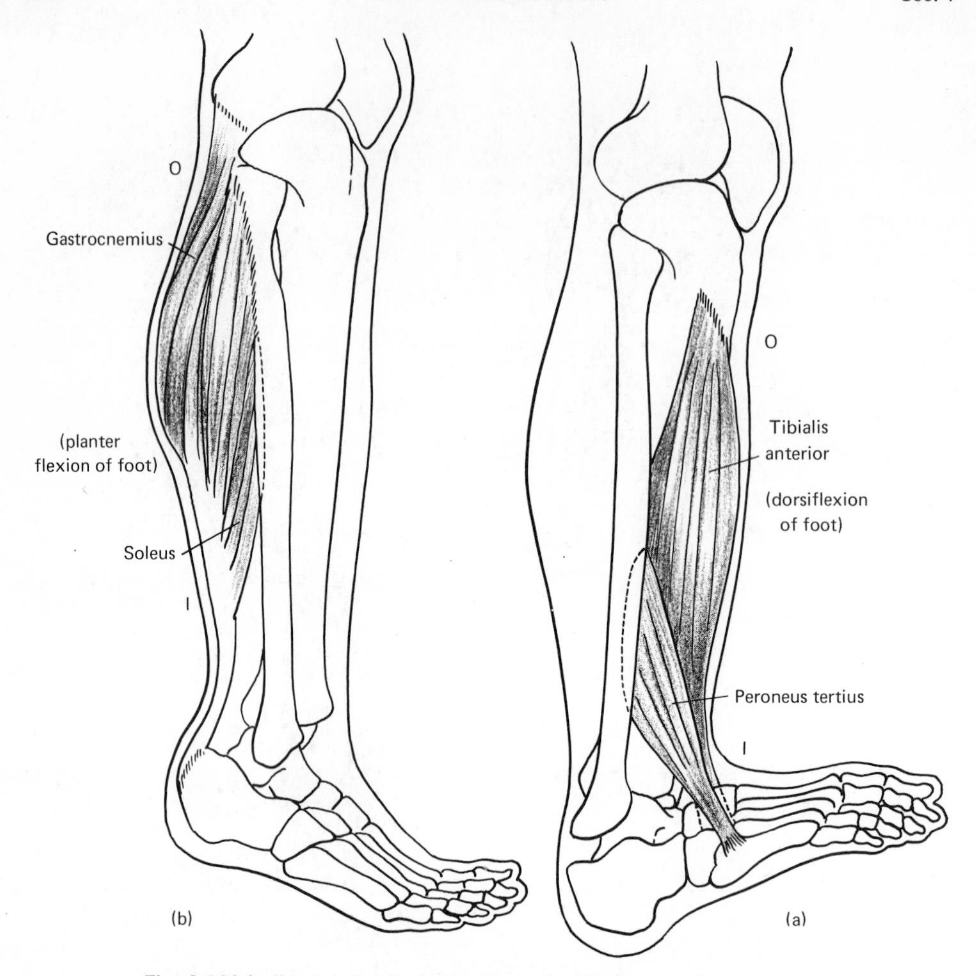

Fig. 4-16(a), (b). (a) Flex foot (dorsiflexion): **tibialis anterior**—*origin* on outer surface of tibia; *insertion* on first metatarsal. **Peroneus tertius**—*origin* on fibula and *insertion* on fifth metatarsal. The gastrocnemius and soleus of the right leg, seen from the right side. (b) Extend foot (plantar flexion): **gastrocnemius**—origin on medial and lateral condyles of femur. **Soleus**—origin on fibula and tibia. Insertion of both on calcaneus via the common tendon of Achilles. The tibialis anterior and peroneal muscles of the right leg, seen from the right side.

side; the *vastus medialis* is medially placed; and the *vastus intermedius* is in between these two on the shaft of the femur.

Direct comparison of the movements of the hand and the foot may be somewhat confusing at first. The sole of the foot corresponds to the palm of the hand; therefore, moving the foot upward at the ankle corresponds to extending the hand at the wrist. The top of the foot is the *dorsum*, and this movement is called *dorsiflexion*. The sole of the foot is the *plantar surface*. The movement of bending the foot downward at the ankle is called *plantar flexion* (Fig. 4-16). The *gastrocnemius* and the *soleus*

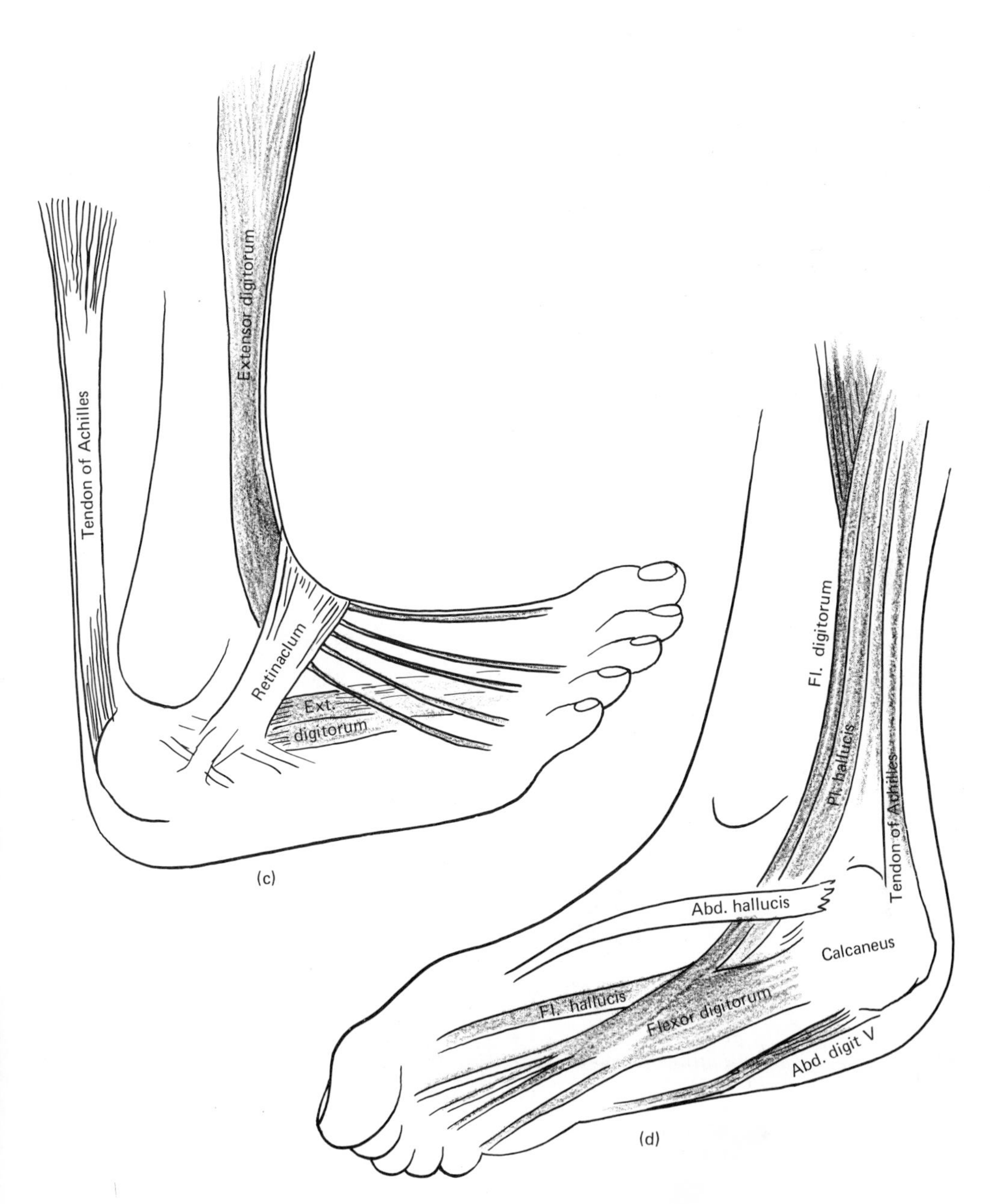

Fig. 4-16(c), (d). (c) Dorsiflexion of foot (extension of toes). (d) Plantar flexion of foot (flexion of toes).

muscles are the most powerful plantar flexors. These muscles insert together on the *calcaneus*, or heel bone, by way of the *tendon of Achilles*. Gastrocnemius means "belly-point" in reference to the round shape of the muscle ending in a point at Achilles' tendon. The muscle has two origins on the medial and lateral condyles of the femur. The gastrocnemius is, therefore, a flexor of the knee as well as a dorsiflexor of the foot. The soleus, originating from the surfaces of the tibia and fibula, is shaped somewhat like an old slipper.

Peroneus refers to fibula, and the several *peroneal muscles* originate mainly on the fibula. The fibula is on the lateral side of the leg, and the peroneus muscles, which insert mainly on metatarsals, are important in turning the foot laterally—that is, everting it. The tibial muscles, originating mainly on the tibia and inserting mainly on metatarsals, turn the foot inward—that is, invert it. *Tibialis anterior* and *peroneus tertius* also dorsiflex the foot, while *tibialis posterior* and *peroneus longus* and *brevis* plantar flex the foot. The toes are flexed by the *flexor digitorum longus* and extended by the *extensor digitorum longus*. Both these muscles originate on the tibia and insert by long tendons onto the appropriate surfaces of the toes. Like the corresponding muscles of the forearm, the long tendons of muscle inserting into the ankle

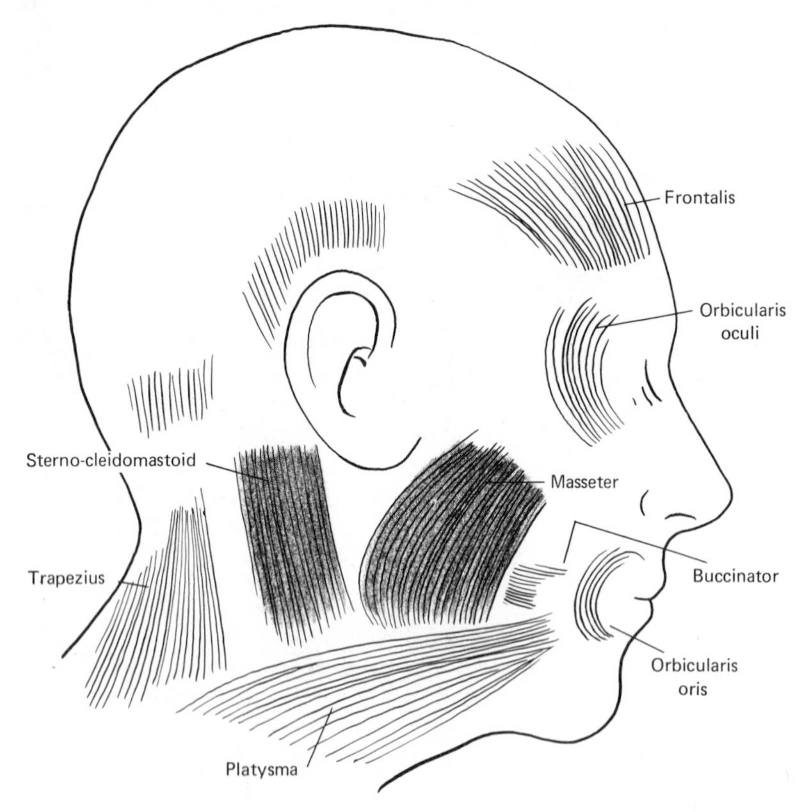

Fig. 4-17. Face muscles.

and toe bones are held in place by *flexor* and *extensor retinacula*. Like the hand, the foot has many small muscles intrinsic to it, especially muscles concerned with adduction and abduction of the toes.

Movements of the face

There are several small but important muscles of facial expression that move the lips, nose, eyes, ears, and scalp. The scalp, a heavy layer of connective tissue that covers the top of the head like a tight-fitting cap, is moved by the *epicranius*, which means on top of the cranium. Actually, the muscle fibers are just above the eyebrows and over the occipital region of the skull. The *corrugator supercilii* draw the eyebrows together. The name refers to the resulting furrows and the location of the muscle above the eyebrows. The eyes are narrowed and the eyelids drawn down by the *orbicularis oculi*, a name that refers to the location of the muscle encircling the eye. Another encircling muscle is the *orbicularis oris*, which puckers up the lips as in whistling. The side of the neck and the corner of the mouth are drawn taut by the large flat *platysma*. The side of the mouth is held flat against pressure exerted, as in blowing a trumpet, by the *buccinator* muscle. (See further Fig. 4-17.)

Section 4: *Questions, Problems, and Projects*

1. You may more easily remember anatomical as well as physiological terms if you understand the root meanings of the words. These roots are used not only in scientific terminology but also in words found in general literature and in ordinary speech. You may need a dictionary in order to find the elemental meanings of the prefixes, suffixes, and roots italicized in the words below: Write those meanings and an example or two of more familiar words having the same origins. Find other words to treat similarly.

   *os*teo*clast*, osteo*blast*, *epiphys*eal, *acro*/*mega*ly, *a*/*chond*ro*plastic*, sternum, *meta*carpals, *syn*/*arthr*osis, *di*/*arthr*osis, *syn*o*vi*al, *quadra*tus, lumborum, *serra*tus anterior, *trapez*ius, *supin*ator, *pron*ator, *extens*or *digit*orum *commun*is, *ab*/*duct*or, *ad*ductor, *quad*riceps.
2. Carry out the following movements and describe the muscles involved. Name synergistic muscles, antagonistic muscles, related to each action: abduct the arm at the shoulder, flex the elbow, flex the wrist, abduct the thigh, flex the knee, stand on the toes.

3. Choose some complicated movement involving several parts of the body and describe in sequence the muscles you would have to bring into action and those you would have to keep relatively inactive in order to carry out the movement at each joint.
4. Make a list of muscles that are named according to origin and insertion. Make a list of those named according to their action. How would you classify the rest?

5

MUSCLE MECHANICS

5.1 Lever Systems, Muscle Force, and Work

Like other machines, muscles can apply *force* and do *work*. So indeed can other organs whose functions may appear to be less spectacular. Certainly the manifestations of cellular activity are most dramatic in muscle, in which invisible chemical energy may be transformed into energy of motion.

Force implies a push or a pull. Perhaps the most familiar example is the *force of gravity*. When you hold a bowling ball in your hand, your muscles apply a force equal to but in the opposite direction to the force of gravity. If you let the ball fall, it will move downward toward the earth with increasing velocity until it is stopped by collision with some object. Without your support, the ball accelerates toward the earth. You provide an acceleration equal in magnitude but opposite in direction and thereby keep the ball motionless.

Subjectively you feel you must exert greater effort—that is, apply more force—to hold several objects of the same kind instead of one; but if you were on the moon, you could hold a dozen bowling balls with the same effort as you hold one on the earth. If you drop a ball in the lunar bowling alley, its acceleration downward will be about one-twelfth the acceleration it has on the face of the earth.

We have been describing a quality of a thing that varies directly as the amount of material in the thing (m) and its acceleration (a). This quality we call *force* (F). In mathematical symbols, we express the relationship as F (force) $= m$ (mass) $\times\ a$ (acceleration).

Work implies productive effort involved in application of a force. You do useful mechanical work on the ball if you lift the bowling ball upward a distance against the force of gravity. The work done is defined as $W = F \times D$, where D is the distance through which the force F is exerted.

Strictly speaking, you do no *useful mechanical work* when you hold an object motionless in a particular position. The work done is of a more subtle kind, involving chemical reactions and movement of electrical charges in your muscles and nerves.

Application of muscle forces in levers of the body

A muscle generally does not apply its force directly in a straight line to the object being moved, but indirectly, as a weight on one side of a seesaw opposes a weight on the other side. Such a simple mechanical device is a *lever*. A lever consists of a rigid beam to which are attached forces that may move the beam up or down clockwise or counterclockwise around a pivot, the *fulcrum*. A lever allows a small force to balance a large force. It also allows a small force moving through a long distance to move a large force through a short distance (Fig. 5-1). A small weight placed on the lever at a suitable distance from the fulcrum will just balance a larger weight closer to, but on the opposite side of, the fulcrum. The distance between the fulcrum and the point of application of the force is called the *lever arm*. In Fig. 5-2 one lever arm is designated L_1, the other L_2. The force at distance L_1 is called F_1, the force at L_2 is F_2. If F_1 is moved farther away from the fulcrum, or if a larger force is substituted, it can balance a larger F_2. Otherwise stated, the effectiveness of the force on the lever depends on the size of the force and its distance from the fulcrum. Effectiveness of the force applied to the lever is called the *torque* (T). The torque is directly proportional to the magnitude of the force and the length of the lever arm; that is, $T_1 = F_1L_1$ and $T_2 = F_2L_2$. If two forces are balanced on a lever, then the torque tending to turn

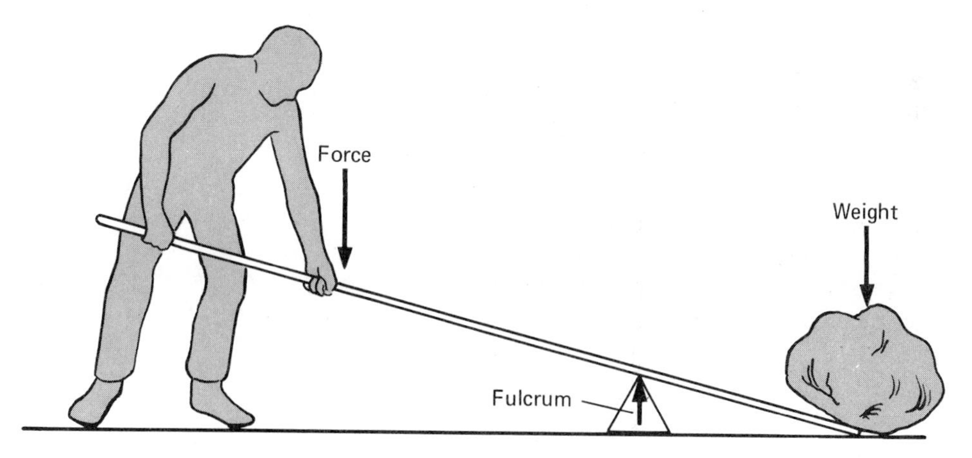

Fig. 5-1. Use of a lever.

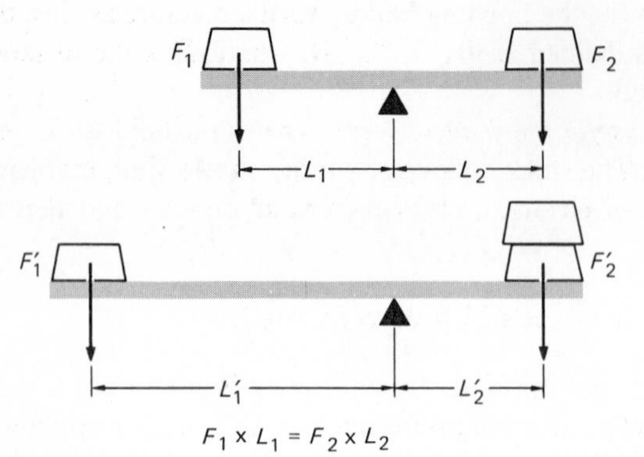

Fig. 5-2. The principle of the lever.

the lever clockwise equals the torque tending to move the lever counterclockwise, and there is no net movement. Then $T_1 = T_2$ or $F_1L_1 = F_2L_2$.

In Fig. 5-2, when one lever arm is increased in length (L_1 to L'_1), the force balancing the weight it bears must decrease proportionately if the system is to remain in balance (F_2 to F'_2). In other words, the lengths of the lever arms are inversely proportional to the forces they support. That is,

$$\frac{L_1}{L_2} = \frac{F_2}{F_1}$$

which is another way or writing

$$F_1L_1 = F_2L_2$$

The attachment of the head on the neck is an example of a simple lever system (Fig. 5-3). If the head is tilted slightly forward, its weight provides force F_1 working through lever arm L_1. The head may be prevented from moving further downward—that is, may be balanced—by F_2, the force of contraction in the semispinalis muscles at the back of the neck.

The body has other arrangements of forces with relation to the fulcrum, but they may be similarly analyzed. In the most common type—for example, in the arm—both forces are on the same side of the fulcrum (Fig. 5-4). This type of lever, compared with that of the previous example, seems to be upside down. However, the forces function in exactly the same way, one tending to move the lever arm clockwise, the other counterclockwise, around the fulcrum, which in this instance is the end of the humerus where it articulates on the ulna. If a weight is held in the hand, the arm being horizontal, the force required to be exerted by the brachialis muscle may be calculated as shown. In the example, the humerus is oriented so that the muscle pulls vertically. When the arbitrary values shown are used in the equation, a force of 100 kg is required to support a weight of 10 kg.

The force that a muscle must exert to balance a particular weight depends on the extent to which the muscle has shortened. In Fig. 5-5 the brachialis muscle is repre-

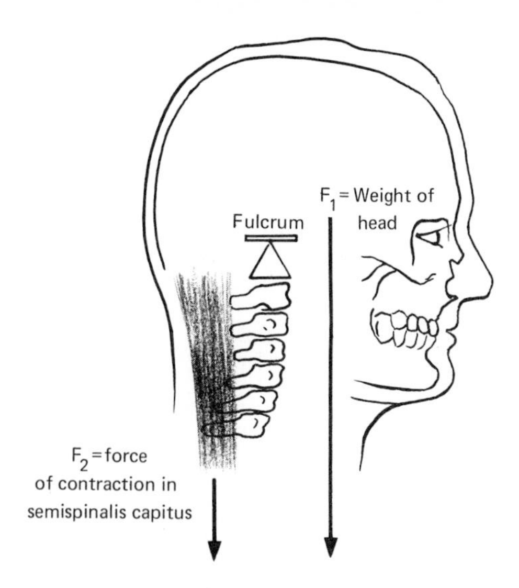

Fig. 5-3. Forces on head as lever.

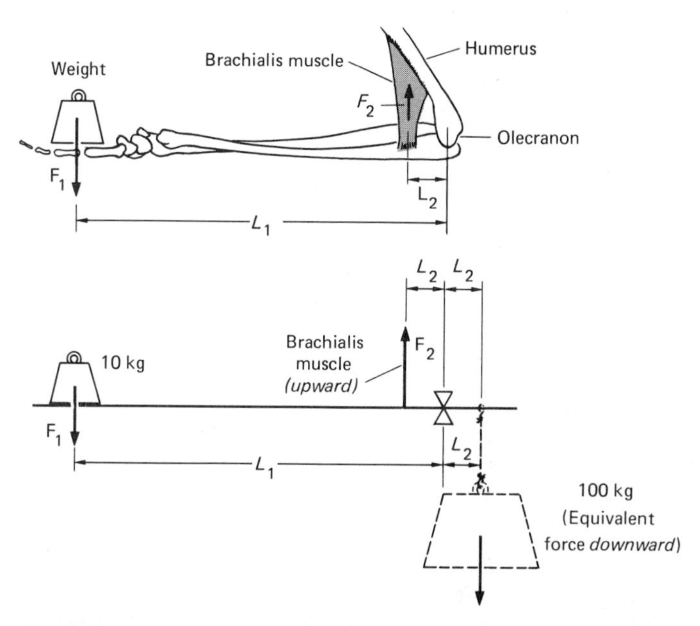

Fig. 5-4. The arm as a lever. (a) The arm is held horizontal; the brachialis muscle pulls directly vertically. The fulcrum is the articulation of the ulna on the humerus. (b) The lever system equivalent to (a). The torque counterclockwise exerted by the 10-kg weight equals the torque clockwise exerted by the muscle. If L_1 = 25 cm, L_2 = 25 cm, then—since $F_1L_1 = F_2L_2$—F_2, the force exerted by the muscle, is 100 kg. Note that for simplicity of analysis, an equivalent downward force may be substituted for the upward force exerted by the muscle.

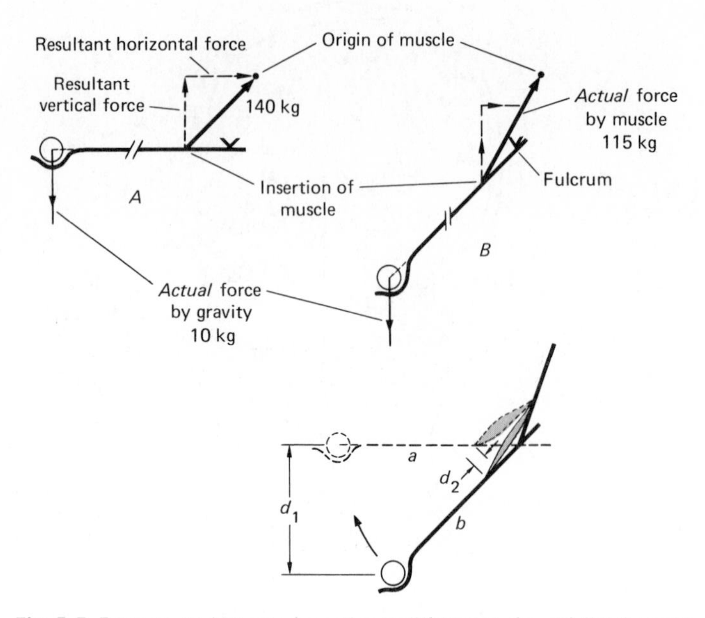

Fig. 5-5. Force exerted by muscles acting at different angles. (a) For the same arrangement as in Fig. 5-4, if brachialis muscle pulls obliquely rather than vertically, greater force is required to hold the weight. In *A*, the muscle force is exerted at 45° to the vertical, while the forearm is held horizontal. In *B*, the muscle force is exerted at 30° to the vertical, and the forearm is held at 45°. In both instances, since the ratios of lever arms is assumed to be 1 : 10, the resultant vertical force is the same as in Fig. 5-4, 100 kg. However, it can be calculated that the resultant horizontal forces are unlike, and the *actual* muscle force required to hold the 10 kg in *A* is about 140 kg, as compared to about 115 kg in *B*. (b) Work done may be calculated for movement of weight vertical distance d_1 when muscle shortens amount d_2 during contraction (see text).

sented as holding the ulna at two successive positions as the weight is lifted. The *direct upward force* necessary to hold the weight *is the same* in either position. If the force were less than that necessary to just balance the weight, then the arm would move down; if it were greater, the arm would move upward.

The force that must be exerted *by the muscle will vary*, however, as the muscle shortens. In each instance in the figure, the arrow, F_2, represents the direct upward force necessary to balance the weight. This force is the same in both (a) and (b). In both instances, the muscle provides this upward force. It also provides a force *horizontally*. In (a), the horizontal force must be greater than in (b). Evidently, as the arm is moved toward the horizontal, more force is required in the muscle to balance the weight.

The muscle is stretched when the limb is brought toward a vertical position. Under these conditions the muscle is actually capable of exerting more force than when it is not stretched. Stretch increases the effectiveness of muscle contraction because none of the contraction force need be used to take up external slack—the force can therefore be exerted more fully on the external attachments.

Work during muscle contraction

A lever is said to provide a mechanical advantage because it allows the operator of the lever to move a large force while applying a small force. As previous examples have shown, the muscles of the body are generally arranged in such a way that they must apply a large force to balance a small force. The mechanical advantage usually provided by muscles is thus an advantage in terms of distance rather than of force. Skeletal muscles are so arranged that their forces are usually exerted close to the fulcrum. However, the opposing force (a weight moved) generally has a long lever arm. Consequently, a small movement of the point where the muscle inserts (a small amount of muscle contraction) will provide a considerable displacement of the opposing force.

In the previous examples we considered the muscle forces necessary just to balance external forces. If we exert force to hold an object at a certain place, we normally say that we are doing work. In a general way, *work* is described as the expenditure of energy. In this sense, all the cells of the body do physiological work, even when they appear to be doing nothing at all. Metabolism involves physiological work. In a simpler and more precise way, work is said to be done if a force acts through a certain distance.

In the example illustrated in Fig. 5-5, the brachialis muscle shortens a small amount d_2 and moves the weight upward against the force of gravity a net large distance d_2. The work done on the weight by lifting it the distance d_1 is equal to the external work done by the muscle in shortening the small amount d_2. Work accomplished is proportional to the force applied (or weight moved) and the distance through which this force is exerted. The force in this instance, as in previous examples, is due to the acceleration of gravity on the mass involved, and the distance is therefore the vertical distance in opposition to the force of gravity. Work equals force times distance: $W = Fd$, and $W = F_1 \times d_1 = F_2 \times d_2$.

In gravitational units, we equate force and weight. Thus if the weight is 10 kg and the distance is 10 cm, then 10×10^3 g $\times$ 10 cm $= 10^5$ g cm of work. If we express the force in absolute units, then the force exerted by the 10-kg weight is 10×10^3 g $\times$ 980 (dynes of force per gram) $= 9.8 \times 10^6$ dynes, and the work done by the muscle is therefore 9.8×10^6 dynes $\times$ 10 cm $= 9.8 \times 10^7$ dyne-cm or 9.8×10^7 ergs of work. In doing this work, the muscle would have expended energy equivalent to a much greater amount of work. As we saw, the applied force is required to be much greater than the weight lifted, because of the poor mechanical advantage of the lever system on which the muscle acts. In addition, the muscle is not particularly efficient, and a great deal of energy is utilized without yielding useful external work. The amount of force the muscle must exert in various positions is shown in Fig. 5-5.

5.2 Regulation of Muscle Force

We can control with great precision the force applied by our muscles. With delicate touch we pick up a tiny insect or with powerful grip lift a 100-lb weight. The amount of force the muscles apply is controlled by the nervous system, but it is the muscles themselves that apply the force. The muscle fibers normally contract only in response to nerve messages (impulses) from the central nervous system to which each and every skeletal muscle is attached by means of a branch of a nerve fiber.*

The motor unit

Each nerve fiber is connected to several muscle fibers. Small muscles used for fine movements may have 10 to 20 muscle fibers attached to branches of one nerve fiber (e.g., the small muscles of the fingers, and those that move the eyes). Large muscles used for coarser movements (muscles of the arm or leg) may have 100 or more muscle fibers attached to the branches of the nerve fiber. A group of muscle fibers innervated by branches of one nerve fiber is called a *motor unit* (Fig. 5-6). An entire muscle consists of many motor units.

* The nature of the nerve impulse (and the muscle impulse) is described in Chap. 6.1.

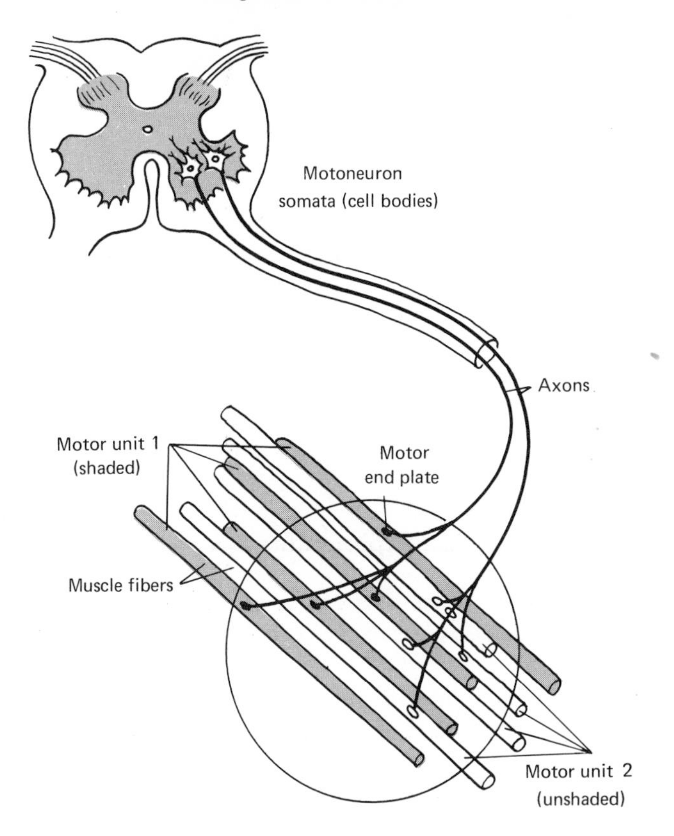

Fig. 5-6. The motor unit. Two motor units are encircled. Muscle fibers of one are shaded; the fibers of the other are unshaded.

The simple twitch

The muscle fibers in a motor unit ordinarily contract in response to a signal coming to them along the nerve. This signal, called a *nerve impulse*, travels over the nerve at about 100 meters per second, and lasts only about $\frac{1}{1000}$ sec (1 msec). Following a nerve impulse to a motor unit, there occurs a momentary shortening of the muscle fibers, lasting less than $\frac{1}{10}$ sec (100 msec). This simplest kind of contraction, which is completed so quickly, is called a *simple* (or single) *twitch* (Fig. 5-7). If several motor units of a muscle are stimulated simultaneously in this fashion, the contraction is still called a simple twitch. The force developed during a single twitch of a particular muscle depends on the number of motor units that are activated.

The knee jerk that follows a tap to the *patella* tendon of the quadriceps muscle is very nearly a simple twitch. A small tap will evoke a small number of motor units and yield a weak kick, whereas a strong tap will provoke a powerful kick.*

* The knee jerk is really controlled by the central nervous system. See Chap. 6.3.

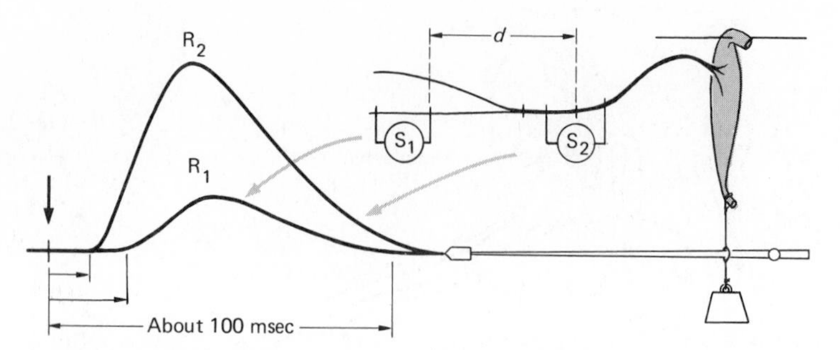

Fig. 5-7. The single twitch. (a) Some of the motor units are activated (some of the nerve fibers are stimulated). (b) All of the motor units are activated (all the nerve fibers are stimulated). *Note* moment of stimulation. Latency for small response is greater than for large response. Why?

Muscle contraction experimentally controlled

A graphic record of the time course of a muscle contraction may be obtained if one end of the muscle is attached to a lever arranged to write on paper moving along at a known velocity (Fig. 5-7). The record of a single twitch will show a period of development of tension and a period of relaxation. The height that the record reaches depends on the amount of shortening *or* the amount of force developed in the muscle.

Experimentally, the *simple* twitch can be most easily and reproducibly obtained by means of an electric shock of short duration. This shock serves merely as a trigger, like the nerve impulse, to start the contraction machinery. The response does not begin immediately, but after a short *latent period* during which the impulse travels along the nerve to the muscle, then goes from nerve to muscle and over the surface of the muscle. By means of suitable electronic devices, the electric shock may be carefully controlled in its strength and duration, as well as in the frequency with which it is applied. Thus a tiny amount of electric current may be applied to a muscle or to the nerve that innervates the muscle. If the stimulus is so small that no contraction at all is produced, the stimulus is said to be *subthreshold* (Fig. 5-8). A slight increase in stimulus intensity may be enough to start contraction in a few fibers close to the electrodes. This might correspond to the delivery of impulses from the central nervous system in a very few of all the nerve fibers coming to a muscle. If the contraction is barely sufficient to be detected, it is called a *threshold response*, and the stimulus itself would also be described as of threshold intensity—that is, sufficient to produce a threshold response. By means of a stronger shock, other fibers having higher threshold would be activated. As the stimulus is increased, so the number of contracting fibers increases. When the current is sufficient, all the muscle fibers will contract, a result that would also be obtained if the central nervous system sent impulses along all the motor nerve fibers going to the muscle, as in maximum effort to move a particularly heavy object. In this situation, the response is said to be *maximal* (i.e., it cannot become greater). The stimulus is also said to be maximal so far as the muscle

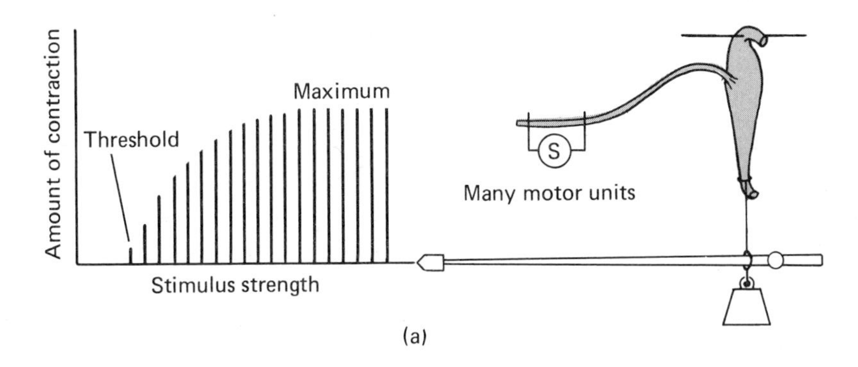

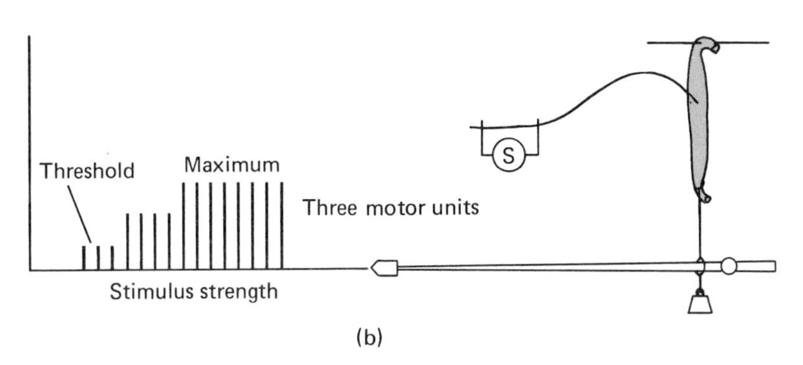

Fig. 5-8. Contraction depending on number of active motor units. The magnitude of contraction is in each instance recorded and the recording paper is moved between contractions. (a) Stimulus-response curve for a muscle with many motor units. This record actually consists of a large number of small steps, which, overall, result in a smooth curve. (b) Stimulus-response curve for a muscle of three motor units. The amplification of the lever is greater in the lower records in order to reveal the steps.

response is concerned. If we now use as a standard stimulus one that will provoke a maximal response in the muscle, we will know by the height of the record whether something we do to the muscle makes the muscle contract more or less effectively.

Isometric and isotonic contraction

Insight into the contraction process may be gained by changing the force that the muscle has to pull against—that is, the *load* on the muscle. If this load is sufficiently great, the muscle cannot shorten. Yet force is certainly developed, as may be seen if a device sensitive to force is attached to the lever. Such a contraction, which occurs without change in length of the muscle, is said to be *isometric* (same length)

(Fig. 5-9). Muscles contract isometrically when we attempt to lift a weight that is beyond our strength. Since a muscle contracting isometrically against a force does not move the mass, it does not *accomplish* any work during the contraction. Energy is expended, but the energy appears entirely as heat rather than as useful work. You may *work* very hard indeed, according to the common usage of that term, but in the physical sense no work has been done. The muscles of posture function isometrically when antagonistic muscles contract to keep a joint stiff. Shivering is another example of isometric contraction.

When a muscle is allowed to shorten during contraction, useful work may be done if the muscle moves a resisting object, such as a weight against the force of gravity. When the muscle is allowed to shorten as it contracts, the contraction is called *isotonic* (same tension) (Fig. 5-9). It should be clear that the force a muscle can usefully apply to its external attachments will depend (up to a point) on the extent to which it is stretched. Thus a flexor muscle at a joint where there is complete flexion or excessive extension can exert less useful force than the same muscle when only moderate flexion exists at the joint.

The dependence of useful force on muscle stretch can be understood when it is realized that ordinarily only part of the force the muscle exerts can be applied directly to the external attachments (Fig. 5-10). Some of this force is used inside the muscle,

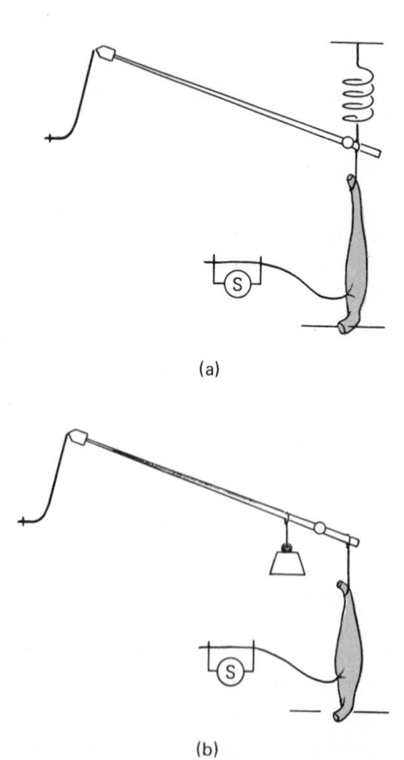

Fig. 5-9. Measurement of force and of shortening in muscle contraction. (a) *Isometric contraction* (shortening prevented). The length of the muscle remains practically constant, but the lever is arranged to give a large amplification of the small amount of shortening that does occur. In this instance the muscle pulls against a very strong spring. The force exerted by the muscle can be measured by this method. (b) *Isotonic contraction* (shortening allowed). The force against which the muscle contracts remains essentially constant. The length of the muscle decreases during the contraction.

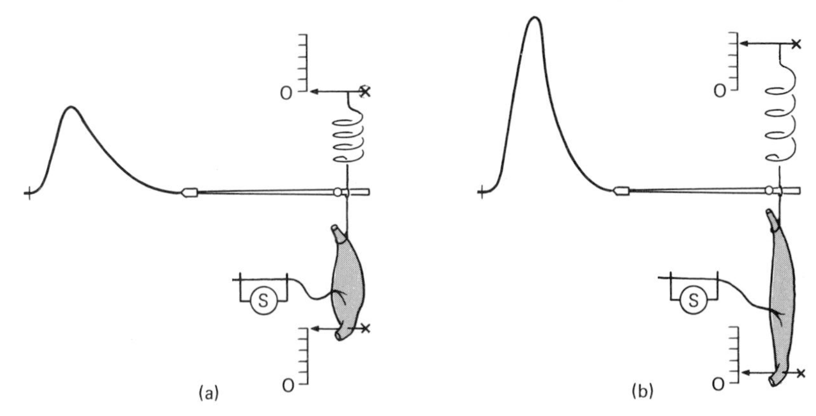

Fig. 5-10. Effect of stretch on muscle force. Force in the muscle is noted by measuring the isometric contraction. (a) The muscle is relatively slack and the force developed on the lever is small. (b) The muscle has been stretched out. This stretch pulls up the tip of the lever, and the spring attachment of the muscle must be moved upward until the lever tip is at zero again. Thus stretched, the muscle exerts greater force when it contracts.

pulling out the stretchable part of the muscle substance. This stretchable part is said to be *elastic* because it will return to its original length when the stretching force is released. When the resting muscle is slack, most of the contraction force will be used in pulling out the elastic component of the muscle. When the resting muscle is stretched, the elastic component is already pulled out, and force can then be applied directly to the external attachments.

However light the load, a muscle can shorten only to about two-thirds of its resting length. Maximum isotonic shortening will occur with zero load, and the useful work is zero in this instance. In isometric contraction, the work is also zero, for there is no displacement of the load although the force developed may be considerable. Somewhere between these extremes, a combination of force exerted and amount of shortening will provide the maximum work of which the muscle is capable under the given circumstances (Fig. 5-11).

The rule that there is an optimal length at which maximum work can be accomplished also applies to heart muscle. The volume of blood the heart will pump depends on how much the heart is filled—that is, the extent to which the heart muscle fibers are stretched.

Sustained contraction of muscle

Muscular effort is ordinarily sustained for a few seconds at least and can usually be maintained for several minutes without difficulty. Force is maintained in a muscle by the arrival of many nerve impulses one after the other (*trains of impulses*) from the central nervous system. When several impulses come in quick succession, the

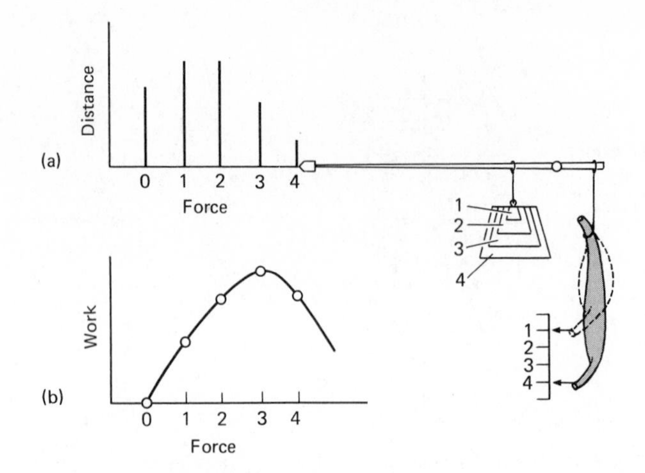

Fig. 5-11. Work done depending on the load on the muscle. (a) As the load lifted by the muscle is increased, the force exerted by the muscle is increased, and the work done becomes greater. (b) The work accomplished by the muscle is maximal for a particular force. Force here is equal to the load lifted by the muscle. Work is the product of load times the distance the load is lifted.

muscle does not have time to relax from one contraction before it must start another. In this way, muscle force may be sustained as long as the impulses continue to arrive.

Like the effects of a single stimulus, the results of trains of stimuli on muscle contraction are best understood by studying a muscle under experimental conditions, removed from the influence of the central nervous system. Experiments demonstrate that not only does the contraction last longer but it also exerts more force if each of a succession of stimuli arrives before relaxation from the response to the preceding stimulus has occurred. The force of the first twitch moves the external load and stretches out the elastic component of the muscle. The second twitch, coming during the stretch (before relaxation), applies more force to the load. Within limits, the greater the frequency of stimulation, the greater the tension. To understand how this situation may occur, consider first the effect of two shocks on a muscle that is under moderate stretch. The contraction that follows the first shock pulls out the internal elastic component in the muscle substance and in the connective tissue, and the remaining force is exerted on the external attachments. If the second shock arrives before relaxation has occurred from the first, the internal elastic component is already stretched out, and the force of the second contraction may be exerted more fully on the external attachments. External force continues to increase with successive contractions thereafter (Fig. 5-12).

Depending on whether the recording is made isometrically or isotonically, the result of trains of stimuli will be recorded either as increased force or increased shortening of the muscle. In either case, sustained contraction is called a *tetanus*. A special feature of the *disease* tetanus (with which *physiological* tetanus is not to be confused) is a sustained muscle contraction that occurs when the nerve-muscle junction, due to poisoning, is kept in a constant state of excitation. In the normal response to trains

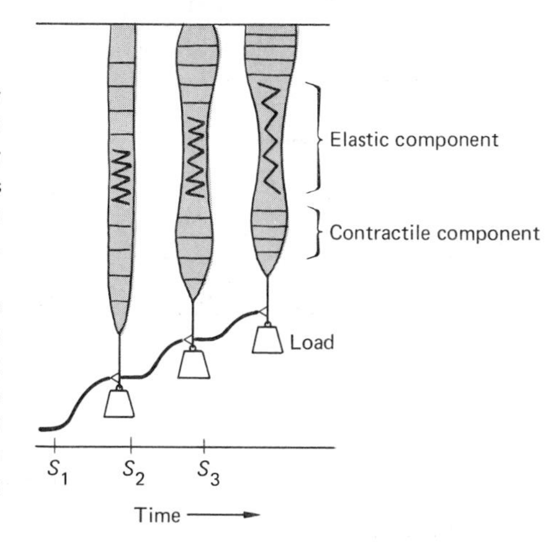

Fig. 5-12. Contraction force exerted by muscle increased by succession of stimuli (summation of contraction, or tetanus). The elastic part of the muscle is actually distributed throughout the length of the muscle, but is here shown localized, to simplify presentation of the ideas.

During the first twitch, the muscle shortens but the elastic components of the muscle are stretched out by the restraining force of the external attachments. This allows the muscle to provide additional external force when another contraction occurs before relaxation from the previous twitch is completed.

of stimuli, if the sustained contraction is smooth, it is considered a *complete tetanus*, whereas it is called a *partial tetanus* if the effect of individual twitches can be seen in the record. A partial tetanus is sometimes called *clonus*, but this term may be more suitably reserved to a rhythmical muscle contraction present in certain diseases of the central nervous system.

The normal frequency of impulses in muscle nerves may be only 10 to 20 impulses per second, at which rate the summation of successive contractions in individual muscle fibers may be of little importance. Each fiber that contracts pulls on the tendon of the entire muscle, however, and summation of contraction force in the whole muscle follows from contractions in the several motor units. As more and more motor units are brought into play, the whole muscle pulls harder. Twitches occurring at 10 to 20 per second could be sensed as definite vibrations, although we seldom note vibrations in our muscles because the various motor units do not contract synchronously but are out of step with one another. Even though several motor units may be stimulated at the same frequency, one may be contracting while another may be relaxing. The overall force on the muscle attachments is therefore smoothly maintained by *asynchronous* motor unit activity during prolonged voluntary effort. This kind of summation of contraction may be called *wave summation*, for a reason that is made clear in Fig. 5-13. It is to be distinguished from *quantal summation*, which refers to the increase in force of contraction that results when the number of active motor units is increased.

Strength of muscles

Besides frequency of stimulation and amount of initial stretch, other factors influence strength of muscles. Most obvious is muscle size, specifically the diameter. When a muscle is used a great deal during the life of a person, it tends to increase in diameter, especially if used near the limit of its strength. This growth (*hyper-*

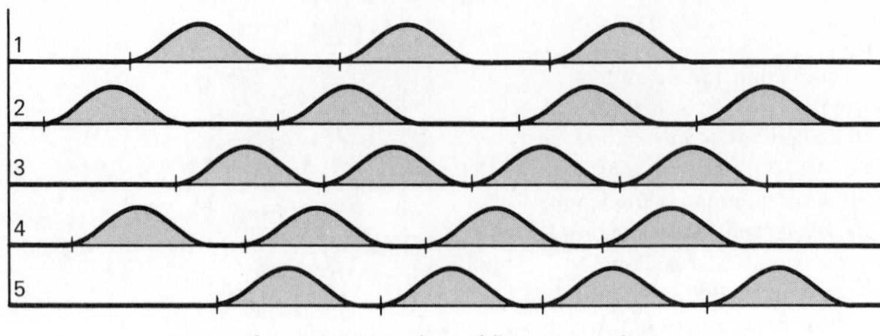

Separate contractions of five motor units

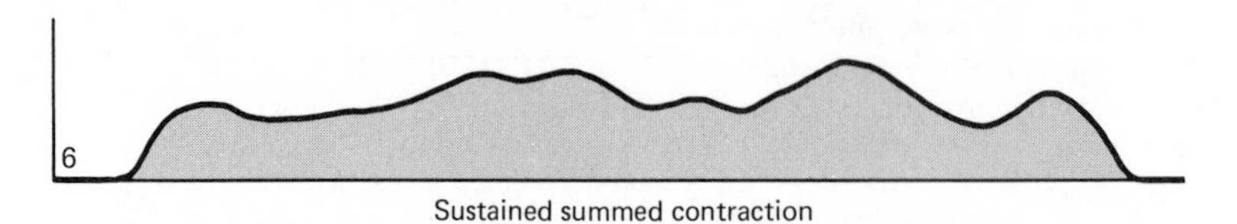

Sustained summed contraction

Fig. 5-13. Wave summation in muscle. (1) through (5) is the force provided by several individual motor units contracting at different frequencies. Vertical marks indicate moment of stimulation. (6) is the overall force provided by the sum of the individual motor units acting on the same attachments.

trophy) is due to an increase in diameter of the individual muscle fibers, in each of which the amount of contractile substance increases. The number of muscle fibers is obviously also important, for a large muscle of many fibers is able to exert more force than a small muscle having few fibers. From one person to another, differences in muscle strength are due mainly to size of muscle fibers, as influenced by prolonged exercise—for the number of fibers in any particular muscle does not change in the adult. The number of muscle fibers and the size of each fiber determine the cross-sectional area of a muscle. Whether stronger muscles are merely larger can be ascertained by comparing the force that would be exerted by portions of the muscles of equivalent cross-sectional area. On this basis, a muscle that is twice the diameter (but of the same length) compared to another, will be four times as strong because the mass is four times as great. The *muscle tension* (force per unit of cross-sectional area) may differ little.

A comparison of the jumping abilities of insects and of men provides an instructive lesson in the interpretation of muscle effectiveness. A man 2 m tall might jump upward a distance approximately equal to his own height. An insect 1 cm long could easily jump upward 10 times his own length. However, the man accomplishes 100 kg (or 1,000,000 g/cm) of work compared to 10 g/cm for the insect. In addition, the cross-sectional area of muscle involved in the man may be 10,000 times as great as in the insect, but the weight lifted may be about 50,000 times as great. In this particular instance, the muscle tension (the force per unit area) will then be about five times as great in the muscles of the man as in the insect. In *maximum* effort for insect or man, the *muscle tension* is probably about equal.

5.3 Microscopic and Molecular Machinery of Muscle

The change in length that occurs in a muscle during contraction is evidence at a visible level of shifting relations between molecules of the muscle substance. So, too, is the force generated during the contraction. Muscle contraction involves expenditure of *energy*. Anything that can do work is said to have energy. Muscle is a machine that can do mechanical work. Thus muscle has energy. Energy at the chemical or molecular level in the muscle is transformed into visible mechanical work when a muscle contracts.

We obtain some insight into the muscle machinery where this energy resides by looking at anatomical changes in form at the microscopic and molecular level.

The contractile machinery of a muscle resides in tiny fibrils oriented lengthwise in each muscle fiber or cell. In smooth muscle, the fibrils are rather similiar to one another, and the cell, examined by light microscopy, reveals only a delicate, *longitudinally striated* appearance. In skeletal and heart muscle, the different types of fibrils are arranged in a regular way that forms a cross-striated pattern.

The transverse bands of striated muscle show up especially well when polarized light is used. One major band is light, the other dark. The light comes "straight through" the *i*sotropic or *I* band, which is therefore light in appearance. The light does not come "straight through" the *an*isotropic or *A* band, and this band therefore appears darker. During contraction, when shortening is allowed, the *I* bands become narrower. The boundaries between the repeating bands is a line (or, more accurately, a plane, since the muscle is a three-dimensional structure) in the middle of the *I* band.

The early German histologists called it the Z line (from Zwischen = between). The sarcomere is the histological repeating unit of bands, from one Z line to the next. (See Figs. 4-7 and 5-14.)

Morphology of the proteins in muscle

Histological stains are differentially effective on the A and I bands, one staining more darkly than the other, thus indicating a different chemical structure. By suitable extraction procedures, the A band can be removed and chemically characterized. It is a protein named *myosin*. The I band remaining may be similarly extracted and characterized as another protein, *actin*. Still a third protein, *tropomyosin B*, seems to make up the structure of the Z zone in the middle of the I band, bounding the ends of the sarcomeres.

Using the electron microscope, we can see an additional wonderfully regular arrangement of the fibrillar actin and myosin material of the muscle bands. The actin filaments of the I band show in cross section an orderly hexagonal array (Fig. 5-14). These filaments continue into the A band, which is in the middle half of a sarcomere. In the A band, a large myosin filament appears within each hexagonal array of I band actin filaments.

Small projections can be seen perpendicular to the long axis of the myosin filaments and attached to the surrounding actin filaments. Even the electron microscope has not told us what really goes on here during contraction. One assumption is that the relation of the myosin to the actin filaments may be compared to a man climbing a rope hand over hand. Similarly, the projections successively make and break connections between the myosin and actin, and the filaments slide farther along among one another.

Actin and myosin extracted from the muscle have some interesting properties that may have some relation to the normal physiological events in the muscle.

Myosin can be separated into two parts designated *light* and *heavy* meromyosin (the prefix *mero* indicates "part," and light and heavy refer to relative molecular weights). The light meromyosin is rodlike and will combine with the heavy form, making clublike structures that will aggregate side by side, all pointing in the same direction (Fig. 5-14). These aggregations, in turn, will arrange themselves tail to tail in a form very much like the myosin filaments seen in electron microscopic pictures of normal muscle. The side "hooks" can be seen at each end, but the middle section is smooth, just as are the real filaments in the muscle. The actin filaments can be separated into globular molecules of actin (*g actin*), which can aggregate to become spirally arranged filaments (*f* actin) very much like the native actin filaments. The interesting general lesson from these observations is that the formation of a relatively large morphological structure occurs apparently spontaneously from the random collisions of the smaller molecules. No special "tools" are necessary to put them together. They fall together in a particular fashion, presumably because they are in the least stressed arrangement. This process may also reflect the way the molecules fall

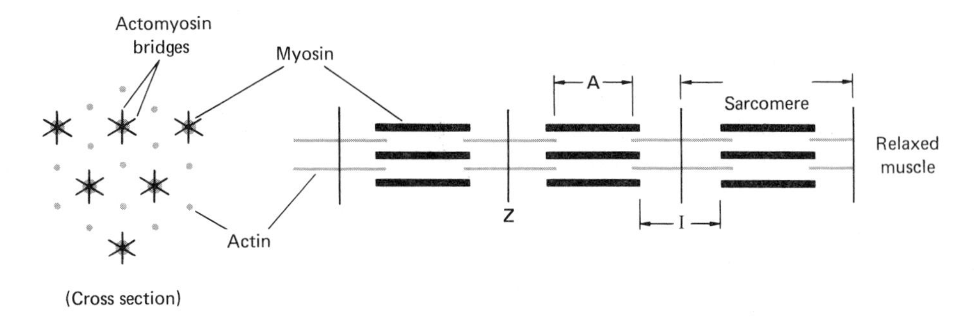

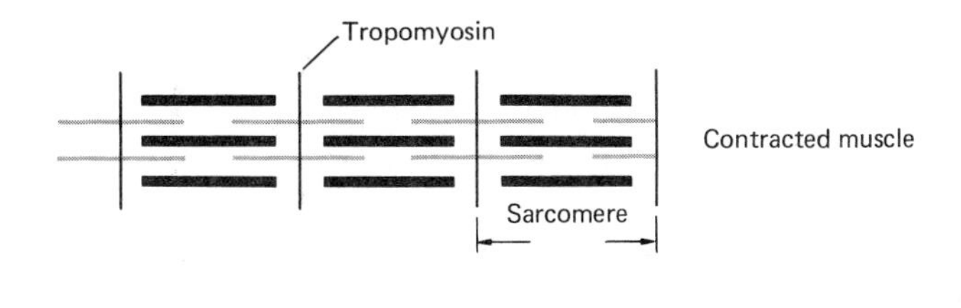

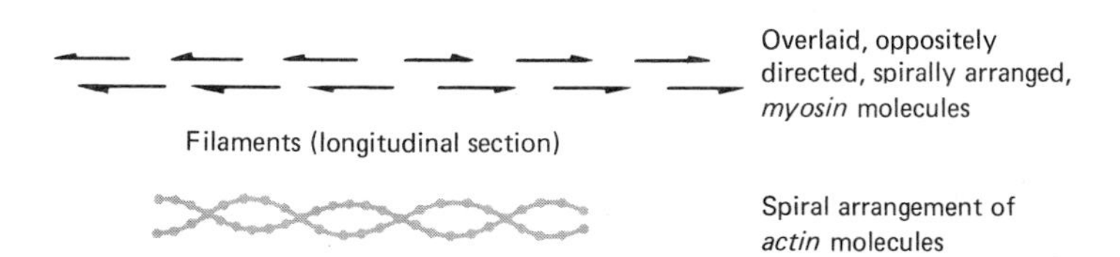

Fig. 5-14. Structure of skeletal muscle *fibrils*.

together during cellular differentiation. In the muscle cell there is still another order of complexity in the assembling process, when the molecules of myosin line up side by side as the *A* band and the molecules of actin line up similarly as the *I* band.

Excitation and contraction

The command for the muscle to contract is a decrease in the electrical potential difference existing between the inside and the outside of the plasma membrane of the sarcolemma, the bounding surface of the muscle cell. Ordinarily this command is delivered through a nerve impulse that activates the muscle along the muscle fiber in a manner essentially like the propagation of an impulse along the nerve fiber, discussed later (see Chap. 6.1).

A response localized to a single sarcomere may be seen if a microscopically small electrode tip is brought into contact with the muscle membrane, and a reduction of the membrane potential is produced at a small spot. The electric pulse will be effective only at certain places along the muscle fiber. A pulse of sufficient intensity will cause a local distortion of the banded pattern of the striations (Fig. 5-15).

In some muscle fibers of man and of mammals generally, the lowest intensity of pulse is effective when the electrode is placed in the middle of a sarcomere—that is, in the middle of the *A* band. Then the *Z* lines at opposite ends of the sarcomere move toward one another, and the sarcomere shortens, as shown in the figure.

Surprisingly, the same experiment tried on the striated muscle of frog gives a different result. The lowest intensity of pulse is effective at the *Z* lines in the middle of

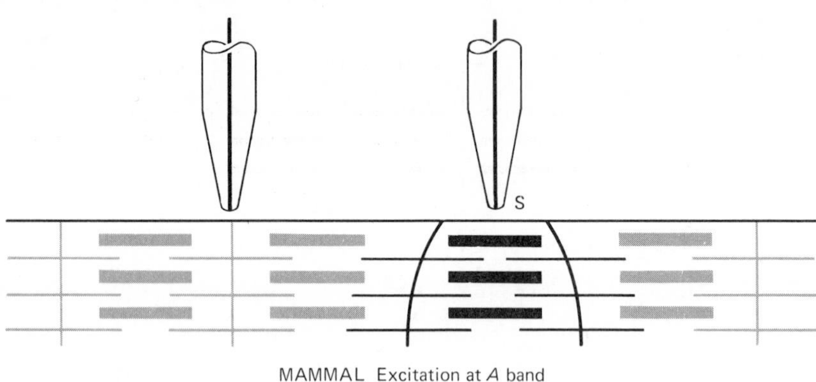

MAMMAL Excitation at *A* band

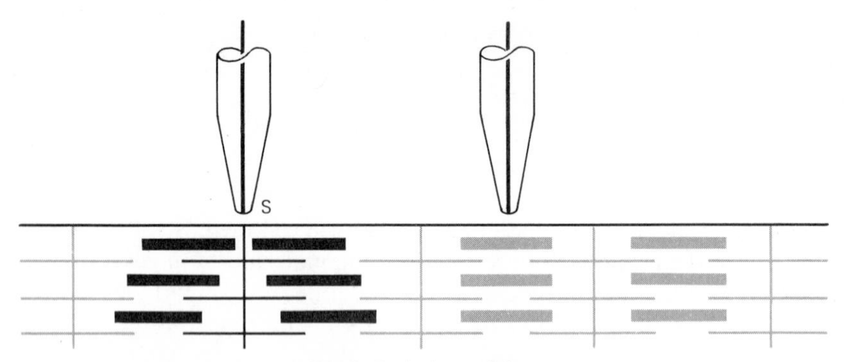

FROG Excitation at *Z* line

Fig. 5-15. Direct stimulation of muscle by means of electric pulses. Experiment showing, by indirect means, the location of the transverse tubular system in striated muscle of mammal as compared to frog. A micropipet electrode is placed at the surface of the muscle fiber. A stimulus pulse is effective at the *A* band in mammalian muscle, and at the *Z* line in frog.

the *I* band. The *Z* lines do not change in position, but the clear *I* band on each side of the *Z* line becomes narrower.

If we examine the fine structure of the muscle fibers between the myofibrils by means of an electron microscope, we will discover the reason for the stimulus spot selectivity. At the region of the *Z* lines in the frog muscle fiber, and between the *Z* lines in mammalian muscle, the sarcolemma dips inward in numerous, long, small pockets. A tangled tracery of transverse tubules at this level provides communication of the interior of the fiber with the outside. Through this transverse tubular system, the command to contract is evidently delivered (Fig. 5-16).

Into the transverse tubules, the muscle fiber impulse carries an electrical charge that alters the interaction between the actin and the myosin, and the myosin filaments of the *A* band move in farther among the actin filaments of the *I* band.

The changing relations of the actin and myosin fibrils during muscle contraction are shown in Fig. 5-17. Here the length-force diagram, previously described for the whole muscle, is shown for a single sarcomere. Note that there is an optimal length at which force exerted is maximal because the maximum number of cross connections are made. If the muscle, and hence the sarcomere, is stretched beyond the optimal length, the number of connections, and hence the force, is decreased. When the muscle is kept slack, and the actin and myosin are allowed to overlap, the force is decreased, perhaps due to the arrangement illustrated; that is, side bridges on a myosin fibril may pull in opposite directions on one and the same filament. The force in one direction cancels the force in the other direction, and the net effective force is thus reduced.

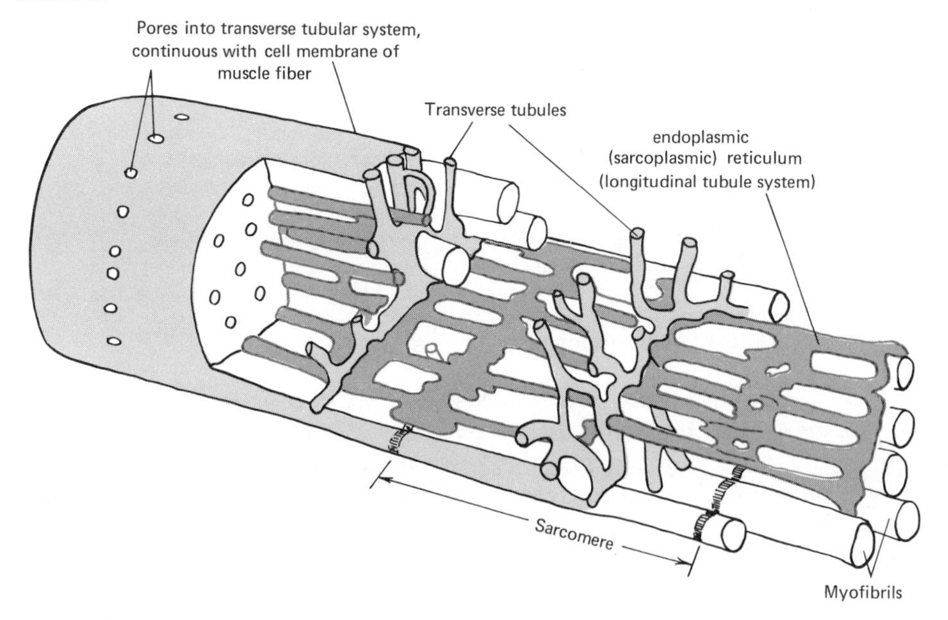

Fig. 5-16. Tubular system of skeletal muscle fiber.

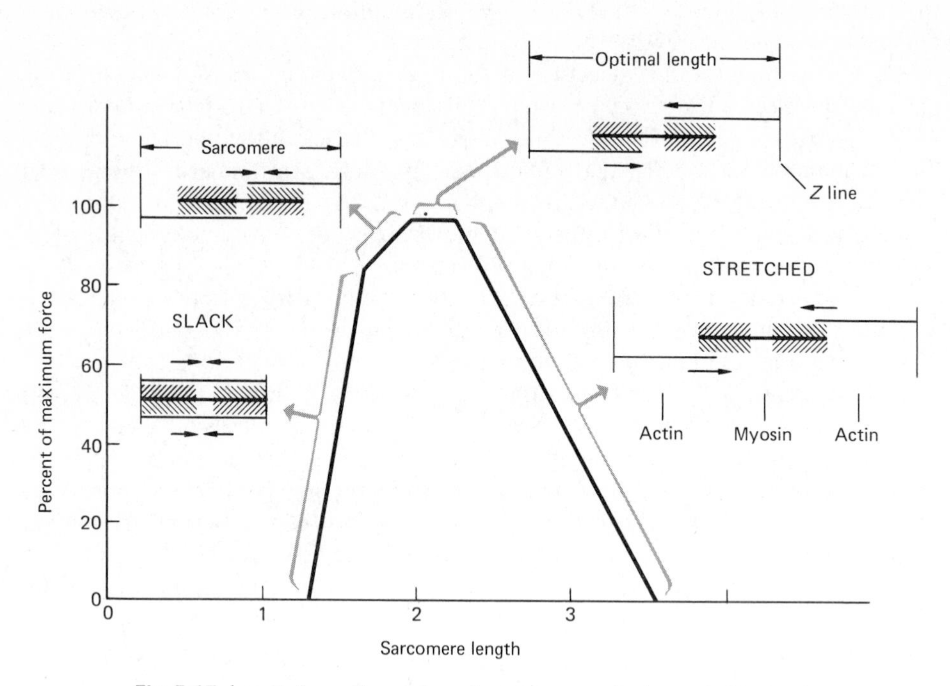

Fig. 5-17. *Length-force diagram for a single sarcomere.* Each myosin filament is actually surrounded by six actin filaments at each end, with which it has cross-bridge attachments. To make a less confusing picture, only an actin filament is shown at each end of the myosin filament. At the optimal length, all the side-bridges of the myosin are in use, at each end pulling each actin filament toward the other. When the muscle is stretched beyond the optimal length, fewer of the side-bridges are attached, and the force is therefore less. When the muscle fiber is slackened, the actin filaments overlap and the force exerted by the myosin side-bridges pulling in one direction on an actin filament is neutralized by side-bridges meeting in opposite directions in the same filament. When the muscle is pulled together excessively, the force is reduced as the thick filaments collide with the Z membrane at the ends of the sarcomere.

ATP energy in muscle contraction

Energy for cellular activities is stored in the chemical bond that holds together a phosphate (PO_4) to an organic molecule. It is called an energy-rich bond and is designated by a squiggle (~). ~Ⓟ means an energy-rich phosphate bond. Details of production and utilization of these energy-rich compounds are described in Chap. 14.2, but here we will mention a few examples relevant to muscle.

As in cellular processes generally, a special organic molecule, *adenosine triphosphate* (ATP), has an important role in muscle function. Exactly how the chemical energy in ATP is converted into cellular work is not particularly clear in any cell, and neither is it clear for muscle, which transforms chemical energy into mechanical work.

The involvement of ATP is suspected from several observations. Myosin is an *ATPase*, an enzyme that facilitates the splitting of ATP; it can thereby make available the phosphate bond energy (see Chap. 14.2).

The ATPase activity of myosin is localized to the heavy meromyosin—the part that forms the bridges connecting the myosin and the actin filaments. ATP, on the other hand, seems to be bound to the actin, and perhaps the breakdown of ATP provides the energy for contraction by union of myosin with actin.

A classical object of study in muscle physiology is the glycerol-extracted muscle. The muscle is soaked in glycerol in order to remove electrolytes, lipid substances, and soluble proteins. The actin and the myosin fibrils remain, more or less in their normal relationship, but the structure cannot be made to contract by an electrical pulse. It can, however, be made to contract and relax by chemical means. ATP added to the solution bathing the preparation will cause shortening if the Ca^{++} concentration in the bath is sufficiently high. Relaxation will occur if the Ca^{++} concentration surrounding the fibrils is reduced to a very low level, about 10^{-6} moles/liter. In an intact, normal muscle, local injection of Ca^{++} ion solution into the fiber will have an effect similar to the effect of an electrical pulse and will produce a local contraction.

If muscle is ground up, the sarcoplasmic reticulum can be separated from other fractions of the muscle substance. This material will eagerly take up Ca^{++} until the surrounding concentration of Ca^{++} is lowered to approximately that required for relaxation of the fibrils.

The picture that emerges from these observations is somewhat as follows: The electrical currents associated with the muscle impulse flow into the transverse tubules and cause the release of Ca^{++} from a bound condition. With a relatively high concentration of Ca^{++} in the environment, the myosin ATPase is activated, thereby making the energy of ATP available to form actomyosin bridge connections so that the myosin slides along the actin. When the released Ca^{++} is taken up by the sarcoplasmic reticulum, ATPase is no longer activated and the bridges separate.

Muscle analogized to a spring

A muscle may be compared with a spring. In Fig. 5-18 the chemical energy in muscle is represented as the mechanical energy of a spring. Energy must be expended to pull the spring out to the hook. This represents the energy in glucose and oxygen that must be used to synthesize ATP during recovery in the muscle. When the spring is released, it provides energy that can do work. Figure 5-18(a) shows the spring analogy to the muscle set to contract. A small amount of "activation energy" in the form of a stimulus, such as a slight electric shock or an impulse, is necessary to trigger the release of the contraction process. Once the trigger has been released (the spring is pushed off the hook in the model), the contraction, involving splitting of ATP, proceeds spontaneously (the spring pulls together spontaneously).

As the isotonic contraction occurs, a force is applied through a distance and work is accomplished [Fig. 5-18(b)]. If the model is now to recover all its lost energy,

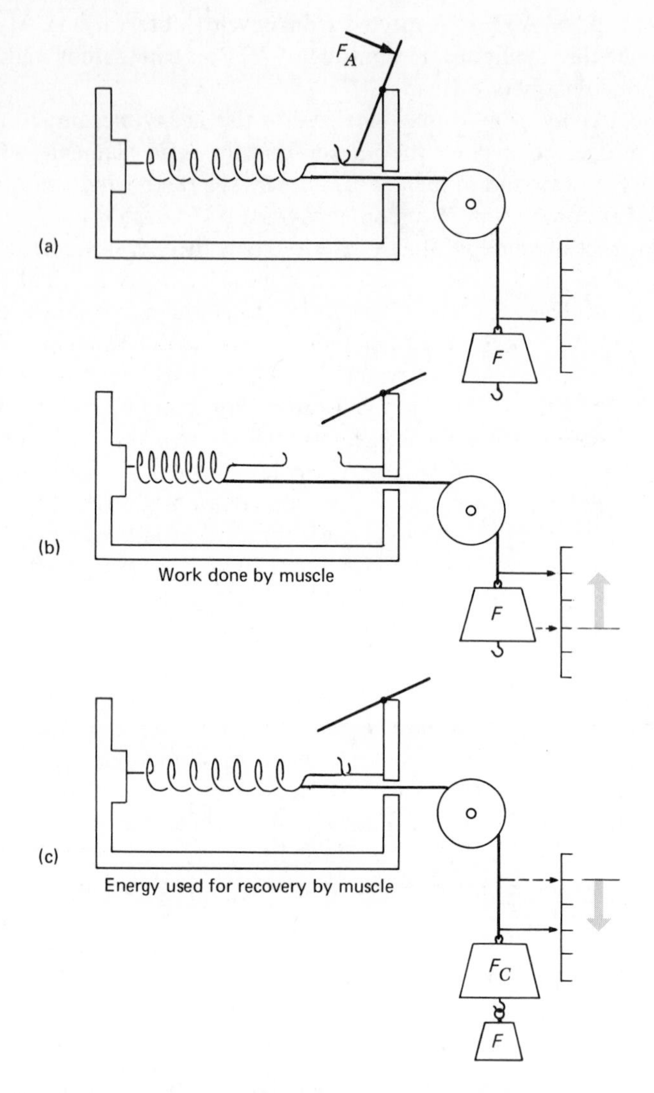

Fig. 5-18. A spring as an analog of a muscle. (a) Energy is held locked in the muscle, ready to be released by the stimulus F_A. (b) Some of the energy is made available—work is done—if the muscle shortens when it contracts. (c) Work is done on the contractile machinery as ATP provides chemical energy to bring about relaxation.

more energy than was released must be put into it to put the spring back on the hook (because some energy has been lost as heat). The product of the force used and the distance through which the mass is moved equals the work that must be done (the energy that must be expended) during the recovery process. If the trigger now releases the spring once again, the spring cannot pull up the weight that has just pulled it out.

It can only pull a smaller weight. Thus the system during contraction can do less work—that is, it makes available less energy than was put into it during the recovery process, for some of the energy is lost as heat.

Muscle metabolism

Useful work, according to the physicist's definition, can be accomplished by a muscle only during an isotonic contraction. Lifting an object against the force of gravity accomplishes useful work during isotonic contraction of muscles. Holding the object requires isometric contraction and may feel like work but accomplishes no work in the physical sense.

During isometric contraction no useful work is done, but the muscle loses energy as *heat*. This energy must be replaced if the muscle is to continue capable of contracting. To say that the energy is lost is not to say that it is destroyed. As a matter of fact, energy cannot be destroyed—it can only be transformed into energy of another kind. This is the implication of the *First Law of Thermodynamics*,* a generalization which states that the total energy in the universe remains constant, whatever changes the energy may undergo.

Whether a muscle contracts isotonically or isometrically, energy is expended. It is not destroyed; it is merely changed into some other less available form in accordance with the *Second Law of Thermodynamics*, which states that the total energy in the universe becomes continually less available—that is, less able to provide work. In the case of the muscle, only part of the energy used for a contraction is actually transferred to the object being lifted. Only about one-third of the energy consumed by a muscle during isotonic contraction can be used to accomplish useful work, even when the muscle is allowed to shorten. The other two-thirds of energy used is transformed into heat.

The heat released during contraction momentarily increases the temperature of the muscle. Blood circulating through the muscle takes up this heat and distributes it throughout the body. Heat generated by contracting muscle, therefore, helps to keep the general body temperature up to a normal level. During isometric contraction, virtually *all* the energy released during the contraction appears as heat.

In certain circumstances the energy as heat can be used to perform useful work (many practical machines use heat as a source of energy). However, when heat from a system at a particular temperature is so used, the temperature of the system decreases unless it is maintained by energy supplied from another source. The fact that muscle tends to *increase* in temperature when it contracts shows that the muscle does not use heat as a principal source of energy; that is, it is not a heat engine.

During muscle contraction, energy is made available to do work when the actin and myosin filaments move closer in among one another. The capacity of the muscle to do this rests in the arrangement of molecules that can use the forces of attraction

* Thermodynamics is the study of the interrelations of work and heat. Its laws apply to the entire universe so far studied.

between atoms to bring about the relative movement. In other words, the energy of muscle contraction comes from the chemical reactions that go on in the muscle cells.

Although ATP is thought to provide chemical energy for contraction, the amount of ATP remaining in a muscle may be unchanged after many contractions. There is another "energy-rich" compound that does decrease in amount, however. This compound is *creatine phosphate*, CP. As quickly as ATP is split, it is regenerated by combination of the ADP with ~Ⓟ contributed by the creatine phosphate (Fig. 5-19). The creatine phosphate is actually only one (although the most important) of many high-energy phosphate storage molecules called *phosphagens*, a term referring to the ability of the molecules to "eat up" phosphate. Other cells using energy at a less violent rate than muscle do not have phosphagens. For these cells, ATP alone is usually synthesized rapidly enough for their slower processes.

Creatine Phosphate	⇌	Creatine	+	Phosphate
CrPh	⇌	Cr	+	Ph
				↓
ATP	⇌	ADP	+	Ph
Adenosine Triphosphate	⇌	Adenosine Diphosphate	+	Phosphate

The overall reaction is

$$CrPh + ADP \longrightarrow Cr + ATP$$

Fig. 5-19. Relation of phosphagen (creative phosphate) and ATP in muscle metabolism.

The reactions shown in Fig. 5-19 are reversible, and the concentration of energy-rich phosphagen compounds in muscle may increase when there is net synthesis rather than breakdown of ATP. As will be described in further detail in Chap. 14.2, the chemical energy released in the breakdown of glucose may be utilized to make ATP.

There are several steps in this breakdown process. First, glucose is moved from the bloodstream into the muscle cell as glucose 6-phosphate. The glucose molecules may then be coupled together to form glycogen, in which form the carbohydrate may be stored and used as necessary, or they may be split into the three-carbon compounds pyruvic acid and lactic acid. Synthesis of glycogen required ATP energy, but the splitting of glucose to three-carbon compounds provides energy for synthesis of ATP. The pyruvate may be further oxidized in the mitochondria, where it may enter the Krebs cycle (see Chap. 14.2). In this cycle, O_2 is utilized and the pyruvate is split into CO_2 and H_2O, with the generation of many more ATP molecules than is possible under anaerobic conditions.

If the supply of O_2 is limited, then lactic acid tends to diffuse out of the muscle and into the circulation. Much of the lactic acid is then picked up in the liver, where about one-fifth of it is oxidized to provide energy for resynthesis of the remainder to glycogen which is then stored in the liver.

During intense muscle activity, O_2 is inadequate, pyruvic and lactic acids are produced, and the acidity of the muscle consequently increases. The *p*H is brought back to normal when the acid products of metabolism are washed out by the circulation during subsequent rest.

Ordinarily, sufficient oxygen is carried to the muscles by the *hemoglobin* in the red blood cells from which oxygen diffuses into the muscle.

In the muscles, the oxygen can be temporarily stored in combination with *myoglobin*, a pigment similar to hemoglobin but held fixed inside of muscle cells instead of in the blood cells. If the chemical processes of the muscle consume oxygen more rapidly than it can be supplied by the circulation, the supply of oxygen held by the myoglobin will be inadequate for the metabolic needs of the muscle, and lactic acid will accumulate.

In order to study the chemistry of muscle contraction, we may place a muscle in a bath of Ringer solution and observe the effects of various substances added to the solution. For example, iodoacetic acid will poison the chemical machinery that synthesizes energy-rich phosphate compounds. In that case, muscle contraction can continue only as long as there remains a reservoir of previously synthesized CP and ATP to be utilized.

The normally functioning muscle machinery may obtain energy-rich compounds from sources other than breakdown of glucose—for example, by the oxidation of fat (lipid).

If O_2 is used to burn sugar as a source of energy to make energy-rich phosphate compounds, the volume of carbon dioxide (CO_2) produced will equal the volume of oxygen (O_2) consumed. If O_2 is used to burn fat, the volume of CO_2 will be less than the volume of O_2. In muscle, the CO_2 produced during any interval may be about 0.7 the volume of O_2 used. A CO_2/O_2 ratio of 0.7 is typical of fat utilization rather than of sugar utilization and suggests that lipid is utilized to a considerable extent by muscle. Further details about cellular metabolism will be found in Chap. 14.2.

Section 5: *Questions, Problems, and Projects*

1. You need to move a large rock weighing 600 kg. You find a sturdy plank, and a smaller rock to serve as fulcrum. If the plank is 3 m long, what distance must you have from large rock to fulcrum if you are to move the rock and if you weigh 70 kg?
2. Estimate the work done when you carry 50 kg up a stairway 7 m high. If your

efficiency is 10 percent, how much energy would you dissipate as heat in doing this job?

3. What is a motor unit? If there are 300 motor nerve fibers in a muscle nerve and there are 2500 muscle fibers, what is the average size of the motor unit of this muscle?
4. If you stimulate a muscle nerve with an electric shock, what are the events that must occur in the latent period before the muscle contracts?
5. What is an all-or-none response?
   Repetitive activation ordinarily produces a larger response than a single shock to a nerve-muscle preparation, yet the n-m system is said to be all-or-none in its behavior. Explain.
6. A slack muscle can do less work than a moderately stretched muscle. Explain.
7. Draw a curve that shows the relation you would expect between the frequency of stimulation delivered to a muscle and the force generated by the muscle.
8. Would you expect to be able to do more work if your muscles are contracting isotonically or isometrically? Explain.
9. Name the compound that is called the chemical transmitter at mammalian neuromuscular junctions.
   What effect does the transmitter have on the end-plate?
10. By way of what structures is excitation conveyed from outer surface to the interior of a muscle cell?
    What appears to be the function of the sarcoplasmic reticulum in respect to the contraction phase of muscle function?
    What appears to be the role of the sarcoplasmic reticulum in respect to relaxation of the muscle?
11. What inorganic ion appears to be most significant in the contraction process of muscle?
12. The thick filaments of muscle myofilaments consist of myosin molecules. What are the thin filaments?
    During contraction, what happens to the
    (a) distance between Z lines?
    (b) length of myosin filaments?
    (c) length of actin filaments?
    (d) distance from Z line to nearest end of a myosin filament?
    (e) distance from Z line to nearest end of an actin filament?
13. The effectiveness of a force applied to a lever is described by the torque. You know that the torque (T) is directly proportional to the force applied (F) and the length of the lever arm (L). Write the simplest equation that describes this relation. If a lever is in balance, the torque (1) in one direction equals the torque (2) in the opposite direction. Write the equation for this relation.
14. If a muscle fiber is approximately a cylinder about 5 cm long and 0.1 mm in

diameter, (a) calculate the exposed surface area (excluding the ends) in centimeters squared. (b) Calculate the volume of the cell in cubic centimeters.

15. Describe the arrangement of the sarcoplasmic reticulum and the transverse tubular system in relation to one another and in relation to the myofibrils.
16. Comment on the anatomical basis of the sliding filament theory of muscle contraction. How do you distinguish the *A* and the *I* bands of a muscle?
17. Describe the role of *f* and *g* actin in relation to the thin filaments of the *I* band, and the role of heavy and light meromyosin in relation to the thick filaments of the *A* band.
18. What is anaerobic glycolysis? What is the fate of lactic acid produced during muscle activity?
19. What are the structural changes during muscle contraction?
20. Define *excitation-contraction coupling* and describe the relationship in skeletal muscle.
21. Of what significance is the accumulation of Ca^{++} by the cisternae of the endoplasmic reticulum?
22. What happens in the contraction machinery when Ca^{++} is released from the terminal cisternae of the transverse tubular system in muscle?
23. What is the relation of actomyosin and ATP?
24. How is ATP regenerated after breakdown during muscle contraction?
25. Describe the molecular events involved in relaxation of muscle.
26. What is smooth muscle?

6

THE NERVOUS SYSTEM—GENERAL FUNCTIONS

6.1 Nerve Impulses

Function of nerves

All cells are excitable, but none so much as nerve cells. All we know of the world and of our own bodies, as well as all the conscious and unconscious control that we have over our behavior, we owe to our nervous system. Information is brought to and from the *central nervous system* along *sensory nerve fibers*, which are prolongations of nerve cells connected at one end to sense organs and at the other to the central nervous system. Motor nerve fibers carry commands outward from the central nervous system to muscles and glands. These afferent (sensory) and efferent (motor) nerve transmission lines are grouped together as *nerves* of the peripheral nervous system.

Some of the peripheral nerves can be discovered close to the body surface. For example, between the olecranon and the medial head of the ulna a nerve comes close to the surface on the dorsomedial aspect of the elbow. This is the *ulnar nerve*, which supplies part of the lower arm and fingers. If it is pushed with a firm, quick motion against the head of the ulna, a tingling sensation will be felt in the little finger of that hand. Evidently excitation of this nerve along its course is much the same as stimulating the sensory end-organs to which it is attached, so far as our sensation about *place* of stimulation is concerned. Touch the hand, and the information that provides the resulting sensation may travel along the same nerve that was stimulated at the elbow.

A strong stimulus to the ulnar nerve may be followed by movement of the little finger; therefore the same nerve that carries sensory imformation into the central nervous system also conveys messages distally to command muscles to contract.

We can voluntarily contract skeletal muscles almost anywhere in the body. In addition, we perceive pressure, pain, heat and cold at any place on our body surface. It is clear that *nerves are distributed to all regions of the body* and *carry information both toward and away from the periphery*. The information is carried along *one* pathway coming *from* the sense organs and along another going *to* the muscles.

Structure of nerves

The cellular unit of nervous tissue is the *nerve cell* or *neuron* (Fig. 6-1). A neuron consists of a *nerve cell body* having one or more protoplasmic prolongations called *nerve fibers*. In the cell body resides the nucleus of the cell, and over the nerve fibers pass the *nerve messages* or *impulses*, which are carried to other excitable cells. Many nerve fibers in a single bundle constitute a *nerve*. In most nerves there are both *afferent* or *sensory* nerve fibers that carry information into the central nervous system and *efferent* or *motor fibers* that carry information out to the muscles.

The entire central nervous system and the peripheral nerves connected to it are composed of aggregations of nerve cell bodies and bundles of nerve fibers. A group of nerve cell bodies within the central nervous system is usually called a *nucleus*, a term implying an organization center, just as it implies an organization center of another kind when it refers to a structure within a cell. Any group of nerve cell bodies found along the peripheral nerves is called a *ganglion*. A ganglion is a group of cell bodies outside the central nervous system. The cell bodies of afferent fibers are generally in ganglia outside the central nervous system; the cell bodies of efferent fibers or *motoneurons* are within the central nervous system. A motoneuron has one main fiber, the *axon*, along which nerve impulses move away from the cell body. Nerve impulses are brought toward the cell body along the *dendrites*, which are many small fibers branching off from the cell body.

A nerve fiber is a long, thin tube consisting of a cell *membrane* enclosing a *core* of cytoplasm. The end of a fiber may be as much as a meter away from the cell nucleus, as, for example, a motor fiber of which the cell body is in the spinal cord and the termination is in a muscle of the foot. As in all cells, the nucleus is necessary for prolonged life of the cell, but if a fiber severed from its cell body is kept in a suitable physiological salt solution, it can carry impulses for several days.

If a nerve is severed, say in a limb, the part distal to the cut is thereby separated from the cell bodies. This portion will degenerate, and the fibers will finally disappear. The proximal portions of the fibers, still attached to their cell bodies, will undergo some deterioration at the cut ends, but the tips will become reorganized and begin to grow into the channels left by the degenerating distal stumps. The growth is at the rate of about 2 mm per day, a rate that corresponds to a velocity of movement of nerve substance observed to occur in intact, functioning fibers. Normally there is a steady transport of nerve material longitudinally in the fibers away from the cell bodies. Presumably this substance is used in the metabolism of a fiber at a great distance away from the cell nucleus in the cell body. Perhaps material also passes from the

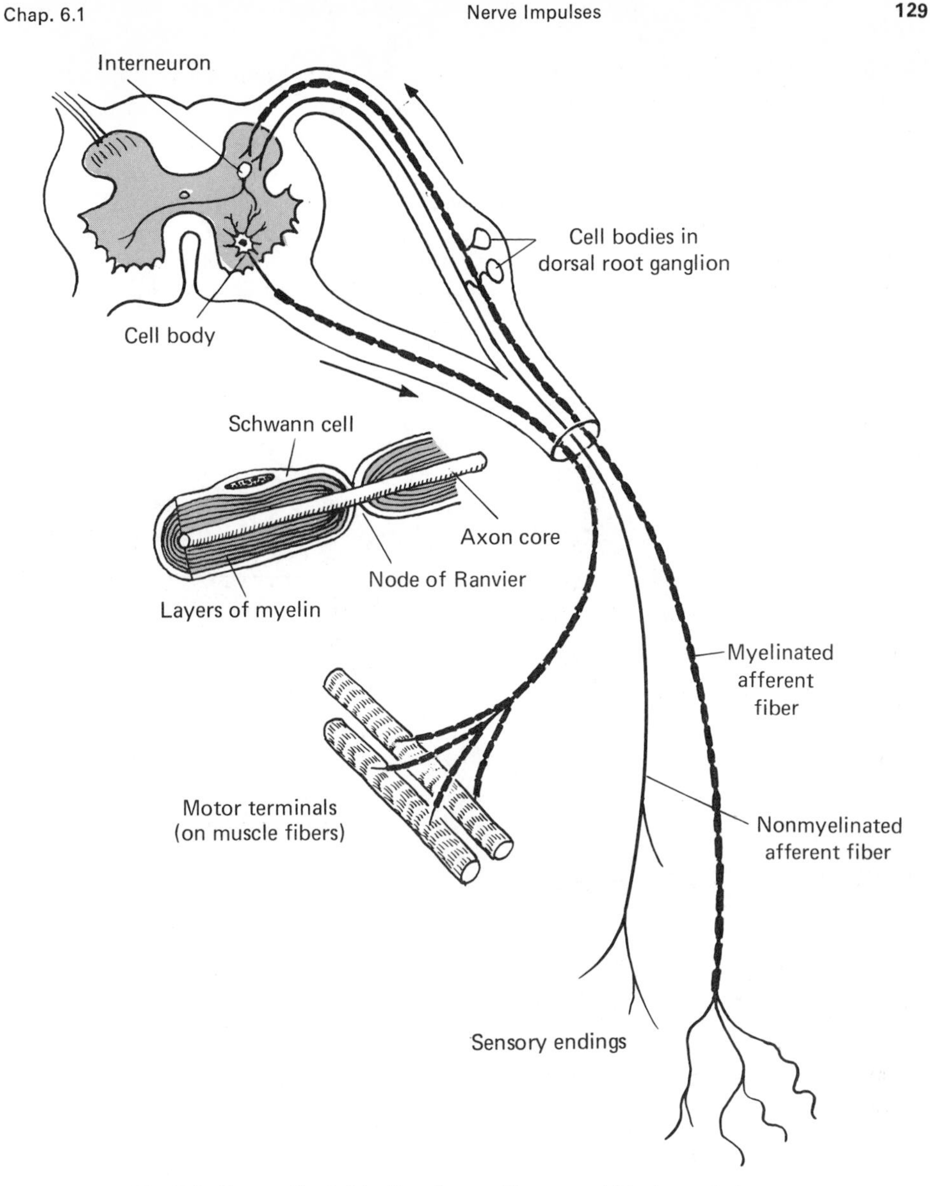

Fig. 6-1. Nerve cells. Cell bodies of motor fibers are within the central nervous system. Enlargement shows the structure of the myelinated fibers in section. Larger nerve fibers are heavily myelinated, smaller ones lightly (and are called unmyelinated). Ordinary direction of impulse conduction is shown by arrows.

nerve terminals into the next excitable cell. A muscle cell, for example, will degenerate if its nerve supply is cut. The muscle's requirement for an intact innervation may be interpreted to mean that the muscle receives some essential substance from the nerve.

Nerve cells and fibers are completely enclosed by special supportive and nutritive *satellite* or *glia* cells. *Schwann cells* are satellite cells that are wrapped in thin layers around the nerve fibers. On *myelinated fibers*, the satellite cells are spirally wound many times, forming thick sleeves of fatty substance, the *myelin*. Each sleeve is separated from the next by a *node of Ranvier* (see Fig. 6-1). At the nodes, the nerve fiber membrane has free access to the surrounding fluids, but between, along the internodes, it is insulated. On *nonmyelinated fibers*, the Schwann cell may be wrapped around only once, and there may be several nerve fibers enfolded in one satellite cell (Fig. 6-2).

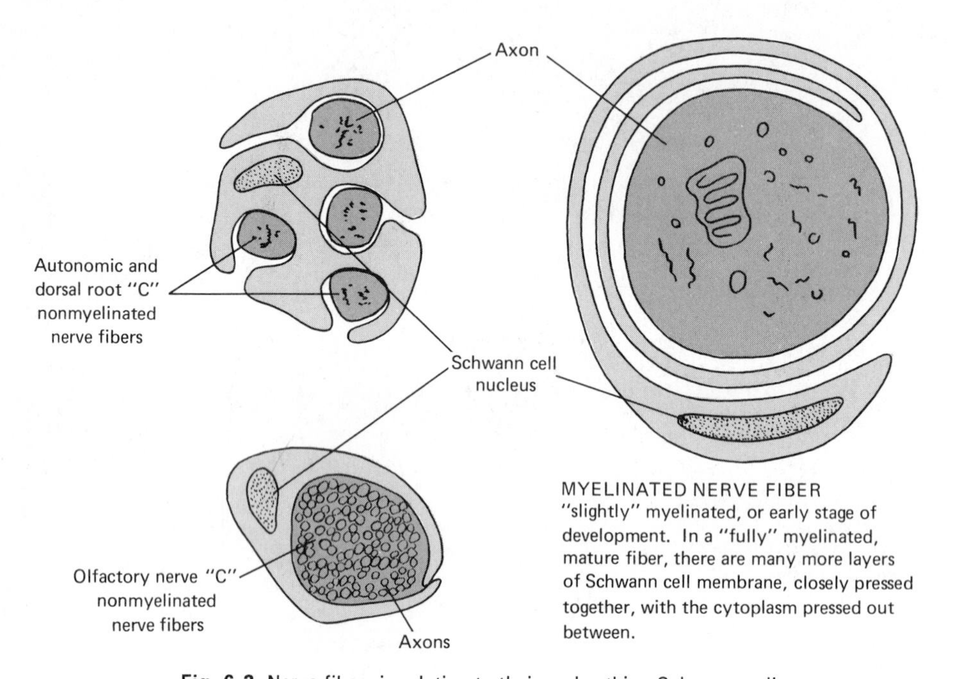

Fig. 6-2. Nerve fibers in relation to their ensheathing Schwann cell.

The nerve impulse

Among the millions of neurons in the nervous system, the *nerve impulse* is the means of communication. It is basically the same in all nerve cells, and if we understand what it is in a peripheral nerve, we may better understand its role in the central nervous system. The nature of the nerve impulse will, therefore, be described before the functional anatomy of the entire nervous system is examined.

The impulse is a momentary change that moves along the fiber away from the region where it started. In a *motor nerve*, this *propagated alteration* moves from the central system to a muscle. Near the muscle, the nerve fiber branches, and each branch terminates on a muscle fiber. The impulse proceeds down each nerve fiber branch and ends at the *neuromuscular junction*, where it acts as a trigger to start an impulse in the muscle fiber. The impulse in the muscle activates the contraction machinery.

A nerve may be stimulated to start an impulse by almost any environmental change if that change is sufficiently intense, rapid, and localized. A quick and intense change in temperature, a sudden application of pressure, a strong salt solution, or any of several other types of stimuli applied directly to the nerve will start the nerve alteration. An electric shock is most convenient, however, because it may be precisely controlled and because it mimics to some extent the effect of the impulse itself traveling along the nerve. An electrical stimulus may be easily and exactly adjusted in intensity or strength, it may rise to its maximum intensity slowly or rapidly, as we may wish, and it may be applied to as large or as small a part of a nerve as may be desired.

Velocity of a nerve impulse

The velocity with which impulses travel in a motor nerve may be measured in terms of the duration of time between the moment when the nerve is stimulated and the moment when contraction starts in the muscle (Fig. 6-3). If the moment of stimulation is marked on the same record as is the muscle contraction, the *latent period* may be determined. The latent period is the length of time between stimulation and response. It includes the time necessary for the impulse to travel to the muscle from the point of stimulation on the nerve. It also includes the time required for *neuromuscular transmission* to occur and for the impulse to move along the muscle and to start the contraction. It includes, as well, the time that may be necessary to get the impulse started in the nerve. The total latency is thus greater than the time necessary for the

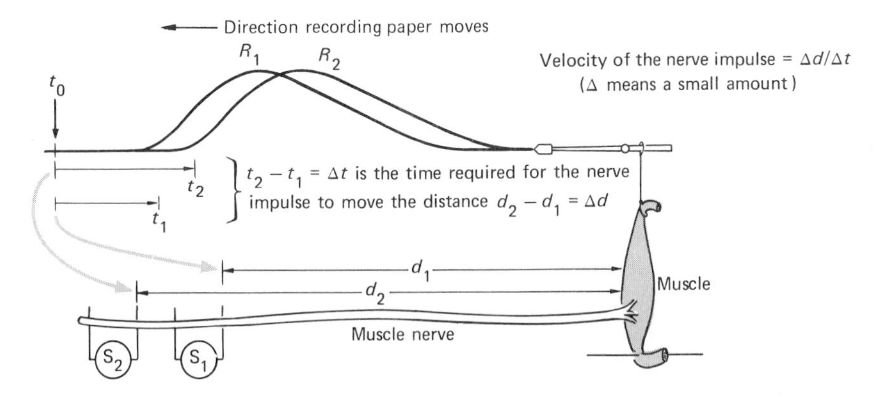

Fig. 6-3. Velocity of nerve impulse in motor nerves. The pointer on the lever makes a line on the paper moving along at a known speed. The two contractions differ essentially only in latency. When the stimulus is applied at s_1, time t_1 passes before the muscle contracts. At s_2, stimulation, contraction occurs at time t_2.

impulse to travel over the nerve and cannot be used directly to calculate the velocity of the impulse.

This complication can be taken care of by obtaining two different records—one with the stimulus close to, the other with the stimulus far from, the muscle. The difference in the latent period in the two records is the time the impulse requires to travel from the distant to the near point of stimulation. By means of this technique, impulses in frog motor nerve fibers can be demonstrated to travel 20 to 40 m/sec. The method is not very accurate for mammalian motor fibers in which impulses travel at 80 to 120 m/sec, nor can it be used to determine the impulse velocity in afferent nerves. More information about impulses may be obtained by other methods of analysis.

The electrical nature of the nerve impulse

The impulse in nerve or muscle consists of a localized change in the structure of the cell membrane. In a sense, the impulse moves along as a wave moves in water. As it passes, the region in its wake is quickly restored to normal, and another wave may pass. Unlike the wave, the impulse in a nerve fiber does not become less in intensity as it travels along the fiber. An impulse, once started, will move with undiminished vigor until it reaches the end of the fiber. In this respect, it is like the burning of a fuse, but unlike the fuse, the nerve repairs itself as the cataclysmic event passes. The nerve impulse is a localized moving wave of molecular reorientation and chemical reaction. Its most easily observed feature is a flow of *electric current* that occurs in the nerve and in the surrounding fluid during the impulse. This current is a movement of ions—that is, of electrically charged particles, especially of ions formed from atoms of sodium (Na^+), potassium (K^+), and chloride (Cl^-)—that are always present in the solution in and around the nerve fibers.

Electrical records from nerve

The energy in the electric current, associated with the nerve impulse, can be used to run a device (a galvanometer) that indicates how much and in which direction current is flowing. The indicator is arranged to point to the cathode, because by general agreement the direction of current flow is considered to be the direction that ⊕ charges move (see Chap. 3.1).

In Fig. 6-4 one electrode from the indicator is connected to the cut (injured) end of a nerve, the other to an intact part of the nerve. Current (*injury current*) flows toward the cut end. The inside of the nerve fiber is electrically negative compared to the outside. The resting nerve membrane is therefore like an electrical cell, the inside being the cathode. The electrical potential between inside and outside of the membrane, corresponding to the separation of charges that exists across the membrane, is called the *membrane potential.*

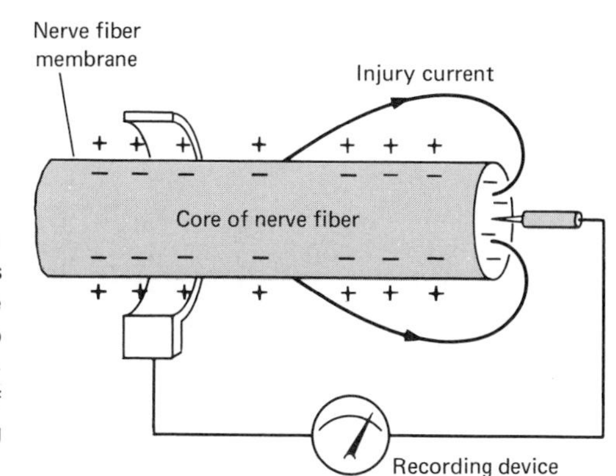

Fig. 6-4. How a nerve cell is similar to an electric battery. The cut end, continuous with the interior of the cell, behaves like the cathode of a battery—current flows to it from uninjured parts of the cell. Externally, the current flows in a thin layer of physiological salt solution surrounding the nerve.

Origin of the resting membrane potential in nerves

No one really knows how the resting membrane potential of a nerve fiber comes about, although much is known about how it can be changed. In the long run, it depends on cell metabolism. The cell membrane is permeable to sodium and potassium ions (Na^+ and K^+), but only K^+ ions stay within the cell. According to one hypothesis, the metabolism acts as a pump to push Na^+ ions out through the membrane as fast as they come in. In an arrangement of this kind, K^+ will exist in high concentration inside compared with the outside of the cell. The membrane potential may then be thought of as the tendency of the K^+ to diffuse outward, away from negatively charged organic ions (A^-) that are too large to pass through the cell membrane. The smaller K^+ ions diffuse as far as they can, held back by the opposing negative charges, and form an electrical double layer with the smaller K^+ ions on the outside and the larger A^- ions on the inside. The energy for maintaining the inequality of distribution of charges is provided by ATP, presumably by affecting the concentration of negative organic ions, and the structure of the membrane.

When metabolism is inhibited by certain poisons, the membrane potential will decline, although rather slowly, because so little energy is required for this cellular function. The size of the electric potential difference between inside and outside of the cell is reflected in the concentration of K^+ ions held within the cell. More accurately, the resting membrane potential can be estimated from the ratio of the concentration of K^+ in the cell compared with that present in the solution surrounding the cell. An increase in the concentration of K^+ outside the cell will reduce the tendency of K^+ ions to diffuse outward, and thus the membrane potential will be reduced.

Getting an impulse started

Excitation (stimulation) of a nerve to start an impulse is most easily accomplished by means of the cathode of an electrical cell. Figure 6-5 shows why this is so. The *cathode* provides a negative charge that attracts the existing charge on the mem-

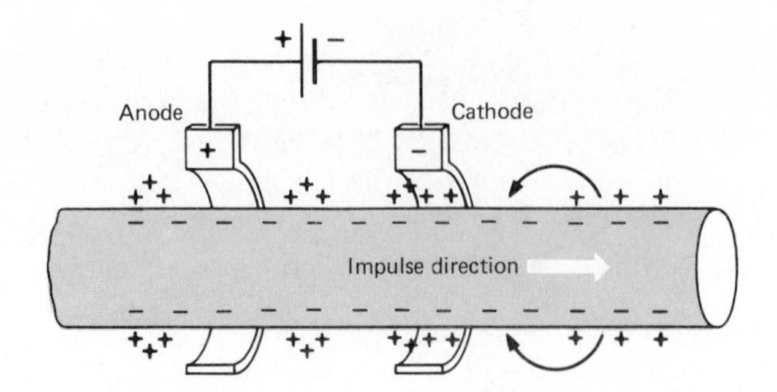

Fig. 6-5. Electrical excitation of nerve. Charge of membrane is removed near cathode where excitation starts. Arrows show flow of current during start of impulse propagation along the nerve fiber. Arrow in core shows direction in which impulse will move. Near the anode, the charge across the membrane is increased, and the excitability of the nerve becomes less.

brane and effectively neutralizes it. In this situation, the membrane is said to be *depolarized*, compared with its resting *polarized* condition. Current from resting membrane will flow to the depolarized region, and the depolarization will spread until the impulse is moving under its own power. Whether the excitation of a nerve results from stimulation by electrical, chemical, or mechanical means, or via a sense organ, or by another nerve impulse at a *synapse* (where one nerve fiber ends and another begins), the impulse has its beginning in the same way—a depolarization of the nerve fiber membrane. This is the crucial beginning of the excitation process.

At the *anode* of the stimulating electrodes, positive charge is supplied, augmenting that already present on the outside of the nerve fiber. Thus the potential difference between inside and outside of the membrane is increased, and more than the usual amount of stimulus energy would be necessary to depolarize the membrane to the critical level and initiate an impulse. The stimulus intensity that is just sufficient to get a nerve impulse started is said to be of *threshold* intensity. The threshold for stimulation is increased at the anode and decreased at the cathode of electrodes through which an electric pulse is delivered to the nerve.

Recording the impulse by electrical means

When a moving impulse reaches the first of two recording electrodes on a nerve, electric currents of the impulse may be recorded. The region of the impulse behaves momentarily like the negative pole of an electrical battery. From this simple comparison, we may call the electrical aspect of the nerve impulse, recorded in this way, an *action current*. Action in this context refers to the existence of an impulse moving along the nerve. The entire impulse travels past a point on the myelinated nerve in less than 2 msec. Most ordinary electric meters take longer than this to get started and therefore cannot give an accurate record of the time course of the action currents.

The *cathode-ray oscilloscope*, an instrument much like a television receiver, is rapid enough to record the action currents faithfully (Fig. 6-6). In the cathode-ray tube, a beam of electrons (the cathode ray) is shot from a cathode at the back of the tube. The inside of the front of the tube has a coat of material that glows momentarily under the impact of these electrons. The point of light thus produced on the tube face can be forced to move from side to side as the electron beam is accelerated first to one side and then to the other by means of an electrical charge alternately plus and minus in sign, placed on plates at opposite sides of the beam.

The electric charge associated with the impulse is very small indeed. It may, however, be sent through a suitable electronic device that will produce a much larger current, which appears and disappears with a time course corresponding to that of

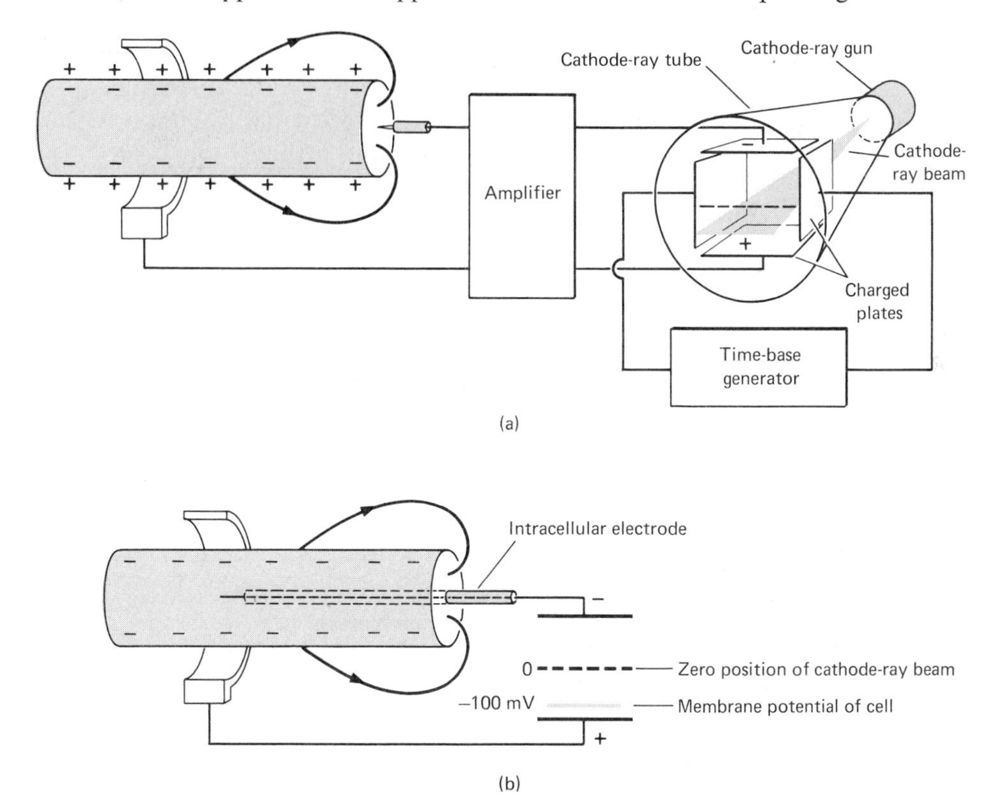

Fig. 6-6. Measurement of injury potential or membrane potential by means of cathode-ray oscilloscope. (a) The beam of electrons is moved back and forth as the charge on the vertically oriented plates of the cathode-ray tube is made alternately positive and negative by means of the time-base generator. The negativity of the cut end of the nerve is applied, after amplification, to the upper horizontally oriented plate and forces the beam of electrons downward. It behaves, therefore, as would the negative pole of a battery. (b) Somewhat more diagrammatic presentation, showing position of the beam when a fine electrode is thrust into the core of the nerve fiber.

the action current in the nerve. Thus sufficiently amplified, the charges are conducted to a pair of plates above and below the electron beam of the cathode-ray oscilloscope. As the beam moves horizontally with a known velocity, it also moves up or down an amount corresponding to the currents detected by the electrode on the nerve.

The action current of nerves is not merely an incidental happening that allows impulses to be easily followed. It is, in fact, the means by which an impulse keeps moving along the fiber. The moving impulse is a sink into which nearby positive charges flow. Thus the impulse moves along as regions ahead of it are depolarized by means of the current that flows from that region and into the center of the impulse at successive regions during its progress along the fiber.

Figure 6-7 is a diagram of the record of the electric potential change that occurs in the nerve fiber membrane as an impulse moves along the fiber. When the impulse arrives at the region of the inside electrode, the record of the membrane potential suddenly moves away from its negative value toward zero and continues on to become positive in relation to the outside. In other words, the inside of the membrane becomes more positive than the outside resting membrane is. To put it another way, the region of the membrane occupied by the center of the moving impulse undergoes a reversal in sign of potential (the inside becoming positive compared to the outside).

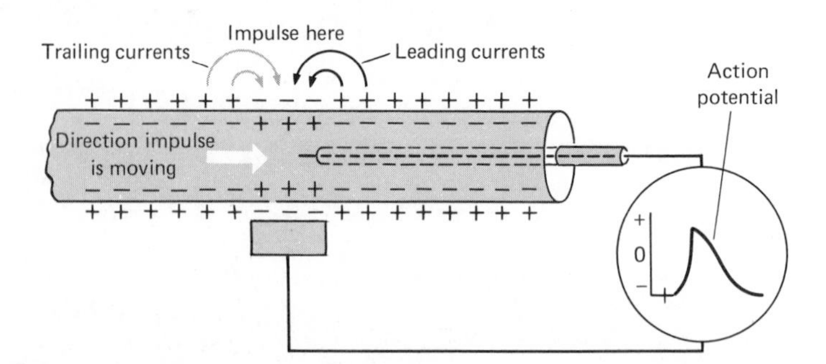

Fig. 6-7. Measurement of action potential of a nerve fiber. The outside electrode sees a momentary negativity as the impulse arrives under it. The inside of the nerve membrane becomes momentarily positive. The cathode-ray beam is moved in accordance with the potential thus placed on the plates of the tube.

Nerve membrane potential and permeability to ions

Just as the resting membrane potential may be explained in terms of permeability of the resting membrane to potassium ions, so the active membrane potential may be explained in terms of permeability of the altered membrane to sodium ions. *Na^+ is present mainly outside* of cells, and although it can penetrate the cell membrane, it is quickly expelled and effectively excluded. It does not, therefore, contribute appreciably to the size of the resting membrane potential. How can it be relevant to the action potential? The concentration difference for Na^+ across the cell membrane is oppositely directed compared to that of K^+, and the concentration of Na^+ outside is perhaps ten times the concentration of Na^+ inside the nerve fibers. The *outside* of a *resting nerve membrane* is *positive* because the K^+ ions that determine the potential and that are ⊕ ions tend to diffuse toward the outside.

In the nerve membrane occupied by the action potential, the permeability to Na^+ is perhaps 300 times as great as the permeability to K^+. In the active nerve, the Na^+ plays the tune—the membrane potential is determined by the relative concentration of Na^+ on the outside compared to the inside of the nerve fiber. The flux of ⊕ ions (Na^+) is now *inward,* so that the *inside* of the nerve fiber becomes relatively *positive* compared to the outside during the *action potential.*

When the resting membrane potential is decreased during the process of excitation, the permeability of the cell membrane to the Na^+ ion increases. The more the membrane potential decreases, the more easily the Na^+ is able to move through the membrane. Finally, the membrane permeability to Na^+ becomes tremendously greater than its permeability to K^+. This is the condition in the fully propagating nerve impulse. In this situation, the size of the membrane action potential can be predicted mainly from the relative concentrations of Na^+ outside and inside the cell. Direct measurement of the active membrane potential confirms this prediction. Figure 6-8 suggests the ionic basis for the resting and active membrane potentials. It also suggests the distribution, along the fiber, of the permeability changes involved in the impulse as it exists at one point in its passage along the nerve. With depolarization, a dramatic increase of Na^+ permeability occurs and then declines during the millisecond or so of life of the action potential spike at a point. Return of the membrane potential to the resting level is hastened by an increase in membrane permeability to K^+. The contribution of K^+ to the membrane potential therefore increases in excess of the resting situation. The sign of this contribution is opposite to that of the Na^+, and the membrane potential moves back toward the resting level. The return is made even more swiftly because the increased permeability to Na^+ is rapidly terminated. This concept of the resting and action potential depends on the formation of an electrical double layer accompanying the diffusion of ions of one sign while those of the opposite sign are restrained (see also Chap. 3.1).

The underlying mechanisms that govern the changes in membrane permeability are not understood, but the sizes of the potential changes are reflected in the ionic currents, which can be calculated. A really accurate description would involve the contributions of K^+, Cl^-, and Na^+ ions to both the resting and action potential.

Sometimes it is necessary to prevent particular nerves from functioning, as in the case of tooth extraction. Nerve impulses can be blocked or stopped by various means. If the increase in Na^+ permeability is prevented, the resting membrane potential may remain unchanged and yet propagation will cease. Anesthetics like cocaine and its derivatives seem to act in this way. If a large region is kept entirely depolarized, it cannot carry an impulse. A sufficiently high concentration of K^+ outside a nerve will have this effect.

In myelinated fibers the permeability changes described occur only at the nodes of Ranvier—not in the regions of membrane covered by the myelin. For this reason, the impulse in myelinated fibers is said to jump from node to node. *Saltatory conduc-*

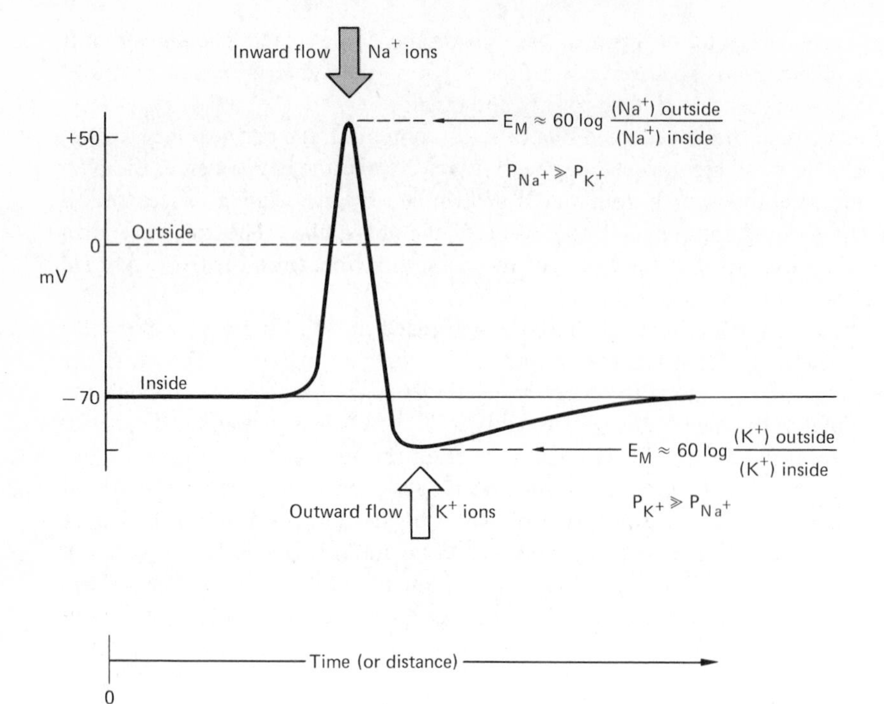

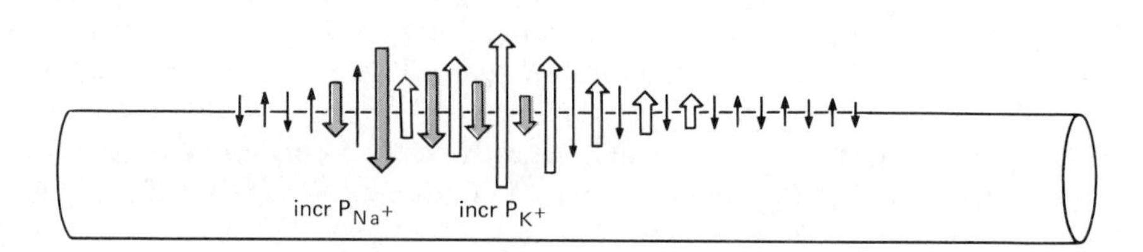

Fig. 6-8. Ion movements through axon membrane during nerve impulse. (Top) Initial spike approaches membrane potential (E_M) predictable from concentrations of Na^+ outside (high) compared to inside (low). Inside of membrane is ⊕ to outside during this spike. In later phase of the impulse, inside of membrane becomes ⊖ to outside, and E_M approaches a level predictable from concentration of K^+ ions outside (low) compared to inside (high). (Below) Arrows through axon membrane suggest the time course of change in membrane permeability responsible for the two extremes described above. Depolarization results in rapid development of high permeability to Na^+ ions (see $P_{Na^+} \gg P_{K^+}$ above) (solid arrows), which tend to move inward, and a more slowly developing high permeability to K^+ ions (see $P_{K^+} \gg P_{Na^+}$ above) (open arrows) which tend to move outward. The figure might also be considered to show the longitudinal distribution of permeability change in an axon with impulse frozen at one moment in time.

tion of this type allows the impulses to travel more rapidly than they do in nonmyelinated fibers.

At any one time, only one impulse can exist at a particular point on a nerve fiber. Having once been adequately stimulated, the fiber cannot again be stimulated until the membrane has recovered sufficiently. During a period of time corresponding to the early part of the action potential, the nerve is thus *absolutely refractory* to any stimulus. The *relative refractory period* follows. During this interval a stimulus will excite if it is somewhat stronger than is ordinarily required. Subsequent changes in excitability reflect the return to normal membrane permeability and the restoration of the membrane potential.

The compound action potential

The size of the action current recorded from a whole nerve depends on the number of active fibers in the nerve. When a small stimulus is applied, there may be no fiber sufficiently stimulated to generate an impulse. The stimulus is said to be of *threshold intensity* when sufficient fibers are active to produce a detectable action current. When all the fibers are active, the action current is maximal and does not increase during increase of stimulus intensity. Under a particular set of conditions, *the size of the externally recorded action potential moving along in each kind of fiber* depends mainly on the size of the nerve fiber carrying it and is *independent of the stimulus intensity*.

The record obtained from a whole nerve is a record of action currents from a large number of nerve fibers. The potential difference registered by the recording device is called a *compound action potential* (Fig. 6-9). The *velocity* of the impulses in any nerve may be determined from the electrical record. The electric current that serves as the stimulus to start the impulse also provides a small electric charge that is instantaneously recorded far from the point of application. This *artifact* serves to mark zero time, when the stimulus is applied. From the moment of the artifact until the arrival of the impulse at the electrode, the electrical record shows nothing happening. That interval mainly represents the conduction time of the impulse in traveling from its origin near the stimulating electrodes to the recording electrode. If several records are taken at different distances, the average velocity of the impulse may be determined by plotting the different combinations of distance and time.

The separate fibers in a nerve trunk are of different diameters; the impulses travel at various velocities according to the diameters of the fibers. Large fibers (10–20 μm to muscle) carry the impulses most rapidly (80–100 m/sec in man); small fibers, most slowly (0.1 μm fibers have impulses traveling only a few centimeters per second). The record from a whole nerve shows these differences (Fig. 6-9). In a record obtained a short distance from the point of stimulation, the action currents of all the fibers are nearly simultaneous. As the impulses travel farther, they become more dispersed because the slow ones lag behind. Having a long conduction time, the slow impulses turn up late in the record of the compound action potential.

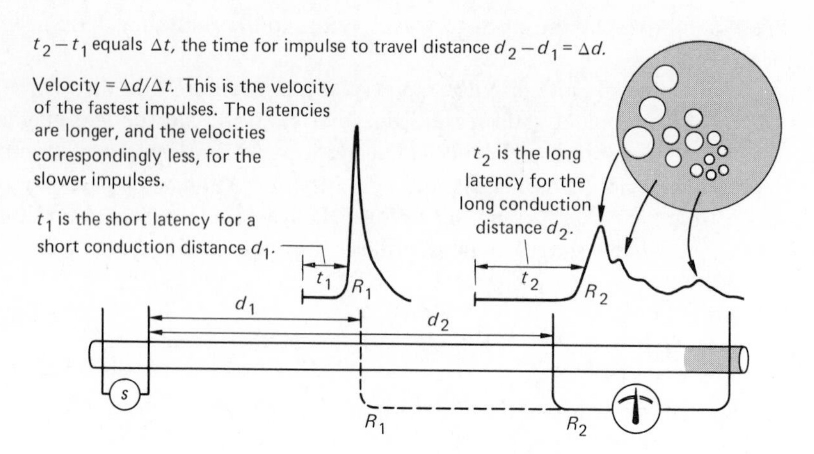

Fig. 6-9. Compound action potential. Traveling the distance d_2, the impulses that start at S and arrive at R_2 become dispersed. The latency t_2 is for the fastest impulses (in the largest fibers). Slower impulses (in smaller fibers) appear as later elevations in the record. The distance traveled to R is small enough that the impulses have not become greatly dispersed, as compared to those travelling to R_2.

Impulses are not usually synchronized in the several fibers of a nerve trunk. In any particular motor nerve fiber, impulses ordinarily follow one another at 5 to 50 times per second, but when one fiber is active, another may be resting. The same story is true in sensory nerves, but because of the many thousands of nerve fibers, there is a continuous traffic of impulses into and out of the CNS.

6.2 Sensory Receptors —The Input Side of the Reflex Arc

Sensory receptors are special physiological devices that permit afferent nerves to be excited by means of stimuli so weak as to be inadequate if applied to the nerve directly. Mechanical pressure, for example, will stimulate a nerve directly, as you are well aware when you accidentally bruise your elbow. The tingling sensation in the finger results from direct stimulation of the ulnar nerve at the elbow. Far less pressure is necessary to arouse a sensation if the pressure is applied directly to the fingers, however. Since less energy is required to produce action potentials in a nerve when a receptor is stimulated than when the nerve is stimulated directly, the receptor is said to lower the threshold to a particular form of stimulus. The receptor is the intermediary between stimulus and nerve; it is a *transducer* that receives stimulus energy and starts impulses in the afferent nerve fibers.

The receptors respond to only a small part of all the energy changes occurring in the environment, and any particular stimulus must be of some adequate or threshold intensity if it is to be detected at all. We do not have specific receptors for radio waves or X rays. Some snakes, but not humans, have infrared detectors, and insects see much farther into the ultraviolet than we do.

Sensory receptors may be classified according to the nature of the stimulus. *Mechanoreceptors*, on which the senses of touch and hearing depend, are stimulated by mechanical displacement. *Chemoreceptors* stimulated by particular chemical substances are found on the tongue and in the nose, as well as in some blood vessels. *Osmoreceptors* in the brain are sensitive to changes in osmotic concentration of the blood. *Photoreceptors* of the eye respond to light impinging on them. *Thermoreceptors* in the skin provide information about temperature changes. *Nociceptors* detect stimuli that are harmful, noxious, or injurious.

Another way of looking at receptors is in terms of where the stimuli are in relation to the body. *Exteroceptors* detect stimuli that impinge on the body from outside. *Proprioceptors* detect changes of the body itself—for example, stresses in mesenteries, in fascia, in tendons, and in muscle. *Interoceptors* detect changes in the internal fluid environment within the body.

The most distinctive pressure receptor about which the most is known is the *Pacinian corpuscle*. This tiny organ has been studied particularly in mesenteries, the thin, double-layered sheets connecting the viscera to the walls of the body cavities, but it is also found in the deep layers of the skin, in mucous membranes, and in connective tissue generally. Each corpuscle, having several concentric layers of connective tissue, is constructed like an onion, through the center of which runs the nerve ending. Impulses seem to start at the first node of Ranvier back from the tip of the nerve fiber imbedded in this corpuscle. The tip itself may not carry an impulse, but it becomes depolarized and sets up a flow of current from the excitable part of the nerve, which is then depolarized to the threshold level for impulse generation.

Effect of receptors on the central nervous system

The impulses from most receptors may have a double role in the central nervous system. They can bring about certain specific reflex movements, and they may arouse in the receptive brain a sensation of a specific quality. For example, afferent fibers whose terminals are "injury receptors" in the skin of the appendages connect in the spinal cord in such a way as to produce flexion of the limb, whereas those in which the terminals respond to pressure on the soles of the feet are arranged to provoke extension. Carrying impulses to the brain, these fibers evoke sensations of pain associated with the flexion, and of touch, associated with the extension.

Generally a single stimulus to any receptor produces not one but several impulses during a few milliseconds, or several seconds, depending on the particular receptor. In all instances, the time interval between the impulses is very small at the start of the response and becomes greater as the discharge continues. When a hair is bent, only a few impulses are produced, even though the hair is kept in the bent position. A stretch receptor may discharge impulses for many seconds while stretch is maintained on a muscle.

The frequency of impulses in an afferent nerve declines during application of an apparently constant stimulus to the receptor. This feature is called *adaptation*, a term that implies a decrease in sensitivity of the receptor system to the apparently constant stimulus (see Fig. 6-10). A receptor that ceases discharging relatively soon, even though the stimulus is continued, is said to adapt rapidly; a receptor that continues firing for a long time is said to adapt slowly. In the case of mechanoreceptors, the

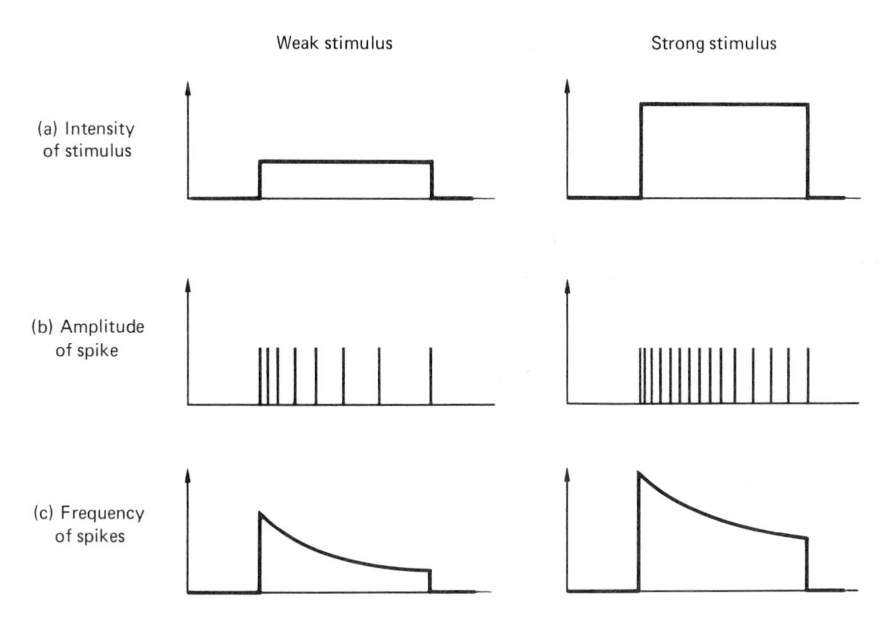

Fig. 6-10. Relation of stimulus and response in a single unit of a receptor. (a) A constant stimulus to a receptor. (b) A declining frequency of impulses in the afferent nerve (adaptation). (c) A plot of frequency of impulses versus time, and a representation of the time course of the intensity of the sensation.

rapid adaptation may be merely a rapid readjustment of stresses within the tissue; slow adaptation may imply that the tissue tensions do not rapidly become dissipated. A slowly adapting receptor will provide a prolonged sensation, whereas the sensation from a rapidly adapting receptor will be quickly terminated.

The intensity of effect produced in the central nervous system by stimulation of a receptor depends on the number of receptors and the number of nerve fibers activated. It also depends on the frequency of impulses traveling in the nerve. This relation is similar to the dependence of muscle force on the number of active motor units and the frequency of impulses in each unit. It is easy to see how *intensity* of a stimulus is sensed by the nervous system, but it is much more difficult to understand how different *kinds* or qualities of stimuli of are discriminated by the central nervous system. Generally speaking, the impulses causing different kinds of sensations are carried in different nerve fibers. For example, muscle stretch receptors are attached to large fibers, while sensation of certain kinds of pain may be carried in very small fibers. The rates of adaptation may vary, as may characteristic firing frequencies. More important for recognizing kinds of stimuli are the places to which the impulses go in the central nervous system. Separation of fibers carrying different kinds of information is relatively clear at spinal and brain stem levels. How *place* of projection of fibers to the cortex may be interpreted by the brain as a particular *quality* of sensation is at present completely unknown.

Receptors in the skin

The entire skin is liberally supplied with touch, pressure, temperature, and pain receptors. By touching the skin with suitable small probes, one may discover that receptors for touch, for warmth, for cold, and for pain are distributed in a punctate fashion. Between the points, the probe provides no sensation when low–intensity stimuli are used. Beneath each sensory spot, there is a nerve ending. On hairless areas of the skin (forehead, lips, palms of the hands, and soles of the feet), the endings have a more or less specific structure for each sensory quality. The endings are twisted and coiled into characteristic forms that are separately sensitive to touch, heat, and cold. The different forms presumably reflect some fundamental differences in molecular structure of the different nerve fibers.

In hairy areas of the skin, places may be found that are sensitive to specific stimuli but where only simple nerve endings can be seen after histological examination. In these instances, one may assume that molecular differences in the nerve fibers are not accompanied by anatomical distinctions. Alternatively, the determinant of the specificity may be the nature of the accessory non-neural cells that are associated with the nerve endings.

The skin is also liberally provided with free nerve endings that appear to be stimulated only by injurious stimuli and that provide a sensation of pain.

A mechanoreceptor exists at the base of each hair in the skin, as may be determined by bending a single hair on the back of the hand. The hair serves as a lever to apply a distorting force to a bundle of nerve endings that surrounds the hair follicle. This stretch probably increases the membrane permeability in the nerve terminals. The terminal region then becomes sufficiently depolarized to start an impulse.

The organization of groups of receptors is analogous to the organization of groups of muscle fibers on the motor side of the nervous system. The dendrite from a dorsal root ganglion cell divides a number of times, and each branch is a receptor terminal. This system—the afferent neuron and the structures it innervates—may be called a *receptor unit*, analogous to a motor unit (Fig. 6-11).

Receptors in muscle

Mechanoreceptors in skeletal muscle and in tendons give the central nervous system information about the condition of stress in the muscle. In order to study these receptors in an animal, we would select spinal roots that innervate the particular muscle we wish to study. Now, if the ventral root is stimulated, the muscle will contract. If recording electrodes are placed on filaments of the dorsal root, impulses will be recorded when mechanoreceptors of the muscle or tendon are stimulated. There are two simple ways to manipulate the muscle to excite the mechanoreceptors. The muscle may be passively stretched, or it may be stimulated to contract. Two classes of responses giving information about these two kinds of manipulation may be recorded from filaments of the dorsal root (Fig. 6-11).

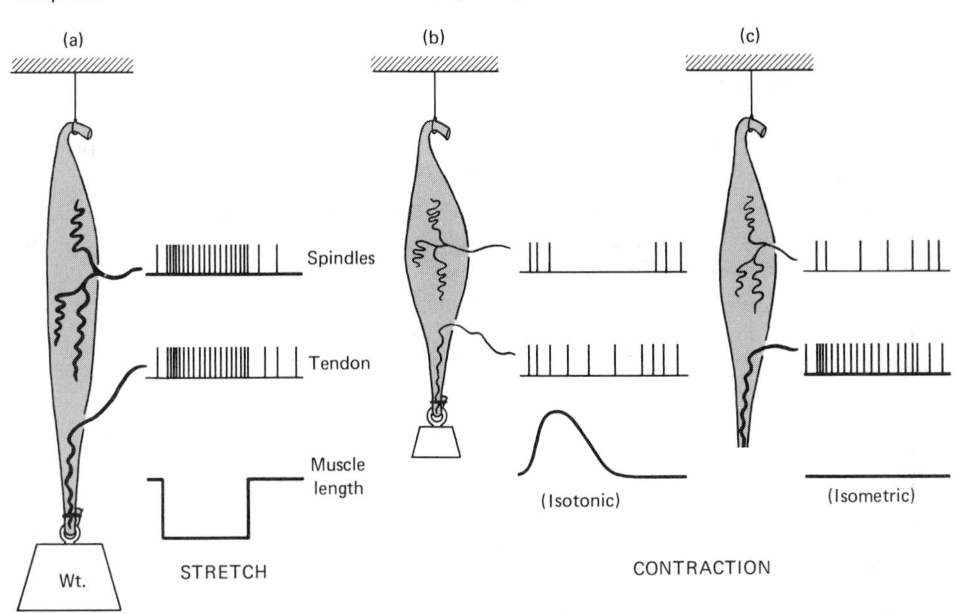

Fig. 6-11. Muscle and tendon stretch receptors. (a) Muscle spindles and tendon receptors both discharge impulses at increased frequency when muscle is greatly *stretched.* (b) During *isotonic muscle* contraction, muscle shortens, stretch is taken off spindles, and spindles discharge impulses at decreased frequency. (c) During *isometric contraction,* tendon receptors are strongly stretched and discharge impulses at increased frequency.

The weight shown in Fig. 6-11 will stretch the muscle. It also stretches the muscle spindles that lie parallel alongside the ordinary muscle fibers. The stretch is a stimulus to the small afferent nerve fibers of the muscle spindle. On the other hand, when the ordinary muscle fibers shorten during contraction, the spindles are not stretched, but instead stress on them is relieved [Fig. 6-11(b)]. The tendon organs that lie in the tendon are pulled somewhat by the weight stretching the muscle, but they are stimulated even more when the muscle contracts and pulls the tendon [Fig. 6-11(c)]. Figure 6-11(b) shows the increase in impulse discharge in the tendon afferent fiber during muscle contraction. Figure 6-11(a) and (b) shows the increase in impulse discharge during passive stretch of the muscle, as well as the decrease of impulse discharge during muscle contraction.

The receptors discussed so far are mainly exteroceptors distributed generally in the skin, plus the proprioceptors found in skeletal muscle. These receptors will serve to give us some understanding of the input side of the reflex arc. Several sets of receptors are organized into large, complex sense organs of the head. These organs, which have large portions of the brain devoted to them, will be discussed following consideration of brain structure and function (Chaps. 8.1 and 8.2).

How we judge stimulus intensity

We know that a weak stimulus to a particular sense organ will produce a low frequency of impulses, and a weak sensation, whereas a stronger stimulus will produce a more intense sensation. Figure 6-10 shows that the strong stimulus generates a higher frequency of impulses in the afferent nerve. One might then reasonably expect a relation between frequency of impulses and intensity of sensation.

Impulses from muscle stretch receptors are used when we try to compare different weights. If we hang different weights from an isolated muscle, we can see how the frequency of firing in the afferent nerve fibers of the muscle depends on the amount of weight [Fig. 6-12(a)]. A minimum threshold weight is necessary before any increase in frequency of impulse discharge can be noted. Above a certain maximum weight, there can be no increase in discharge frequency because the nerve fiber is firing as fast as possible under the circumstances. In between, there is a range over which impulse-discharge frequency seems to be approximately directly proportional to stimulus intensity. It seems probable that a simple relationship of this kind must be characteristic of the early stage of impulse generation by all receptors.

Let us start at the other end of the problem to relate the stimulus to the sensation. If we make judgments of weights, we may ask how different can the weights be and still seem to be the same? When weights are large, this Just Noticeable Difference (JND) is also rather large, while at moderate weights the Just Noticeable Difference is a smaller weight. When we say that one weight is just noticeably different from another, we are reporting a similar kind of *difference in sensation* whether the weights or the pair are light or heavy. The simplest hypothesis to account for equal differences in sensation is that the difference in the frequency of impulses is the same—that is, that a JND is due to a particular increment in nerve impulse frequency.

According to this idea, if the weights are light, they produce a low frequency of impulses, but the *change in number of impulses* that is just noticeable from one weight to the next is the same as between heavier weights. In the latter case, however, the reference weight produces a higher frequency of impulses. If this assumption is true, then if we consider the whole range of weights that we can handle, from very light to very heavy, we will also cover the whole range of change in frequency that we can sense, from low to very high. Of course, in a particular case, when we try to detect differences in weights, we do not have a record of the impulse frequency. We have only a record of the weights and the differences that are just noticeable. We find that the larger the reference weight (W), the larger the increment (ΔW) must be before we can decide that the weight is heavier. If the detectable increment is directly proportional to the reference weight, then $\Delta W/W$ is the same for all values of W. A plot of $\Delta W/W$ vs. W would be a straight line parallel to the abscissa. In constructing such a curve, we find that generally the line is relatively flat in a middle range but rises at each end, as in Fig. 6-12(b). In the flat middle range, weights are most easily discriminated.

Curve *B* could be constructed from the muscle receptor response curve (*A*). The ordinate was divided into eight JND intervals. The dashed lines show how the ΔW

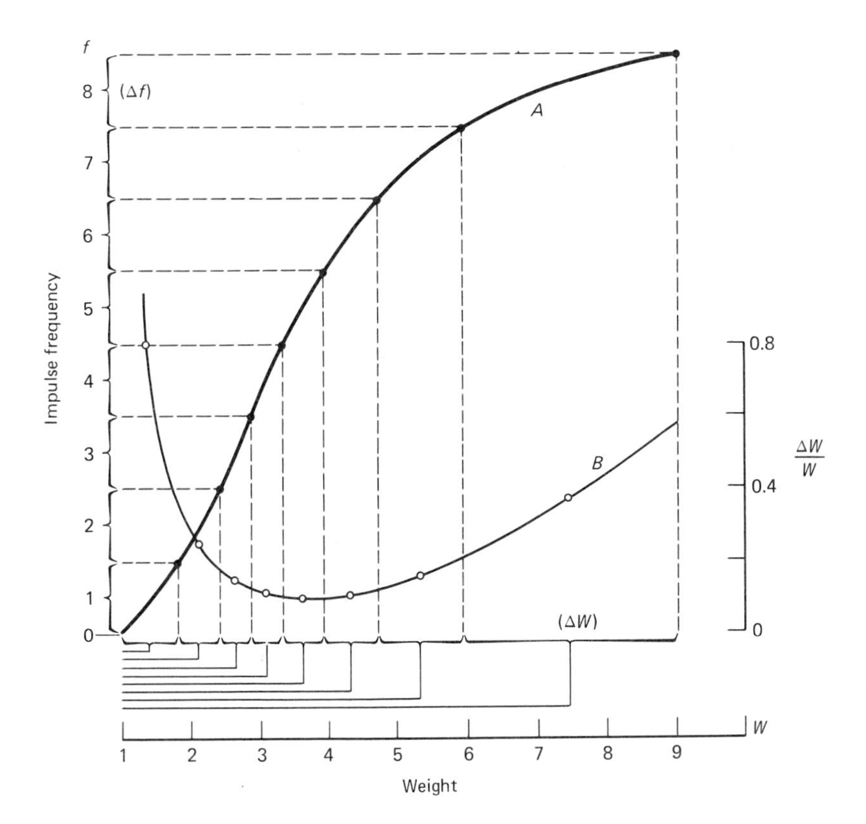

Fig. 6-12. Weight discrimination. *Heavy line:* impulse frequency (ordinate axis at *left*) in afferent nerve from muscle, as a function of weight (abscissa) stretching muscle. *Thin line:* Weight discrimination predicted from above curve, as follows: Abscissa at left is divided into equal increments of frequency (Δf). The increment of weight (ΔW) corresponding to each Δf is shown by the dotted lines. The relative weight differences (*right-hand ordinate*) $\Delta W/W$ corresponding to each Δf, ΔW combination is calculated and plotted as a function of weight.

The diagram shows that there is a middle range of weights over which the ratio of JND to the standard is relatively constant. In other words, the relative differences in weight required to obtain equal increments in impulse frequency are the same. At the ends of the curve, where the reference weights are rather small or rather large, the relative weight differences necessary to obtain equal increments of impulse frequency are larger.

corresponding to each successive JND was determined. Each ΔW so determined was divided by the relevant W. The similarity in the curves derived directly from isolated muscle and indirectly from sensation lends support to the idea that intensity of sensation is directly related to the frequency of impulses that reach the brain and that just noticeable differences in sensation correspond to equal increments in impulse frequency.

Changes in the basic form of the JND curve may be due to adaptation mechanisms that, in effect, alter the stimulus intensity. There is, for example, a reflex that increases the tension on the eardrum when the sound is excessively loud. This reflex may decrease the impulse frequency generated by loud sounds. A similar mechanism exists for the eye in relation to light.

6.3 Reflexes—Structure and Function of the Spinal Cord

Anatomically, it is easy to distinguish the central and peripheral parts of the nervous system. The *central nervous system* is a compact mass of interconnected nerve fibers and cell bodies constituting the spinal cord and brain. The *peripheral nervous system* consists of bundles of *afferent* and *efferent* nerve fibers—the nerves connecting the central nervous system to the periphery of the body. Some of the nerves interweave as *plexuses*, for example, where the arms and legs attach to the trunk of the body (Fig. 6-13).

Another kind of plexus is present in the viscera, where groups of cell bodies exist as complex interconnected networks of neurons of the peripheral nervous system.

It is also helpful to describe broad functional divisions of the nervous system, according to whether the part in question regulates skeletal muscle or smooth muscle, such as in the visceral organs of the body cavity. The nerves and central connections of the spinal cord and brain that are concerned with control of the skeletal (somatic) musculature may be called the *somatic nervous system.* This is a functional subdivision of the nervous system not easily distinguished anatomically, except in certain instances, from the nervous system as a whole. The part of the nervous system concerned more directly with control of visceral functions, such as the beat of the heart, the contraction of smooth muscle in blood vessels and digestive tract, and the activity of glands, may be designated the *visceral nervous system.* The somatic nervous system is easier to study than the visceral, and we know more about it.

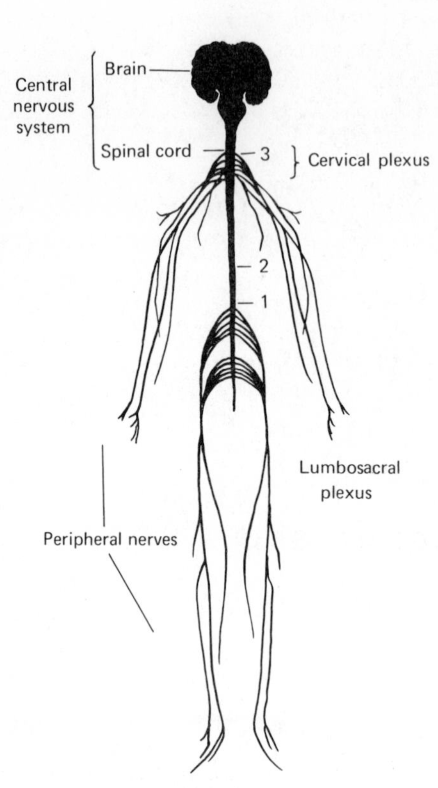

Fig. 6-13. Dorsal view of central nervous system (together with nerve roots and nerve trunks). Only some of the main peripheral nerves are roughly indicated. The entire system is of course far more extensive than this diagram shows.

General organization of the central nervous system

In the CNS, the cell bodies are grouped according to the function of their fibers. The groups are interconnected by their fibers, which travel together as *tracts* in the central nervous system. Some groups of cells are connected to the periphery of the body by way of the *peripheral nerves.* A peripheral nerve consists of hundreds of fibers that carry impulses toward the central nervous system, plus other hundreds that carry the impulses away from the central nervous system.

In the nervous system, the fatty covering or myelin surrounding many individual fibers provides the color to the *white matter*, whereas *gray matter* consists of cell bodies and small fibers, both of which lack this fatty layer. The central nervous system may be divided into the *spinal cord*, where the nerves to the limbs and trunk originate, and the *brain*, the origin of the nerves to the head (Fig. 6-13). In the spinal cord, the gray matter is at the center; in the brain there is, in addition, an outside layer or *cortex* of gray matter.

The nerve fibers make connection with the central nervous system in bundles called *roots* (Fig. 6-14). These roots occur in pairs, segmentally along the CNS. Each segment has a *dorsal* and a *ventral* root on each side. In the dorsal roots are the afferent fibers. Almost all fibers of this type have their cell bodies located outside the central nervous system along the roots, in the *dorsal root ganglia*. The efferent fibers in the ventral root originate from cell bodies located in the central gray matter. Fibers from dorsal and ventral roots join to form the mixed bundles (nerve trunks) of the peripheral nerves.

The impulses traveling into the central nervous system stop at the ends of the afferent nerve fibers that carry them. The impulses traveling up and down within the central nervous system and out the efferent nerve fibers to the muscle are new impulses. The transformation of the incoming impulses into the new efferent pattern takes place at the *synapses*, the places where neurons touch in such a fashion that the impulses in the *presynaptic fibers* can excite (or inhibit) the production of impulses in the *postsynaptic neurons*. In some instances, afferent fibers make synapses directly with efferent fibers, but in most cases they are separated by one or more *interneurons*.

The reflex

Most of the nerve cells of the body are entirely within the central nervous system, a complex network that receives information from sense organs and sends information to the muscles. The information going to the muscles is followed by patterns of movement that are usually appropriate responses to the messages arriving from the sense organs. Reproducible patterns of movement that result from particular kinds of peripheral stimulation entering the central nervous system are called *reflexes*. The sudden withdrawal of the hand from a hot object is an example of a reflex. The response is such as to remove the offended part of the body from the stimulus. The effect of the stimulus seems to travel into the central nervous system and be reflected back to the appropriate muscles. For this reason, this type of reaction is called a *reflex*. Any specific, reproducible reaction following stimulation of a receptor and involving the central nervous system, but dissociated from voluntary control, may be called a reflex. The pathway involved in a reflex includes *receptors*, *afferent fibers* carrying impulses *into* the central nervous system, *central connections*, and *efferent fibers* carrying impulses *out* to muscle or to other effectors. This pathway is a complete *reflex arc* (Fig. 6-14).

Reflexes involving the spinal cord can proceed without the presence of the brain as long as the peripheral connections are intact. Ordinarily, considerable control may be exerted over the spinal motoneurons by impulses coming from the brain, but the rules governing the functioning of most of the rest of the nervous system may be established by a study of the relatively simple arrangements in the isolated spinal cord.

Structure of the spinal cord

The spinal cord is contained entirely within the vertebral canal; the brain is housed in the skull. Nerves from the spinal cord exit between the vertebrae; those of the brain come out through special apertures—*foramina*—in the skull wall.

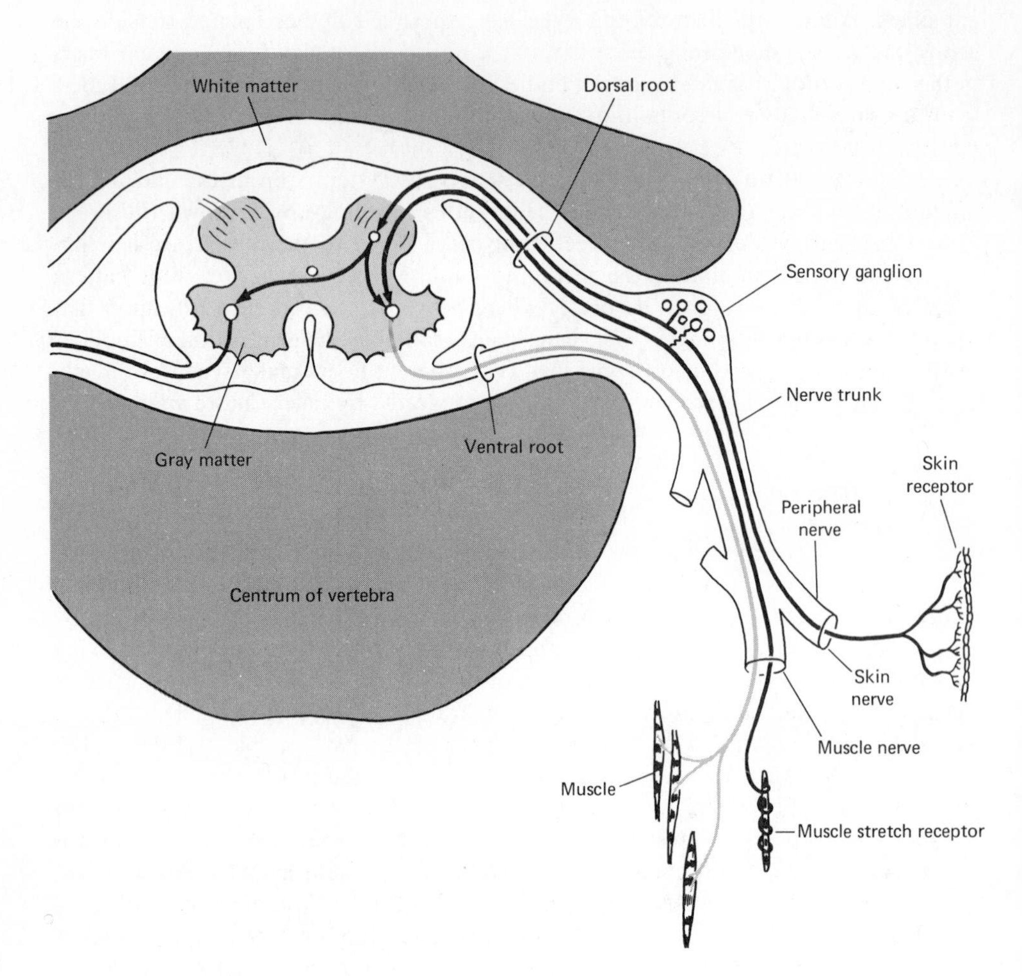

Fig. 6-14. Some relations of peripheral nerve, dorsal and ventral root fibers at one segment of spinal cord. G = gray matter; W = white matter of spinal cord.

The *nerve roots* of the spinal cord are within the central canal of the vertebral column. In the cervical and thoracic regions of the vertebral column, the roots are short and go out between the nearby vertebrae. In lumbar and sacral regions, the roots are longer; the spinal cord ends at about the level of the first lumbar vertebra, but the roots continue to their exits lower down between vertebrae of the lumbar and sacral regions. The relatively great length of the roots results from the fact that the more caudal vertebrae grow faster than the spinal cord and pull the roots along with them.

The regions of the cord are named according to the regions of the vertebral column where their roots exit. The cervical and lumbar regions of the cord are enlarged, compared with the thoracic region, because of the presence of many motoneurons that control the large masses of limb muscles. At all levels of the cord, the afferent fibers of the dorsal roots carry information from receptors; the efferent fibers in the ventral roots carry impulses out to the muscles. Outside the vertebral column, the nerve fibers are assembled into nerves or nerve trunks, in which the afferent and efferent fibers are mixed.

Within the spinal cord, the cell bodies of motoneurons are aggregated ventrally in the gray matter, while cell bodies of interneurons are in the more dorsal parts of the gray matter. The gray matter of the cord appears butterfly-shaped in cross section. The dorsal wings of the gray matter are designated *dorsal horns*, while the ventral portions are called *ventral horns*. Actually, of course, the aggregations of cell bodies form continuous columns along the length of the spinal cord.

Afferent fibers from the dorsal roots are distributed to several places in the spinal cord. In general, each fiber divides several times and *diverges* from its point of entry into the cord to form synapses with several neurons (Fig. 6-15). Impulses in one fiber alone are usually unable to excite neurons with which they synapse. However, presynaptic fibers from many afferents converge on each neuron, and the synaptic effects in these several fibers summate to provide sufficient effect for excitation of the postsynaptic cell. The fibers connect not only in the segment at which they enter but also at adjacent segments.

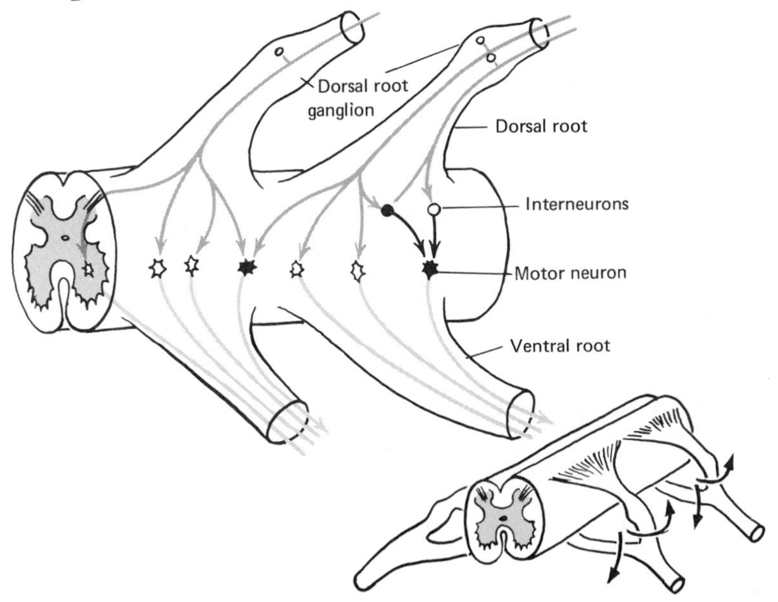

Fig. 6-15. Connections between nerves at successive levels in spinal cord. Two afferent fibers, each in a separate root, converge to supply one motoneuron. Another pair of afferent fibers, both in one root, converge to supply one interneuron. All three afferent fibers diverge to supply several motoneurons and interneurons. Two interneurons converge to supply one motoneuron. Note that the afferent nerve fibers fan out from the root as they approach the cord. (Below) Dorsal and ventral roots have been cut and turned aside (arrows) to provide a transparent side view.

Stretch reflexes and regulation of posture

The striated skeletal muscles of the body are known as *voluntary muscles* because their contractions can be consciously controlled. Yet much of the activity of this somatic or body musculature is reflexly regulated. The afferent impulses that initiate the reflex movements do not have to reach consciousness in order to be effective. When we do become aware of the stimuli, the information may be used for voluntary adjustment of muscle tension.

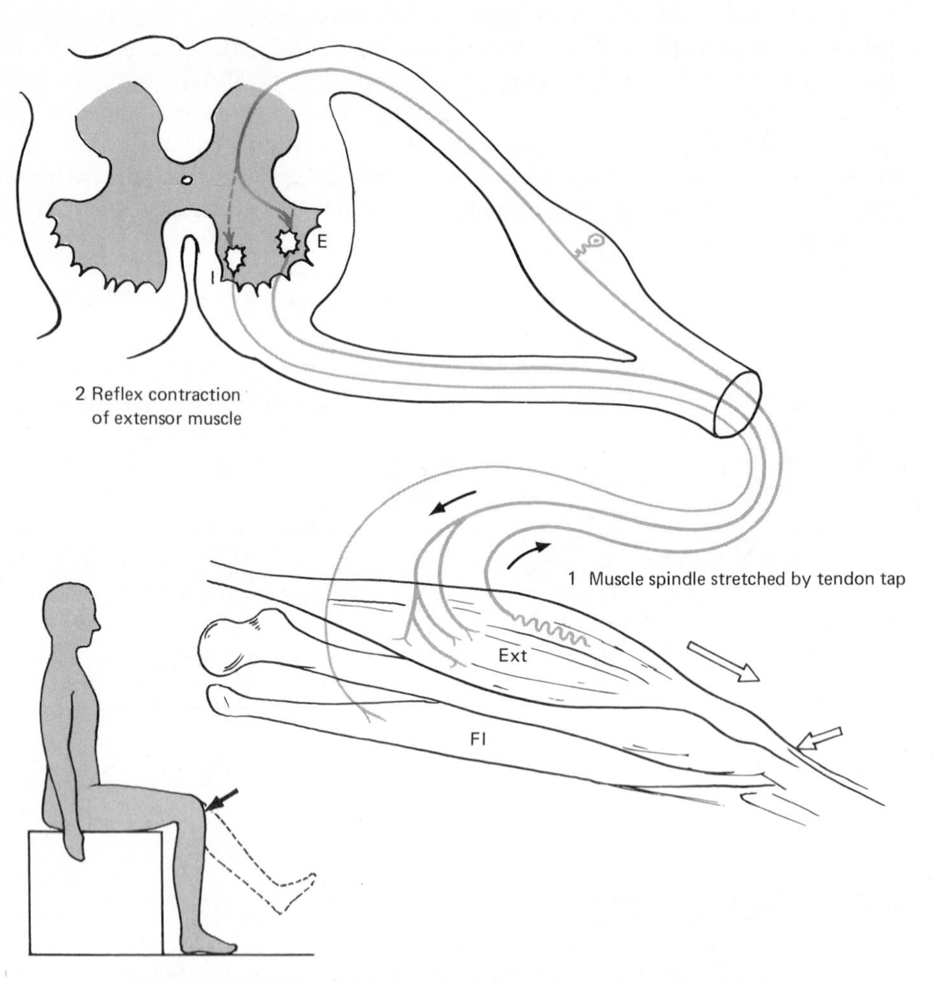

Fig. 6-16. The stretch reflex—a monosynaptic reflex. Afferent impulses from stretch receptors result in reflex excitation of the muscle stretched and inhibition of the antagonistic muscle. Muscle is stretched by a tap applied to the tendon. Numbers indicate sequence of events.

One of the simplest reflexes may be obtained by stimulating the stretch receptors of a muscle. The knee jerk is a good example, readily elicited by striking the freely suspended but flexed leg just below the patella. When the patella is tapped, the quadriceps femoris muscle is momentarily stretched. A slight pull is thereby applied to the stretch receptors among the muscle fibers (Fig. 6-16). These receptors consist of coiled nerve endings ramifying over the surface of certain muscle fibers that are smaller than the ordinary muscle fibers and that do not contribute to the tension of muscle contraction. These special muscle fibers of the stretch receptor are enclosed in a spindle-shaped sheath and are called *intrafusal fibers*. When the intrafusal muscle fibers are stretched, the small nerve fibers surrounding them are stimulated, and impulses travel into the central nervous system. A strong pull to the quadriceps femoris muscle stimulates a large number of such intrafusal fibers, and a large synchronous volley of impulses, traveling in many afferent fibers, enters the central nervous system. Centrally, these fibers have direct connections with the motoneurons that innervate the quadriceps femoris muscle fibers.

Several motoneurons are more or less simultaneously activated by the relatively synchronous volley of impulses coming from the stretch receptors. Impulses travel out to the quadriceps femoris muscle, and the leg is kicked forward. This reflex contraction of a muscle is called a *stretch reflex*.

Stretch receptors have been found in most voluntary muscles. They are even present in the extrinsic muscles of the eye and are especially important in the muscles that control posture. We exert very little conscious effort in ordinary standing. If the knees tend to buckle from the weight of the body, they are automatically straightened. Bending of the knee pulls the stretch receptors of the quadriceps slightly, and reflex contraction of that muscle straightens the leg. If the head falls slightly forward, it is quickly readjusted by a similar mechanism. Skeletal *muscle tone* is thus continually regulated in the muscles to a degree appropriate for maintaining normal posture. The muscle stretch receptors are responsible for this regulation and for the tone in skeletal muscles. Tone in skeletal muscles depends on continual asynchronous contractions in the motor units. The excitation of the motoneurons depends on the continual play of impulses coming into the central nervous system along afferent nerves from the stretch receptors.

EFFERENT CONTROL OF STRETCH RECEPTORS. The frequency of impulses sent into the CNS by the stretch receptors depends on how much the muscle is stretched, as described in Chap. 6.2. It also depends, in a curious way, on the intensity of tone (frequency of impulses) in special efferent fibers bringing, from the CNS, impulses that increase the excitability of the stretch receptors. Figure 6-17 shows the arrangement. The afferent fiber innervates the middle section, and the efferent fiber innervates each end of the intrafusal muscle fiber. The efferent fibers to the spindle are smaller than the large Aα fibers that innervate the regular muscle fibers. They are myelinated fibers of the γ (gamma) subgroup. When they are active, a small amount of force is apparently produced at each end of the intrafusal muscle fibers. The force is directed away from the middle section, which is therefore put under strain. The mus-

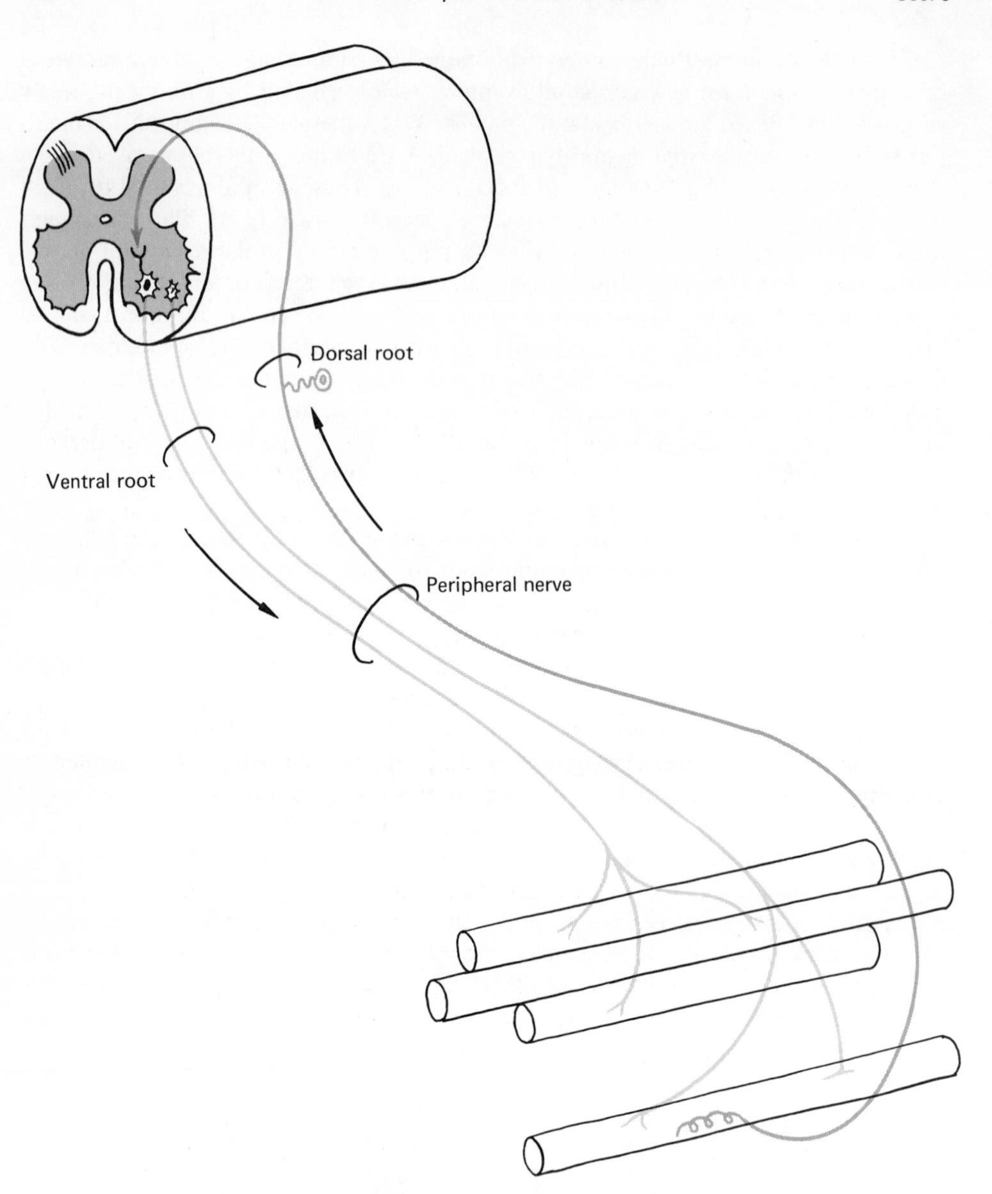

Fig. 6-17. (1) Reticulo-spinal influence, from higher levels of CNS, determines frequency of impulses going along γ efferent fibers to the intrafusal muscle fibers. (2) Frequency of impulses in the spindle afferent fibers depends on stretch of the spindle produced by (a) stretch applied to the tendons of the whole muscle (the extrafusal fibers are also thereby stretched) and (b) tension developed within the extrafusal fibers according to γ efferent tone. (3) Tension in the muscle fibers depends directly on (a) stretch applied to the muscle, and (b) frequency of impulses generated in the motoneurons. Spindle fiber afferent tone is one factor controlling motoneuron discharge rate.

cle-spindle nerve terminals are thereby stretched, and impulses go along the afferent fiber into the central nervous system and excite the α motoneurons of that muscle. The γ efferent motoneurons are regularly excited by impulses coming from upper regions of the nervous system.

Reciprocal organization of reflexes

The same afferent impulses that reflexly cause an increase of impulse frequency (*excitation*) in the motoneurons to the quadriceps muscle cause a reflex decrease of impulse frequency (*inhibition*) in the motoneurons innervating the antagonistic hamstring muscles. Therefore, at the same time that tension in the quadriceps increases, hamstring tension decreases. All the antagonistic pairs of muscles are arranged in this way. They are said to be *reciprocally innervated.* That is, the central connections of the neurons are such that whenever any particular muscle is reflexly stimulated to contract, its antagonist will be reflexly inhibited and will relax. How *presynaptic* fibers may in one instance excite the *postsynaptic* cells, and in another inhibit them, is not fully understood. There are indications that differences in anatomical relationships may be significant.

Presynaptic fibers seem to be of two types. Those that are excitatory apparently end mainly on the cell body. In a manner not entirely understood, they depolarize the cell body membrane either by electric current or by the secretion of special depolarizing substances. The depolarization spreads to the place where the axon leaves the cell body and the impulses start. Usually the depolarization produced by one presynaptic ending is insufficient to produce excitation, and impulses arriving at the cell over several presynaptic pathways are generally necessary. Other presynaptic endings that seem to terminate mainly on the dendrites are inhibitory in nature. The flow of electric current in a suitable direction through the cell membrane, or special repolarizing substances associated with these endings, prevents the growth of the depolarization that excitatory endings produce.

An impulse at a single synapse is rarely sufficient to provide enough effect to excite or inhibit the generation of an impulse in a postsynaptic neuron. When several impulses arrive more or less simultaneously along several different presynaptic endings, their collective action may be effective. This is the situation in the stretch reflex, in which the impulses in many afferent fibers impinge on the several motoneurons. There are many synapses in this instance, but all are at the same level between afferent and efferent fibers, and there are no neurons between afferent and efferent fibers. This arrangement of a direct connection is typical of a *monosynaptic reflex arc.*

Polysynaptic reflexes

Contraction of skeletal muscles to cause movement is superimposed on the resting postural tone. When a painful stimulus is applied to the skin, say, of the foot, the injured member is immediately withdrawn. This movement may begin even

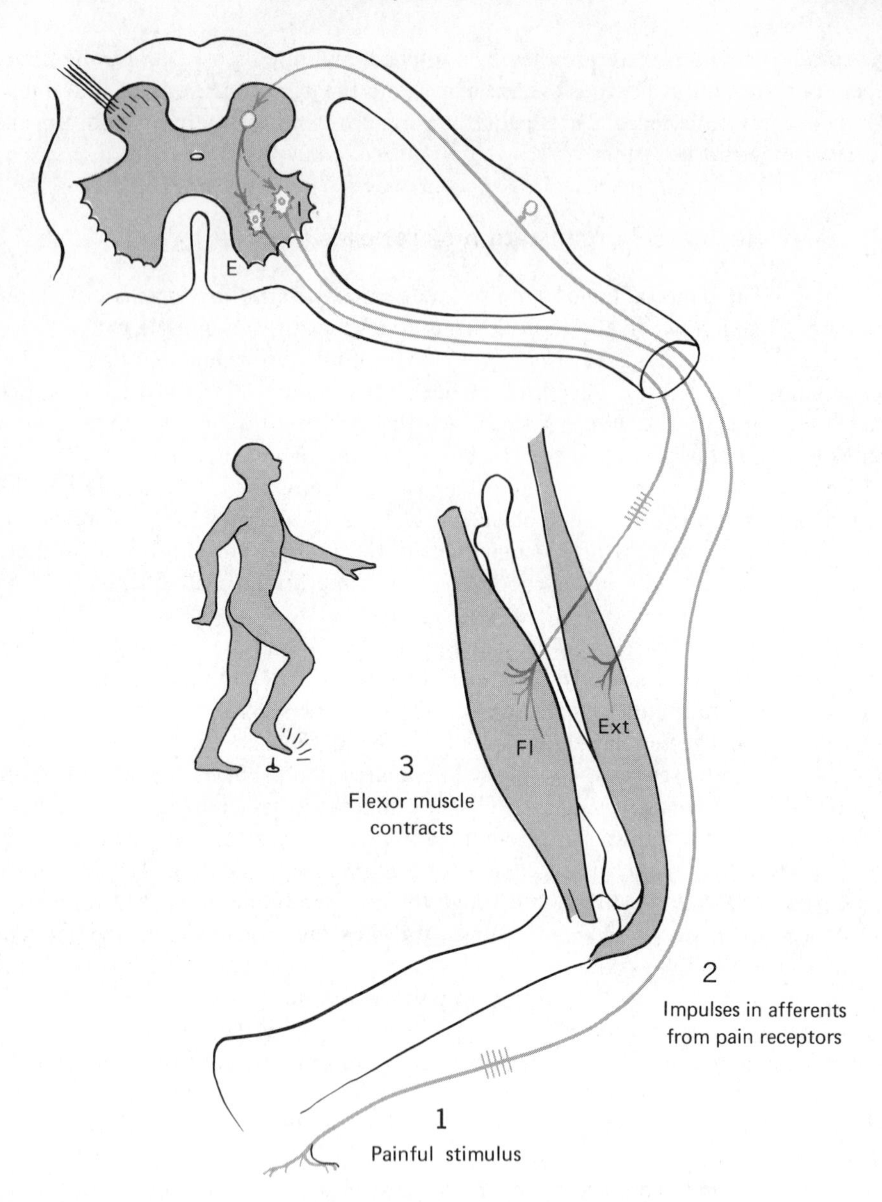

Fig. 6-18. The flexion reflex—a polysynaptic reflex. A painful stimulus provokes excitation of motoneurons of flexor muscle and inhibition of motoneurons of the extensor muscle.

before there is conscious awareness of the pain. Like the muscle stretch reflex, it will occur even when the brain is cut off from the spinal cord. In this *flexion reflex*, there are neurons (interneurons) *between* the afferent and the efferent fibers (Fig. 6-18). The flexion reflex arc is thus at least disynaptic and is better described as *polysynaptic*.

During withdrawal of the foot from a painful stimulus, the quadriceps must relax to allow flexion at the knee when the hamstring muscles contract. When this reaction occurs during standing, the reflex muscle tone in the quadriceps decreases, while the frequency of impulses arriving at the hamstring muscles increases. Thus arrangements of reciprocal innervation apply also in this example, involving a *polysynaptic reflex*.

The body weight is normally distributed on both legs. If one leg is flexed, the body will topple over unless the weight is redistributed and the tension in the muscles of the other leg is readjusted. Such readjustment does, of course, occur. In fact, the same stimulus that provokes flexion of the leg results in extension of the opposite limb (*crossed-extension reflex*). This is reciprocal innervation involving muscles on opposite sides of the body.

Presynaptic endings on any particular motoneuron have several origins; consequently, the motoneuron is under the influence of impulses from several different places. Generally, however, the strongest connections are with regions close to the muscle governed by the motoneuron. This anatomical fact is reflected in the phenomenon of *local sign;* that is, a reflex response is related to the region stimulated. The connections with other segments account for the *irradiation* or spread of the response beyond the local region. A reflex response often lasts much longer than the original stimulus (e.g., the flexion and crossed extension following a painful stimulation to the toe). This prolongation in time results from the many interneurons between afferent fibers and the motoneurons. These interneurons provide a long path with additional delays at each synapse. In addition, many interneurons carry impulses back upstream to restimulate neurons and thus increase the duration of the discharge to the muscles.

Mechanisms of synaptic transmission

What is happening at the synaptic junction when sensory impulses go into the CNS and cause reflex movement of a limb? We can get some insight into the mechanism of the reflex if we simplify the system we are looking at. Ordinarily, impulses come into and go out of the CNS over innumerable pathways and produce complex movements. The relation between input and output can be made clear by decreasing the overall activity and by paying attention to one particular input-output pathway through the CNS.

A simple reflex

The favorite experimental animal for neurophysiologists is the cat. From the study of the nervous system of the cat, we have obtained a large part of our understanding about how the CNS of man functions. If a cat is suitably anesthetized, we can surgically expose the spinal cord at the lumbosacral level and discover the long dorsal and ventral spinal roots that run caudally, join to form spinal nerves that exit from the vertebrae and become peripheral nerves. If we stretch a suitable muscle—for

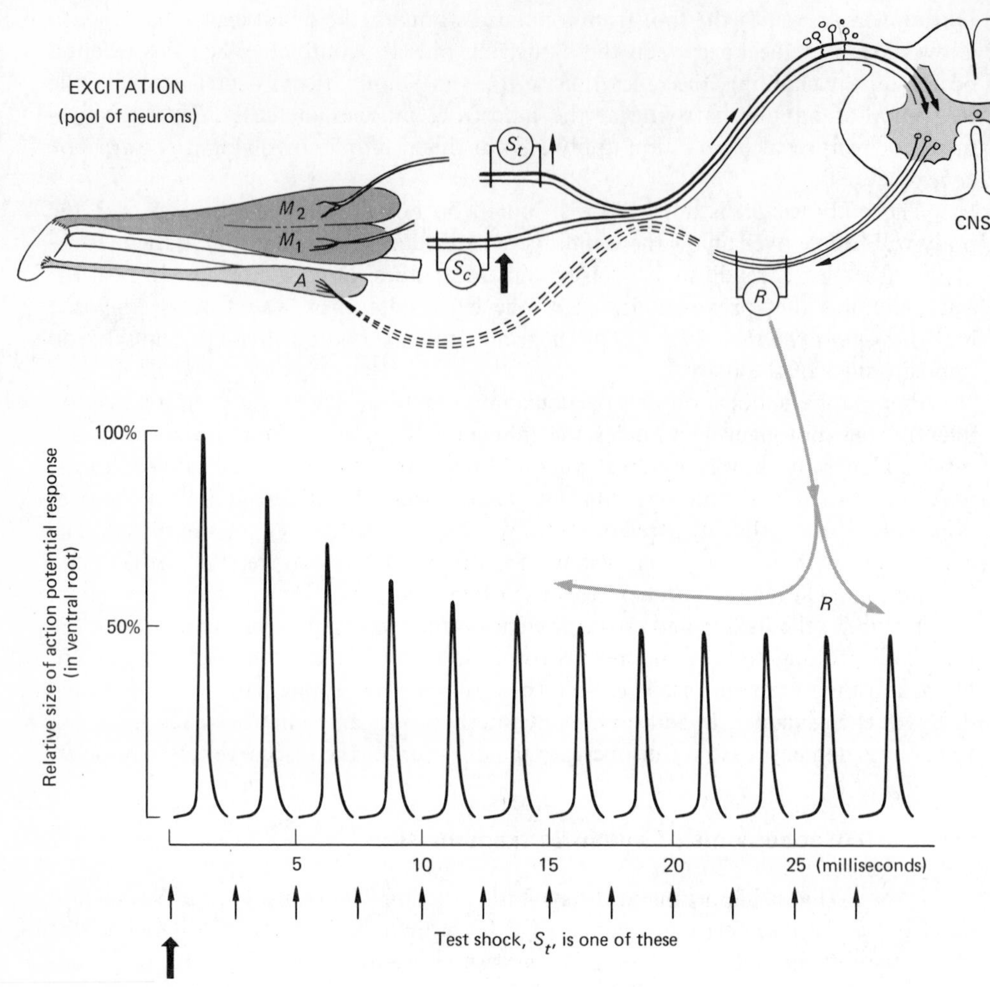

Fig. 6-19. Central excitatory state. Upper diagram shows muscle heads M_1 and M_2, which are synergistic on one side of the joint (A is the antagonistic muscle). The nerves to the synergists are cut and set up on electrodes. The ventral root which carries efferent fibers to these muscles is cut. The muscle-nerve contains both afferent (sensory) and efferent (motor) fibers, but during stimulation of the central part of the cut nerve only the afferent impulses reach the spinal cord, because the motor root is cut.

During test stimulation, S_t, to nerve from muscle M_2, record R is seen in ventral root motoneurons innervating muscle M_2. Size of R depends on antecedent conditioning stimulus S_c to nerve from *synergistic* muscle M_1. Impulses generated by S_c have a central excitatory effect on M_2 motoneurons. Effect of S_c impulses lasts about 15 msec. That is, when the conditioning volley precedes the test volley, by 15 msec or less, the test response is greater than it is when the conditioning shock occurs much earlier or not at all.

example, if we tap the tendon of the gastrocnemius—we can observe a *stretch reflex*. If we cut the appropriate *ventral roots*, the reflex will fail, but we can record action potentials coming to the cut end of the central portion of the sectioned root (Fig. 6-19 *R*). The muscle is flaccid because the impulses cannot reach it. Now, it we place electrodes on the nerve as it enters the gastrocnemius muscle, we can use an electric shock to stimulate the nerve and make the muscle contract. The same stimulus will also send impulses into the spinal cord, for the nerve to the muscle contains both afferent and efferent nerve fibers. If we cut the nerve and stimulate only the central end (Fig. 6-19 S_t), we will have produced a relatively simple arrangement on which to study the relation of the input to the output of the spinal cord. As we increase the stimulus intensity from a low level, we excite more nerve fibers. If the input is adequate, the size of the compound action potential recorded from the motoneuron axons will also increase as more motoneurons are brought into action by the impulses in the sensory nerve fibers.

Conditioning—testing procedures; central excitation

What will happen if, at the same time, we stimulate another muscle-nerve (S_c)? If the nerve is from a muscle synergistic to the first, the size of the ventral root action potential generated by S_t will be increased. If the impulses along the two afferent lines arrive simultaneously, there will be a very large increase in response. The later the second shock comes in relation to the first, the less will be its effect. Suppose that we adjust one shock so that the response is about one-half the maximum possible that can be evoked. We will call this the *Test Shock* (S_t). Then we adjust another shock to an intensity not quite sufficient to give a perceptible response. We will call this the *Conditioning Shock* (S_c). Now, the response to the test shock will be larger if the conditioning shock precedes it by a short interval. The conditioning shock is too small to generate an output in the motoneurons, but it has an effect that shows up as an increased response to the test shock. We plot the time course of this effect in Fig. 6-19.

Evidently impulses produced by the conditioning shock, while seemingly ineffective alone, have generated some effect—a *central excitatory state*—to which the impulses produced by the test shock can add to bring the affected motoneurons to their excitation threshold. The neurons so affected are represented by the additional increment of compound action potential above the dotted line in Fig. 6-19.

Inhibitory processes

If S_c and S_t are delivered to nerves that come from antagonistic muscles, quite a different picture is seen. Take, for example, the *tibialis anterior muscle* vs. the *gastrocnemius*. Suppose that we produce a stretch reflex in the gastrocnemius muscle by stretching that muscle. At a certain moment we try to produce a stretch reflex in the tibialis anterior muscle. If we try to produce both at about the same time, one muscle may not reflexly contract, or its response will be much reduced compared to the other one. This behavior is an expression of *reciprocal inhibition* in the spinal reflex. We can study it more closely by preparing muscle nerves and spinal roots for stimulation and recording, as we did for the study of the excitation process. S_c, the conditioning shock, will be delivered to the nerve from tibialis anterior, S_t, the testing shock, to the gastrocnemius nerve. We select, for recording, a ventral root that has a good output to gastrocnemius but little or none to tibialis anterior at the stimulus intensities used (Fig. 6-20).

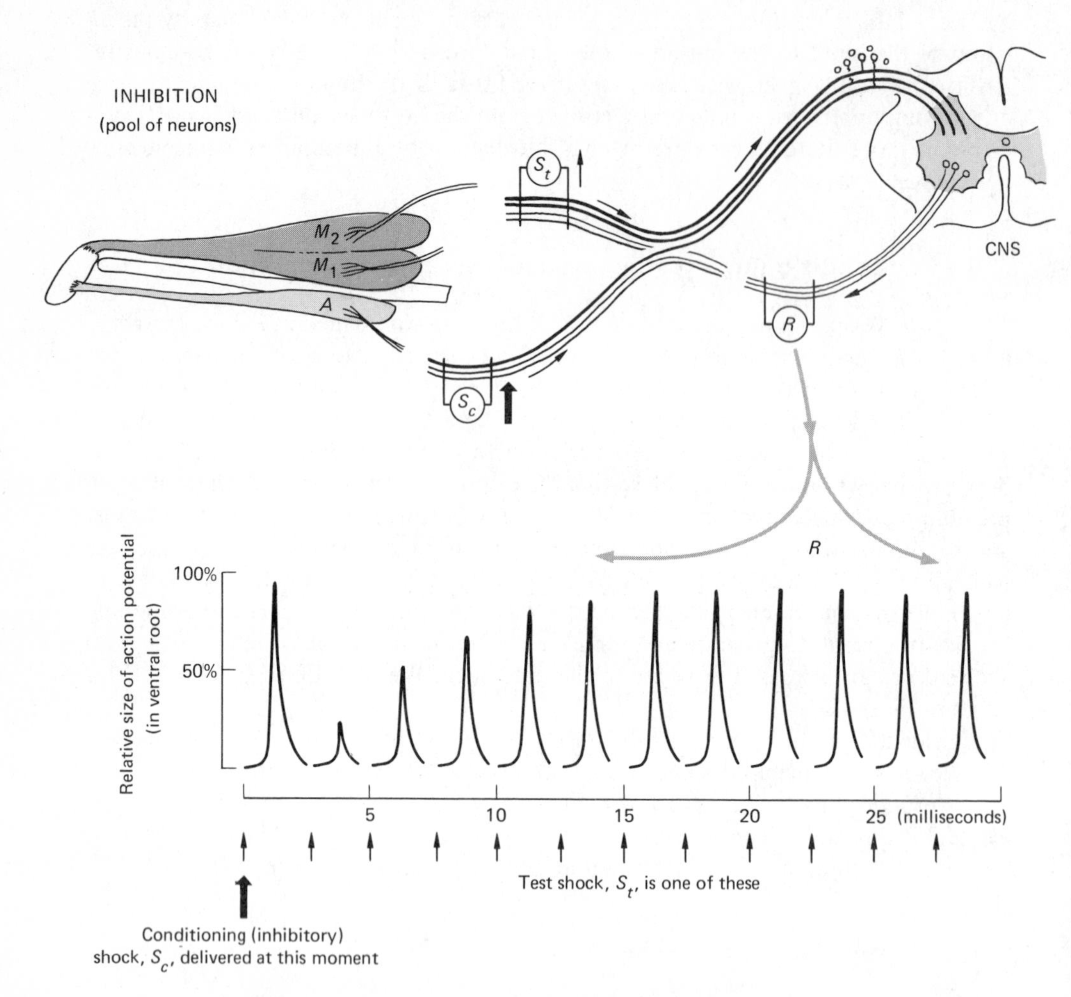

Fig. 6-20. Central inhibitory state. During test stimulation, S_t, to nerve from muscle M_2, record R is seen in ventral root motoneurons innervating muscle M_2. Size of R depends on antecedent conditioning stimulus to nerve from *antagonistic* muscle A. Impulses generated by S_c have a central inhibitory effect on M_2 motoneurons. Effect of S_c impulses lasts about 15 msec. Note nerves and roots cut.

We note that in this arrangement the conditioning shock (delivered to a nerve from an antagonistic muscle) provokes a decrease in the response to the test shock delivered to the gastrocnemius nerve. If the two shocks are simultaneous, excitation will dominate, but if the conditioning shock from tibialis anterior precedes the test shock by an appropriate interval of time, the response will be diminished. The impulses generated by the conditioning shock have had an *inhibitory action* on the gastrocnemius neurons. A *central inhibitory state* has been produced having about the same duration as the central excitatory state.

Intracellular synaptic recording

The action potentials illustrated in the excitatory and inhibitory curves represent the activity of a large population, perhaps thousands of neurons. Means have been devised for studying the behavior of individual neurons in the central nervous system. A tiny glass tube is drawn out until it is only about 0.5 μm in diameter at the tip, and it is then filled with a salt solution. A wire thrust into the tube is connected to an electrical recording device. The microelectrode is plunged into the spinal cord, and with luck it will penetrate the membrane of a motoneuron cell body. Now we can measure the electric potential changes that occur between the inside and outside (across the membrane) of the motoneuron cell body. We can stimulate afferent muscle nerves, and we may now be able to understand the synchronized activity of the large population in terms of events in the single cell.

Synaptic facilitation (EPSP)

In an arrangement to look at the excitation or facilitation process, when the conditioning shock alone may be too weak to provoke an output in the ventral nerve root, we can see a change by using our microelectrode. If S_t is sufficiently strong, enough afferent fibers will be stimulated to generate action potentials that will be propagated out the ventral root and the microelectrode will record a single spike occurring in the impaled cell. When S_t is made weaker, only a "local" change is observed. That change is a decrease in resting potential of the impaled cell (Fig. 6-21). It lasts about as long as the central excitatory state. It occurs in connection with synaptic excitation, in a postsynaptic cell, and it is called an *excitatory postsynaptic potential* (EPSP). It is very similar to the *end-plate potential*, which may be recorded by means of a microelectrode thrust into a muscle cell, at the end-plate zone where the nerve enters. Evidently a small volley of afferent impulses decreases the membrane potential to a small extent, and the change then fades away to disappear in about 15 to 20 msec. A second small volley during this time can add its effects to the first to bring the membrane to its threshold for discharge. If we were to look closely at the neuron, we would find that the cell body and its dendrites are covered with hundreds of tiny nerve endings that are the terminations of axon branches of interneurons and of afferent fibers.

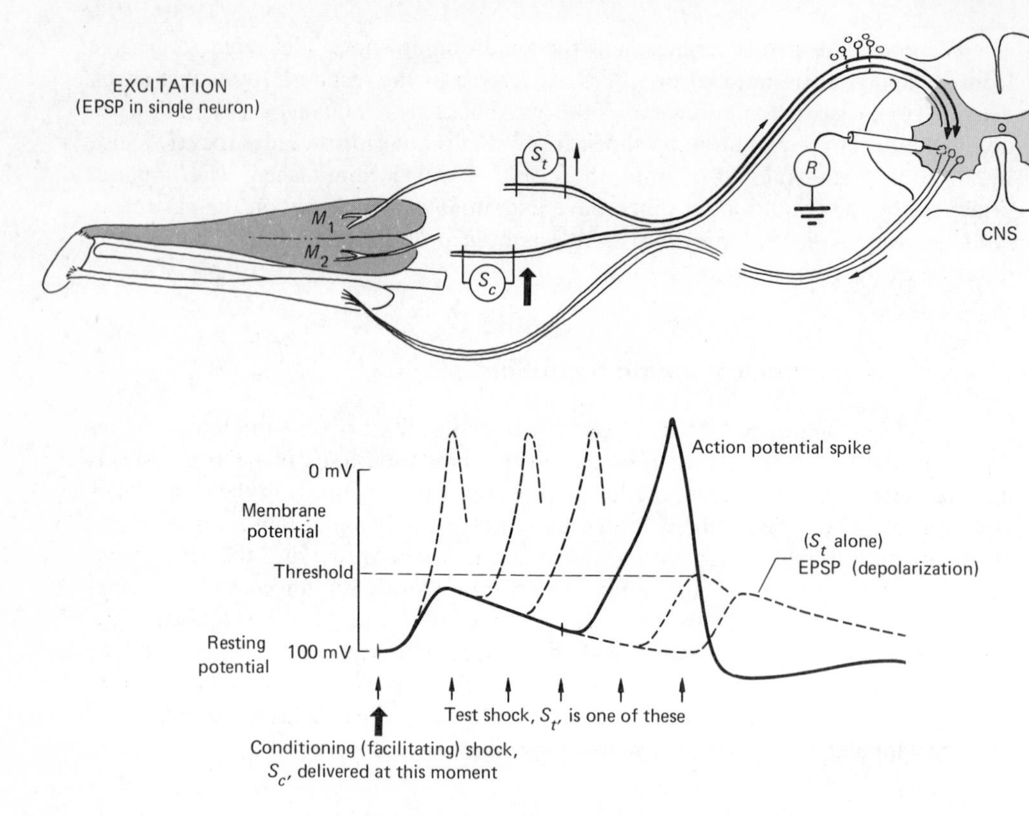

Fig. 6-21. Excitatory postsynaptic potential (EPSP). Test stimulation, S_t, alone to nerve from muscle M_2 is adjusted in intensity so that only a non-propagated depolarization is recorded by the intracellular microelectrode in a motoneuron innervating muscle, M_2. The volley from S_c is just barely insufficient to generate an action potential. Antecedent stimulation to nerve from synergistic muscle M_1 adds further depolarization, bringing postsynaptic membrane to threshold for propagated action potential. Depolarization from conditioning volley lasts about 15 msec.

We can imagine a small excitatory afferent volley producing a small amount of membrane depolarization because only a small number of synaptic knobs, and hence a small portion of the total postsynaptic neuron surface, is involved. When another volley is added, more synaptic endings are activated, and the postsynaptic depolarization is more intense. *Spacial summation* of the depolarizing action has occurred over the surface of the neuron. We see the result in terms of *temporal summation* of the depolarizing process as we adjust the interval between conditioning and testing shocks.

Synaptic inhibition (IPSP)

We can examine the inhibitory process in the same way. Although previously we saw no response in the ventral root when stimulating the inhibitory afferent fibers, we may see a change in the membrane potential recorded by means of the intracellular electrode. That change will be in the direction to increase the resting membrane potential. The small excitatory volley, then, does not decrease the membrane potential sufficiently to excite the motoneuron. The *inhibitory postsynaptic potential* (IPSP) has about the same duration as the EPSP. The weak excitatory volley (S_t) used in Fig. 6-22 does not yield spikes until the IPSP is terminated. A stronger shock (a large excitatory volley) would be able to overcome the inhibitory effect before the termination of the IPSP. The control of impulse discharge in most, if not all, of the neurons of the central nervous system is exerted by adjusting the magnitude of EPSPs and IPSPs through variations in the source and magnitude of afferent volleys. Changes in composition of tissue fluids, whether natural secretions or drugs, will alter the effectiveness of the two opposing synaptic actions. The depolarizing EPSP

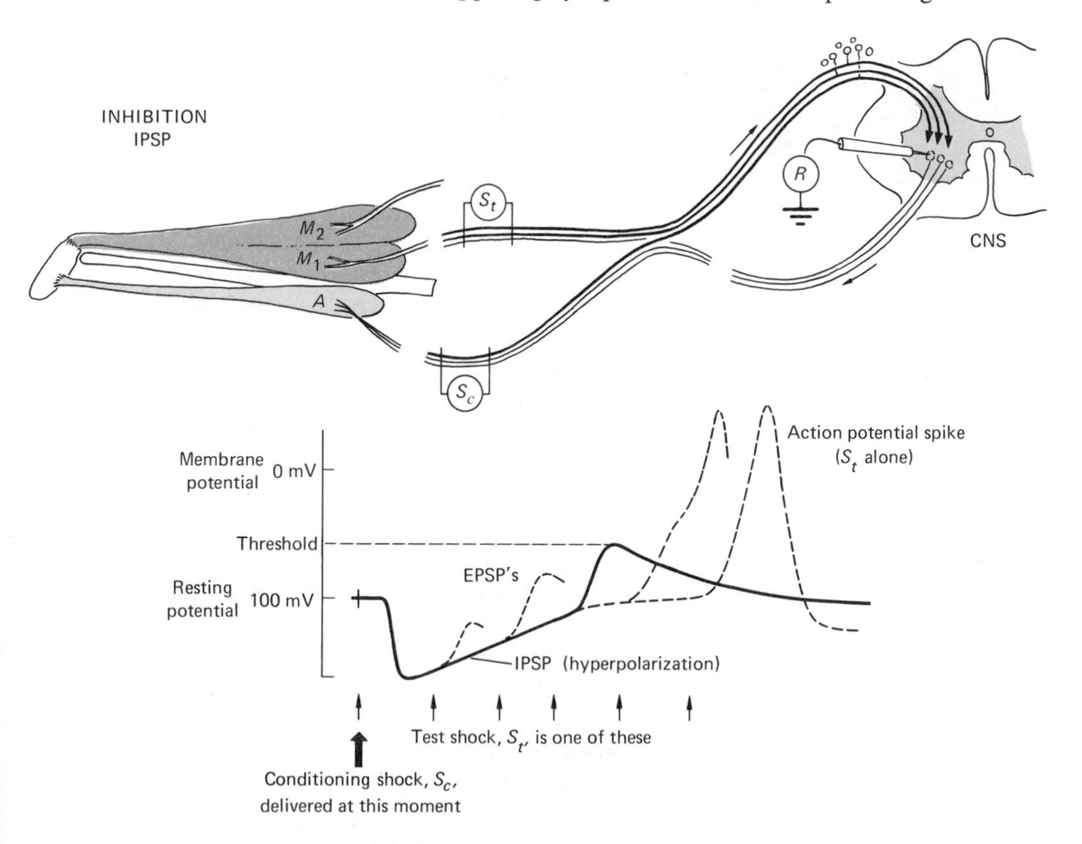

Fig. 6-22. Inhibitory postsynaptic potential (IPSP). Test stimulation, S_t, alone to nerve from muscle M_2 is adjusted so it is just sufficient to generate an action potential in a motoneuron impaled by the intracellular microelectrode and innervating muscle M_2. Antecedent stimulation to nerve from antagonistic muscle A prevents propagated action potential from arising and allows only the EPSP to be seen. The inhibitory action evokes the hyperpolarization which is the IPSP.

and the hyperpolarizing IPSP may be due in the first instance to secretion of synaptic transmitters that alter the membrane permeability of the postsynaptic membrane, in a direction to enhance or diminish excitability.

Structure of the spinal cord—connections to higher centers

The information brought into the spinal cord and brainstem by the afferent nerve fibers is not only used to carry out spinal and brainstem reflexes. It may slso find its way upward to the brain, where it may be felt as specific *sensations*, or it may provide information for the brain to use in other ways. In the pathways leading to the cerebral cortex at the top of the brain, there are at least three synaptic levels. The afferent neurons are the *primary* or *first-order neurons* in the sequence. The *second-order neurons* receive the information from the afferent fibers, take it across to the other side of the spinal cord, and then carry the information to the thalamus of the brain. The *third-order neurons* carry the information from the thalamus to the cerebral cortex.

In the central nervous system, nerve fibers are grouped into bundles called *tracts* (Fig. 6-23). Many tracts are named according to the regions between which they carry impulses. Most of the white matter of the spinal cord consists of tracts carrying impulses upward, called *ascending tracts*; those carrying impulses from the brain down to the motoneurons are the *descending tracts*. The arrangement of fiber tracts at various regions or *levels* of the spinal cord is shown in Fig. 6-23.

The *lumbar region* (1) supplies nerves to the lower limbs. The ventral horn gray matter where the motoneuron cell bodies are located is therefore relatively prominent, but since little of the body is caudal to this level, the fiber tracts are not large.

The *thoracic region* (2) supplies the middle of the trunk. The ventral horns are poorly developed, because of the small amount of musculature to be served, but there are accumulations of cell bodies on the sides of the gray matter, from which the fibers of the sympathetic nervous system arise (see Chap. 9.1). Large tracts are ascending and descending through this region. Most of these tracts are found at all levels of the spinal cord.

In the *cervical* or neck region of the spinal cord, the gray matter is well developed, for it supplies the motor control of the arm and hand muscles. The ascending and descending tracts of the white matter are large because they include fibers concerned with the entire body caudal to this section.

Ascending tracts of the spinal cord

The dorsal white matter in the spinal cord consists of axons of afferent neurons whose cell bodies are in the dorsal roots and whose dendrites innervate muscle stretch receptors. These fibers also connect to the spinal motoneurons and are re-

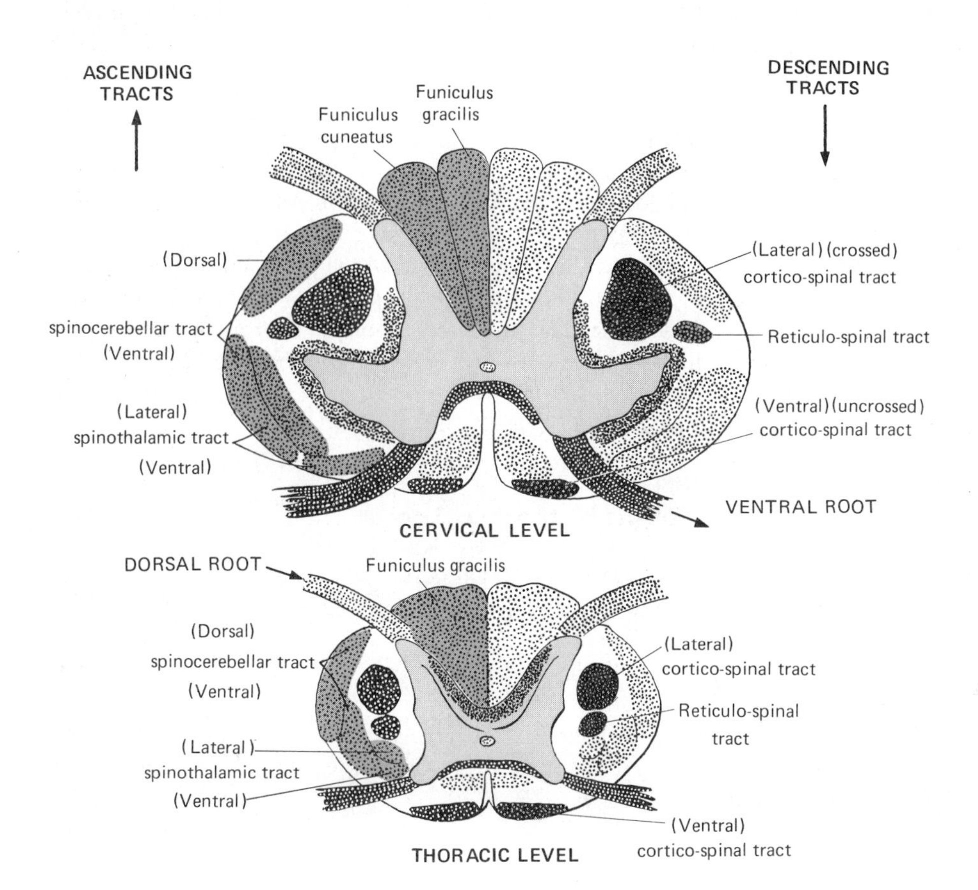

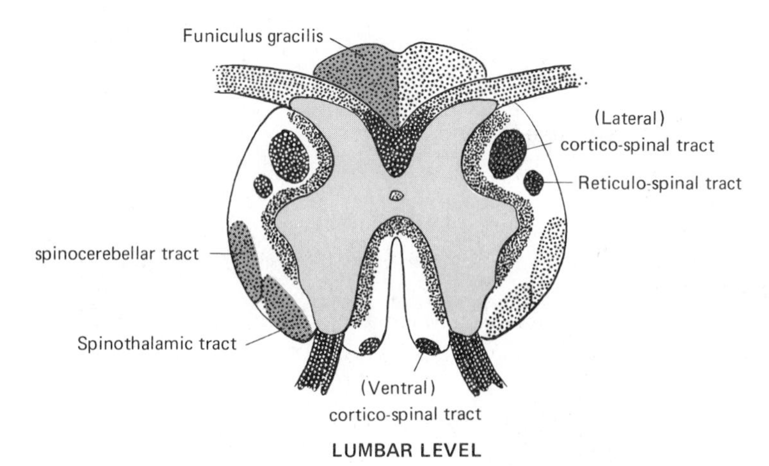

Fig. 6-23. Some tracts of the spinal cord.

sponsible for the stretch reflex. They run upward to the brain on the same side that they enter the spinal cord.

Almost all other ascending tracts of the spinal cord arise from second-order neurons in the gray matter of the cord itself, and similar pathways exist in the brainstem. The stretch receptor afferent fibers, for example, also synapse on the side of their entry into the cord with second-order neurons sending axons into the *spinocerebellar tracts*, which ascend to the cerebellum. Information related to skin senses of pain and temperature is not only used in local reflexes but is also delivered to nearby second-order neurons in the dorsal horn gray matter of the spinal cord. These second-order neurons immediately cross to the other side of the spinal cord and ascend in the *spinothalamic tracts* of the other side. Like second-order neurons concerned with muscle sense, they reach the *thalamus* of the brain, from which the information is relayed to the cerebral cortex via the *third-order neurons*. Connections in the brain are described in the next chapter (7.1).

Descending tracts of the spinal cord

The *descending tracts* of the spinal cord are shown in Fig. 6-23. The largest of these tracts are lateral to the spinal cord gray matter and consist of large, high-speed fibers that are a direct line from cerebral cortex to spinal motoneurons. These are the *corticospinal tracts*. Other descending pathways come from other areas of the brain, especially from a network of cells and fibers called the *reticular system*. The cell bodies of this system are the source of the *reticulospinal tracts*, which include fibers from the *vestibular centers* that govern posture by making use of information from the proprioceptors of the middle ear.

Section 6: *Questions, Problems, and Projects*

1. (a) What is the general meaning of (1) afferent? (2) efferent?
   (b) What is the meaning of each of those terms (1 and 2) as applied to the peripheral nerves?
   (c) Can you make a distinction between afferent and sensory nerves?
   (d) Where are the cell bodies of (1) somatic motor nerves? (2) afferent nerves?
   (e) Distinguish between (1) nerve root and (2) nerve trunk.
2. (a) In which direction are impulses ordinarily carried: (1) in the dorsal root fibers? (2) in the ventral root fibers?
   (b) What kinds of nerve fibers are in the (1) dorsal root? (2) ventral root?

(c) In what part of the spinal cord will you find (1) the cell bodies of the motoneurons? (2) the central processes of the afferent neurons?
(d) What are the end organs of (1) somatic motoneurons? (2) of somatic sensory neurons?
(e) In the vertebrates, into which side of the spinal cord (1) do afferent signals enter? (2) do efferent signals leave?

3. (a) Where do you localize the sensations resulting from (1) stimulation of the ulnar nerve at the elbow? (2) stimulation of the skin of the elbow?
(b) Stimulation of a nerve in the leg may result in a sensation referred to the foot. Explain.
(c) What is your sensation (1) at the moment when, and (2) shortly after a hair is bent and held in the bent position?
(d) Explain (1) and (2) of part (c) above.
(e) How would you characterize the response of the touch receptor at the base of the hair, as compared to a receptor in which the sensation continues for a longer time?

4. (a) Distinguish between (1) proprioceptor and (2) exteroceptor.
(b) To which cerebral lobe are impulses from (1) proprioceptors of the left side mainly projected? (2) skin receptors of the right side?
(c) Draw a diagram to show a dorsal view of the spinal cord and medulla oblongata. Draw the pathway showing location of cell bodies and of synaptic junctions and label: (1) skin sense; (2) muscle sense fibers.

5. (a) What are the equivalents in the terminology of human anatomy of (1) dorsal and (2) ventral aspects of the spinal cord?
(b) Distinguish between (1) dorsal horn and (2) dorsal column of the spinal cord.
(c) In which direction along the spinal cord do the pyramidal fibers carry impulses? Where do they start?
(d) Where are the cell bodies of the pyramidal fibers? of a motor unit?
(e) What is meant by an inhibitory action of a nerve?

6. (a) What happens to the force exerted by a muscle when the tendon of the muscle is stretched, if:
   (1) The spinal cord has just been severed?
   (2) Recovery from spinal shock has occurred after spinal cord section?
   (3) After (2), the ventral root supplying the muscle having been cut also?
   (4) After (2), the dorsal root having been cut?
   (5) After (2), both roots having been cut?
(b) What happens to the force exerted by a contracting flexor muscle in a spinal preparation (after recovery from spinal shock) if:
   (1) An antagonistic muscle is stretched (the spinal roots being intact)?
   (2) Pain receptors of the limb are stimulated?
   (3) Pain receptors of the opposite limb are stimulated?
   (4) The stretch receptors of the muscle are stimulated?
   (5) The motor nerve to the muscle is stimulated?

7. (a) A nerve to a skeletal muscle contains at least three types of fibers, classified and having functions as follows (list them):
   (b) Name the two kinds of end organs innervated by the efferent nerve fibers of muscle.
   (c) Suppose that you are recording from a sensory nerve fiber innervating an isolated muscle, what happens when you (1) stimulate the muscle to contract? (2) stretch the muscle?

7

STRUCTURE AND FUNCTION OF THE BRAIN

7.1 Organization and Pathways of the Brainstem

General structure and function of the brain

The cervical region of the spinal cord is continuous upward into the *brainstem*, where the *cranial nerves* originate. This part of the brain is organized basically like the spinal cord, but it is concerned with reflex control of the muscles of the head and face. In addition to the groups of nerve cell bodies (nuclei) associated with the cranial nerves, the brainstem contains great fiber tracts that carry impulses between spinal or cranial nuclei and other parts of the brain. There are, moreover, numerous nuclei in the brainstem besides those of the cranial nerves. Some of these nuclei organize complex reflexes involving the cranial nerves and are also the source of long tracts into the spinal cord. Important nuclei in the brainstem exert control over functioning of the circulatory, digestive, respiratory, and other systems of the body.

Superimposed on the brainstem, and having been developed from it in the course of evolution, are the great *cerebral* and *cerebellar lobes* or *hemispheres*. By means of impulses arriving at the motoneurons from the higher centers of the brain, especially from the cerebrum, the simple reflexes of the spinal cord and brainstem are organized into more complicated patterns of behavior than they are able to produce alone.

There are four primary subdivisions of the brainstem: *medulla oblongata* (together with the *pons*), *mesencephalon* (or *midbrain*), *diencephalon*, and *telencephalon* (Fig. 7-1). The cerebellum and cerebral lobes rise above the brainstem but are not part of it. The medulla oblongata is most like the spinal cord, with which it is continuous. The narrow *spinal canal* opens from the spinal cord into the medulla, where it becomes a wide chamber, the *fourth ventricle* of the brain. The central cavity of the

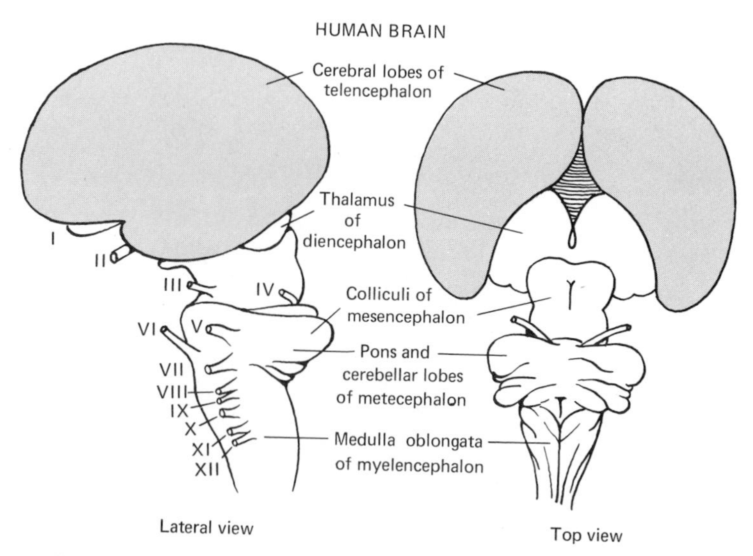

Fig. 7-1. Human brain at about three months embryonic development. These views of the human brain show major subdivisions more clearly than is possible in the fully developed brain in which the great cerebral lobes overwhelm the rest of the brain. At this early stage the cerebral cortex is smooth, as in primitive adult mammals.

nervous system is called the *cerebrospinal canal.* The roof of the fourth ventricle is very thin but is well supplied with small blood vessels. Cerebrospinal fluid, which fills the cerebrospinal canal, is formed by the highly vascular roof (*choroid plexus*) overlying the four cerebral ventricles.

It is mainly the cerebral hemispheres that give man such a relatively large head compared with other animals. In man, more than in any other vertebrate, the information coming to the cerebral hemispheres from all the body senses and from the special senses of the head is pooled, analyzed, and used as a basis for adjusting patterns of muscle tension in the whole body—to a degree impossible at the simple reflex level. The cerebral hemispheres are the ultimate controllers of voluntary movement, the final regulators of behavior, particularly those complicated and subtle aspects of behavior that distinguish human beings from lower animals.

Caudal to the cerebral hemispheres is the smaller *cerebellum.* The main role of the cerebellum is to provide smoothness to the movements initiated by the cerebrum and to assist in unconscious postural adjustments.

The outstanding anatomical feature of both the cerebrum and the cerebellum is the tremendous extent of the outer surface, composed not of fibers such as form the outside of brainstem and spinal cord, but of millions of cell bodies. This outside layer of gray matter is the *cortex*. Great bands of fibers beneath the cortex make up the *white matter* of the cerebral and cerebellar lobes. Deep within the cerebral lobes, other

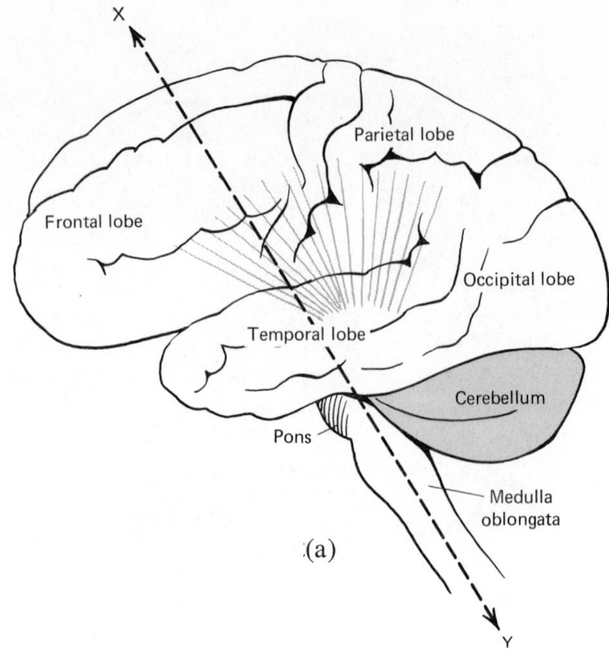
X
Parietal lobe
Frontal lobe
Occipital lobe
Temporal lobe
Cerebellum
Pons
Medulla
oblongata
Y

(a)

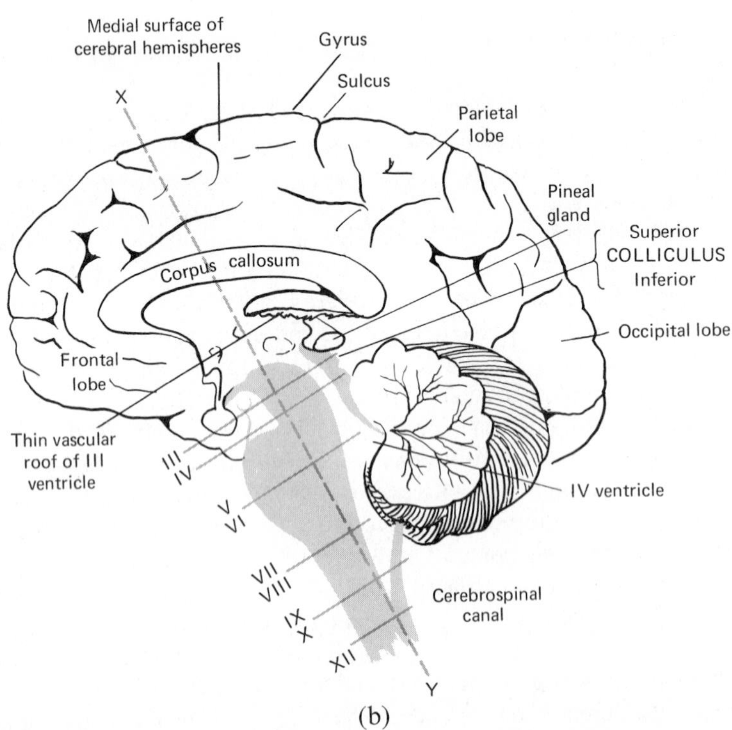
Medial surface of
cerebral hemispheres
Gyrus
Sulcus
X
Parietal
lobe
Pineal
gland
Superior
COLLICULUS
Inferior
Corpus callosum
Occipital lobe
Frontal
lobe
Thin vascular
roof of III
ventricle
III
IV
V
VI
IV ventricle
VII
VIII
IX
X
XII
Cerebrospinal
canal
Y

(b)

masses of gray matter have developed into large nuclei, the *basal ganglia* of the brain.*

Figure 7-2(a) shows side (A) and sagittal (B) views of the brain of an adult. In Fig. 7-2(b) the brainstem is indicated in black. Gray lines show the levels of sections shown in Figs. 7-7 and 7-8. The roman numerals designate the cranial nerves, which are approximately at the levels indicated. These nerves are shown in Figs. 7-3 and 7-4. Pathways that ascend and descend through these levels are shown in Figs. 7-7 and 7-8.

Cranial nerves of brainstem†

The cranial nerves are serially continuous with the spinal nerves. Several of the cranial nerves serve rather specialized sense organs, however, and the central connections of these nerves are correspondingly specialized. In the spinal cord, the dorsal and ventral roots are more or less equally developed. In some nerves of the brainstem, however, one root may be lacking, and there may be a correspondingly greater development of the other root. The cell bodies of the motoneurons are assembled into more distinct *motor nuclei* than is the case in the spinal cord (Fig. 7-3).

The cell bodies of the primary afferent fibers entering the brainstem are, in most instances, located in ganglia, as are the corresponding fibers of the spinal roots. These afferents make connection with second-order neurons assembled into distinct groups, the *sensory nuclei*. Figures 7-3 and 7-4 show the location of the motor and sensory nuclei of the brainstem.

PONS AND MEDULLA OBLONGATA. Central connections of cranial nerves VII through XII are in the medulla oblongata. Nerve XII, the *hypoglossal*, and XI, the *spinal accessory*, correspond to spinal nerves in lower vertebrates, which during the course of evolution have come to lie in the skull. They innervate muscles of the

* Except for the ganglion cells of the retina, this is the only general usage of the word "ganglion" to refer to cell bodies in central nervous tissue.

† The reader may find the following mnemonic device useful in remembering the cranial nerves in numerical sequence:

On	Old	Olympus'	Tow'ring	Tops	A
Olfactory	Optic	Oculomotor	Trochlear	Trigeminal	Abduscent
I	II	III	IV	V	VI
Finn	And	German	Viewed	Some	Hops
Facial	Auditory	Glossopharyngeal	Vagus	Spinal accessory	Hypoglossal
VII	VIII	IX	X	XI	XII

Fig. 7-2. Adult human brain. (a) Lateral view. Note enormous expansion of cerebellar and cerebral lobes compared to embryo (Fig. 7-1). Masses of fibers radiating upward to and coming down from cerebral cortex are indicated in transparency. (b) View of sagittal section of brain. Diagonally lined are structures cut in the sagittal plane. *Corpus callosus* is white matter consisting of enormous mass of fibers connecting the two cerebral hemispheres. Numbers indicate levels of sections shown in succeeding figures, 4-6 in Fig. 7-7, and 7-9 in Fig. 7-8, X-Y, plane of section in Fig. 7-11.

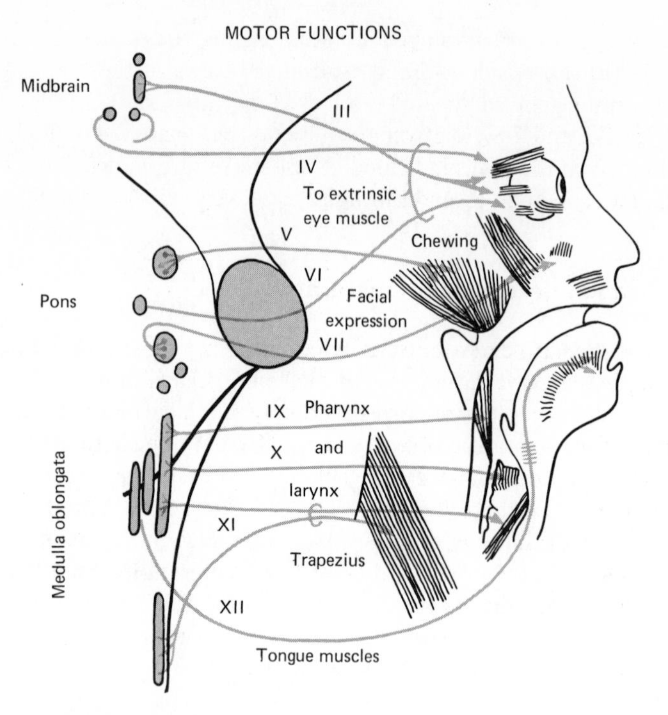

Fig. 7-3. Components of cranial nerves of brainstem and the regions supplied by these nerves. (a) The somatic motor nuclei are groups of nerve cell bodies from which arise motor fibers of the cranial nerves. Visceral motor nuclei (parasympathetic) are also shown (without fibers). The somatic motoneuron fibers go to muscles of the head and neck.

tongue (XII) and of the neck (XI). The only afferent fibers they carry are from stretch receptors of these muscles. Nerve X, the *vagus*, is a most important part of the *visceral nervous system.* It contains afferent fibers from the viscera and efferent fibers that control smooth muscle and glands of the viscera.

Nerve IX, the *glossopharyngeal*, is also most important as a part of the visceral nervous system. For example, it carries afferent fibers from the taste receptors of the tongue and efferent fibers to the salivary glands. It also has some fibers concerned with muscles of the throat.

In relation especially to the central connections of nerves IX and X, there are regions of the medulla oblongata that exert rather direct control over many of the vital functions of the body, such as respiration, heartbeat, and blood pressure. The cell bodies concerned in each function are not assembled as separate nuclei, however, and the term "center" is used to suggest a general region of the medulla oblongata in each instance.

Nerve VIII, the *auditory*, is the nerve of the inner ear. The acoustic division provides information concerning vibrations of sound, while the vestibular component informs us about movement of the head. Nerve VII has two important components. Like the glossopharyngeal, it innervates taste buds and salivary glands. In addition,

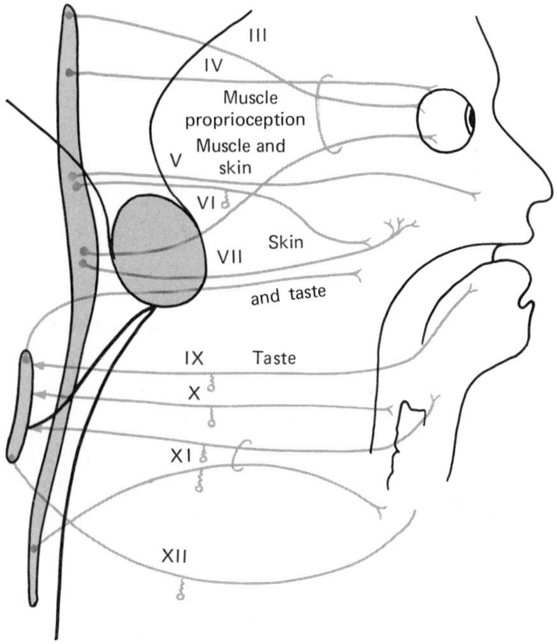

SENSORY FUNCTIONS

Fig. 7-4. Shows the location of the groups of second-order neuron cell bodies to which go somatic sensory nerve fibers of the cranial nerves. The cell bodies of the sensory nerve fibers are located mostly in ganglia outside the brainstem. Certain proprioceptive fibers of V are an exception.

it has components of the somatic nervous system, particularly in relation to muscles of the face.

The rostral end of the medulla oblongata is greatly enlarged by the presence of a mass of fibers connecting the two sides of the cerebellum by way of deep nuclei in the region. Because these fibers appear to form a bridge between the two sides of the brain at this point, the region is called the *pons*. Nerve VI, the *abduscent*, is in the pons. It contains afferent fibers from stretch receptors of some of the muscles that move the eyes. It also carries motor fibers to these same muscles.

Nerve V, the *trigeminal*, is remarkable in that its sensory component is essentially an upward extension of the dorsal horn of the gray matter of the spinal cord. It supplies the receptors of skin sense of the face and head. The central connections of this nerve run from the medulla oblongata through the pons and into the *mesencephalon*, or *midbrain*.

MESENCEPHALON: CRANIAL NERVES IV, III, AND II. The cerebrospinal canal is very narrow in the mesencephalon, while the surrounding walls are thick and the roof heavy. Here may be found reflex centers for movements of the eyes. Nerve IV, the *trochlear*, and III, the *oculomotor*, like nerve VI, innervate voluntary muscles of the eye and contain both somatic motor and sensory components. In addition, the oculomotor is part of the visceral nervous system in that it controls smooth muscle of the iris of the eye.

The first two cranial nerves are quite different from the others. Nerve II, the *optic*, which brings visual information to the midbrain, is not actually a nerve but a central tract. The retina already provides several layers of synapses between the light receptors and the optic tract and is comparable to cerebral cortex in its structure.

The large *optic tracts* travel partly to the same side, partly to the opposite side of the brain. In vertebrates other than the mammals, the *optic lobes*, corresponding to the roof of the mesencephalon, may be more important than are the cerebral lobes, for organization and execution of basic behavior patterns.

DIENCEPHALON. On the way to their connections in the brainstem, about half the fibers of the optic tracts cross below the *diencephalon*. The largest part of the diencephalon is the *thalamus*. This is a large group of nuclei including and related to the third-order neurons that receive information from the receptors of the body and project the information to the cerebral cortex. Immediately below the thalamus in the diencephalon is the *hypothalamus*, a group of nuclei that controls many aspects of visceral functions. An organ of the endocrine system, the *pituitary body*, is suspended from and controlled by the hypothalamus.

TELENCEPHALON: CRANIAL NERVE I. The most rostral subdivision of the brain is the *telencephalon*. Out of this part of the brain has developed the great *cerebral lobes* or hemispheres. At the rostral end of these hemispheres, but lying underneath because the hemispheres have grown so large, are the small olfactory bulbs. Olfactory tracts connect these bulbs to the cerebral lobes, and the nerve fibers of the olfactory nerve, cranial nerve I, reach the bulb from the olfactory epithelium after passing through the cribriform plate. The primitive function of the cerebrum is the organization of reflexes related to the sense of smell. Connections exist here among all the sensory systems; in higher vertebrates these connections are retained and bloom into the great expanse of the cerebral cortex.

Control of lower CNS centers by the brain

In their simplest relations, the sensory and motor nerve fibers of the brainstem function in terms of simple reflex arcs, in the same manner as the nerves of the spinal cord. They also have a similar relation to the higher centers of the brain. Superimposed on these lower brainstem and spinal centers are higher levels of control.

In the spinal cord, the aggregations of neuron cell bodies in the dorsal horns of the gray matter correspond to the sensory nuclei of the cranial nerves, whereas the ventral (and lateral) horns correspond to the motor nuclei of the cranial nerves. The intermediate zone, between the sensory and motor nuclei, is developed to an enormous extent in the brain as compared to the spinal cord. Much of it forms the diffuse fabric of fibers and cells known as the *reticular formation*. The nuclei of the reticular formation are masses of cells and fibers that form a sort of central core of the brainstem, especially prominent in the midbrain or mesencephalon. They appear as a relatively unchanged fabric throughout the vertebrates and control primitive behavioral functions. The thalamic nuclei make up the thalamus, which is in the diencephalon.

Large and numerous nuclei have become elaborated in the brain, and their connections allow complex interactions among pathways associated with effectors and with receptors of the head as well as the body generally. They provide a higher level of control of the brainstem and spinal cord centers.

Making possible the control of the lower by the higher centers are systems of pathways of *ascending fibers* that carry impulses from spinal cord and lower brainstem centers upward to the control centers (Fig. 7-6) and other systems of *descending* pathways that carry impulses from the control centers ultimately down to the motoneurons (Fig. 7-5).

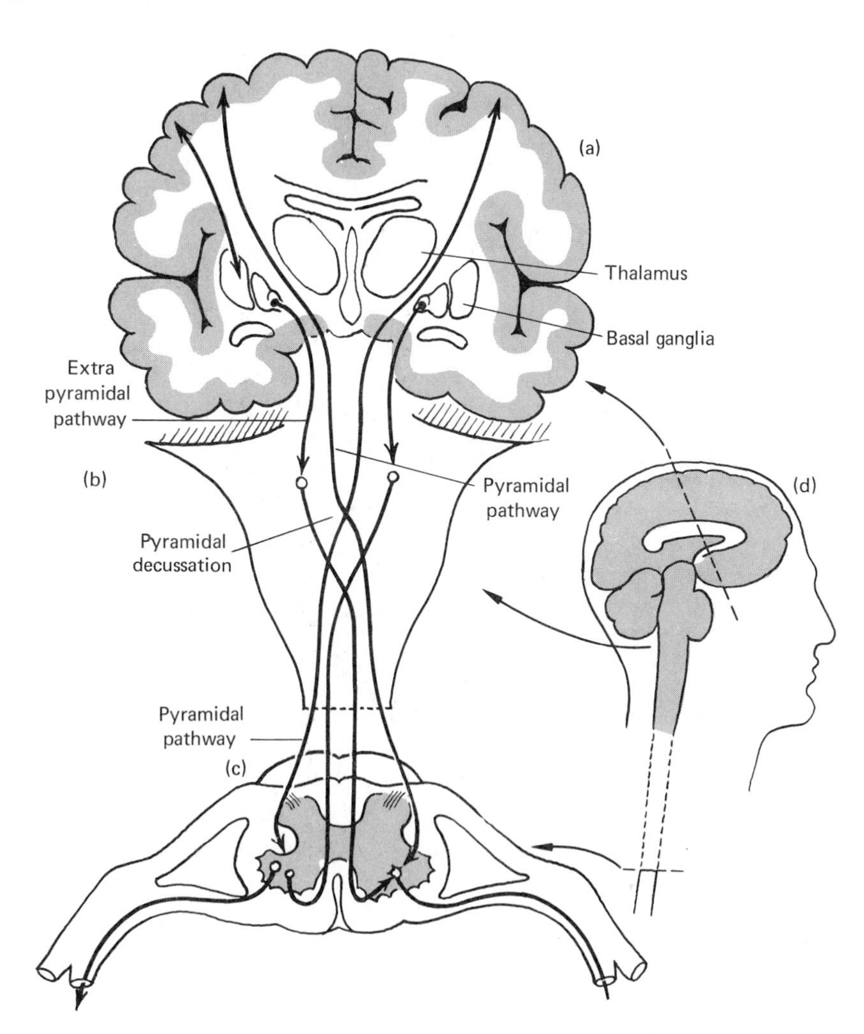

Fig. 7-5. Some tracts of nerve fibers descending from the brain to the spinal cord.

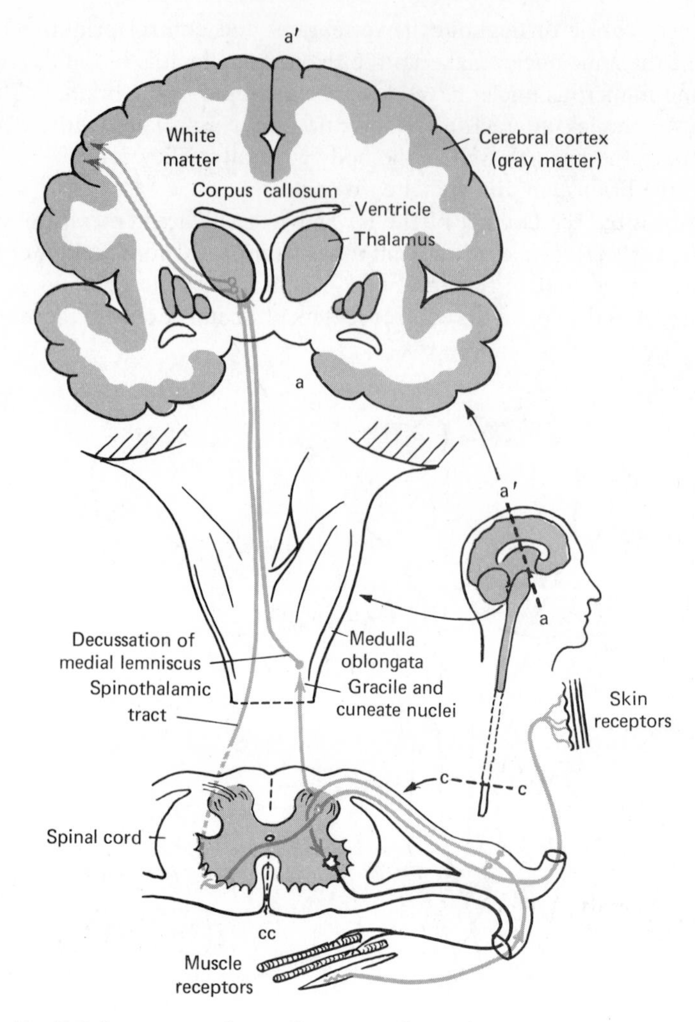

Fig. 7-6. Some tracts of nerve fibers ascending to the brain from spinal cord.

The locations of some of the major ascending and descending pathways or tracts in relation ot one another at various levels of the brainstem are shown in Figs. 7-7 and 7-8. The several cross sections in Figs. 7-7 and 7-8 are taken at levels noted in Fig.

Fig. 7-7. Some tracts and landmarks of the medulla oblongata. This figure continues upward into the brainstem, the series of sections of the spinal cord are gray (ascending tracts) and light gray (descending tracts). Cross-sections through the medulla oblongata are at three levels:

1. *Motor decussation:* the corticospinal tracts cross over from the ventral to the lateral region of the brainstem.
2. *Sensory decussation:* the second-order fibers from the gracile and cuneate nuclei of each side begin to cross over to form the medial lemniscus of the brainstem.

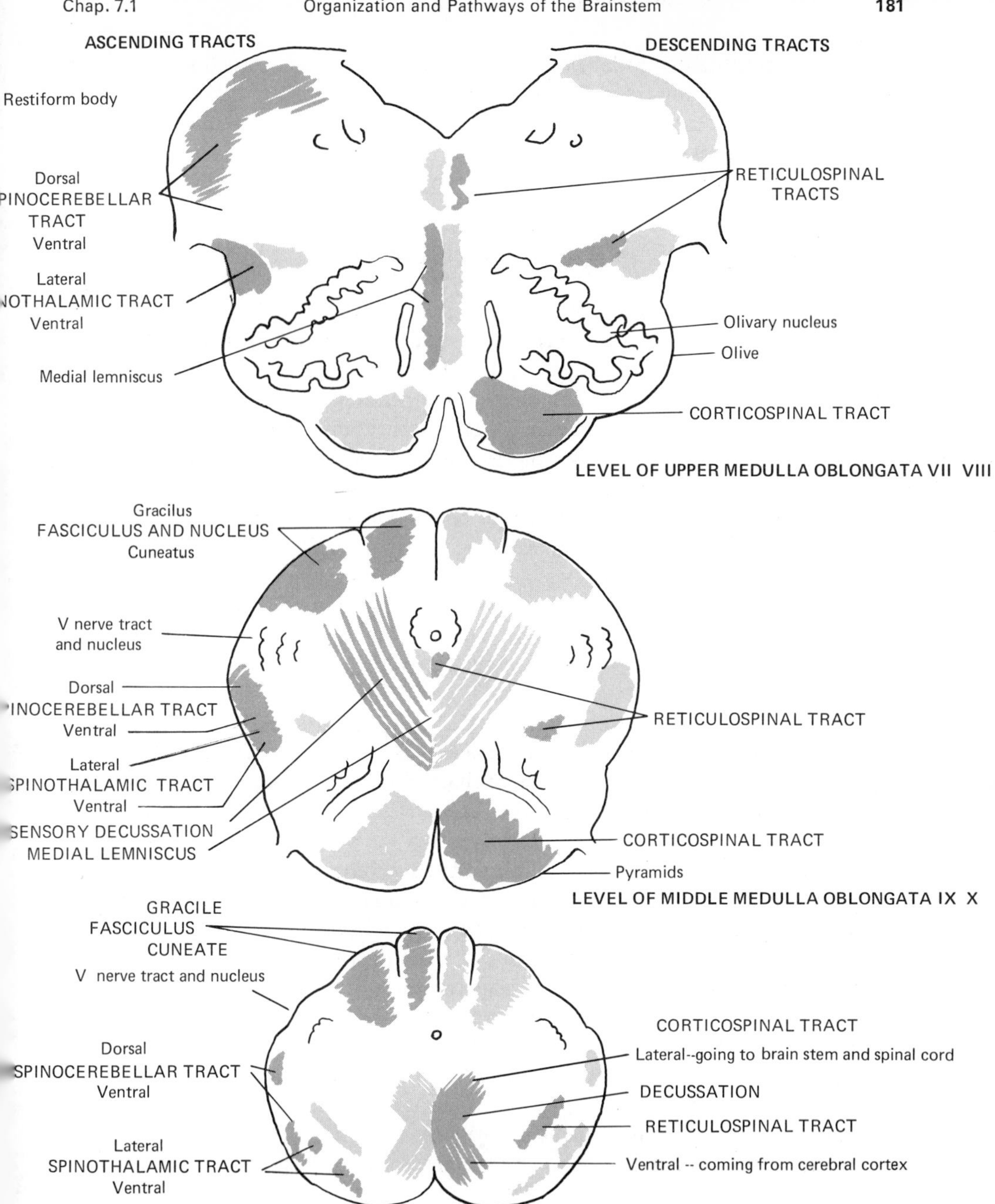

3. The level of the olive, which bulges out the side of the medulla oblongata. Internally, the olivary nucleus is seen as a convoluted mass of gray matter. The medial lemniscus is on its way to the thalamus, the dorsal spinocerebellar tracts begin to enter the cerebellum, and the second-order fibers from the gracile and cuneate nuclei have nearly all crossed over. There remains the lateral cuneate nucleus at this level.

7-2(b). Ascending tracts are labeled at the reader's left side. Descending tracts are named at the right. Only the major tracts are shown, but these tracts, together with various nuclei, fill most of each section and are labeled at each section to show how they go from one level to the next. The course of the pathways may therefore be traced from spinal cord through brainstem and from upper brainstem down into spinal cord.

Pathways up to cerebrum and cerebellum

All the information the cerebral cortex receives from the receptors, except the olfactory, arrives via the thalamus of the brainstem. In the thalamus are the third-order neurons that project to the cerebral cortex. In the spinal cord, impulses from receptors concerned with the senses of pain, temperature, and touch are carried in relatively small fibers that send branches short distances up and down the spinal cord immediately after they enter. They synapse relatively soon, however, with the *second-order neurons* whose cell bodies are in the dorsal horn of the gray matter. Axons of second-order neurons cross to the opposite side and ascend toward the thalamus in the *lateral* and *ventral spinothalamic tracts.*

The impulses of muscle sense, which tell us the position of a limb, travel in a somewhat different pathway. The *primary (first-order) fibers* from the receptors are rather large and ascend in the cord on the same side they enter. They travel all the way up the spinal cord to the brainstem and finally synapse with *second-order neurons* of nuclei in the medulla oblongata of the brainstem (Figs. 7-6 and 7-7). The *gracile nuclei* receive fibers coming from the posterior limbs, while to the *cuneate nuclei* come the fibers from the anterior limbs. In the medulla oblongata, fibers from both pairs of nuclei cross (*decussate*) to the other side of the brainstem and ascend in a great tract, the *medial lemniscus*, which travels to the thalamus.

Nerves from the numerous general receptors of the head synapse in the brainstem on the side they enter, and the second-order neurons cross to the opposite side of the brainstem. All the fibers carrying sensory information finally arrive together at the *thalamus*, where they make contact with *third-order neurons* that ascend (project) to a particular part of the cerebral cortex, the *postcentral gyrus* (Fig. 7-6).

At the sides of the fourth ventricle, massive tracts of fibers lead upward from the spinal cord to the cerebellum. These fibers carry to the cerebellar cortex information from muscle stretch receptors. Impulses then travel in other fibers from the cerebellar cortex to nuclei within the cerebellum. Fibers arising from these central nuclei pass out in large tracts directed rostrally toward the *reticular* and *thalamic nuclei* of the brainstem. Here these impulses may alter the pattern of impulses traveling to or from the cerebral cortex. As a result, the cerebral cortex is supplied with additional information to use in regulating the spinal motoneurons.

Fig. 7-8. Ascending and descending tracts in the pons and midbrain.

1. *The pons:* Fibers cross the field, from dispersed pontile nuclei, and head for the cerebellum via the middle cerebellar peduncle (not shown). These fibers break the corticospinal tracts up into many bundles. The medial lemnisci have been pushed upward and outward. The ventral spinocerebellar tract moves into the cerebellum along side the brachium conjunctivum (superior cerebellar peduncle).

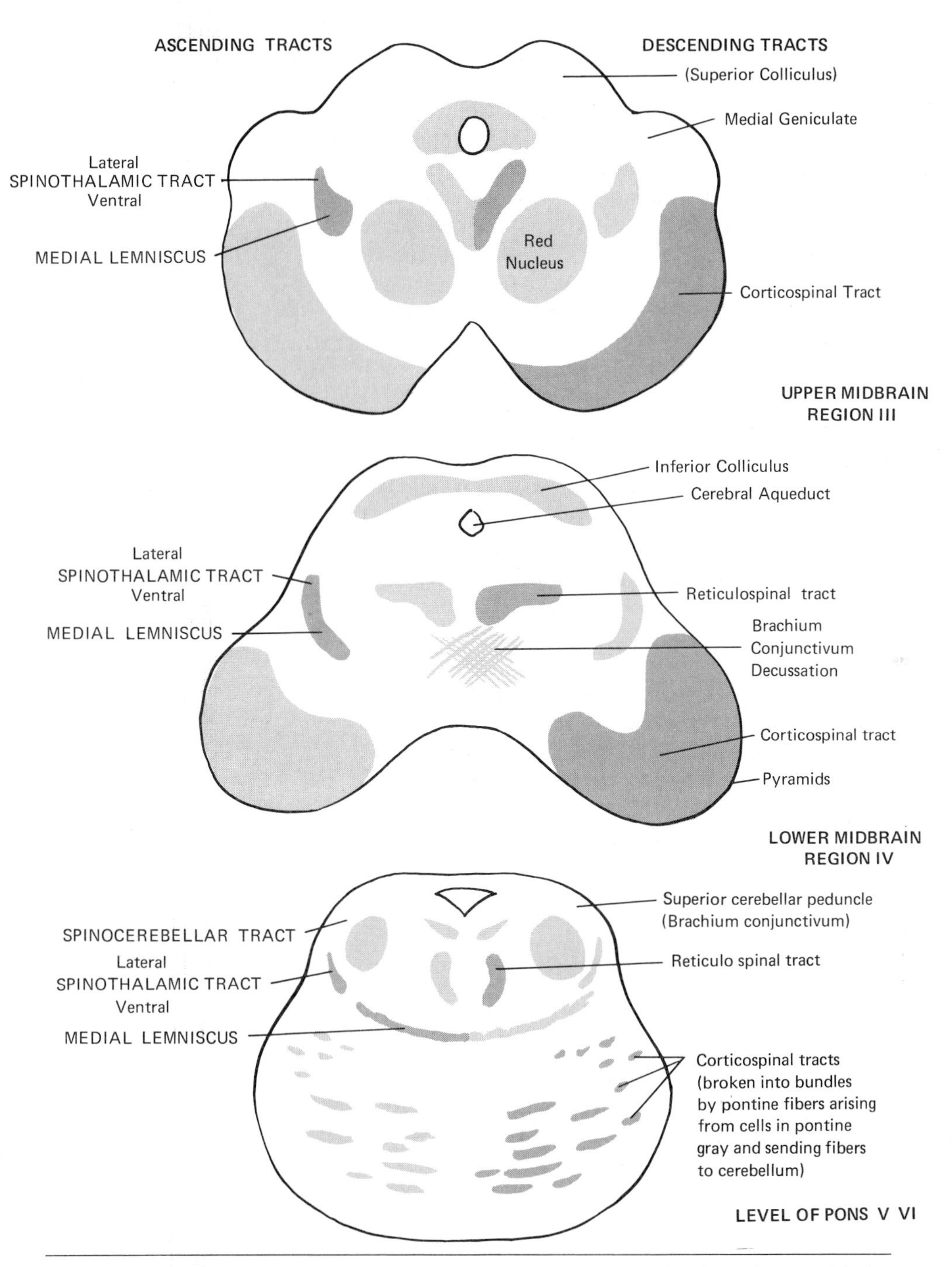

2. *The midbrain:* The cerebrospinal canal is the narrow aqueduct. Medial lemniscus is pushed further laterally by the decussation of the brachium conjunctivum whose fibers come from the cerebellum and are distributed to nuclei of the brainstem (arrows).

Descending pathways of brainstem and spinal cord

Having received information from lower brainstem and spinal cord as to what is going on, the higher centers of the brain can send down impulses in the descending pathways to command the motoneurons to carry out appropriate action. These descending pathways are of two general kinds. The most direct are high-speed pathways from cerebral cortex to motoneurons. These are the *pyramidal tracts* or the *pyramidal systems*.

The pyramidal tracts, which arise from large cell bodies in the cerebral cortex of the opposite side, are mainly in the lateral parts of the spinal cord. A smaller number of pyramidal fibers, coming from the cerebral cortex of the same side, are found ventro-medially in the cord. However, these fibers cross to the other side just before they innervate the motoneurons.

The other less direct, descending pathways are lumped together as the *descending reticular system*. These pathways involve a variable number of synaptic levels. From an evolutionary standpoint, they constitute an older system than the pyramidal. Fibers of this system arise in various brainstem nuclei—the vestibular, the olivary, the red nucleus, and from cells of the midbrain roof or tectum—and innervate the motoneurons of brainstem and spinal cord. These nuclei, in turn, are governed by fibers from yet other nuclei. The highest levels of control, from the cerebral cortex, are exerted by extrapyramidal fibers that arise from neurons of the cerebral cortex but are other than, are outside of, or are "extra" with respect to the pyramidal system. The extrapyramidal fibers travel along with the pyramidal fibers to get to brainstem levels. At all synaptic levels all the way down to the motoneurons, the systems of neurons are organized into excitatory–inhibitory reciprocally functioning arrangements that allow the appropriate movements to be carried out.

7.2 Higher Nervous Functions

Levels of control by the brain

The central nervous system is arranged in a hierarchy of levels of control over the motoneurons of brainstem and spinal cord. Insight into the role of various levels may be gained by several means in experimental animals as well as in man during operations involving the brain.

If part of the nervous system is destroyed, the capabilities of what remains may be observed. Electrical activity at a point in the nervous system may be observed during application of electrical stimuli at other points. These methods are used together with direct microscopic study of slices of the brain to provide a picture of the way in which the various parts of the brain are functionally interconnected.

Vestibular control of posture

One lower brainstem nucleus of considerable importance is the *vestibular nucleus.* This nucleus is continually bombarded by impulses arising in the organ of equilibrium of the inner ear. The excitatory effect of this input is transmitted to the spinal motoneurons along the fibers of the *vestibulospinal tract,* which arises from the cell bodies of the vestibular nucleus. This excitatory action is normally held in check by impulses coming to the nucleus from higher brain centers. The role of the vestibular nucleus in spinal motoneuron control is dramatically shown if a cut is made across the brainstem just rostral to the vestibular nucleus in an experimental animal—

for example, in a cat. This operation (called decerebration) releases the vestibular nucleus from all control by impulses from higher centers. The excitatory effects of the impulses from the inner ear then generate, unhampered, an intense and continuous play of impulses that descend to the spinal motoneurons. Here they provide a sustained excitatory effect that strikingly intensifies the stretch reflex. An animal placed on its feet after such a cut will stand with pillar-like rigidity of the limbs for several hours. The slight stretch that is imposed on muscles during standing stimulates the stretch receptors. The excitatory effects of the resulting afferent volleys sum with the excitatory action of the vestibulospinal impulses to provide an intense excitatory action on the motoneuron.

In the normal, intact animal, the excitatory and inhibitory actions are suitably balanced to provide the appropriate postural adjustments characteristic of normal behavior.

The cerebellum

Subdivisions of the Cerebellum. Overlying the pons in the region of the vestibular nucleus, and, in fact, evolving out of neural tissue related to the vestibular nerve, is the *cerebellum* (Fig. 7-9). Part of the cerebellum, close to its attachments to the brainstem, receives fibers directly from the vestibular nerve. This most ancient part of the cerebellum, the *archicerebellum*, includes the *flocculus*, extending out on each side on a large stalk, and the small *nodule* and *lingula* lobules caudal and rostral to the origin of the flocculus. From an evolutionary standpoint, this portion corresponds to the cerebellum of fish. An intermediate portion of the cerebellum corresponds to the cerebellum of amphibia and includes the anterior lobe, the *tonsillar* lobules, and the *uvula*. Most of the mass of the lateral lobes of the cerebellum is *neocerebellum*, newly evolved and having no homologs in the lower vertebrates. The cortex of the mammalian cerebellum is thrown into an enormous number of parallel folds covered by the gray matter atop the dense white matter, which, in a sagittal view of the cerebellum, branches like a tree and is called the *arbor vitae*.

Tracts and Nuclei. Almost all the incoming fibers to the cerebellum go directly to the cerebellar cortex. A large part of this afferent inflow is along the *spinocerebellar* tracts that carry information about muscle tension of the limbs. (Fig. 7-10). The *ventral spinocerebellar tract*, particularly supplying the older parts of the cerebellum, arrives via the *superior cerebellar peduncle*. The *dorsal spinocerebellar tract* is more newly evolved and arrives via the *inferior cerebellar peduncle*. An enormous number of fibers come to the cortex by way of the middle cerebellar peduncle. These fibers arise in the diffuse nuclei of the pons. The pontile nuclei receive information brought by fibers coming from the cerebral cortex. The largest and most obvious cells of the cerebellar cortex are the enormous *Purkinje cells*, which are arrayed in all the folia of the cortex. From the cell body of each Purkinje cell extends an enormous dendritic tree; the branches of each cell are all in a plane perpendicular to the long axis of the folium. Around these branches entwine the terminations of the afferent fibers to

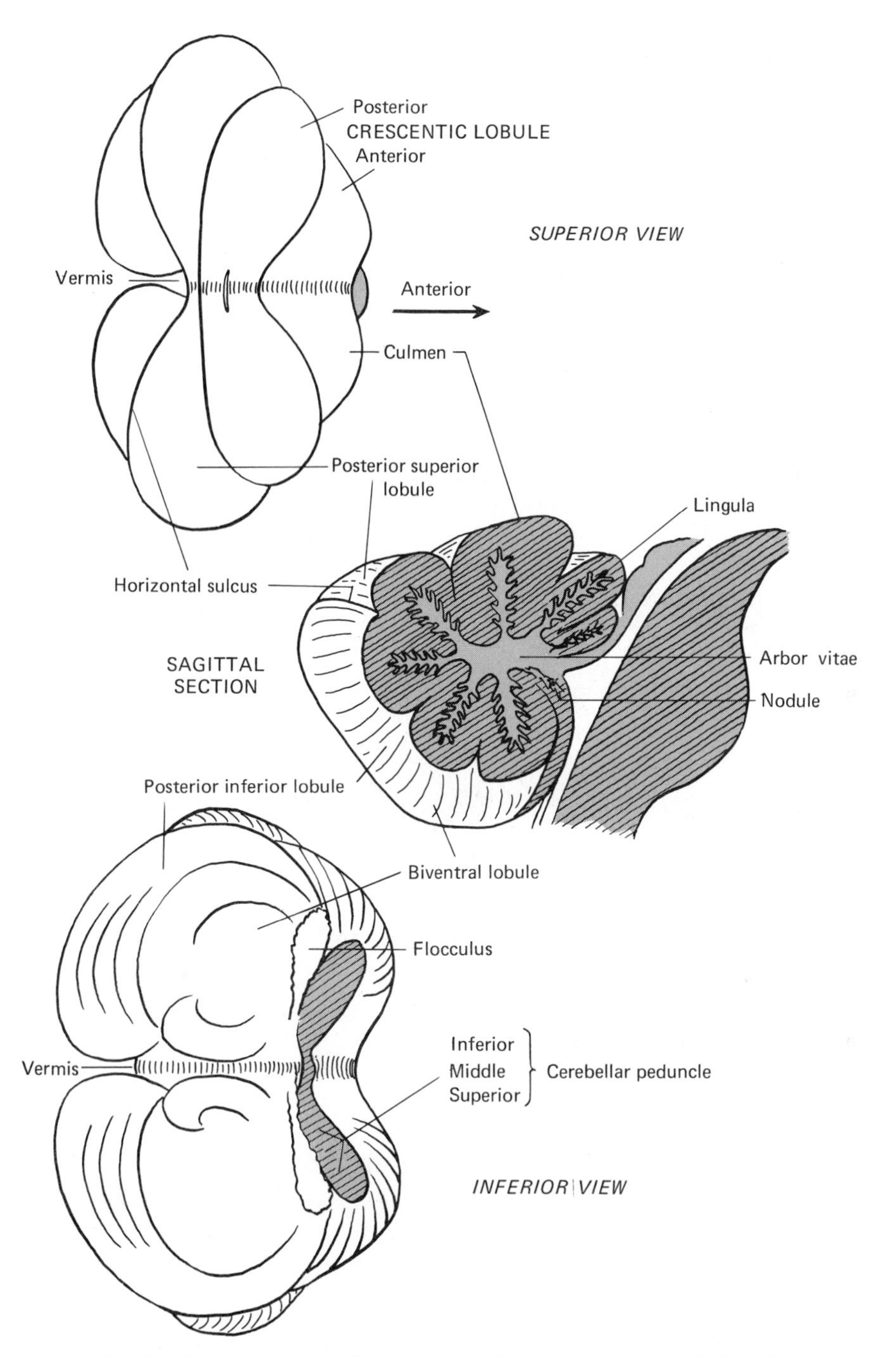

Fig. 7-9. External anatomy of the cerebellum. The sagittal view (middle figure) oriented as in the erect head, may be seen in relation to the rest of the brain by comparing with Fig. 7-2(b). It rises over the fourth ventricle, above the pons.

the cerebellar cortex. The axons of the Purkinje cells are the sole efferent path from the cerebellar cortex. These fibers all travel to the deep nuclei of the cerebellum, which overlie the fourth ventricle. The archicerebellum makes connections to the more medial nuclei (particularly the *fastigial*). The neocerebellum connects particularly to *dentate* nuclei, which are more lateral in position. From these nuclei, cerebellar information is brought to red nucleus or to thalamus.

CEREBELLAR FUNCTIONS. A simple scheme shows the participation of the cerebellum in the control of body movement (Fig. 7-10). Each line represents many

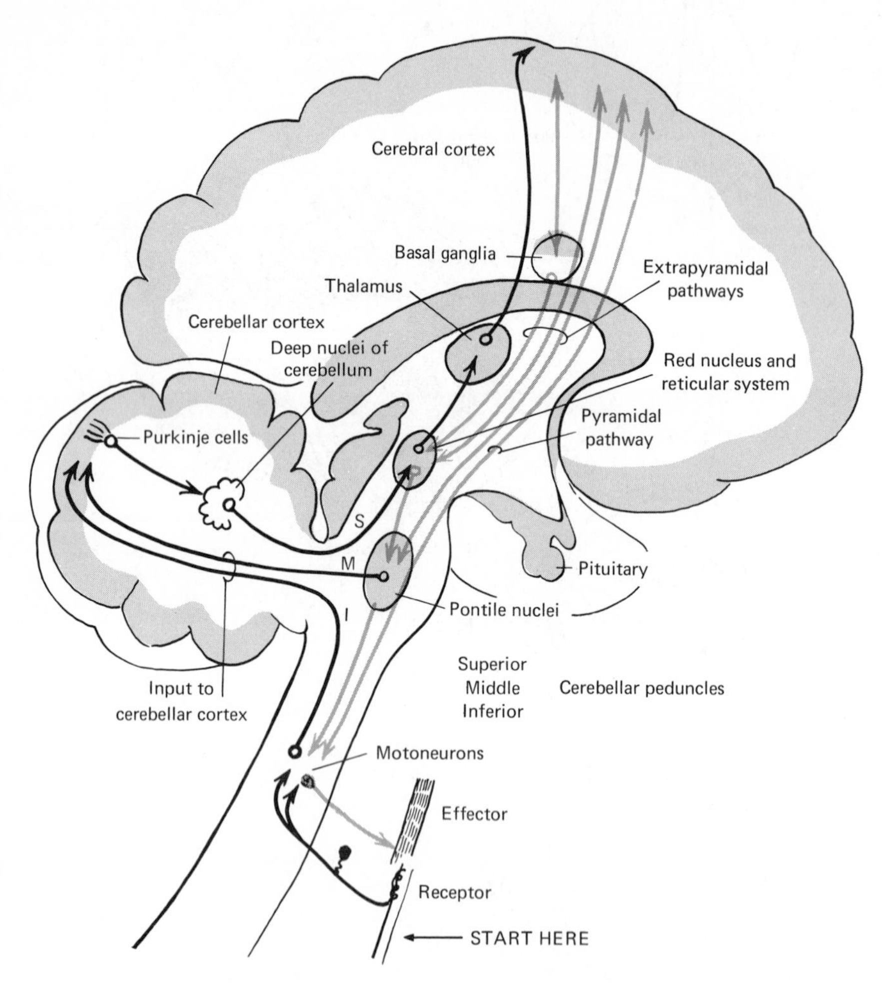

Fig. 7-10. Cerebellum and cerebellar control of motoneurons. Section shown in Fig. 7-2(b). Pathways of one side shown. Black arrows, some pathways rather directly related to cerebellum. Gray arrows, some pathways rather directly related to cerebrum. Follow arrows to realize direction of impulse traffic.

thousands of fibers in the various pathways. Begin at the receptor and follow the arrows on the neuron pathways. The information from the receptors goes not only to segments of the spinal cord and up into the cerebral lobes but also to the cerebellar cortex. The Purkinje cells of the cerebellar cortex provide the output from the cerebellum. The temporal pattern of firing in these fibers depends on the synaptic influences arriving from receptors directly, plus those arriving from various brainstem nuclei, particularly the cerebral cortex. The cerebral cortex outflow, in turn, is affected by the information coming from the cerebellum as well as numerous other sources, including the direct projection of afferent information from the cortex. The cerebellar information reaches the cerebral cortex via the *thalamus* and the *red nucleus*, as the figure shows. The red nucleus and the associated reticular system are also part of the extrapyramidal pathway from cerebral cortex down to motor neuron in brainstem or spinal cord.

The cerebellar output may be to the extrapyramidal system or to the reticular nuclei of that system in the brainstem. The net output by the motoneuron to its muscle fibers depends on the extrapyramidal and pyramidal influences. In the case of a voluntary effort, pyramidal activity occurs in a background of extrapyramidal "tone" generated by the cerebellum. More important, the cerebellum seems to act as a computer to help calculate the appropriate amount of activity—the frequency of impulses—in the appropriate fibers necessary for the smooth and accurate accomplishment of particular actions.

The cerebral lobes

THE PYRAMIDAL SYSTEM. The cortex of the cerebral lobes is the highest governing level of the nervous system. It controls the motoneurons of brainstem and spinal cord by means of two general groups of fibers: one route runs directly from cell bodies in the cerebral cortex to the motoneurons; the other, an indirect route, goes to the brainstem, from which fibers of other neurons carry the information to the spinal cord. From an evolutionary standpoint, the direct-line system from cortex to cord is a relatively new arrangement. It permits a more rapid response than is possible when the impulses must cross several synapses, each of which introduces a delay and increases the time between command and response. The direct route from cortex to motoneurons consists of fibers that originate in pyramidal-shaped cells of the cerebral cortex (Fig. 7-11) and that constitute the *pyramidal system*.

The larger pyramidal cells are in the *precentral gyrus*. Many smaller pyramidal cells are scattered in various *association areas* of the cortex. Most of these fibers travel together down through the brainstem, where some innervate the cranial nerve nuclei, and others continue on into the spinal cord via the *pyramidal tract*. In the region of the lower brainstem, fibers of each side cross to the other side and come to occupy a large portion of the lateral white matter of the spinal cord. As they descend into the cord, fibers leave the main tract and end close to the spinal motoneurons. Movements of particular body regions follow stimulation of the precentral gyrus

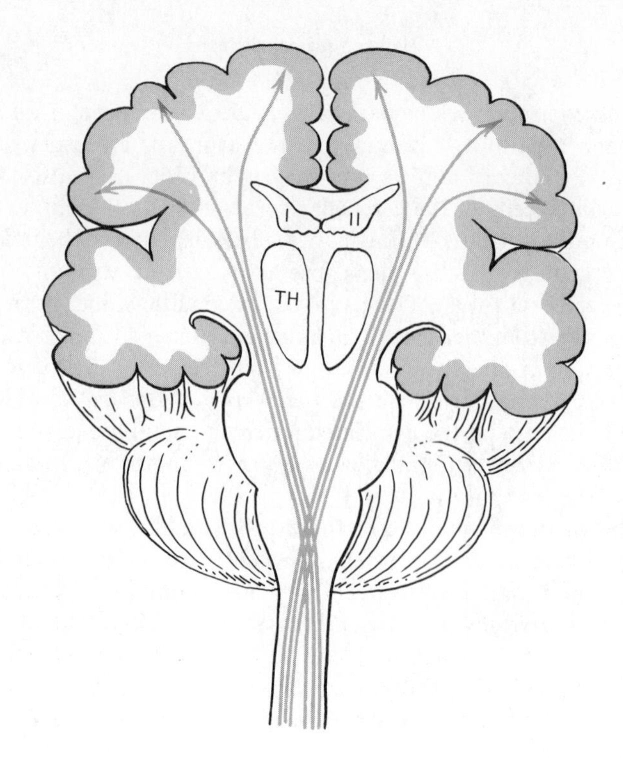

Fig. 7-11. Pyramidal tract. Horizontal slice through brain in plane shown in Fig. 7-2(a). Gray arrows show pathway of pyramidal fibers from cerebral cortex, through brainstem and crossing in medulla oblongata.

where the larger pyramidal cells are located. Stimulation of the topmost region will produce movements of the posterior limbs and lower trunk. Movements of the upper trunk and face occur during stimulation of more lateral and lower parts. The precentral gyrus is called the *somatomotor area,* in reference to its control of body movements. The extensiveness of the subdivisions in the somatomotor area reflects the importance, rather than the dimensions, of the body regions to which they are related. The muscles of the thumb and index finger have relatively few muscle fibers in each of the many motor units. In the large trunk muscles there are many muscle fibers for relatively few motoneurons. The relative sizes of the cortical subdivisions reflect this difference (Fig. 7-12).

Somatosensory Area of Cerebral Cortex. Electrical stimulation of the *postcentral gyrus* in the conscious patient provides vague sensations referred to the body surface, but pain is not felt. If the stimulus is applied to the topmost region of this strip of cortex, the patient will experience vaguely defined sensations that seem to involve his lower trunk and limbs. In the lower parts of this band of cortex, sensations from the upper trunk and face are evoked. Because sensations from the entire body seem to be thus projected to the cerebral cortex, this special zone is called the *somesthetic* (body senses) *projection area* (Fig. 7-12).

The subject will report flashes of light and other visual effects if the *occipital* lobe of the brain is stimulated. This lobe is devoted to the retina of the eye and is designated

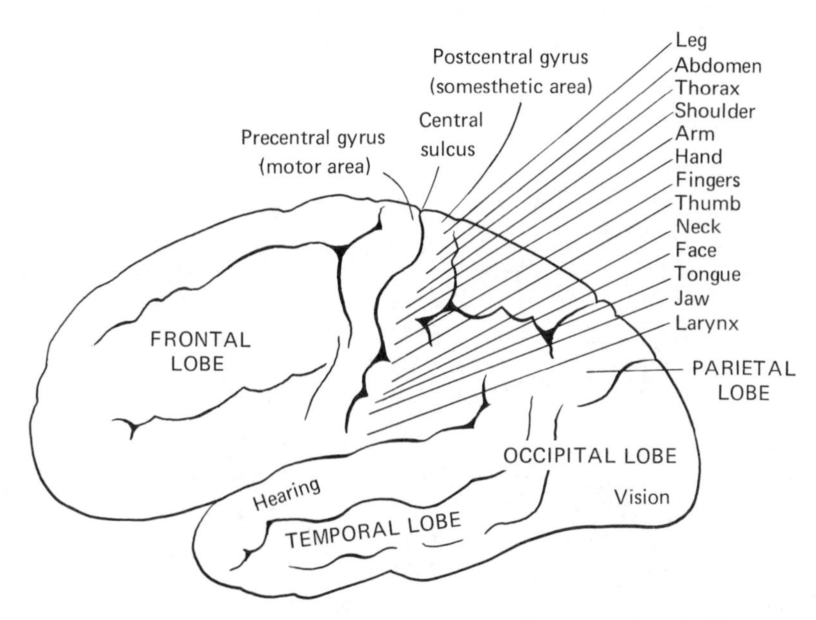

Fig. 7-12. Areas of the cerebral cortex. Lateral view of cerebral hemispheres.

the *visual projection area.* Electrical potentials may be recorded here when light shines in the eye. Blindness follows extensive injury to this area. Another special sense organ, the *cochlea* of the ear, projects to the *temporal lobe* of the cerebral cortex. The olfactory system is most closely connected with an area of the cerebral cortex underneath the frontal lobes. The relative proportions of cortex functionally connected to particular parts of the body reflect the density of receptors and not the sizes of the various body regions. Thus the high sensitivity of the tongue and the finger is revealed in a correspondingly large cortical area devoted to these regions, compared with places such as the back, where the sensitivity is low.

THE EXTRAPYRAMIDAL SYSTEM. The indirect route from cortex to brainstem and spinal cord motor nuclei is the *extrapyramidal system.* "Extra" in this context means "outside of." The *extrapyramidal system* arises from small cells sending fibers from the cerebral cortex to the *basal ganglia* and the brainstem. The descending components of the *reticular system* of the brainstem and spinal cord are of special importance in the extrapyramidal system. In higher vertebrates, numerous distinct nuclei have developed from the pervasive reticular system, an undifferentiated network in the lower vertebrates (Fig. 7-13).

The basal ganglia send fibers to the reticular areas of diencephalon and mesencephalon, while other fibers of the extrapyramidal system run from the cortex to the reticular areas of the medulla. As elsewhere in the nervous system, this control is in terms of balanced patterns of excitation and inhibition.

Control of motoneuron activity may be directly on the motoneuron itself or

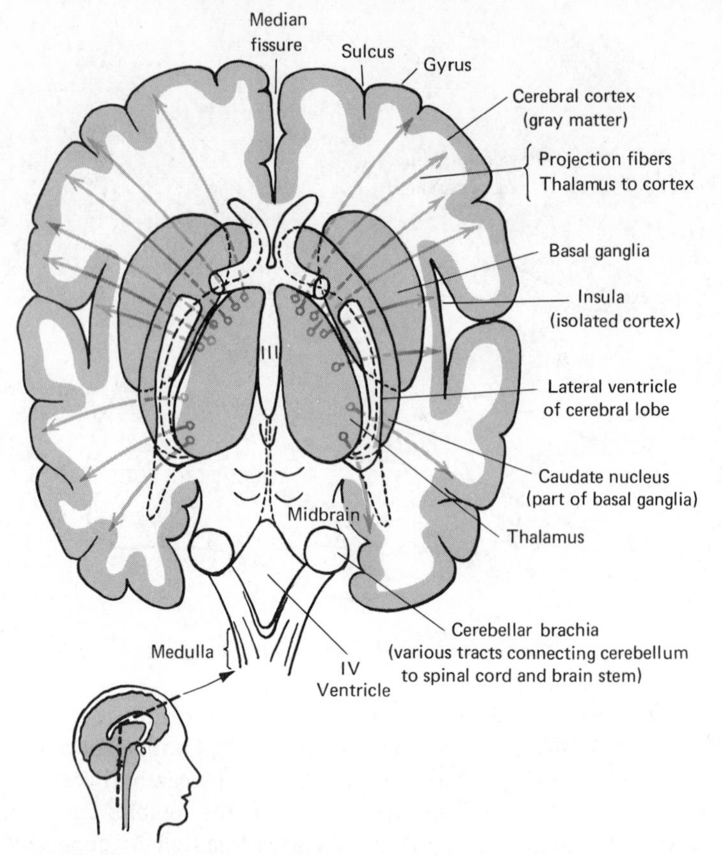

Fig. 7-13. Relation of basal ganglia and thalamus to brainstem. The brain is sliced approximately as shown in the small figure. The top is removed and the large figure is a representation of what may be seen. The lateral cerebral ventricles are shown as tubes. The substance of the cerebral lobes makes their walls. The lateral ventricles are continuous with the ventricle of the diencephalon (III), of which the thalamus provides the walls. The dotted lines show the continuity of the III ventricle through the mesencephalon to the wide IV ventricle, the cavity of the medulla. The lateral ventricles may be imagined as pulled around, up, over, and backward as the cerebral lobes have grown. The basal ganglia are great assemblages of gray matter in the interior of each cerebral lobe, lateral to the thalamus, and beneath the lateral ventricles. The caudate nucleus is part of the basal ganglia, drawn out as a long tail (whence its name) beside the curving lateral ventricle of each side. The ventricle narrows as it passes through the midbrain, and above it are the prominent superior and inferior colliculi, centers for visual and auditory reflexes.

indirectly by way of small efferent fibers that go not to the tension muscle fibers but to the *intrafusal* fibers of the muscle stretch receptor system. In addition to the afferent innervation, discussed in the previous section, the intrafusal fibers are also supplied by a motor innervation. When these efferent fibers are active, the intrafusal fibers contract slightly, thereby applying stretch to the stretch receptors. The stretch receptors, in turn, pour impulses into the central nervous system and activate the large motoneurons governing the tension muscle fibers. Thus reflex contraction of the muscle is maintained even when stretch on intrafusal fibers is removed during shortening of the muscle.

THE RETICULAR SYSTEM. *Sensation*, a quality of the conscious, aware brain, exists in relation to impulses that reach particular parts of the cerebral cortex. The experiencing of specific sensations depends on impulses arriving at the cortical projection areas along the major pathways described above. However, a condition of consciousness must exist if sensation is to be experienced. The conscious, awake condition of the cerebral cortex seems to depend on impulses that reach the brain along other than the major sensory pathways.

Besides the well-defined, large-fiber, high-speed tracts, a network of small fibers in spinal cord and brainstem carries impulses through multiple synaptic pathways up to the cortex. This diffuse pathway is the ascending *reticular system*. Impulses from this system in the brainstem supply virtually the entire cerebral cortex. Apparently the cerebral cortex is maintained in a receptive, awake condition when this flow of impulses comes steadily into the cortex. Under these circumstances, the cortex is set to respond to the more precise information provided by the high-speed, specific afferent systems. In its role of thus keeping the cortex awake, this diffuse system is called the *reticular activating system*.

Electrical activity of the brain

The nerve cells of the brain are always active, but different groups of neurons carry impulses at different times. As a result, the great mass of nerve fibers and cell bodies produces a changing pattern of action currents that flow throughout the cranium. In addition, many cells of the brain undergo fluctuating nonpropagated changes in potential that provide additional flow of current into the surrounding regions. All this electrical activity provides the currents that may be recorded on the surface of the skull as the *electroencephalogram* (EEG) (Fig. 7-14). The potentials produced by these currents are less than 0.1 mV at the surface of the skull, although they arise from membrane potential changes a thousand times as great. The neurons of the brain carry impulses according to a rhythm that seems to be determined partially by an innate self-excitatory quality of the neurons and partially by the impulses the brain receives from the receptors by way of the spinal cord and the brainstem. The electrical currents from the brain activity can be recorded from the skull surface, for the bone, in spite of its hardness, has a considerable proportion of salts dissolved in water and will, therefore, like all the body fluids, conduct an electrical current.

The pattern of the EEG depends on the location of the electrodes on the skull or brain from which the records are taken; it changes according to the state of consciousness, alertness, or attention of the individual. During intense concentration the record is relatively flat. In sleep, when attention is at its lowest level, large waves at less than 10 per second can be seen in the record. It is assumed, in the latter instance, that slow changes are occurring synchronously in many neurons. During alert be-

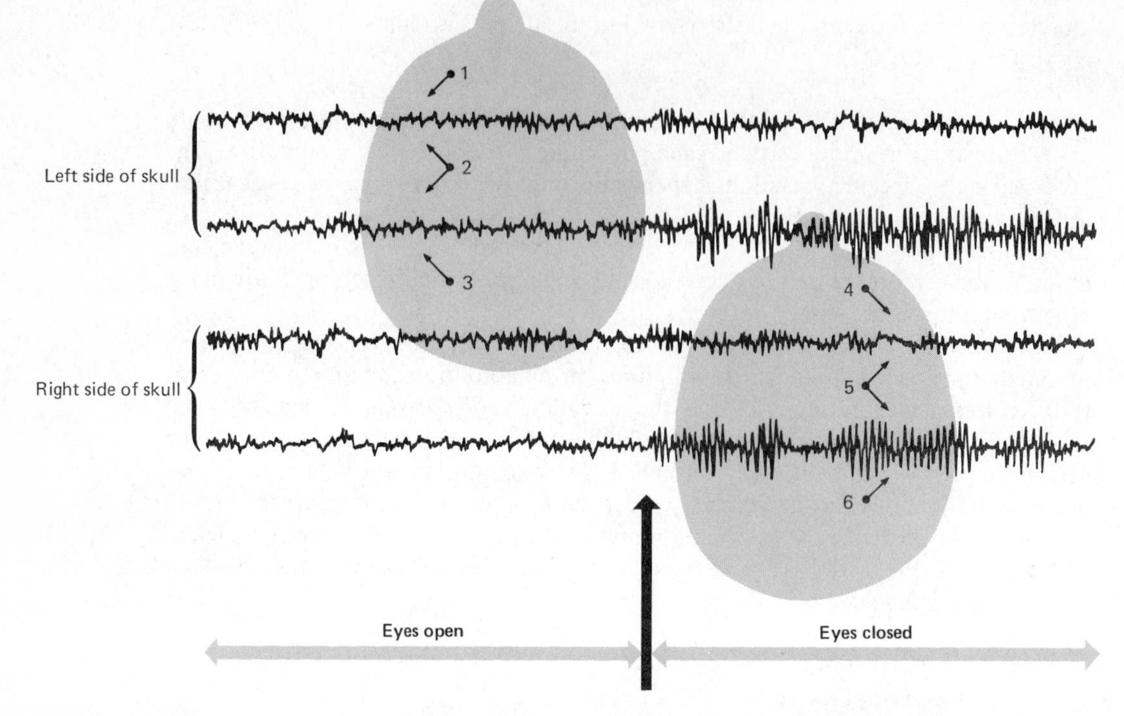

Fig. 7-14. Electrical activity of the brain. Electrodes placed on scalp at numbered points. Top pair of records from points 1, 2 and 2, 3. Bottom pair from 4, 5 and 5, 6. Note increase in amplitude of EEG when eyes are closed. Highest activity at posterior region of skull.

havior the total neural activity is greater, but the impulses move about in a more unsynchronized fashion. The record is said to be of *desynchronized* activity.

The normal patterns of currents flowing throughout the cranium may be changed when alterations occur in the electrical conductivity of the contents of the cranium. For example, the amplitude and frequency of the EEG may be altered by the presence of tumors, the precise location of which can be determined by examination of the EEG recorded from various places on the head. When a particular sensory input is provided, a momentary new wave called the *evoked response* occurs in the EEG. This wave signals that impulses from the stimulated sense organs have arrived at the brain.

General functions of the cerebral cortex—hemisphere dominance

The dominating role of the cerebral cortex is undoubtedly a result of its organization in relation to incoming and outgoing pathways. Surprisingly, large parts of the cerebral cortex may be removed with little change in behavior. As in many other organs of the body, there is a great deal of redundancy in its construction. Several cortical areas may be involved in the same processes, and there is more than the minimum amount of structure necessary to do the job.

The primary receiving (somesthetic) and the primary motor areas are essential, and their removal may result in anesthesia and paralysis. Defects from loss of certain other areas may be more subtle and difficult to specify.

The language centers constitute an assembly of several subcenters relating to hearing for oral language, vision for reading, and motor control for writing. Language facility goes considerably beyond the perception of sound, the detection of visual form, or the holding of a pen. Particular areas of the brain are apparently concerned with different aspects of language, for it is possible to have separate and distinct kinds of deficiency result from an injury to the language areas of the cortex.

Experiments on animals have shown that the two cerebral lobes can function more or less independently when the optic tracts and the corpus callosum are cut in the midline. In humans the corpus callosum has occasionally been cut in order to control epilepsy that otherwise radiates from a focus on one side to a corresponding area in the opposite hemisphere. After such section or cut, a skill learned by the use of the left hand, for example, could not then be carried out by the right hand. Evidently the corpus callosum normally serves as an important path for communication between the two cerebral lobes, and in many ways the lobes must certainly be essentially equivalent, exchanging information as equals. It is even conceivable that most of one cerebral hemisphere could be eliminated and the individual would still be able to function more or less normally. However, one side ordinarily controls the other—one side is dominant. If the dominant hemisphere were to be destroyed, recovery would be less likely. Side of cerebral dominance is usually related to hand preference. In right-handed people, for example, the left hemisphere is probably the dominant one. The dominance of the opposite side makes sense when one realizes that the pyramidal fibers innervating the motoneurons of the right hand are in the precentral gyrus of the left hemisphere. A right-handed person can generally do with the left hand whatever he can do with the right, but a good deal less effectively.

The dominance also extends to other aspects of performance and perception. There is some indication that the side of dominance is different for different skills. For example, music seems to be processed more effectively in the right hemisphere, and speech in the left. It is likely that much of the *subconscious* exists in the circuitry of the nondominant hemisphere. An event embarrassing to the dominant hemisphere may be acceptable to the nondominant side.

Higher nervous activity—the cerebral cortex and society

Consider the meaning of behavior: the central nervous system receives a barrage of impulses that must be sorted out and acted on according to their survival value for the organism. Even the simplest reflex is organized so as to optimize the effect of the stimulus giving rise to it. In the case of a noxious stimulus, this means reduce the stimulus; preferably, get rid of it. In the case of a favorable stimulus, this means continue it and increase it so long as it has a favorable effect. The brain takes the rich visual, auditory, chemical, and kinesthetic patterns that it receives, and places them in the context of its past experiences. Reflex behavior is then modulated not merely in terms of present reality but also in terms of future hopes woven from the patterns of the past. Into the genetic capabilities of neuronal circuits, unique for each individual, is fed the information of experience, also unique for each individual. Most of us believe that the world has a substantive reality, but that reality differs for each person in subtle ways. Each of us behaves differently in response to his environment. When the environment elicits what is judged as appropriate behavior conducive to self-survival, we consider the behavior as normal. Insofar as the behavior is judged as inappropriate, we consider the behavior as abnormal.

Derangements of behavior may occur at any level of neural function—sense organ, sensory nerve, central synapses, ascending pathway, cortical association, descending pathways, motoneuron, neuroeffector junction or effectors. And the causes may be genetic errors of development, toxic substances, nutritional deficiencies, or complex maladaptive conditioning, usually called psychological problems.

Human behavior is so enormously rich and varied, as well as unpredictable, that often there is difficulty in assessing the cause or causes of a malfunction or even in ascertaining that a malfunction exists. Normal behavior is more or less consistent and predictable behavior. Most of us behave consistently because we find the world consistent and our reactions are governed by it. Our rational behavior is conditioned by a rational universe to which we can make meaningful adjustments. If we are conditioned by an inconsistent environment, there will be confusion in our cerebral circuits, and the behavior may also seem unpredictable, irrational, and abnormal to an observer. If malnutrition and genetic deficiencies are combined with malconditioning, the probability of successful normal behavior is slim indeed, and self-integrity is easily threatened.

Our human brains allow us to extend the meaning of the term "individual survival" far beyond the wildlife meaning of simple physical aliveness. The human individual is his body, but he is also his thoughts and his self-consistent behavior. For instance, the police may threaten the physical survival of a picket or a protest marcher, but, for the protestor, upholding a certain point of view may be consistent with his whole outlook and not to behave as he is doing would be inconsistent and therefore inappropriate.

The more rational the mind, the more likely that the individual will be able to see the long–term consequences of certain kinds of behavior. The social critic who is often denounced as irrational may, in fact, be more rational than the majority of the members of the society whose unjust institutions he rails against and who judge him to be mad. Only when bizarre behavior reflects a disintegration of the self-identity is it pathological from a biological point of view.

Man's crowning glory is his cerebral cortex. With its help he can try to understand the universe and his place in it, for he is no better than any other animal except in this capacity to use his brain to imagine and to work toward that goal. If such is his uniquely human attribute, then the society that allows each individual to participate in this endeavor to the best of his abilities is the best one.

People feel better and think more clearly when they have adequate diet and exercise and when the body cells are not overwhelmed by toxic agents. However, the conditions of modern life are not, in general, conducive to healthful living, either physically or psychologically, and too often we have allowed inventions that could help us extend our vision to be used merely for temporary comfort and private profit.

From a physiological standpoint, much of the behavior of civilized man is highly irrational. To allow oneself always to be conveyed effortlessly from place to place rather than to use one's own muscles, to breath a polluted atmosphere, to eat too little because of deprivation or too much out of habit, to allow oneself to become accustomed to visual and auditory discord when beautiful as well as useful surroundings are possible—this typical behavior of so many of us in modern Western society is deranged in the extreme. The behavior that allows millions of our fellow humans to starve while a few live in carefree opulence is equally deranged. From a physiological and a social standpoint, those who protest and wish to change the state of the world may be the more rational.

Section 7: *Questions, Problems, and Projects*

1. Distinguish between (a) cell nucleus and (b) brainstem nucleus.
2. Where are the cell bodies and the cell nuclei of a brainstem nucleus? Where do the axons of these cells go?
3. Where are the cell bodies of most afferent nerve fibers? What about those of the olfactory nerve? Where do the axons of sensory neurons go?
4. Where are the cell bodies of the sensory nuclei of the brainstem?
5. What is the relation between the sensory and the motor nuclei of the brain-stem?
6. Distinguish somatic vs. visceral aspects of the nervous system.
7. What great sensory systems are related especially to these parts of the brain: mesencephalon, medulla oblongata, telencephalon?
8. With what part of the brain is the autonomic nervous system particularly identified?
9. What is the relation between the thalamus and the sense organs? Between the hypothalamus and the pituitary?
10. Explain what is meant by first-, second-, and third-*order* neurons, and trace the pathways involving these neurons, from receptor to cerebral cortex, in the in-

stance of: (a) a light touch receptor in the left big toe. (b) A muscle stretch receptor in the extensor muscle of that same toe.

11. Describe the successive changes you would expect in neuromuscular tone if the spinal motoneurons are separated from effects of higher centers by: decerebration, destruction of the vestibular centers, a spinal transaction, destruction of the motoneurons innervating the muscle in question.
12. What is meant by localization of function in the cerebral cortex? Describe the nature of the localization of motor activity and of sensation.
13. Describe behavior that is not reflex in nature, and distinguish conscious and unconscious movement.

8

RECEPTORS AND SPECIAL SENSE ORGANS

8.1 The Ear: Proprioceptive and Auditory Functions

The dramatic expansion of the brain in the vertebrates in general and in the human in particular is forced by the enormous development of the sense organs of the head. Cerebral cortex has evolved from primarily olfactory centers, a large part of the thalamus is devoted to vision and audition, and the cerebellum developed partly to process the information from the proprioceptors of the inner ear. That part of the cerebral lobes, thalamus, and cerebellum not devoted directly to specific sense organs organizes this receptor information and relates it to that coming from the body generally, to provide the basis for intelligent action. This chapter is concerned with the ear.

The inner ear apparatus

A set of special *mechanoreceptors* lies in the inner ear and provides information about movement and orientation of the head. Part of the same apparatus detects vibrations originating outside the body and thus serves the special sense of hearing. The entire apparatus consists of a thin-walled membranous sac of many interconnected, complicated passageways, appropriately called the *membranous laby-*

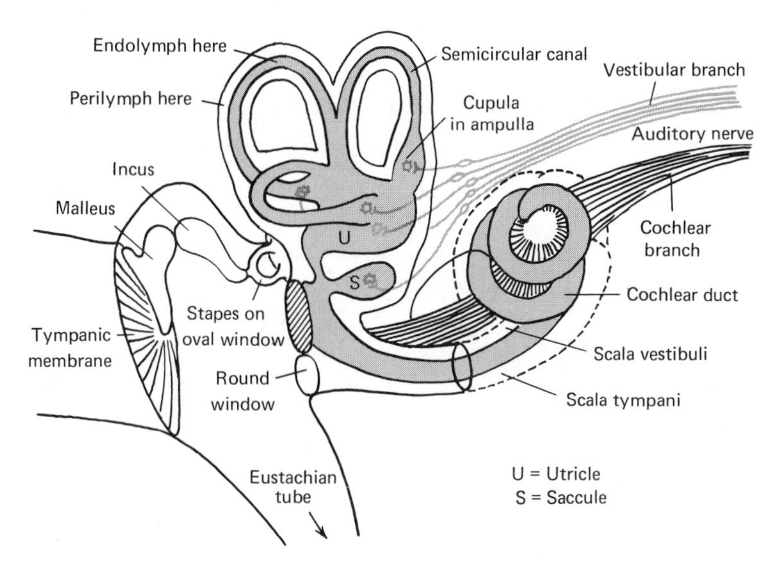

Fig. 8-1. Structure of the middle and inner ears. (Arrow shows plane of section in Fig. 8-6.)

rinth, which is suspended and attached within a slightly larger cavity of similar shape, the *bony labyrinth*, within the temporal bone of the skull (Fig. 8-1).

The entire membranous labyrinth is filled with a fluid, *endolymph*, and is surrounded by *perilymph*, which fills the bony labyrinth. Both fluids resemble blood plasma but differ slightly from one another in composition. The *utricle*, which is the central chamber of the labyrinth, contains receptors that signal information about the *static* or stationary orientation of the head with respect to gravity. The smaller *saccule* may respond to movement of the head in a straight line in any direction (linear movement), whereas the *semicircular canals* provide information about *rotation* of the head. The cochlea is appended from the utricle but responds to vibrations of sound waves. Sound waves ordinarily gain access to the inner ear by way of the *middle ear bones*, through which the vibrations are transmitted after impinging on the eardrum or *tympanic membrane*. The utricle and saccule are collectively called the *vestibule* in reference to their location at the entry beyond the middle ear bones and on the way to the cochlea. The receptors of the vestibular system and the semicircular canals feed impulses into the *vestibular nucleus* of the medulla. The brain uses the information in the regulation of posture. From the receptors of the cochlea, a separate nerve root (the cochlear branch of the auditory nerve) runs to the cochlear nucleus of the medulla oblongata.

In all parts of the labyrinth, the receptor end organs are fundamentally similar. The terminals of the afferent nerve fibers embrace the receptor cells, which have specialized cilia at their apices. These cilia are hairlike in appearance under the microscope, and the cells are therefore called *hair cells*. The cilia are imbedded in an overlying massive structure, and shearing forces applied to them by the displacement of the overlying mass provide stretch to the hair cell. A discharge of impulses along the afferent fibers results. It is thought that the receptor cell membrane increases in permeability and is depolarized during stretch. Current presumably then flows from the rest of the cell to the depolarized zone, and an excitatory effect is provided to the tiny nerve endings at the base of the cell.

Receptors for static position

The hair cells of the utricle are imbedded in a gelatinous material containing concretions of calcium carbonate ($CaCO_3$) called *statoliths* or *otoliths*. (Fig. 8-2). Arrangements similar to this, called *statocysts*, are found in nearly all animals. When the head is in a position tilted toward the right, the hair cells on the right-hand side of

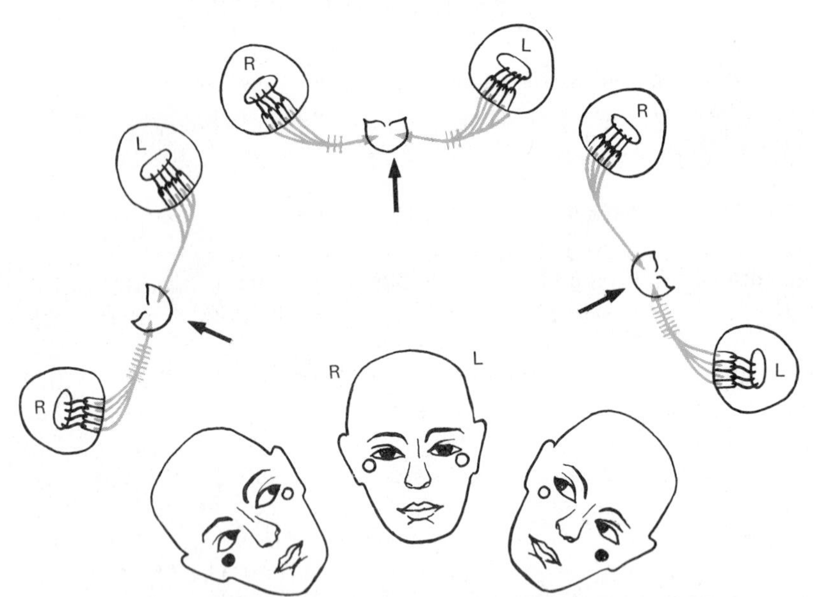

Fig. 8-2. Sensing of head position by utricle. In the normal upright position, the impulse traffic from both sides is equal into the central nervous system. When the head is tipped toward the right, the utriculus of the right-hand side (R) is stimulated more, that of the left (L), less. Displacement of the head to the one side has an opposite effect on the receptors of each side, and the resulting total impulse pattern provides information about the direction of the displacement. The same kind of difference between the two sides exists with respect to the information the semicircular canals provide to the central nervous system concerning direction of rotation of the head.

the head are put under extra stress, owing to the pull of gravity on the statoliths. There is a consequent increase in frequency of impulses generated in the utricular nerve of that side. The receptors are so constructed that when the head is displaced to the right, the stress on the receptors of the left side of the head is decreased, and the impulse frequency is correspondingly diminished. Thus, by means of the different patterns of impulses, the central nervous system is informed of the position of the head. In the normal position, the trains of volleys of impulses from both sides are about equal.

Receptors for rotation

The sensitive regions in the semicircular canals are somewhat different from those in the utricle (Fig. 8-3). Each semicircular canal has a distended portion, the *ampulla*, at one end. Within the ampulla is a small patch of sensory hair cells, whose apices are imbedded in a gelatinous cap (*cupula*) that extends above the hair cells to the wall of the ampulla. The cupula forms a complete boundary across the ampulla. The sides of the cupula are free except at their attachment to the hair cells but are closely fitted to the sides of the ampulla. When the fluid moves in the canal, the cupula is pushed, like a swinging door, by the pressure of the moving liquid. The gelatinous cap is so light that the receptors are not particularly affected by the static orientation of the head, and the fluid will not be displaced by a constant velocity of movement in any direction. The consistency and the elastic properties of the cupula are of such a nature, however, that *changes* in angular velocity momentarily displace the cap and therefore provide a transient pull on the hair cells. The movement of the cupula occurs particularly at the beginning of a rotation of the head and at the stop of a rotation lasting a minute or more, but not during constant rotation. The cupula moves inside the ampulla for the same reason that you, as an autombile passenger, are jolted *back* in your seat when the car suddenly starts *forward* or jolted forward when the car suddenly stops moving.

The principle of this effect may also be illustrated by applying a sudden spin to a bowl of soup. A floating cracker crumb will serve as a marker to show that at first the bowl moves around the liquid, which because of its inertia tends to lag behind. As the rotation is continued at a constant angular velocity, the container and the fluid tend to move together because of the friction between the fluid and the walls. When the rotation is suddenly stopped, the fluid tends to continue moving because of its inertia.

The flexible gelatinous cupula in the semicircular canal tends at first to be held back with the fluid when a rotation begins. When the rotation is constant, fluid and cupula move along together with the rotating head. When rotation stops, the fluid, continuing to move, carries the cupula along with it; the latter is therefore displaced at its attachment to the hair cells. At start and at stop of rotation, stress is exerted on the hair cells, and the impulse frequency is increased.

A clockwise rotation evidently increases the stress in the hair cells of the right-hand horizontal canal and moves those of the left side into a position of less distor-

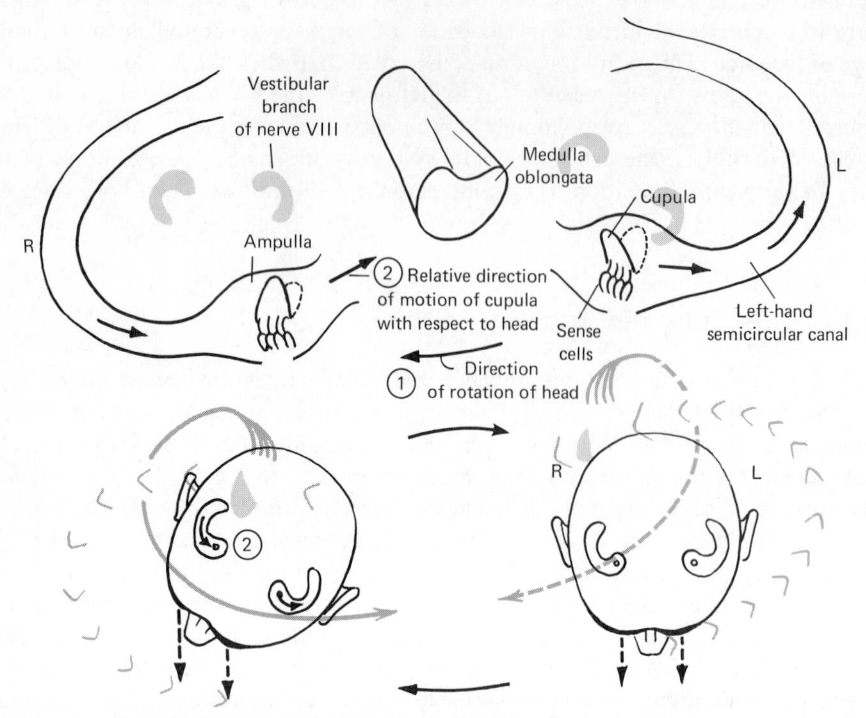

Fig. 8-3. How the semicircular canals signal rotation of the head.

1. At the start of the clockwise spin of the head, the fluid together with the cupula in the semicircular canals tends to remain behind. During smooth, constant rotation, the fluid moves along with the head, the cupula straightens up and no sensation is provided. When rotation stops, the cupula moves with the fluid, which continues moving after the head stops rotating.
2. The relative movement of the cupula with respect to the head is opposite to the direction of the spin of the head.
3. The movement decreases the stress existing on the hair cells of the left-hand ampulla, while increasing the stress existing in the right-hand one. Thus the impulse frequency from the left side decreases while that from the right side increases.

tion, since the frequency of impulses from the right ampulla increases while that from the left decreases. This pattern of impulses provides a sensation of movement toward the right. When the head stops, the relative direction of movement of the fluid with respect to the head is reversed, the frequency of impulses in the left side increases, that on the right decreases, and a sensation of rotation toward the left is felt, although the head is actually stationary.

On each side of the head there are three semicircular canals, each perpendicular to the other two. The horizontal canals (one on each side) respond to rotation in the horizontal plane, and the others are affected by rotation in planes at right angles. Rotation at an intermediate angle will involve participation of more than one pair of canals.

Eye movements controlled by the semicircular canals

The semicircular canals are reflexly connected via the vestibular nucleus to virtually the entire somatic motor system, and postural adjustments are partly determined by the information supplied to the central nervous system by these organs. Control by eye movements is particularly striking.

Reflex connections with the extrinsic (external) voluntary eye muscles are easily demonstrated (Fig. 8-4). Whenever objects move rapidly across our field of vision, we tend to fix the gaze on some point ahead, then move the eyes in the direction the object appears to be moving. After following the object a short distance, the eyes are snapped back in the opposite direction to fixate on another object that has appeared in the visual field. This behavior serves the useful purpose of keeping objects in focus

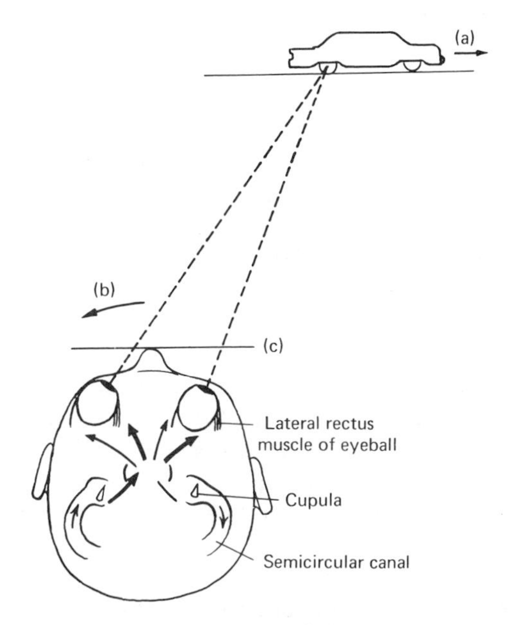

Fig. 8-4. The mechanism of nystagmus. (a) If the environment moves, the eyes will follow to keep a clear image on the retina. If the person moves his head counterclockwise as shown by arrow (b), the eyes will follow the apparent clockwise movement of the environment to keep a clear image on the retina. (c) If the eyes are closed, the same movement of the head provokes similar reflex eye movements through stimulation of the vestibular system. Arrows in canals show displacement of fluid and cupula relative to the head. Heavy arrows show pathways of increased frequencies of impulses (see Fig. 8-3) that result in eye movements.

and preventing a blur on the retina. This type of alternating movement of the eyes is called *nystagmus*. Nystagmus will also occur when the head, rather than the environment, is moving.

Referring to Fig. 8-4, we may note that during counterclockwise rotation of the head the environment would appear to be moving clockwise. (*Clockwise* here refers to the direction that you would sense the movement of a clock's hands if you were standing at the center of the clock face). The eyes would move clockwise to keep the gaze fixed on a particular point. As that point passes out of the field of vision, the eyes snap quickly back to fixate on another point, which is followed in turn. These movements are quite logically arranged to provide a fixed point of visual reference in an environment that appears to move in relation to the head.

Remarkably enough, the same movements occur when the eyes are closed and the moving environment cannot be seen. The receptors of the semicircular canal are thus reflexly locked to the external muscles of the eye, and the eyes will be made to move from side to side, up and down, or even in circles, depending on which semicircular canals are stimulated. These movements occur whether the stimulation is by movement of the head or by means of a temperature change. Warm or cool water applied close to the inner ear may set up convection currents of movement in the fluid.

Central connections of vestibular system

The afferent fibers from the vestibular system of the inner ear evidently make strong connections with the nuclei of the *extrinsic muscles* of the eye. These connections are not made directly but through intermediate neurons. The primary afferent fibers are from bipolar cells of the *vestibular nerve*. The peripheral fibers (the dendrites) come from the vestibular receptors; the centrally directed fibers (the axons) enter the *vestibular nucleus*, which is an important postural control center in the medulla oblongata of the brainstem. The cell bodies that constitute this nucleus send axons rostrally in the brainstem, to synapse with the cell bodies of the motor nuclei controlling the eye muscles and thus establish reflex control of these muscles. The vestibular muscles also send fibers caudally into the spinal cord, to make connections with the motoneurons innervating the postural muscles. Impulses along the vestibulospinal tracts provide information to the motoneurons about the position or movement of the head. This information is used to adjust the frequency of impulses to particular muscles to match the circumstances of body position at the moment. The impulses act on the motoneurons, along with other impulses, particularly from proprioceptors of the muscle innervated by the motoneuron, as well as from related muscles.

Hearing: The perception of sound

CHARACTERISTICS OF SOUND. The word "sound" refers to the sensation provided by the pulsation or oscillation of a material body when that pulsation is transmitted to the cochlea and perceived by the brain. The vibrations themselves are also called *sound.* (Pure tones are heard when the vibration occurs regularly at a definite *frequency.*) Most sounds result from mixtures of vibrations of several different frequencies. When the oscillations are of many different frequencies, mixed together, and not in regular sequences, the result is noise. In general, the frequency of a vibration is sensed as *pitch.* The "low deep" notes in a musical instrument are low in frequency. The "high shrill" notes are high in frequency.

A record of several vibrations may be obtained by allowing a tuning fork to write on a moving smoked drum. The vibrations set the air in motion, and waves of displacement of the air move away from the source in all directions (Fig. 8-5). The most familiar waves are the ripples that move outward when a pebble is dropped into placid water. In these surface waves, the water moves up and down, as the waves move

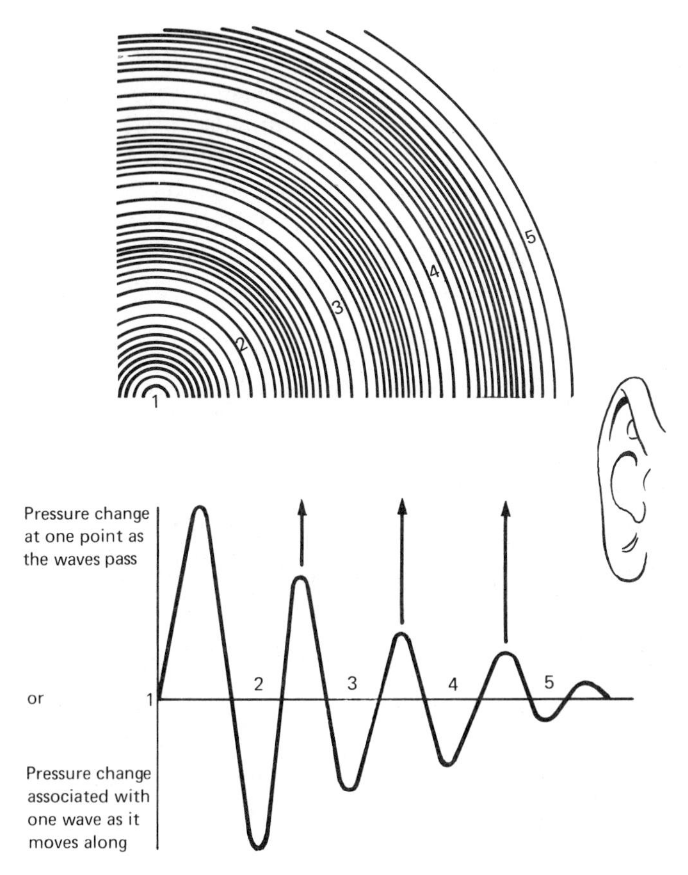

Fig. 8-5. Sound waves. The density of circles represents alternate compression and thinning of the air that occur when the object at the center vibrates. The numbers represent the movement in time of these pressure waves away from the sound source. The damped sine wave represents the pressure changes in the sound wave.

horizontally along the water. Movement of these waves in a certain direction through a material does not involve any significant net movement of the material in that direction. The material merely undergoes a back-and-forth motion. As it does so, it transmits a back-and-forth movement to the nearby molecules of the material. The energy of the vibration in the tuning fork is concentrated in a small volume. The molecules of air near at hand give a momentary kick to the molecules surrounding them; these, in turn, agitate those still farther out. If there were no friction, all the energy of the original vibration would be divided among more and more molecules. Thus the energy in the original small volume is spread out in all directions, and the intensity of sound decreases with distance from the source of the sound. In addition, some of the sound energy is lost as heat by the friction developed when the molecules collide with one another during their movements. The rapidity with which sound energy is lost as heat depends on the material that is vibrating.

RECEPTORS FOR SOUND. A ventral extension of the labyrinth is sensitive to vibration carried to it. This structure is merely a small outpocketing in lower vertebrates, but it becomes a prominent, coiled structure, the *cochlea*, in higher animals, which are sensitive to a great range of sound frequencies (Fig. 8-6). The *cochlear duct* of the membranous labyrinth divides the bony cochlea into two chambers. Each chamber is imaginatively referred to as a *scala* or staircase. The *scala vestibuli* begins close to the vestibule. It ascends to the apex of the coil, where it is continuous with the

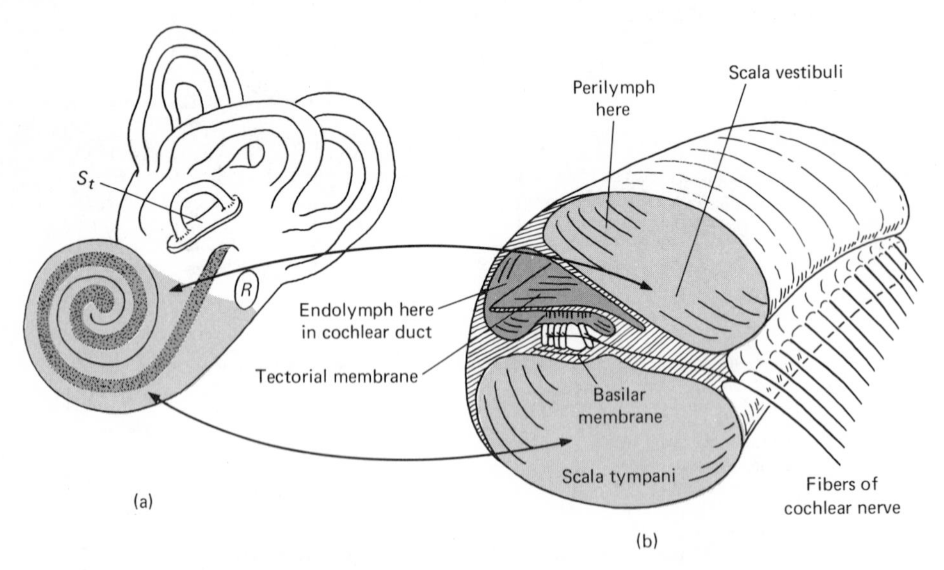

Fig. 8-6. Cochlea. (a) Outline of bony labyrinth within which is contained the membranous labyrinth. Cochlea is shown continuous with the vestibular system. Gray membranous cochlea duct. Light gray: space separating cochlea duct from its bony container.

scala tympani. At its lower end, the scala tympani terminates at the *round window,* a thin but tough membrane separating the scala from the middle ear cavity. The scala vestibuli and the scala tympani wind parallel to one another on opposite sides of the cochlear duct.

EFFECTS OF SOUND ON THE EAR. Movement of the cochlear fluid can occur from vibrations of the surrounding bone—for example, if a vibrating object is placed on the skull. Ordinarily, however, the sound waves are conducted from the air to the cochlea by means of the external and middle ear apparatus.

Sound waves that enter the external auditory canal impinge first on the *tympanic* membrane (Figs. 8-1 and 8-6). On the inner surface of this membrane is attached the first of the three inner ear bones, the *malleus* (hammer), which in turn articulates with the *incus* (anvil). The innermost bone is the *stapes* (stirrup), which transmits the vibrations to the *oval window.* Each of the two chambers of the bony coil of the cochlea opens into the cavity of the middle ear. Each opening is covered by a thin membrane called a *window.* The scala vestibuli begins at the oval window, into which is set the base of the stapes, which pumps back and forth during vibration transmitted to it. Water is virtually incompressible, and the membranous round window moves outward every time the foot of the stirrup pushes the oval window inward. The round window, at the base of the scala tympani, looks out to the middle ear cavity, an air-filled chamber into which the pressure oscillations in the cochlear fluid are dissipated. The sound receptor mechanism is contained within the cochlear duct. The receptor cells rest on the *basilar membrane,* and their apices are imbedded in the *tectorial membrane* within the duct. Both membranes extend the entire length, and from one side to the other, of the cochlear duct. Vibrations of the fluid (the endolymph within the cochlear duct, the perilymph in the surrounding bony cavity) produce a relative movement of the basilar membrane with respect to the tectorial membrane. The apices of the hair cells are bent, and the distortion is translated into nerve impulses in the fibers of the *cochlear branch* of the *auditory nerve.* Small movements of the stapes will provide only small oscillation of the fluid in the cochlea, small displacement of the basilar with respect to the tectorial membrane, and small distortion of the hair cells. Only a few impulses will be generated in the nerve fibers at each cycle of the vibration, and these impulses will be sensed as a weak sound if heard at all.

A higher-vibration *intensity* (greater vibration amplitude) causes more impulses at *each cycle,* and a louder sound is heard. The increase in number of impulses at each cycle involves a higher frequency of impulses in each participating nerve fiber and a greater number of active nerve fibers.

PERCEPTION OF PITCH OF SOUND. As the frequency of vibration is increased, the groups of impulses come more frequently. The change in sound frequency is then sensed as a change in *pitch.* The impulses cannot possibly be produced any faster than 1000 per second, for nerves cannot carry impulses any more often than 1000 per second. Yet we are able to distinguish vibrations 10 to 20 times this frequency. How can we distinguish frequencies of sound that are greater than the frequencies to which the nerves can respond? The answer lies in the manner in which the cochlea responds to vibrations (Fig. 8-7).

If an injury occurs to a restricted area along the cochlea, then the perception of sound near a particular frequency may be profoundly impaired. If the injury is at the *base* of the cochlea, *high*-frequency sensitivity near 10,000 Hz may be diminished. An injury near the *apex* of the cochlea is accompanied by loss of hearing for *low*-frequency sounds of 10 to 100 per second. In an intermediate region of the cochlea, intermediate frequency sensitivity is lost. In general, a sound appears to be of low or high pitch, depending on which area of the cochlea is stimulated or, succinctly, "pitch is which."

The basilar membrane is *narrow at the base* and *wide at the apex* of the cochlea. It was once thought that the fibers of this membrane vibrate sympathetically to the impinging vibrations, as might the strings of a piano. Unfortunately for that theory, the membrane is *not* under tension and the fibers cannot act in this way. The motion

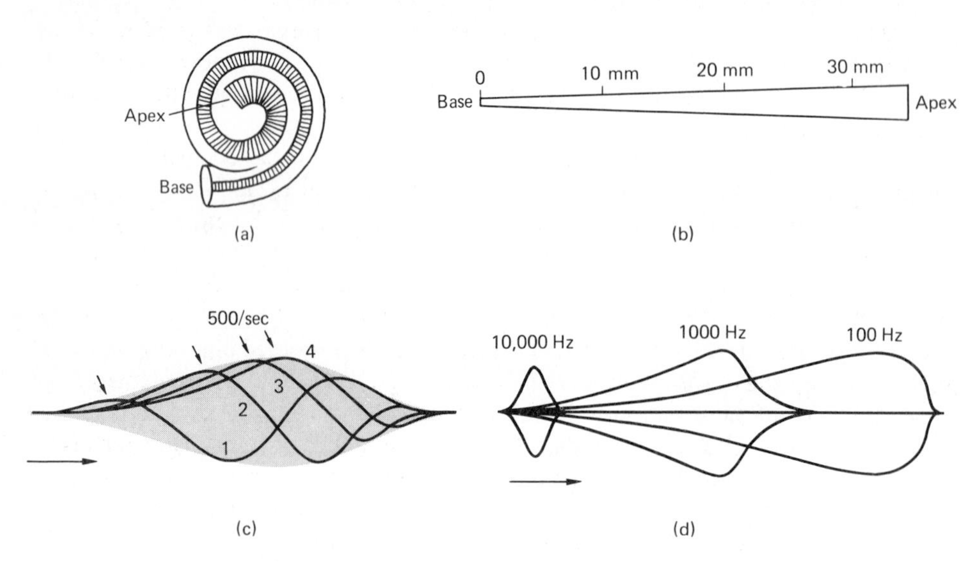

Fig. 8-7. Travelling waves in the basilar membrane. (a) The fibers of the basilar membrane within the cochlea increase in length from base to apex. Only about half the coil is shown. For full coil see Fig. 5-2. (b) Represents the full length of the basilar membrane rolled out flat. (c) The displacement of the basilar membrane produced by a sound wave in the cochlear duct. The amount of motion is exaggerated with respect to the length of the membrane. The numbered lines show the displacement of the membrane along its length during three successive equal time intervals. The small arrows show the successive positions of the maximum displacement upward. The shaded area shows the overall extent of membrane brought into motion for a sound frequency of 500/sec. (d) Overall displacement of the basilar membrane by waves generated in the membrane by three different frequencies of sound.

imparted to the cochlear fluid during each oscillation of the stapes produces a *traveling wave* of displacement of the basilar membrane. This wave starts at the base of the cochlea and travels toward the apex (Fig. 8-7).

The traveling wave increases in amplitude as it moves away from the oval window; then, after reaching a maximum, it declines. The place at which this maximum displacement occurs along the basilar membrane is near the oval window for sounds of high frequency and near the cochlear apex for sounds of low frequency. For some frequencies, especially of loud sounds, almost the entire basilar membrane may be displaced. There will always be one place, however, that will be displaced slightly more than any other, and the central nervous system has ways of emphasizing small differences so that they can be recognized. Thus frequencies of sound are distinguished according to the place of maximum displacement of the basilar membrane during the traveling wave.

Two characteristics of the basilar membrane determine the different places at which different frequencies of sound will be effective. First, high-frequency waves are damped more rapidly along the basilar membrane than are waves of low frequency. Second, the membrane, being wider at the apex than at the base, can be more easily displaced near the apex. As the wave progresses, it moves into parts of the membrane that can more readily be displaced, but at the same time the energy available to do the displacing diminishes. For a particular frequency, somewhere along the route a combination of remaining sound intensity and membrane elasticity provides maximum displacement of the membrane for that frequency.

For sounds of equal intensity, the combination of remaining intensity and membrane elasticity at one particular part of the membrane provides a greater effect for one particular frequency than for any other frequency at any other point. This is the frequency that can be heard at lower intensity than any other frequency.

Intensity-frequency sensitivity of the ear

The sensitivity of the ear may be described by means of an *audiogram*, which is merely a plot of the intensity-frequency combinations to which the ear is sensitive (Fig. 8-8). To obtain this information, each frequency of sound that is presented to the ear by means of a sound generator is adjusted in intensity so that it is barely heard. This is a *threshold* intensity. The least intensity is required for a frequency near 2000 Hz, while greater intensities are required for greater and for lesser frequencies. If a sound is intense enough, it may be felt as a painful sensation. The top curve in the audiogram shows frequency-intensity combinations that are just sufficient to produce pain. Any frequency-intensity combination within these curves can be sensed as sound.

Notice that both scales of the graph are plotted so that each interval is a multiple or power of 10. This is a *logarithmic plot*. On the horizontal axis (abscissa) is shown the *frequency*, in Hertz (Hz)—i.e., cycles per second. The ordinate indicates the *intensity* of the sound. The range of frequencies over which we can detect sound is rather

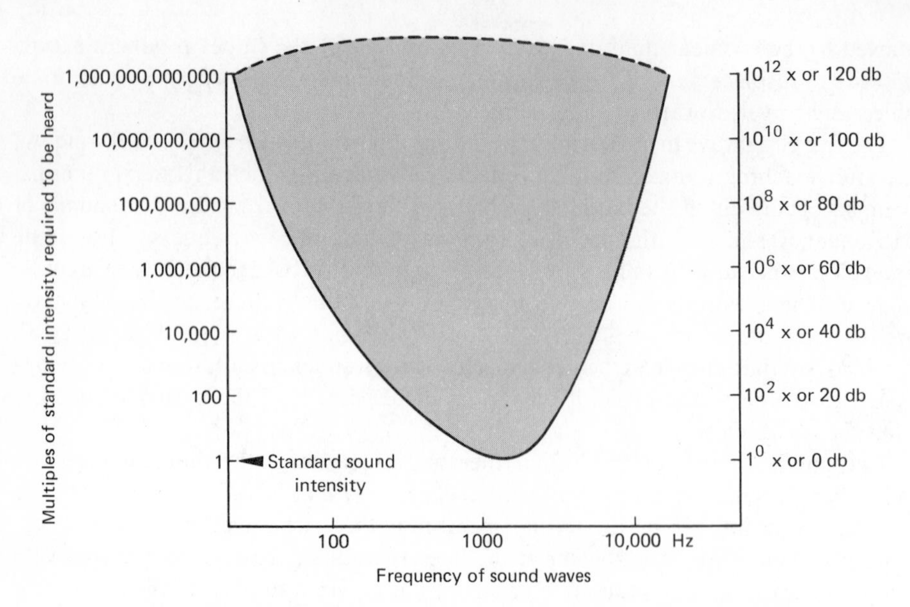

Fig. 8-8. The sensitivity of the ear to sound. The ordinate shows the relative amount of sound energy made available to the ear. The abscissa shows the frequencies of the sound waves that can be heard. The solid line shows the relative amount of energy necessary for any particular frequency of sound to be just heard. The threshold is lowest at about 2000 Hertz. Any point in the shaded space is a combination of energy and frequency that can be heard. Any point outside cannot be sensed as sound. On the right, the sound intensity is shown in terms of multiples of 10 and in terms of decibels.

large, about ten thousandfold, although this range is extended ten times farther on the higher-frequency side in bats and porpoises. The range of intensities over which we perceive vibrations as sound can only be described as enormous by comparison—about a thousand billionfold. If the information in the audiogram is plotted arithmetically, the details of the curve do not show up well at all. The double logarithmic plot shows the sensitivity of the ear in its more, as well as its less, sensitive range. The very large numbers required to compare sound intensities can be made less unwieldly if expressed as powers of 10.

A convenient and simple way to measure the intensity of a sound is to compare it with a standard sound. A particular sound (I), for example, may be ten times or 10^7 times as intense as a given standard (I_0). That is,

$$\frac{I}{I_0} = 10^1, \text{ or } 10^7, \text{ or whatever.}$$

The intensities over a great range are more easily compared by describing them in terms of multiples of 10, which we can do if we take logarithms of both sides of the equation. That is,

$$\log_{10} \frac{I}{I_0} = 1, \text{ or } 7, \text{ or whatever.}$$

This measure is called a *bel.* That is,

$$\log_{10} \frac{I}{I_0} = x \text{ bels.}$$

As an example, let

$$\begin{aligned} x = \log_{10} \frac{500}{I} &= \log_{10} (5 \times 100) \\ &= \log_{10} 5 + \log_{10} 100 \\ &= 0.6990 + 2 \cong 2.7 \text{ bels.} \end{aligned}$$

It turns out to be more convenient to use the *decibel,* defined thus

$$1 \text{ bel} = 10 \text{ decibels} = 10 \text{ db.}$$

In the foregoing example,

$$x = 2.7 \times 10 = 27 \text{ db.}$$

In the audiogram, the intensity of a standard sound is shown by an arrow. It has been chosen to be equal to the sound intensity required for threshold perception of the frequency most easily heard.

Impulses may be recorded from single fibers in the cochlea nerve trunk, when the ear is exposed to sound waves. If the intensity at each frequency is made just sufficient to produce impulses in the nerve (threshold), then an audiogram may be constructed for each such single fiber. Each fiber covers a limited frequency-intensity range, and for each such single fiber there is a particular frequency that is the most effective stimulus. Other frequencies can also stimulate, but at higher intensity. Each fiber presumably innervates the hair cells of a very small part of the cochlea, and this restricted distribution is reflected in the local sensitivity of the basilar membrane to particular frequencies.

Central connections of auditory receptors

The ear is subject to an enormous range of sound intensity. From a faint whisper to a shattering roar of thunder, a millionfold difference in pressure (force per area) is applied to the eardrum. Excessive vibration may injure the ear, but some protection is afforded by a reflex that reduces the effectiveness of loud sounds on the ear. The stapes is supplied with a muscle that contracts, decreasing the movement of that bone, when the sound becomes very intense (Fig. 8-9). This is an example of *feedback*. The central nervous system feeds back to the receptor impulses that control the sensitivity of the receptor. The frequency of impulses fed back to the periphery is determined by the frequency of impulses coming into the CNS from the receptor.

The information from the cochlea travels in the *cochlear branch* of the *auditory nerve* to the cochlear nucleus in the medulla (Fig. 8-10). From here impulses travel both on the same and on the opposite side of the brainstem, to a part of the thalamus

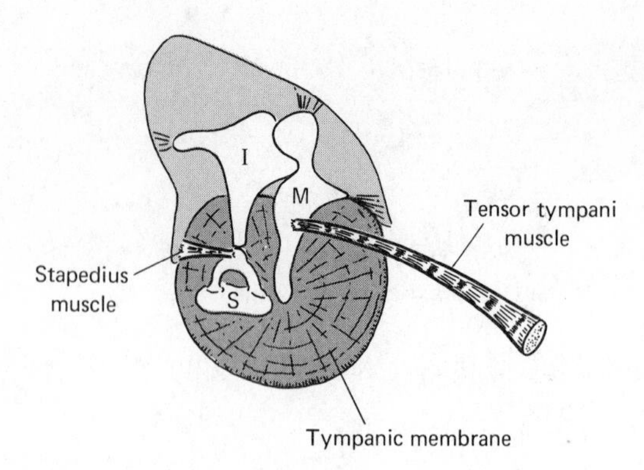

Fig. 8-9. Muscles of auditory ossicles. (View toward inner surface of tympanic membrane.)

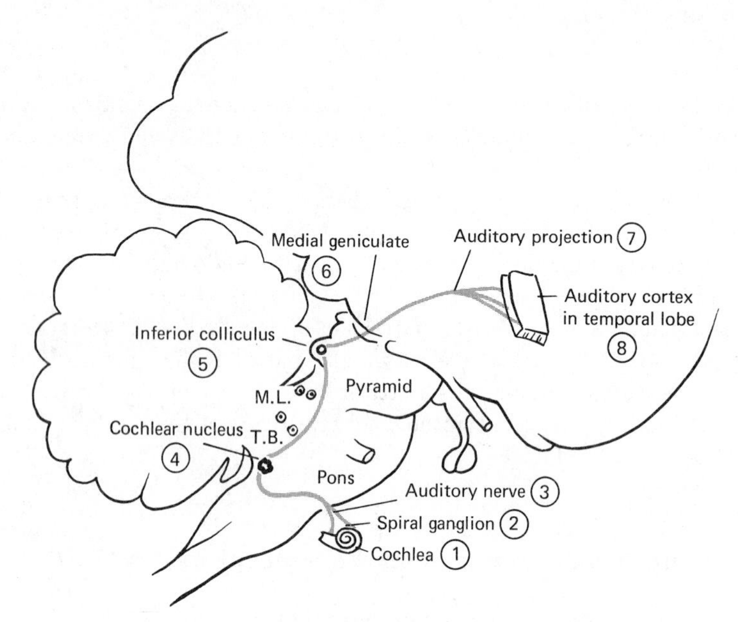

Fig. 8-10. Auditory pathways. Projection of cochlea to the temporal lobe of cerebral cortex. *T.B.: Trapezoid body* where 2° order fibers from cochlear nucleus cross over to the other side of the brain, decussate, and become *Medial lemniscus* (*M.L.*). These fibers ascend to the inferior colliculus. Compare diagram to Fig. 7-2(a), (b) for orientation.

(*lateral geniculate*) and then to the *temporal lobe* of the cerebral cortex. Here the various sound frequencies are serially projected in the way they started out at the cochlea. Perception of pitch depends at least partly on the place to which the impulses signaling a particular frequency of sound arrive at the cortex.

8.2 Vision

Structure of the eye

Light consists of tiny packets of energy (photons) traveling at high velocity in straight lines from their origins. These packets have sufficient energy to bring about changes in some particular molecules of protoplasm that they may hit. In the eye, the photons are absorbed by special pigments of the retina. The energy they provide is used by the light receptors to start nerve impulses that travel from the retina into the central nervous system.

The eye is the most complex of all sense organs. Its complexity is matched by its importance to the body. Information received by the eyes can partly substitute for considerable deficits in many other receptor systems. In man, more area of cerebral cortex is devoted to the eye than to any other sense organ. The information reaching the cerebral cortex has already undergone considerable processing in the retina, which is really CNS tissue in its origin and in its complexity.

Embryonically, the eye begins as an outpocketing from the hollow brain tube or neural canal (Fig. 8-11). This evagination turns in upon itself to form a sort of double-walled cup that is attached to the brain by means of a thin stem. The stem becomes the optic "nerve," more accurately described as the *optic tract*, since it is actually a bundle of fibers between parts of central nervous tissue. In its origin, the deep surface of the lining of this cup is continuous with the inner surface of the brain. The cells lining this inner surface form a thin epithelium (*ependyma*) along the cerebrospinal

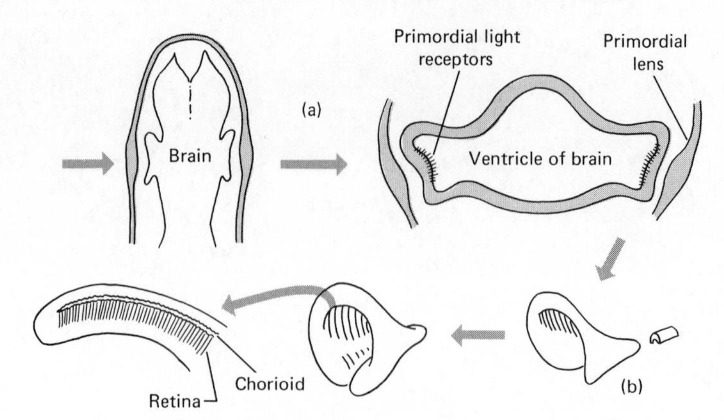

Fig. 8-11. The development of the eye from the central nervous system. (a) At an early embryonic stage, lateral outpocketings develop from the brain, while the overlying ectoderm thickens, and eventually becomes the lens. (b) Meantime, the optic cup becomes folded inward, in a special way. The inner layer becomes the retina, while the outer layer becomes a pigmented layer (the choroid) over the surface of the retina (see Fig. 8-12).

canal of the brain and spinal cord, but in the eye they become transformed into the *light receptors.* The receptor layer is a part of the *retina* that lines the cavity of the eyeball. The rest of the retina consists of nerve cells and fibers organized like the central nervous system, of which it is a part. On the surface of the retina, directed *away* from the center of the eyeball, the receptors project forth like the pile of a rug (Fig. 8-12). The surface of the retina directed toward the center of the eyeball is a layer of nerve fibers that turn and pass through and out of the retina at one spot and continue to the brain as the *optic tract.* There are no receptors at this region, which is, therefore, a *blind spot* in the retina.

The rest of the eye is an optical apparatus for focusing the light rays from the outside world onto the light receptors. We may see many miles of the world before our eyes; thus this optical apparatus must compress a large area onto a receptor surface of only about a square inch. At the same time, the pattern of light in the object seen must be retained in the image cast on the retina. The lens system of the eye (*cornea* plus *crystalline lens*) bends the light rays into a smaller pattern. The cornea, a transparent part of the eye, does most of the bending of the light rays, to project the extensive visual field upon the small area of the retina surface. The crystalline lens is responsible for small adjustments that change the focus as required for close or distant objects.

Optical features of the eye

How the cornea, like any lens, can bend light rays may be understood by considering the successive positions of a photon of light energy moving from air into

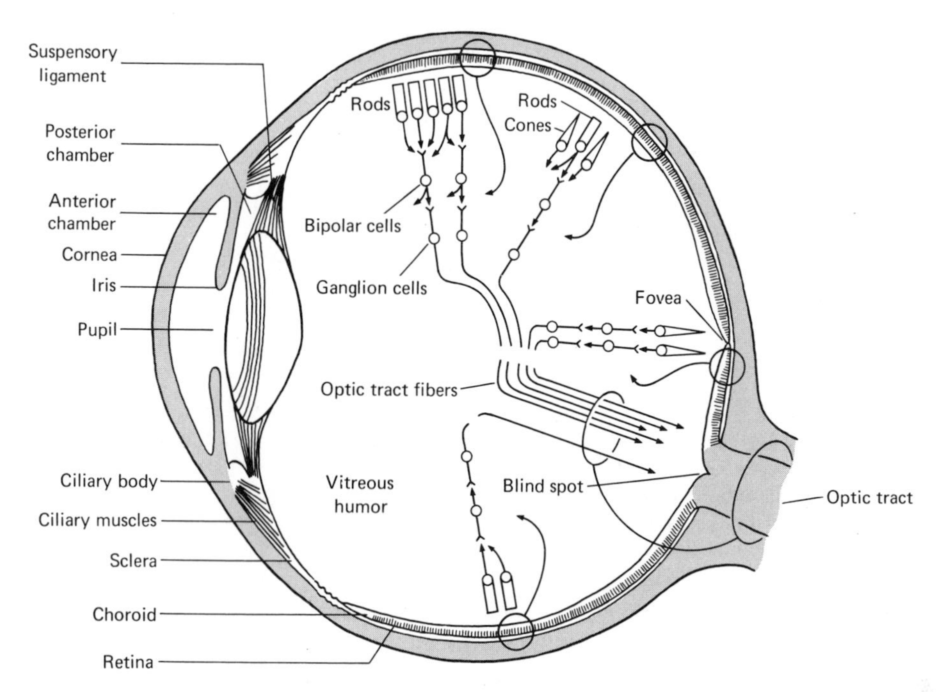

Fig. 8-12. Anatomy of the eye. The eye shown in vertical section. The largest space within the eyeball is filled with the jelly-like vitreous humor. In the figure this space is occupied by enlargements of the retina. Note layer of optic tract fibers covers entire inner surface of retina. The light must pass through several layers of nerve fibers and cells before it reaches the distal segments of the rods and cones where energy from absorbed light generates nerve impulses.

the denser substance of the cornea (Fig. 8-13). The velocity of the light rays is decreased when the rays pass from air into another transparent, but more dense, material. The part of the ray that enters first is slowed down, and the ray swings around to follow a new path. The part of the ray that is last to leave the other surface lags behind, and the ray swings around again to follow another path as it leaves the lens.

Any *visible object* may be thought of as consisting of an infinite number of points of light, each point sending rays out in all possible directions. The eye intercepts a small fraction of the rays from each point, much as the earth intercepts a small portion of the light leaving the sun. The light rays are bent by the lens system of the eye to form an *image* on the retina. If the object observed is a point, the image ought also to be a point. If the point observed is a considerable distance (about 20 ft or more) away from the eye, only a small fraction of the total light reaches the eye; the angle between the rays is very small, and the rays are almost parallel. The amount of bending of the light by the lens is indicated by the angle A in Fig. 8-13. In a normal eye looking at a distant object, this angle is such that the point of light from the object is focused on

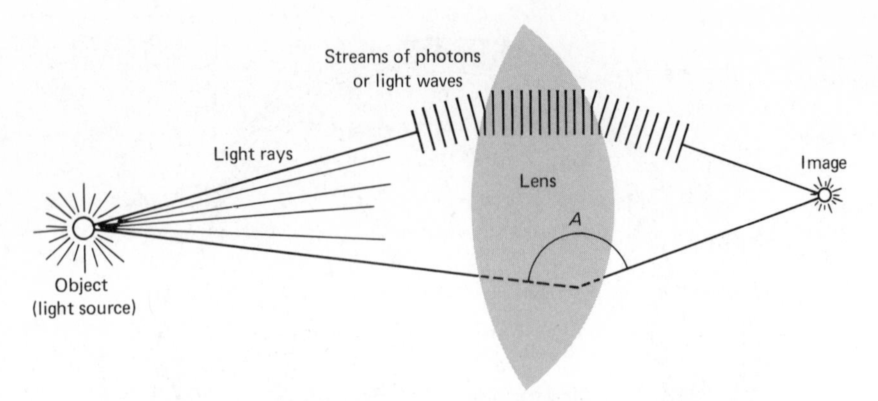

Fig. 8-13. The bending of light rays by a lens. A small fraction of the photons moving away as light rays from a point of light are intercepted by a lens. The photons travel more slowly as they enter the material of the lens. The resulting angle of bending *A* of the light rays depends upon the curvature of the lens on the side the rays enter from the object, as well as the side on which they exit to focus as the image.

the retina as a point image. The rays of light are symmetrically arranged in their course toward the eye. This fact is illustrated by the uniformity of cross sections of the cone of rays. If the lens is of uniform curvature and of homogeneous material, this regularity of distribution will be maintained after the rays are bent, and the rays will come together as a point.

Incident rays that enter corresponding places at the periphery of the lens are equally bent, regardless of the angle at which the rays strike the lens. Figure 8-14 shows the pattern of rays from two points. The rays from the lower part of the object fall on the upper part of the retina. Those from the upper part of the object fall on the lower part of the retina, and the image is oriented upside down (inverted), compared with

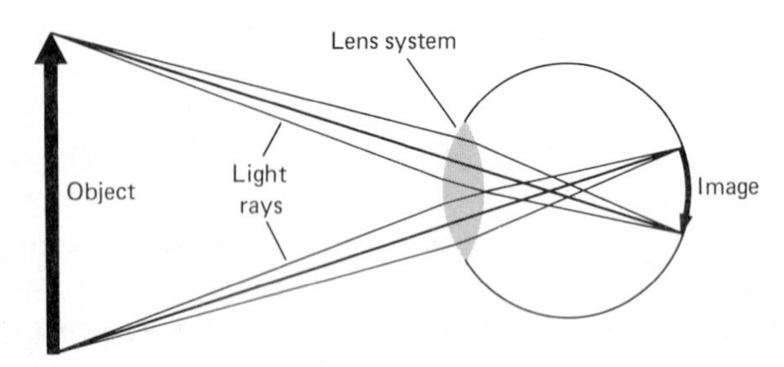

Fig. 8-14. Lens focusing rays from two points of light. The cones of rays from two points of an object approach the lens at different angles, but the rays that strike perpendicular to the lens surface pass straight through the lens. Those striking at any other angle are bent, as shown. Note that the convex lens projects an inverted image on the retina.

the object. Rays striking the lens on a line perpendicular to a tangent of the surface at that point are not bent. Rays not on such a line when they strike the surface tend to be bent toward it, on passing through the lens.

Visual adjustments to distance

An object at a distance will be focused as an image on the retina of a normal, relaxed eye. The crystalline lens in this instance is relatively thin, pulled out by the tension in radially arranged connective tissue fibers of the *suspensory ligament*, which is attached around the rim of the lens (Fig. 8-15). The fibers of the suspensory

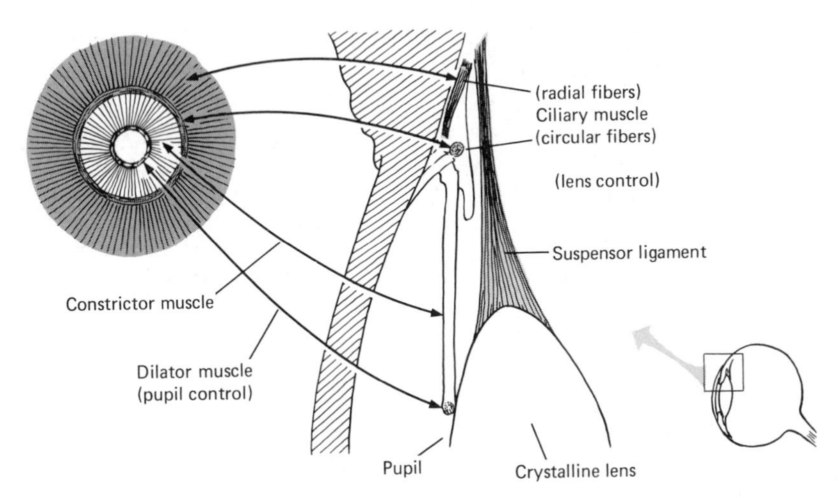

Fig. 8-15. Intrinsic ocular muscles (smooth) controlling crystalline lens and pupil.

ligament extend from the rim of the lens to the *ciliary body*. The ciliary body forms a ringlike thickening on the inner wall of the eyeball, at the junction between sclera and cornea. When a point of light is brought close to the eye, the angle of the cone of rays intercepted by the eye increases (Fig. 8-16). These divergent rays must be bent more than the relatively parallel rays from a more distant point if the point image is to remain focused on the retina. The amount of bending of light rays by the cornea cannot be changed, but adjustments can be made in the thickness of the *crystalline lens*. If the *crystalline lens* is allowed to bulge out, the amount of bending of the light rays will increase.

Paradoxically, although focusing on nearby objects requires a definite muscular effort, the position of *near focus* is one of least tension on the crystalline lens and is the *resting position for the lens*. This curious situation comes about through the existence of special muscles that pull on the ciliary body in such a way as to slacken the fibers

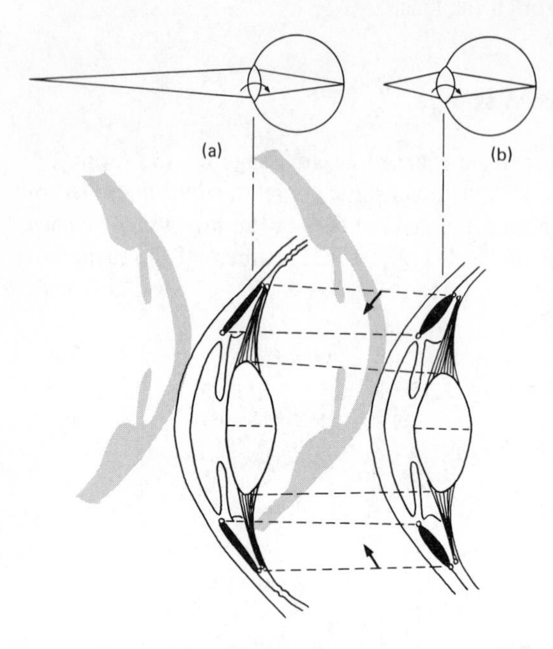

Fig. 8-16. Mechanism of focus adjustment by crystalline lens for objects at different distances. The rays from a near object must be more strongly bent (b) than the rays that come from a distant object (a) if the image is to remain focused on the retina. (a) and (b) below show the corresponding adjustments in the crystalline lens.

Shortening of the muscle fibers is shown by the arrows. (b) relieves tension on the suspensory ligament fibers and the lens is no longer pulled out. As it becomes thicker, the lens bends the light more than it did when it was thinner and flatter (see Fig. 8-12 for key).

of the suspensory ligament. In the absence of tension in these fibers, the elastic crystalline lens is allowed to be restored to the unstretched condition—in which it is relatively thick.

As photons move away from their origin, they diverge like the spokes of a wheel. Therefore fewer photons fall on a given area far from the light than close to the light. Consequently, the total light intensity diminishes with distance from a light source,

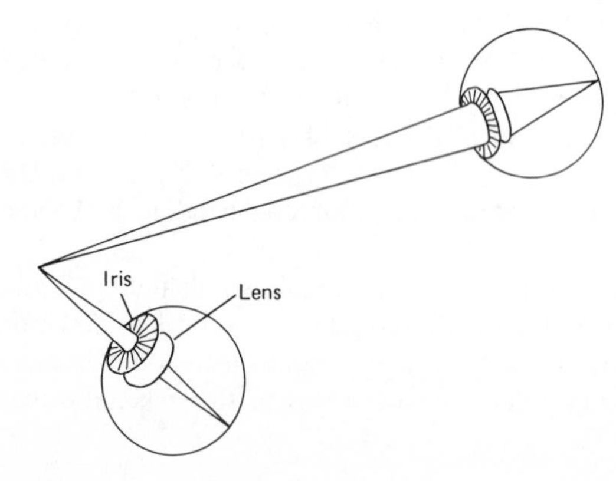

Fig. 8-17. Pupillary adjustment during accommodation for near vision. In addition to lens thickness (Fig. 8-16), the diameter of the iris aperture changes during visual accommodation, being larger when distant objects are viewed.

and the intensity of light is much greater close to the light source than it is far away. If a point of light can be seen at a distance, it can still be seen close at hand even if the total light reaching the retina be reduced. Such a reduction is brought about in the eye by decreasing the size of the pupil when close objects are viewed. In Fig. 8-17 the total light energy reaching the eye from the point of light could be approximately equal in the two instances.

Another advantage of the smaller optical aperture is that the rays are bent more uniformly at the center than at the edge of the lens. This is true because the total amount of bending is least at the center, and the differences in the amount of bending of the different rays are correspondingly small.

Defects of the eye in adjustments to distance

Even though the rays are bent uniformly, a point may appear as a blurred image if the length of the eyeball is inappropriate to the bending power (strength) of the lens system. When an eyeball is abnormally long, relative to the strength of the optical system, a point will be focused in front of, rather that on, the retina. A round spot (rather than a point) then appears on the retina (Fig. 8-18). If the lens cannot be adjusted to bend the rays sufficiently, some other adjustment will have to be made. If the incident rays can be made to diverge a suitable amount, the focal point will move backward (the bending by the lens remaining unchanged) and the image will be focused. This increase in divergence may be accomplished by moving closer to the object or by placing in front of the eye a lens to diverge the rays. The rays will then assume the new position shown by dotted lines in the figure. The abnormal condition described above is *nearsightedness* (*myopia*), so called because near objects are seen more clearly than distant objects [Fig. 8-18(c)].

The opposite abnormality is *farsightedness* (*hyperopia*) [Fig. 8-18(b)]. In this situation, the eyeball is too short, relative to the strength of the lens. The image is focused back of the eyeball, and a round spot appears on the retina. To sharpen the image, the rays must be made to converge toward the eye, so that the amount of bending by the cornea will be sufficient to focus the rays on the retina. Without lenses, this cannot be done. The best a farsighted person can do to improve his vision without lenses is to move away from the object. A hyperope cannot see distant objects more clearly than an emmetrope (a person whose vision is normal), but his own vision is somewhat better for distant than for near objects.

In the defects so far discussed, we have assumed that the image would be clearly focused if the retina could be moved forward or back with respect to the lens. The rays of light are bent uniformly by the lens system of the eye, but they are bent too much or not enough. Suppose that the rays were bent properly in the horizontal plane but insufficiently (or excessively) in the vertical plane. Then a point would focus as a point horizontally, on the retina, but would be smeared out and appear as a line vertically. The failure of the eye to focus a point object as a point image is called *astigmatism* ("condition of not being a point") (Fig. 8-19). Astigmatism is usually due to an irregularity in the cornea. Correction of astigmatism requires a lens that bends

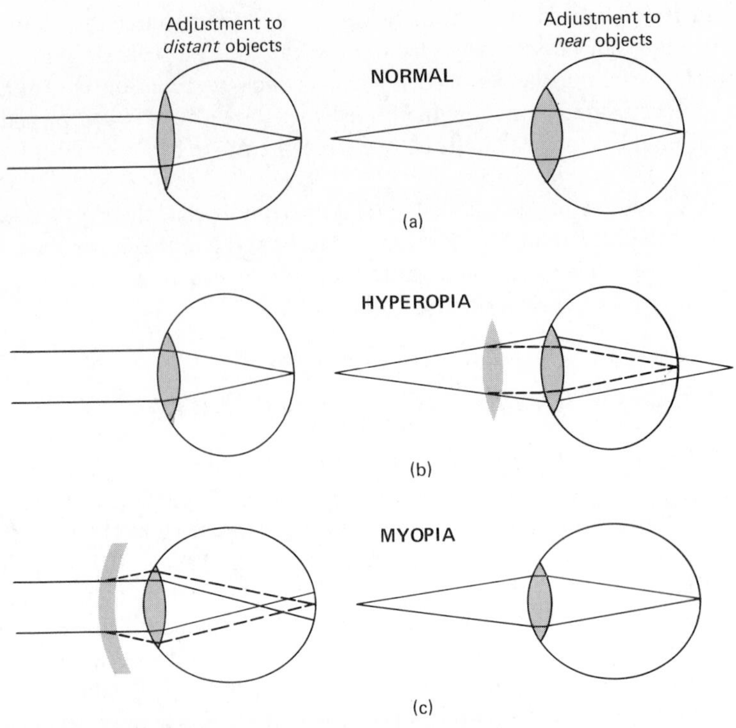

Fig. 8-18. Normal and inadequate visual accommodation.

(a) Virtually parallel rays from a distant object are focused on the retina when the lens is thin and bends the rays minimally. Rays from near objects diverge, but a thickened lens increases bending of light to allow focusing of image on retina.

(b) Lens power is adequate to focus rays from distant objects on retina, but as eyeball is too short relative to lens strength, the rays from near objects are insufficiently bent. A convex lens will bring rays more nearly parallel and improve near vision.

(c) The eyeball is too long relative to the strength of the lens (or the lens is too strong—i.e., it bends the rays excessively) during adjustment to distant objects. The rays may be diverged and therefore focused on the retina if the distance is decreased (hence the condition is nearsightedness). A concave lens will diverge the rays and improve vision for distant objects.

light in one plane more than in another, and such a lens is a portion of a cylinder rather than a sphere.

Binocular vision—use of the eyes together

The lens change and the pupillary change that occur in adjusting focus from distant to close objects are part of *visual accommodation.* A third change involves the simultaneous use of both eyes. When both eyes are used, the image in each

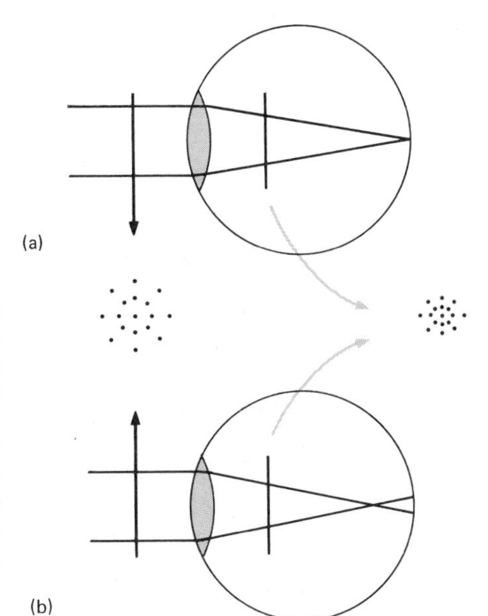

Fig. 8-19. Astigmatism resulting from nonuniform bending of light by lens. When the rays of light are bent uniformly by a lens, both the cone of light entering and the cone of light leaving the lens are round in cross section, and the focus of a point object is a point image. In this illustration of astigmatism, a cross section of the cone of light rays leaving the retina is flattened, and the image is a line rather than a point. (a) and (b) are perpendicular to one another.

eye is slightly different, one from the other, since the eyes are directed toward opposite sides of an object in the visual field. The two images are fused into one in the brain, and the compound image provides the sense of three-dimensional structure of an object. When an object is moved close to the eyes, the total angle of the rays subtended increases. Both eyes move so as to remain fixed on the object; that is, they turn inward (nasally). *Convergence* of the eyes onto a nearby object is, therefore, a binocular (two eyes) aspect of visual accommodation (Fig. 8-20).

Extrinsic Muscles of the Eye. In convergence, corresponding medial muscles of the two eyes contract, but the medial muscle of one eye and the lateral of the other contract during nystagmus in the horizontal plane (Fig. 8-21). The muscles that move the eye are shown in Fig. 8-22. A movement outward from the midline will be brought about by contraction of the *lateral rectus muscle*. Motion away from the midline is *abduction*, and the nerve innervating the lateral rectus muscle takes its name from that fact (*Abduscent*, nerve VI). The *superior oblique* muscle that imparts a rotating motion to the eye is constructed in a curious fashion. Its long, thin tendon passes through a loop as in a pulley, and the name *trochlear* (meaning pulley) is given to its nerve (IV) because of that fact. The other four muscles, *superior*, *inferior*, and *medial rectus*, and *inferior oblique*, are innervated by the oculomotor nerve (III).

Central Fusion of Retinal Image. If the eyes are properly directed by the appropriate adjustment of tone in the extrinsic muscles, the image in one eye will be focused on the retina in a manner functionally complementary to the other. If we have learned how (and most people learn by the age of 4), we can fuse these two im-

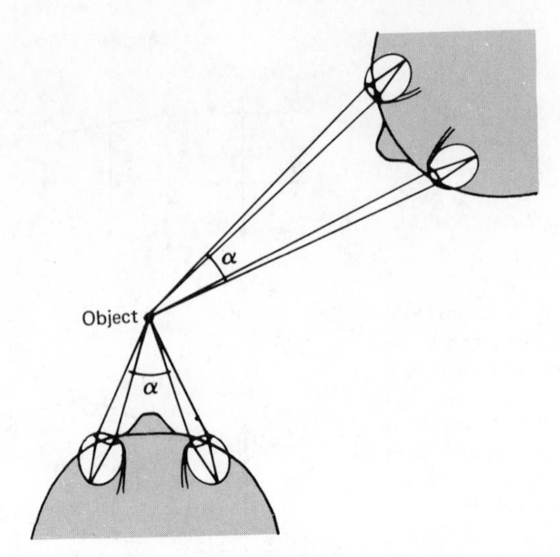

Fig. 8-20. Convergence of eyes during binocular viewing. Angle α is larger when near objects are viewed.

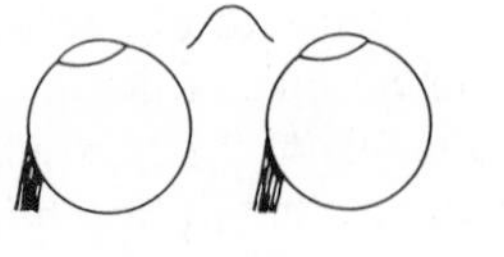

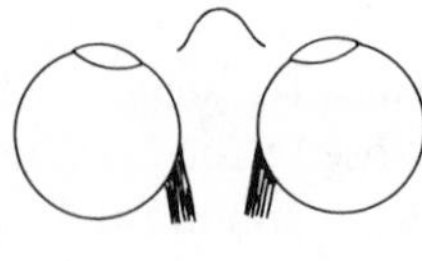

Fig. 8-21. Muscle pairs used in convergence and nystagmus. Tone in particular muscle pairs increases to accomplish the movements illustrated.

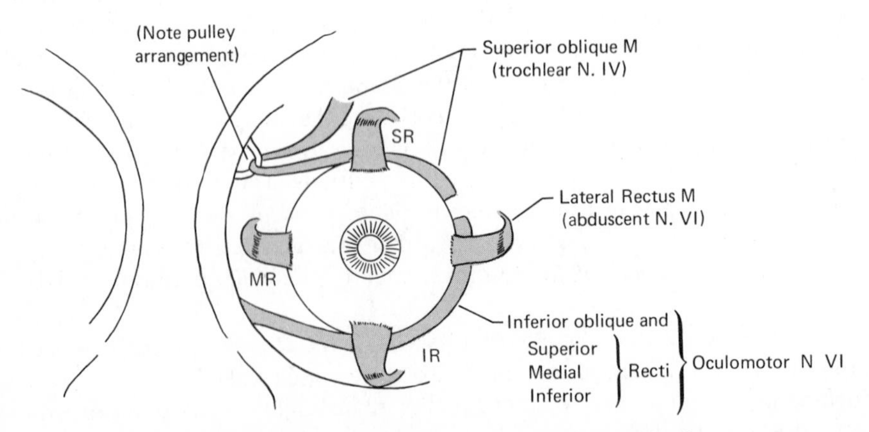

Fig. 8-22. Extrinsic, extraocular muscles (striated, skeletal).

ages into a single three-dimensional picture—we have functional binocular vision. Figure 8-23 shows how an object is focused on each retina. In this example, the gaze is directed toward the circle at the center of the arrow. The arrowhead comes to a focus on the nasal portion of the left retina and on the temporal portion of the right retina. The situation is the opposite for the tail of the arrow. In most animals the

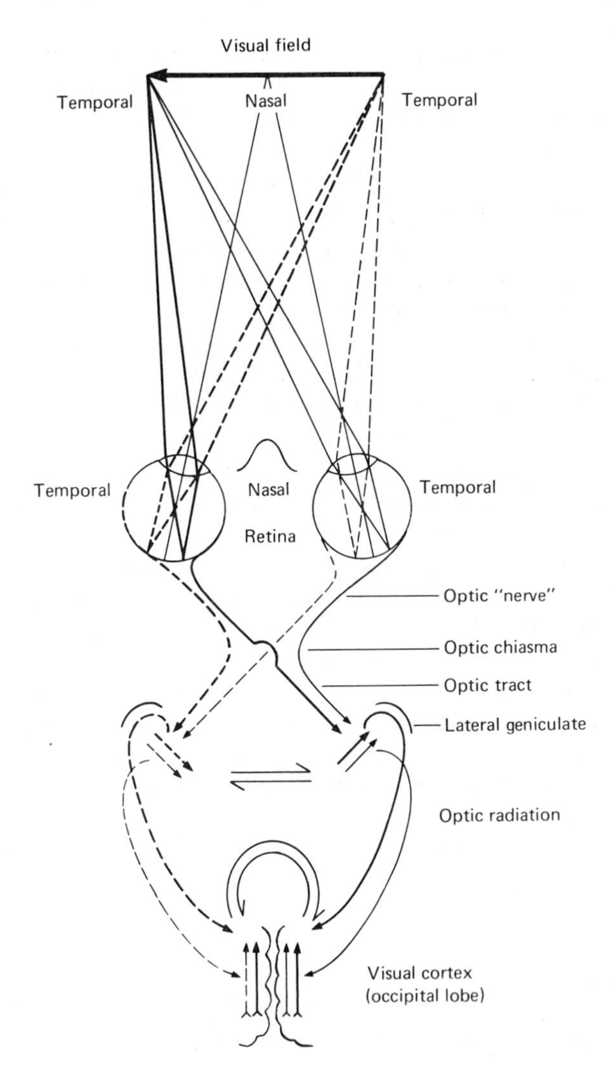

Fig. 8-23. Binocular vision.

fibers of the optic tract cross completely; that is, the fibers from the retina of one eye enter the opposite side of the brain. In the mammals, however, including ourselves, some of the fibers cross and some do not. As the figure shows, the fibers from the nasal part of each retina cross over to opposite sides of the brain. The fibers from the temporal or lateral parts of each retina do not cross over. Those of the left temporal retina go to the left side of the brain; those of the right temporal retina go to the right side of the brain.

It follows that, in this example, the head half of the arrow will be projected to the right-hand side of the brain, the tail half to the left-hand side of the brain. The problem in binocular vision would seem to be to get these two parts of the image assembled into one image.

However, if one eye is blind (or if it is merely shut), half the image is still projected to each half of the brain, as the diagram shows. Certainly there is never any difficulty in joining these two images into one. The brain circuitry is apparently constructed so that the interconnections between the two sides of the brain ensure that projection of the retinal image is whole and complete.

Consider, however, the effect of an optical defect—for example, astigmatism in one eye. The diagram shows that a distorted half-image is projected to each side of the brain. A common consequence of this sort of defect is *amblyopia* or cross-eye. The condition occurs when the image from one eye is distorted and the brain cannot superimpose the images properly. In cross-eye, the defective eye may be excessively turned nasally, and the image is no longer focused on the central fovea, where resolution is best, but instead is focused on the more peripheral part of the retina. This situation reduces still further the definiteness of the image projected to the brain, and attention is paid more exclusively to the clearer image. The unsatisfactory image from the defective eye is said to be *suppressed.* If suppression is inadequate, the two images may be seen separately, and displaced from one another. We say that the image is focused on noncorresponding points in the retina, and therefore it projects to noncorresponding points in the brain. As the diagram shows, the temporal side of one retina corresponds to the nasal side of the other retina, and conversely. *Binocular fusion,* the integration of two images into one percept, must be learned. A defect in one eye may make fusion impossible. If the optical defect is removed by means of corrective lenses, the person may require special training in order to learn how to fuse the images properly.

Light as waves

The streams of photons that are light rays have, like sound, a wavelike character. The waves recur regularly at a particular *frequency*; thus they have a certain duration in time, plus a certain *length* in space, and they travel at a particular *velocity* (Fig. 8-24). These several features are related to one another in a way that holds for both sound and light. If you stand at a spot and watch the waves go by, you may be able to decide that the frequency with which the waves pass by (i.e., how many per second) depends

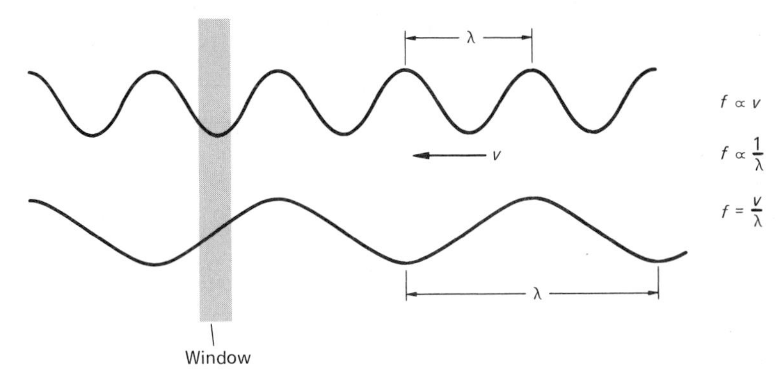

Fig. 8-24. Wave character of light. Relation of velocity, v, wave length, λ, and frequency, f. If the waves are observed through the narrow window, the apparent frequency of waves passing the window depends directly on the velocity and inversely on the wavelength, so we may write $f = v/\lambda$.

on how fast they are moving by (their *velocity*), for the faster they move along, the more you see per second. But the frequency also depends on the *wavelength*, although in an inverse fashion—that is, the greater the wavelength, the less the number passing per second. Since the frequency (f) is directly proportional to velocity (v) and inversely proportional to wavelength (l), we may write $f = v/l$. (In customary symbolism for light, $f = \nu$, $v = c$, $l = \lambda$, so $\nu = c/\lambda$.) Although the *wavelength* has been the most commonly used descriptor of light, the *frequency* is a more natural basis for comparison, for the energy in the photons varies directly as their frequencies, and energy in the photons allows the light to be seen.

Action of light on the eye

Light rays are only a small part of a tremendous spectrum of radiation extending, on the one extreme, from radio waves having relatively long wavelengths, low frequencies, and low energy, to rays such as x-rays at the other extreme, having short wavelengths, high frequencies, and high energy. From a biological standpoint, the special feature about photons of the visible part of the spectrum is that the energy they carry is suitable for bringing about useful chemical changes in living things. In the eye, the energy of the photons is used to generate nerve impulses in the retina. An image focused on the retina consists of a pattern of different light intensities. An object is seen when this pattern of light is translated into a pattern of impulses that travel from the retina to the cerebral cortex. Excitation of the light-sensitive cells depends on the absorption of the light by a visual pigment in the rods and cones.

Nine-tenths of the light energy that strikes the cornea is absorbed by structures in the eye, other than visual pigment. The energy is dissipated as heat or is reflected back out of the eye. Only the remaining one-tenth of the light energy striking the cornea is actually absorbed by the visual pigment and is therefore useful in stimulating the retina.

The effect of light on the visual pigment may be seen in the exposed retina of an animal that has been in the dark for half an hour or more. Such a retina has a reddish color that fades when the retina is exposed to light. The retina of an animal adapted to the light is pale. Evidently the retina contains a pigment that is bleached by light. This pigment is *visual purple*, located in the terminal parts of the receptor cells, specifically in the *rods*. The visual purple molecule undergoes a change in structure and in color when it absorbs a photon. Pigment extracted from the retina will also become bleached in the light.

Dark adaptation of the retina: Duplexity of visual function

All of us have noticed the progressive change that occurs in sensitivity of the eye to light after we have moved from a brightly lighted to a darkened room. Entering a darkened theater, we stumble over the spectators, but after a few minutes we can distinguish figures in the dim light.

This change can be studied more precisely in the following way: An observer looks at a strong light for an arbitrary length of time. When the light is turned off, the observer is in complete darkness. A weak light is then flashed and is increased in

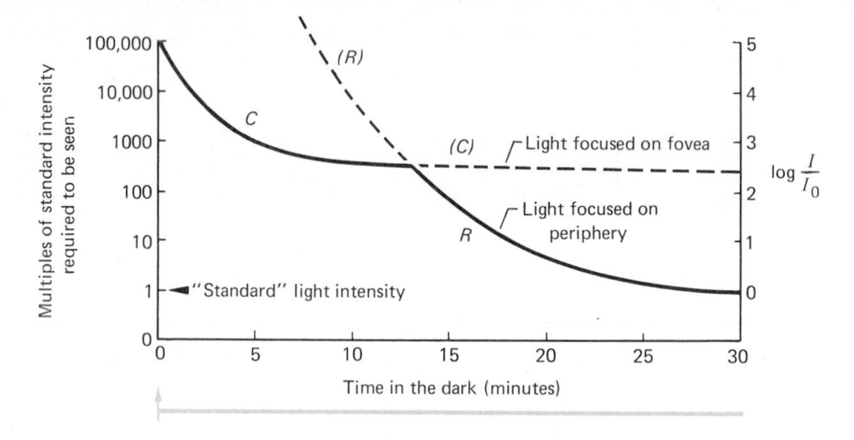

Fig. 8-25. Dark-adaptation curve of the eye (white light sensitivity). At time 0, a strong light is turned off. At first, the subject in darkness can see only a rather strong flash of light and cannot see a weak light (vertical axis of the graph). As time passes, the intensity required becomes less The dotted line shows the least intensity that can be seen when the light is focused on the cones of the fovea. Light focused on the periphery of the retina provides a double curve. The first part corresponds to the cones of the periphery, the second (solid line) is the curve of recovery of rod function. Note that the rods are nearly 1000 times as sensitive as the cones (see also Fig. 8-27).

intensity until the observer can just see it. A few moments later another flash is presented. A record is kept of the threshold intensities of light as a function of duration of time the subject is in the dark. If the test flash of light has been directed to the periphery of the retina, curve *A* of Fig. 8-25 results. If the light has been directed onto the *fovea* at the back of the eye (see Fig. 8-12), curve *B* is obtained. These curves show two sensitivity ranges in the eye. Evidently the fovea is less sensitive than the periphery. The difference can be attributed to the rod and the cone systems of the retina.

This experiment is only one of several showing that the eye is organized into two functional systems that operate under different light conditions. Perception of light and shade, vision under low illumination, and the seeing of objects in relatively coarse detail, all depend on the *rods*, which are distributed peripherally in the retina (see Fig. 8-12). The discrimination of color, vision under high light intensities, and the seeing of objects in fine detail depend on the *cones*, which are concentrated in the fovea from which rods are absent.

Sensitivity of the retina to different wavelengths of light

The series of wavelengths of which ordinary white light is composed can be separated by letting the light pass through a prism. The rays of light will be bent, just as they are by a lens, and for the same reason.

Although the velocity of light in free space is the same for all wavelengths of light, the velocity with which light travels through transparent materials depends on

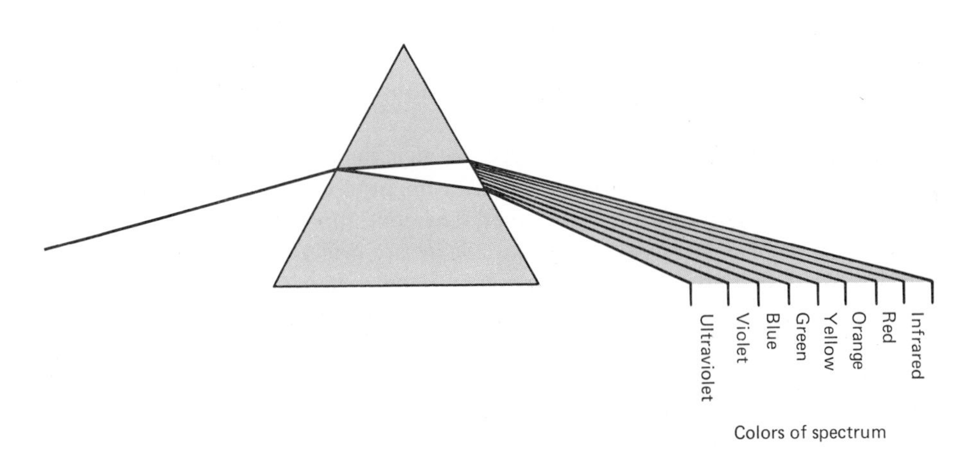

Fig. 8-26. Separation of the wavelengths of the visible light spectrum. Light of shorter wavelength (and higher frequency) is slowed more than light of longer wavelength, when it enters a denser (but transparent) medium. The different wavelengths are thus separated. The cone system of the retina permits us to sense these wavelengths as different colors.

the wavelength. In passing through the prism, the shorter wavelengths will be slowed down more and thus will be bent more than the longer wavelengths. Consequently, the light will be dispersed, with the rays falling in a series of increasing wavelengths called a *spectrum* (Fig. 8-26). The series is a smooth one with no gaps, but we see them as bands of color (red, orange, yellow, green, blue, violet) grading into one another.

When only the *rods* are functioning, or when color vision is absent, such a series may appear as a band of varying intensities of gray. The brightest light would appear to come from the wavelengths corresponding to the blue-green color. This is the wavelength to which the eye is most sensitive.

The photons of light have characteristics of wave motion, and the energy of a photon can be related to its vibration frequency and wavelength. The energy of the photons is least for the long wavelengths, greatest for the short wavelengths. If beams of each color are adjusted to equal intensity, then the peak of sensitivity is still in the blue-green region.

The visual purple molecule in the rods of the retina is of such nature that the change from a structure that is strongly colored to one that is pale is brought about by the least amount of energy when photons of blue-green light are absorbed.

The light-absorption curve of visual purple and the spectral sensitivity curve of the eye are very much alike. It has therefore been assumed that the energy of the photons is used to trigger production of nerve impulses during the bleaching of visual purple. If visual purple has this function, there must be a great reservoir of extra pigment, for the amount bleached by a strong light in the functioning human eye is a barely perceptible fraction of the total pigment present in the retina.

The duplex aspect of the retina is dramatically illustrated by means of sensitivity curves of fovea and of periphery of the eye when different wavelengths of light are used in the completely dark-adapted retina (Fig. 8-27). These curve may be compared with the audiogram. In vision, it is customary to plot excitability against wavelength, rather than threshold against frequency. The curve shows that maximum sensitivity of the cones is at slightly longer wavelength than for the rods and that the sensitivity to light is far lower for cones.

The duplex character of the eye depends on differences in the rods and cones themselves and also on differences in the arrangement of the neurons over which they deliver their information to the optic tract. The cones do not contain visual purple but do contain some other visual pigment. There are evidently three kinds of cones, sensitive to red, blue-green, and yellow colors. Each cone has virtually a private line to the optic tract, but several rods converge (via the bipolar cells) on the ganglion cells that provide the axons of the optic tract. A weak light, by activating several rods, may have its effects summated adequately to excite a ganglion cell (see Fig. 8-12). Conversely, the intensity may be insufficient to permit transmission through the cone systems in which summation is a less prominent feature and in which the threshold for excitation is higher. This same difference in arrangement helps explain the greater detail discernible by means of the cones. Light from several different points falling on several rods tends to give a blurred picture when the several rods activate the same ganglion cell.

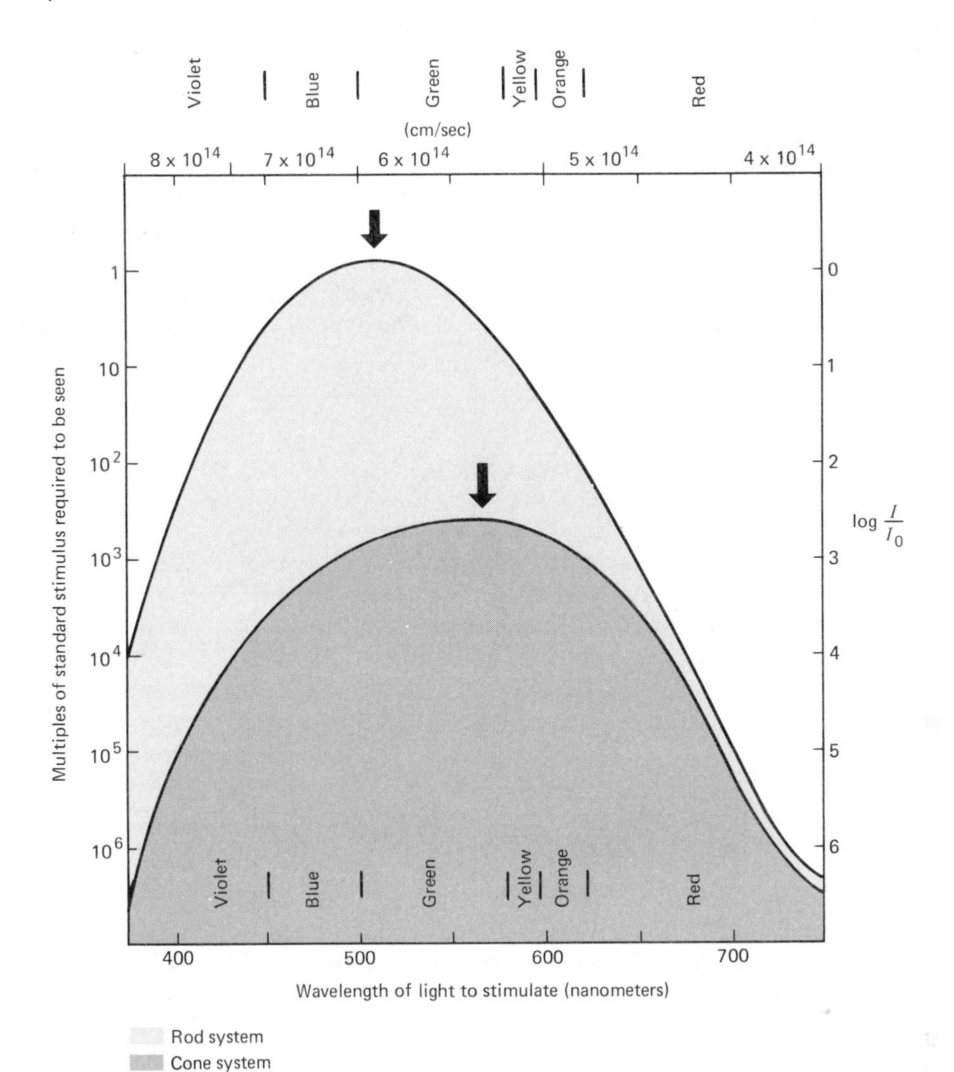

Fig. 8.27. Light sensitivity of the rod and cone systems of the retina. To the dark-adapted subject's eye is presented a small spot of light, controlled in intensity (shown on the ordinate in multiples of 10) and in wavelength (shown on the abscissa in millimicrons). The rods line shows the combinations of intensity and wavelength that can be just seen when the light is directed on the periphery of the retina. The cones line shows the threshold combination when the light is focused on the fovea. Any point in the heavily shaded area will be seen in either instance, but points in the lightly shaded area will be seen only by the rod system. Note the rods are nearly 1000 times as sensitive as the cones. Note also, the peak of sensitivity is shifted toward the red for cones. At the top of the graph are shown the colors corresponding to various parts of the spectrum. The frequencies for the various wavelengths are also shown.

Color vision

In speaking of the "visible spectrum," we mean that the wavelengths of light included in this scale can be detected by the eye. Nothing is implied about the special characteristic that we call color and that most of us take for granted as a real quality of things that we see. Specification of color is a function of the cone system of the eye. Three different types of cones have been described, which differ according to the wavelength of light that they preferentially absorb. The absorption curves for these three types of cones are shown in Fig. 8-28. One type of cone absorbs maximally

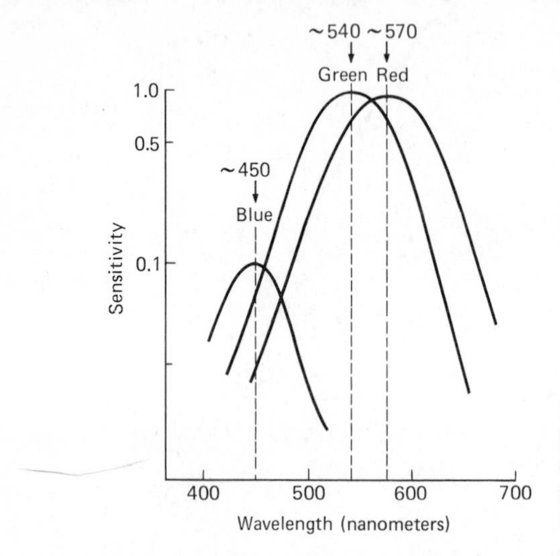

Fig. 8-28. Cones systems of the retina. Ordinate: relative amount of light absorbed at different wavelengths by each type of cone. Each curve is compared only to its own maximum.

at about 570 nm. We call a color of this wavelength *red.* Another type of cone absorbs maximally at 535 nm, a wavelength that appears green, while a third type absorbs maximally at about 450 nm, a wavelength that gives us the sensation of a blue color. Notice that any particular wavelength may be absorbed by two or even three of the different types of cones. What would you see if you looked at radiation of 550nm? At a very low intensity of light, none of the cones might be stimulated—only the rods would respond and you would be aware of a light but be unable to state the color. At a higher intensity, both the red-sensitive and the green-sensitive cones would be stimulated, but you would have a sensation of color that you would call yellow.

The visible spectrum is a graded series of colors conveniently designated red, orange, yellow, green, blue, and violet. Actually, hundreds of gradations can be distinguished in the scale. How can only three kinds of cones provide information so finely discriminating? Each wavelength of light stimulates one kind of cone to a different extent than it does any others. Color distinctions must, therefore, be made according to the degree of stimulation of one kind of cone compared to the other two. Specific colors are sensed according to the relative activity along pathways controlled by the different types of cones.

Deficiencies in one or another of the cones produce color blindness. If we designate the red-, green-, and blue-sensitive systems, respectively, as 1, 2, 3, then a defect of the first system is designated *protanopia,* of the second, *deuteranopia,* of the third, *tritanopia* (Fig. 8-29). In protanopia, since the green-sensitive cones respond to the wavelengths that we call red but the cones that deliver the message "red" are not functioning, red and green colors are confused. Similarly, in the person who is a deuteranope, the red-sensitive cones respond to the wavelengths normally detected by the absent green-sensitive cones, and again red and green are confused. In the tritanope, blue and green are confused. The person is blue-green color-blind. In total color blindness only one type of cone or only rods may be functional. In either case, no color is sensed as such. If everything is seen as *one* color, then there is *no* color, for subjectively a color can be a color only by comparison with another color. If your entire world is various shades of green and you have never seen any other color, then what you see is only shades of illumination from lightness to darkness.

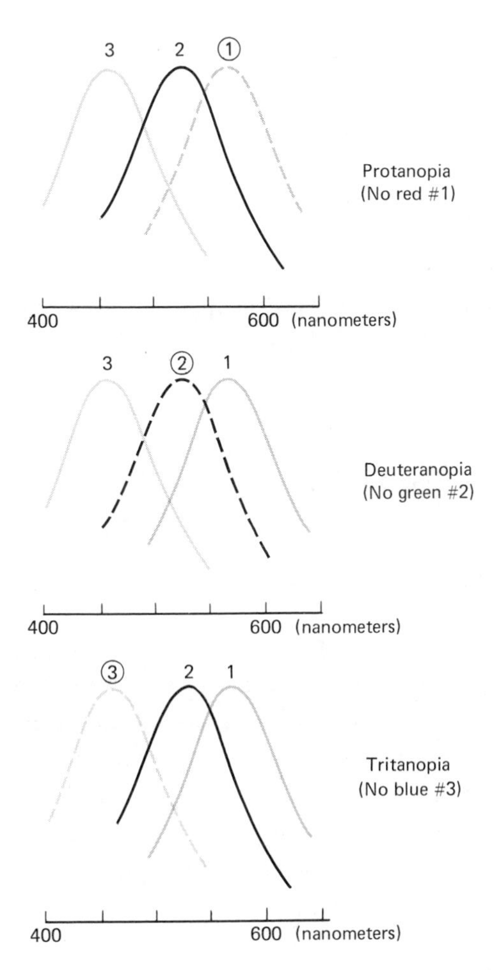

Fig. 8-29. Types of color blindness.

Section 8: *Questions, Problems, and Projects*

1. Draw a longitudinal section through an eye and label these parts: retina, sclera, cornea, crystalline lens, aqueous humor, vitreous humor.
2. What structures must light pass through before hitting the retina?
3. What is the "lens system" of the eye?
4. Show how a lens works. Compare it to a prism (or two prisms).
5. Show how a convex lens bends the light rays from a point object to focus the rays as a point image.
6. What is the focal point of a lens?
7. Show how rays from two points are refracted by a convex lens. Show the action of a concave lens.
8. Show why an image produced by a convex lens is inverted with respect to the object.
9. What is emmetropia?
10. Diagram the optical problem in hyperopia, and show how it can be corrected with lenses. Do the same for myopia.
11. If the lens system refracts the light to the same extent in the eyes of two individuals, but one is hyperopic and the other myopic, what is the basis for the difference?
12. Suppose that a lens bends light in one meridian more than in another. What is the consequence for the focusing, say, of a point of light?
13. What is astigmatism?
14. Show how astigmatism may be a mixture of far and nearsightedness in one lens system.
15. Where do the light rays normally focus in a nearsighted individual? In a far-sighted person?
16. In the structure of the retina, where is the pigment layer with respect to the light entering the eye and with respect to the wall of the eyeball?
17. Distinguish as to location and function in the eye, melanin and visual pigment.
18. What function does the pigment layer serve?
19. Distinguish between (a) rods and (b) cones.
20. In darkness, what happens to the vitamin A made available to the retinal rods?
21. Which cranial nerves innervate the extrinsic eye muscles?
22. In what direction would the lateral rectus muscle move the eyeball? Superior rectus? The inferior rectus?

23. The normal form of the crystalline lens is a somewhat rounded form. What happens to the light-bending power of the lens when it is pulled flat? In which condition may one see near objects best? Far objects?
24. How is focusing of the image on the retina brought about?
25. What is the pupil of the eye?
26. Pupillary diameter can vary from 1.5 to 9 mm. What area change does this represent?
27. What effect does contraction of radial muscle fibers of the iris have on the diameter of the pupil?
28. Realizing that the parasympathetic nervous system plays a protective, conservative role in body economy, what would you predict to be the innervation of the circular, smooth muscle fibers of the iris? Of the radial smooth muscle fiber?
29. What is the pupillary light reflex?
30. What is rhodopsin?
31. What effect does light have on rhodopsin?
32. How do impulses originate in the optic tract?
33. Describe the chemical cycle of rod function.
34. Give an example of dark adaptation. What is the chemical basis for the phenomenon?
35. What is light adaptation?
36. What does it mean when we say that the retinal sensitivity may increase 10^6 times when going from bright light to darkness?
37. About what aspects of the visual world do cones provide information?
38. What are the primary colors? What are the three types of cones, in respect to color sensitivity?
39. How can different wavelengths selectively stimulate different kinds of cones?
40. What determines intensity of sensation of light seen?
41. If yellow is seen when a green and a red receptor are stimulated, the color sensed may be due to the ratio of impulse frequency in the green to that in the red receptor. What might be the effect of changing the sensitivity of the stimulus?
42. If the red cones are defective, what sort of color blindness is present?
43. To which side of the brain does the right half of each retina project? The left half?
44. In what fashion do the temporal halves of the retinae project to the cerebral cortex?
45. Where is the geniculate body in relation to the optic tract and the optic radiation?
46. Where is the visual cortex?
47. Describe the arrangement of fibers from the retinae, as they form the optic chiasma.
48. The visual image is a mosaic of light and dark spots. Explain.
49. Where is the primary visual cortex? The visual association area?

50. What limits the field of vision?
51. Where is the blind spot? What is its origin?
52. To what part of the retina of the right eye does the right part (temporal) of the visual field project?
53. Draw a diagram showing how the optic fibers project from retina to cortex.
54. What part of the visual field will be blind if the left optic radiation is destroyed. If the left optic tract is destroyed?
55. What is meant by: 20/40 vision? 20/15? 20/200?
56. How do you measure visual acuity?
57. Where in the retina is visual acuity greatest? Why?
58. What determines the limits of visual acuity when the optical system is adequate? How may the characteristics of the optical system decrease visual acuity?
59. If two points 1 mm apart can be just distinguished as separate at 10 m, and the length of the eyeball is 20 mm, what is the distance of separation of the two points on the retina?
60. Why is the blind spot not normally noticed?
61. Where in the retina is the greatest density of cones?
62. How do you judge the relative distance of objects from the eyes?
63. If two objects are being observed by both eyes and one is in front of the other, what are these relative positions on the two retinae?
64. How many pairs of muscles are involved in positioning each eye?
65. Make a simple sketch showing the following structures of the ear: external ear, auditory canal, tympanic membrane, malleus, incus, stapes, and cochlea.
66. What produces sound in air?
67. Describe the sequence of movements produced in the parts of the ear when a sound wave strikes the ear.
68. An oval membrane terminates the scala vestibuli where the sound wave enters the cochlea. Which auditory ossicle faces this membrane?
69. Water is virtually incompressible. How then can the oval window vibrate?
70. What function is served by the membranous round window that terminates the scala tympani?
71. How is it possible to hear a tuning fork held to the skull if the sound does not come via the tympanum?
72. Contrast the basilar and tectorial membrane. What function does each serve?
73. Where are the hair cells of the cochlea?
74. The fibers of the basilar membrane are $2\frac{1}{2}$ times as long at one end as at the other. Are the fibers longest at the base or at the apex of the cochlea?
75. What is the relation between length of vibrating string and the resonance frequency? What is meant by resonance frequency? Comparing the time required for a back-and-forth movement of the string and the time for a cycle of vibration

of the air, decide which of two strings, long and short, will vibrate "in tune" with a low, which with a high-frequency sound.

76. When sound waves vibrate the oval window, a displacement of the basilar membrane travels as a wave toward the cochlea apex. Where is the greatest displacement when the sound is of low frequency? Of high frequency?
77. Although any particular frequency of sound will bring into vibration a large part of the basilar membrane, it is the point of maximum displacement that determines the frequency that is sensed. What is the relation between sound frequency and sensation along the cochlea?
78. "Pitch is which." Explain.
79. What determines intensity of sound perceived?
80. How are vibrations of the basilar membrane transduced into nerve impulses?
81. What is the relation between the tectorial membrane and the hair cells?
82. The cell bodies of the cochlea nerve constitute the spiral ganglion along the coil of the cochlea. Where and what are the dorsal and ventral cochlear nuclei?
83. The trapezoid body consists of axons from the cochlear nuclei, which cross to the other side of the brainstem and end in the superior olivary nucleus. What is the proper sequence of the following, on the way to the auditory cortex: superior olivary nucleus, inferior colliculus, medial geniculate body? What are these structures?
84. What is conduction deafness, and how may it occur?
85. What would be the result of immobilizing the auditory ossicles?
86. What is nerve deafness?
87. How can you discriminate conduction and nerve deafness?
88. What is an audiometer?
89. If a sound is 100 times a standard, then it is 10^2a or the change is 2 bels. If another sound is 10,000 times the standard, how many bels change is involved? What is the bel change in each instance? How many fold change is represented by 50 db?

9

COORDINATION AND CONTROL OF VISCERAL FUNCTIONS

9.1 The Visceral Nervous System

The reflexes of the somatic nervous system are concerned primarily with adjustments in tone of skeletal muscle, but the reflexes of the visceral nervous system involve tone of smooth muscle and activity of glands.

Centers of visceral control in spinal cord and brainstem can be distinguished from the centers of the somatic nervous system. The peripheral efferent part of the visceral system is quite distinct in organization and is called the *autonomic nervous system*. The afferent fibers are related especially to receptors for pressure, pain, and chemical substances. Visceral reflexes are, of course, also influenced by somatic sensory information. Some of the impulses from viscera reach consciousness as sensations, but we are unaware of the existence of most of them. Their main function is to control reflex adjustments of visceral structures over which we are not ordinarily able to exert voluntary control. Thus the system appears to function in a way that is independent of the rest of the nervous system. It is this autonomous kind of behavior that provides the name autonomic nervous system to the *efferent* side of the visceral nervous system.

General visceral afferent information and visceral sensations

When general visceral afferent impulses reach consciousness, the sensations they arouse are generally vague impressions of pressure or pain. These sensations, often poorly localized, may be sensed as coming from a region on the surface of the body. Pain originating in viscera but sensed as localized elsewhere is therefore designated *referred pain* to emphasize that the painful stimulus is applied elsewhere than at the place where it seems to be felt. A region of reference may have originated embryonically close to the actual site of stimulation. For example, the fact that a painful stimulus to the diaphragm may be felt as a pain in the region of the shoulder and neck emphasizes the embryonic origin of the diaphragm from slips of muscle that migrate down from the upper chest area. Through visual and tactile aids we learn to recognize places stimulated on the body surface. For the internal organs, we have no such clues. It is possible that the impulses from visceral regions end centrally near somatic sensory projection areas, and the sensations are therefore *referred* to these areas.

Special visceral receptors

The senses of taste and smell are part of the visceral nervous system. The afferent nerves innervating the receptors for these senses are called *special visceral afferents*. Unlike the general visceral afferents, these fibers ordinarily project their information to the level of conscious sensation. However, reflexes involving them are primarily related to the functioning of the digestive tract, a visceral function. Means of sensing the chemical environment are phylogenetically very old among animals. Among vertebrates, certain free nerve endings may be stimulated by strong solutions of salts, acid, and other substances, but our own knowledge of chemical environment comes mostly from the sense organs of smell (olfactory receptors) and partly from the sense organs of taste (gustatory receptors).

Gustation: Taste

Although taste receptors are liberally distributed over the body surface in some fish, these structures are restricted to the tongue and posterior walls of the oral cavity in other animals. Thus they are affected by soluble substances taken into the mouth.

The taste receptors are special epithelial cells (Fig. 9-1). As in all epithelial cells of the mouth, cell division replaces worn-out cells at the surface. The taste receptor cells differ from the ordinary epithelial cells, however, in that they occur in tiny groups as *taste buds*, in which they are arranged like the sections of an orange. Small nerve terminations surround the cells, as in the sensory epithelium of the internal ear apparatus. In addition, each cell terminates externally in tiny hairlike processes on which the tasteful substances are presumably absorbed. The taste bud is probably the only sense organ that is continually renewed by cell division. Different regions of the tongue have different thresholds for detection of various taste substances. It is possible that these variations in sensitivity are determined by differences in the life span of the taste cells, due to unequal accessibility of the cells to the environment in the mouth. A population of young cells may have a sensitivity different from that of older cells.

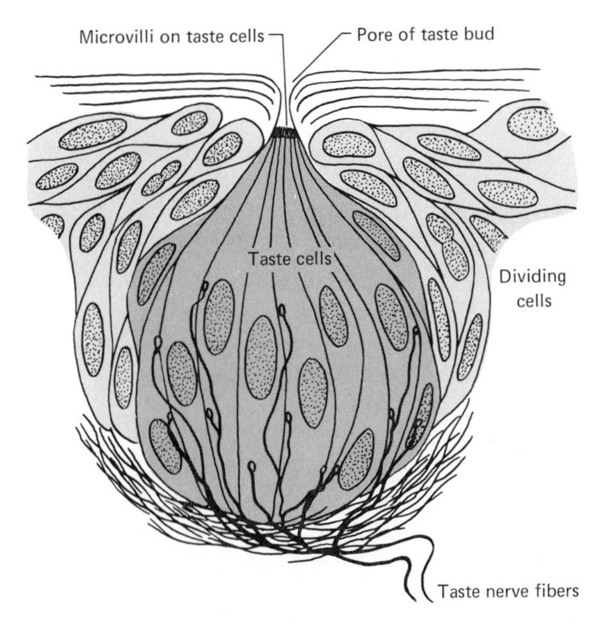

Fig. 9-1. Structure of taste bud. Note how cells of germinative layer of the ordinary epithelium seem to move into the taste bud from the sides and become taste receptor cells.

A substance absorbed on the taste cell perhaps provokes a change in membrane permeability of the cell, so that ions move more easily through the membrane. Current flows from the base of the cell and through the tiny nerve terminals, which are thereby excited to discharge impulses that the afferent nerve carries into the brainstem. Central connections of taste fibers and reflex effects of gustatory stimuli are further described in the chapter on the digestive system.

Olfaction: The sense of smell

The olfactory organs allow us to sense volatile substances in the air we breathe. Modern man depends very little on his nose, although it contributes to his enjoyment of food. Among primitive peoples, olfaction is a very useful sense, especially for tracking. Animals other than man have a keenly developed olfactory sense that is often of primary importance, particularly in the pursuit of the opposite sex, in the hunting of prey, and in the avoidance of enemies.

The zone of olfactory receptors is located far inside the nasal cavity, beneath the front of the brain. It is continuous with the epithelial lining of the nose and lies on the surface of the ethmoid bone through which fibers of the olfactory nerve pass on their way to the brain (Fig. 9-2).

Air coming into the nose passes through complicated passageways lined by a richly vascular epithelium in which there are many cells that secrete mucus. The air

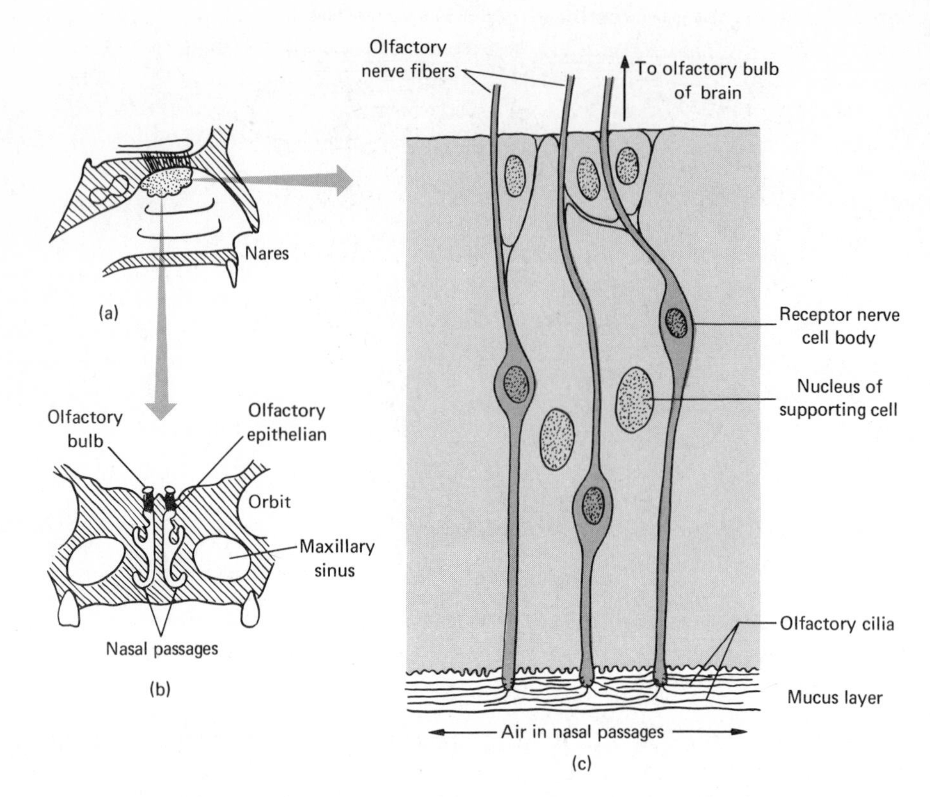

Fig. 9-2. Location of olfactory receptors. (a) Sagittal section of skull in region of olfactory mucosa. (b) Frontal section of skull in region of olfactory mucosa. (c) Enlarged section of olfactory mucosa.

traveling into these passageways is quickly saturated with water vapor by means of the watery mucus and is brought to body temperature by the liberal blood supply. Only a small part of the inspired air passes over the olfactory epithelium during ordinary inspiration. Sniffing is a means of bringing more air rapidly over the area without greatly altering the air exchange in the lungs.

There are about a half-million olfactory receptor cells in a limited patch of epithelium lining each of our two nasal cavities. The olfactory is the only sense organ of which the cell bodies of the primary sensory neurons are located at the surface of the body. Proximally, each cell sends a fiber 0.2 μ in diameter into the olfactory bulb of the brain. Distally, the cell body is prolonged into a short process, from the swollen apex of which a dozen tiny filamentous olfactory hairs float in the overlying layer of mucus. Presumably on these hairs the airborne molecules provide the excitatory effect that is preliminary to the production of nerve impulses that travel centrally from the olfactory epithelium during inspiration. The number of active cells and the frequency of the impulses increase when the concentration of odorous material is increased. Each breath is a separate stimulus to the receptors, and, with moderate concentrations, there is no decline in response of the receptors if equal volumes of air move over the olfactory surface at each breath. The intensity of response in the olfactory nerve does not decline during successive breaths, although the intensity of *sensation* that an odor produces does decrease after a short time. Evidently this olfactory sensory adaptation is primarily in the central nervous system.

Intense olfactory stimuli, as well as stimuli to other receptors, may alter the response by reflexly changing the volume of airflow through the nasal passages. There is a very rich blood supply in these regions. When the blood vessels are engorged with blood, the passageways are blocked, and the airflow is reduced. Emptying of the vessels results in reduction of the impediment to airflow. The flow of blood in these vessels is controlled by the central nervous system; the reflex control may be provided by sensory information from the nose or from other receptors almost anywhere in the body. Free endings of the trigeminal nerve in the nose are stimulated by odorous materials, although their sensitivity to most substances is less than that of the olfactory system.

Odors have been classified in many ways, but no satisfactory theory has been developed to explain the central perception of different qualities of odor. However, the total pattern of impulses leaving the olfactory epithelium is different for different substances, perhaps because different molecules are not absorbed to the same extent on the receptors. Thus a multitude of different odors may be distinguished by means of a homogeneous population of receptors. There is no evidence of anatomic differences that would account for selective sensitivity among the olfactory receptors.

On their way to the central nervous system, the *olfactory fibers* pass through the apertures in the cribriform plate (see also Figs. 9-2 and 4-5) and make synapses with neurons in the *olfactory bulbs* that rest on the inside of this plate. The two bulbs are connected to one another by fibers going from one side to the other and to more central portions of the brain by way of the *olfactory tract*. In primitive vertebrates, the entire anterior end of the brain is concerned with olfactory connections. Impulses from this region influence lower brainstem centers, whereas reflex effects of other afferent systems are taken care of in the lower brainstem. Most of the great area of cerebral cortex in man has evolved from a small region of cortex originally concerned mainly with olfactory connections.

The visceral efferent system (autonomic nervous system)

Afferent visceral information is received by the central nervous system via dorsal roots of the spinal cord and via several of the cranial nerves. The information constitutes the afferent side of the visceral reflexes. The efferent side is the autonomic nervous system that arises from cell bodies in the brainstem and in the lateral horn of the spinal cord (Figs. 9-3 and 9-4). The autonomic nervous system, unlike the somatic motor system, has synapses between neurons *outside* the central nervous system. These synapses are in the autonomic *ganglia*—groups of nerve cell bodies located outside the central nervous system. As in the case of the somatic motor system, myelinated efferent fibers with cell bodies in the central gray matter travel out toward the periphery, but instead of innervating the effector cells directly, these fibers (designated *preganglionic*) form synapses with other neurons in the ganglia. *Postganglionic* nonmyelinated fibers, arising from the other neurons, then travel from the ganglia to the effectors. The effectors of the visceral nervous system are generally smooth muscles, glands, and the heart.

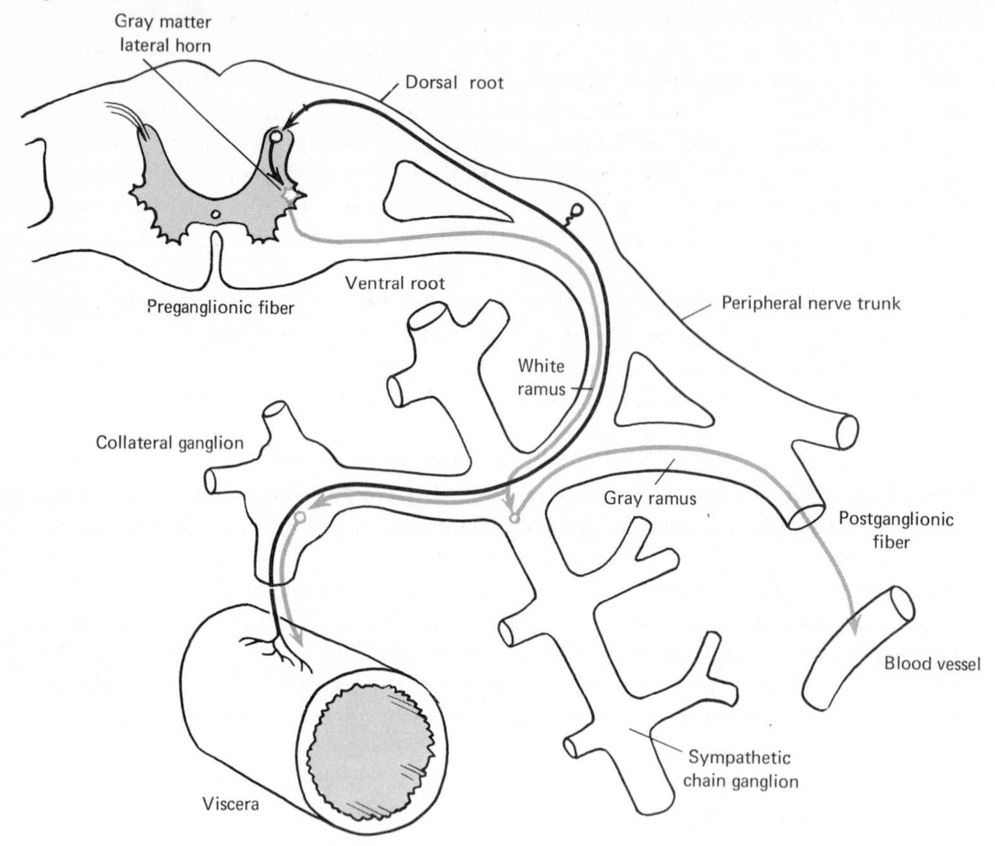

Fig. 9-3. Organization of sympathetic nervous system. The arrangement of fibers of the sympathetic division of the autonomic nervous system is shown at one spinal segment. The ganglionic chain runs outside of and parallel to the vertebral column. The gray ramus is shown running from paravertebral ganglion back to the nerve trunk.

The autonomic nervous system is conveniently divisible into two parts, the sympathetic and the parasympathetic. In the *sympathetic* or *thoracolumbar division*, the cell bodies of the preganglionic fibers are in the thoracic and lumbar regions of the spinal cord. These neurons are restricted to the *lateral horn* of the gray matter, where they may also be influenced by somatic sensory as well as by visceral sensory impulses. Thus autonomic reflexes may arise from stimulation of somatic as well as of visceral structures. Preganglionic fibers of each segment of the thoracolumbar division travel to the nearest *paravertebral ganglion*, located just outside the bodies of the vertebrae. Here they may form synapses with the ganglionic neurons, or they may send fibers upward or downward to nearby ganglia. Some fibers continue through the collateral ganglia located among the abdominal viscera. Postganglionic fibers carry impulses to the effectors. In the digestive tract, there are two more layers of ganglia: the *submucosal plexus*, supplying the gland cells, and the *myenteric*, reaching the smooth muscle of the digestive tract wall.

In the second part of the autonomic nervous system, the *parasympathetic* or *craniosacral division*, preganglionic fibers arise from cell bodies located in the brainstem and in the sacral segments of the spinal cord. The fibers of cranial origin travel with several of the cranial nerves to ganglia located close to the organ innervated. In the craniosacral division there is a rather direct pathway between the preganglionic fibers and the organ innervated. In contrast, the preganglionic fibers of the thoracolumbar division may control several effectors by way of the interconnected paravertebral ganglia.

Chemical features of the autonomic nervous system

The preganglionic autonomic nerve fibers, like the somatic motor nerve fibers, produce *acetylcholine* at their terminals. This depolarizing substance is an aid to, if not the cause of, synaptic transmission at the preganglionic terminations in both sympathetic and parasympathetic ganglia. At the postganglionic terminations, where the fibers innervate the effectors, there is a chemical difference between the two systems. In the craniosacral division, the postganglionic fibers, like the preganglionic fibers, secrete acetylcholine, which acts on each effector in a specific fashion. In the heart, for example, parasympathetic stimulation is followed by hyperpolarization of the cardiac cells and inhibition of the heartbeat. The enzyme *cholinesterase* quickly destroys acetylcholine, which therefore has effects only locally—where it is secreted.

In the thoracolumbar division, an adrenaline-like chemical, *noradrenaline*, is secreted and acts on the effector cells. Noradrenaline is destroyed more slowly than acetylcholine. Its longer life in the circulation enables it to act on sympathetically innervated effectors in various parts of the body. Thus both anatomical and chemical differences conspire to restrict parasympathetic activity to particular organs, but they encourage a general effect of sympathetic activity.

The effect of sympathetic stimulation is made even more general by the fact that sympathetic preganglionic fibers innervate the *medulla* (central part) of the *adrenal gland*, which secretes *adrenaline* into the circulation. Adrenaline has a general effect similar to noradrenaline. The adrenal gland is a part of the endocrine system.

Autonomic effects

In a rough way, the effects of autonomic stimulation can be described in terms of adjustments of the body to the two opposing situations of rest versus "flight or fight" (Fig. 9-4). Generally speaking, the effects of parasympathetic stimulation

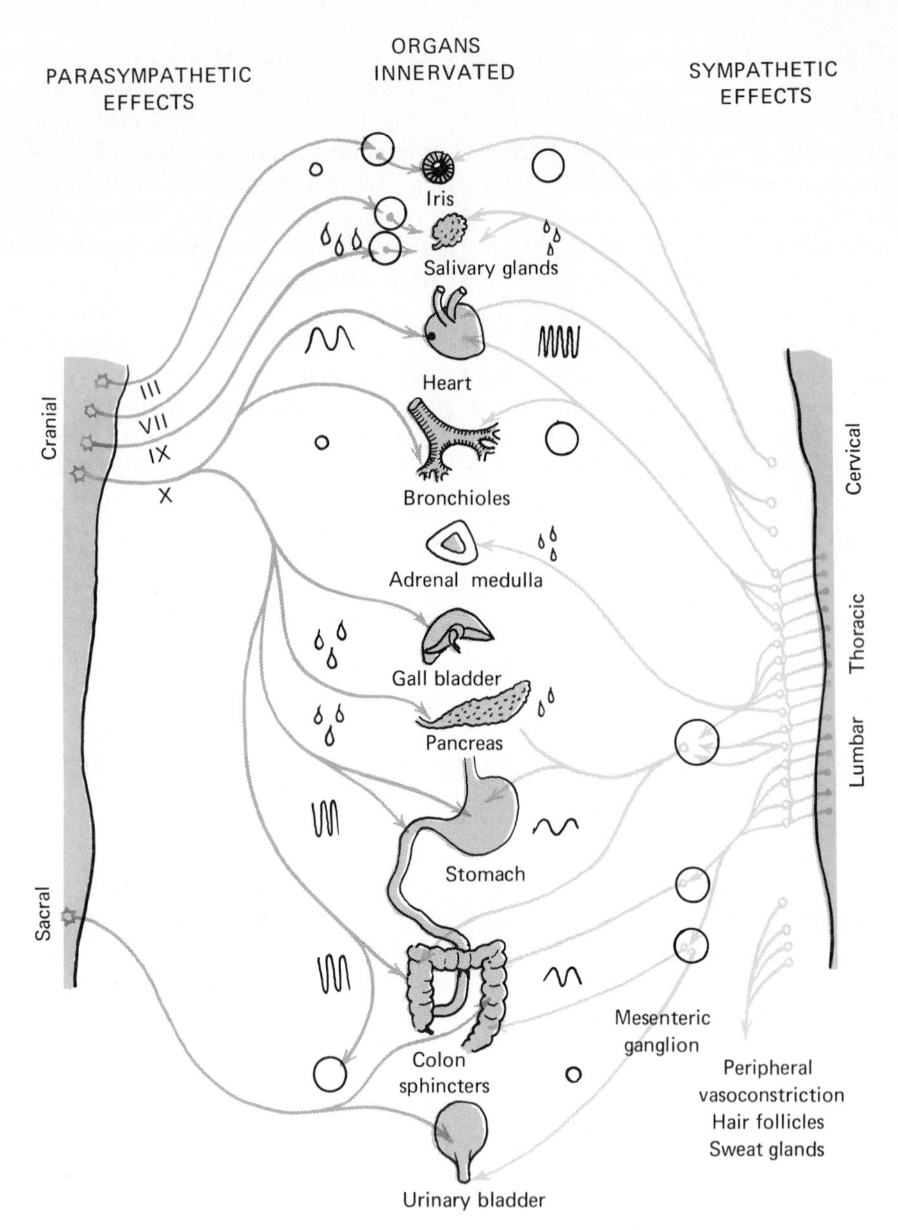

Fig. 9-4. Distribution and effects of autonomic nerves. Organs innervated are shown schematically down the center. On each side the influences of nerve impulses in augmenting or depressing secretory activity or contraction amplitude and rhythm are shown in terms of droplets, circles and wavy lines. Parasympathetic effects are on the left, sympathetic on the right. For example, impulses in parasympathetic fibers of nerve III constrict the iris, in contrast to the dilated pupil present during sympathetic activity.

are most obvious during conditions of calm and rest. Parasympathetic effects may be remembered in terms of "usefulness" to that end. For example, fibers in the oculomotor nerve stimulate the smooth circular muscle of the iris, thus protecting the eye from excess light. Fibers in the facial and glossopharyngeal nerves stimulate secretion of saliva and thus aid digestion. Digestion is an activity most suitably carried out during rest, especially because it requires a large supply of blood that during exercise must be diverted to skeletal muscle. Digestion is further aided by the vagus parasympathetic fibers, which stimulate secretion of digestive juices along the alimentary canal and increase the activity of the smooth muscle of the digestive tract. Impulses in other efferent vagus fibers decrease the rate and force of the heartbeat during rest. This is a reasonable adjustment, for the skeletal muscles are inactive, and the body's demand for blood supply is relatively low at that time.

Acting to prepare the body for stressful situations, the sympathetic nervous system has effects mainly opposite to those described for the parasympathetic. Fibers traveling in the *cervical sympathetic chain* innervate the radially oriented smooth muscle fibers of the iris, and contraction of this smooth muscle increases the size of the pupil. Thus sensitivity to dim light is improved by the increase in total light striking the retina. The influence of sympathetic innervation on glands may be mainly indirect, via the circulation. That is, the sympathetic impulses may control the tone in the smooth muscles of the walls of small blood vessels and thereby alter the blood flow. Since secretions of glands are ultimately derived from the blood, secretory activity will be affected. Under stressful situations, the saliva is likely to be sparse and sticky. This is a result of sympathetic activity.

Muscular exertions require that good supplies of blood be provided to the skeletal muscles. It is reasonable, then, that sympathetic impulses stimulate the smooth muscle of tiny blood vessels of the digestive tract and cut down the flow of blood to stomach and intestine. In addition, digestion is decreased by the increase in frequency of sympathetic impulses, which have an inhibitory effect on the smooth muscle of the digestive tract wall. These impulses tend to decrease the contractions of the stomach and intestine. The requirement of greater blood supply to the skeletal muscles is further aided by cervical sympathetic impulses, which, arriving at the heart, cause an increase in rate and force of heartbeat. When an adjustment of autonomic activity in response to stress occurs, the effectiveness of the adjustment is increased by the presence of reciprocal innervation in the central connections of the two systems. Increase in parasympathetic activity to a particular organ may be accompanied by decrease in frequency of sympathetic impulses, and vice versa.

Higher-level autonomic control

Like the peripheral somatic motor system, the peripheral autonomic nervous system is controlled by levels above the spinal cord and above the motor nuclei of the lower brainstem (Fig. 9-5).

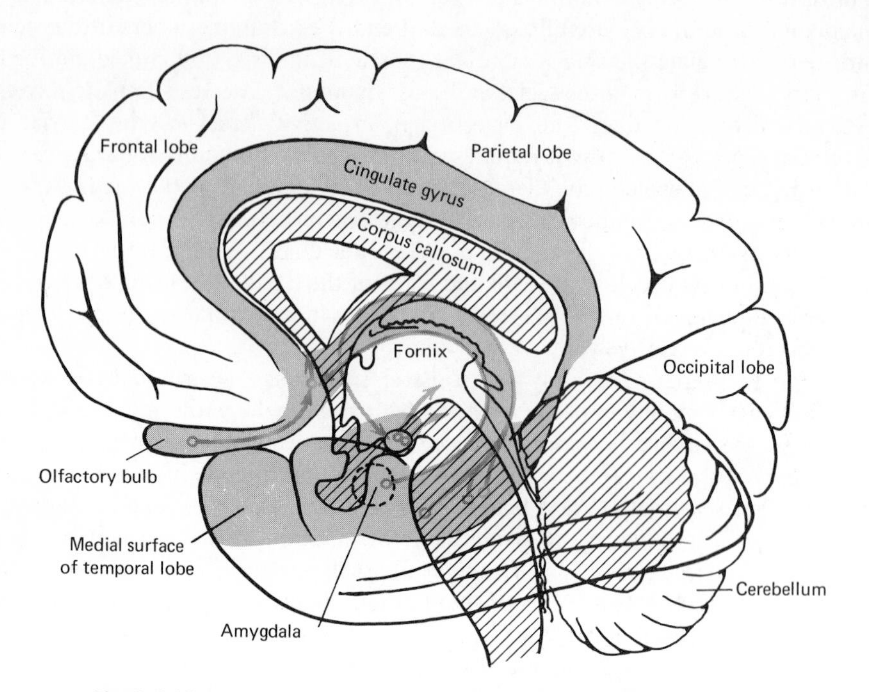

Fig. 9-5. Higher centers of the visceral nervous system. This diagram is a view of the medial surface of the brain after most of the brainstem has been removed. The brainstem and cerebellum are shown in light lines and can be seen in Fig. 9-6. Cut surface of medial sagittal plane is diagonally lined. The shaded area is the rhinencephalon or visceral brain. The limbic lobe is cerebral cortex closely surrounding the brainstem. Starting at the olfactory region, it grows up over the corpus callosum dorsally and then returns anteriorly, enfolding the brainstem in its embrace. Pulled along also is the fornix, connecting hypothalamus nuclei to the limbic lobe, as well as other connections between olfactory areas and the amygdala, at the end of the caudate nucleus. The whole complex has control over visceral functions and relates visceral functions to somatic.

An important part of the cerebral hemispheres is concerned with regulation of autonomic activity. In the *precentral gyrus* of each cerebral lobe, the body is topographically represented in terms of movement. Stimulation of a particular spot with short electric pulses will be followed by contraction of the skeletal muscles in a particular region of the body. This rule of topographical representation applies as well to the autonomic effectors, for the same stimulus that produces a particular limb movement will produce contraction of smooth muscle of blood vessels, or of glands, or the smooth muscle associated with hairs in the same body region where the skeletal muscle response is seen. Stimulation in the cortical face area may produce salivation, and stimulation nearby may activate other digestive tract activities. These observations show the close relationship of somatic and autonomic neural control. Another region of the cerebral cortex devoted more completely to visceral functions is the *cingulate gyrus*, the area of cortex most closely surrounding the brainstem. It may be found around the base of the cerebral lobes and can be partly seen in a sagittal section of the brain. The two cerebral hemispheres are separated by a deep fold, the *medial fissure*. At the bottom of the fissure, a large mass of white matter, the *corpus callosum*, provides one of the pathways between the two hemispheres. Along the sides of this fissure, and extending rostrally and caudally, the cerebral cortex is folded under and wrapped around laterally. Below the corpus callosum, it embraces the outside of the thalamus and extends ventrally to meet the anterior areas of the cortex. This entire ring of cortex is the *cingulate gyrus*. When certain associated nuclei are added, the entire ensemble may be called the *visceral brain* (Fig. 9-5). This descriptive term is justified by the fact that many of the main connections relate to the autonomic nervous system that controls the viscera. A large part of the visceral brain is composed of nuclei and pathways related to the olfactory system; therefore the term *rhinencephalon* (or "nose-brain") is also used to describe it. Particularly among lower mammals, but even in man, olfactory stimuli evoke autonomic responses, and the rhinencephalon provides a special emphasis to the coordination of visceral-somatic-sexual functions.

Among the stimuli that have the most profound influence on the entire body are those related to sexual activity (concerned with reproduction of the species) and those related to the securing of food (concerned with individual survival). Both activities involve powerful activation of both visceral and somatic effectors and, primitively, are closely related to olfaction. In lower animals, often more than in man, the fulfilling of both needs involves alert, searching behavior, mobilizing the entire body resources, especially (as may often be the case) when the pursuit of food or of sex involves struggle.

Whether the struggle is actual physical combat or an emotional crisis involving conflicting demands, the visceral nervous system takes a most active part. On the basis of afferent information, particularly as this information relates to such basic drives as food and sex, patterns of impulses leave the visceral brain and travel to centers of both somatic motor and visceral motor control. Thus visceral-somatic behavior patterns result. Electrical stimulation of certain areas of the rhinencephalon result in facial and body movements and in visceral effects (especially in circulation and digestion) associated with strong emotion. Particular facial expressions and positions of the limbs follow from patterns of impulses traveling over the extrapyramidal system. The visceral effects are via the *hypothalamus*, with which the visceral brain has strong connections.

The hypothalamus

The hypothalamus surrounds the base of the third ventricle in the diencephalon, and it is a coordinating center having several nuclei that receive olfactory impulses, which are then related to other information received from the thalamus, basal ganglia, and elsewhere (Fig. 9-6). A particularly rich network of capillaries

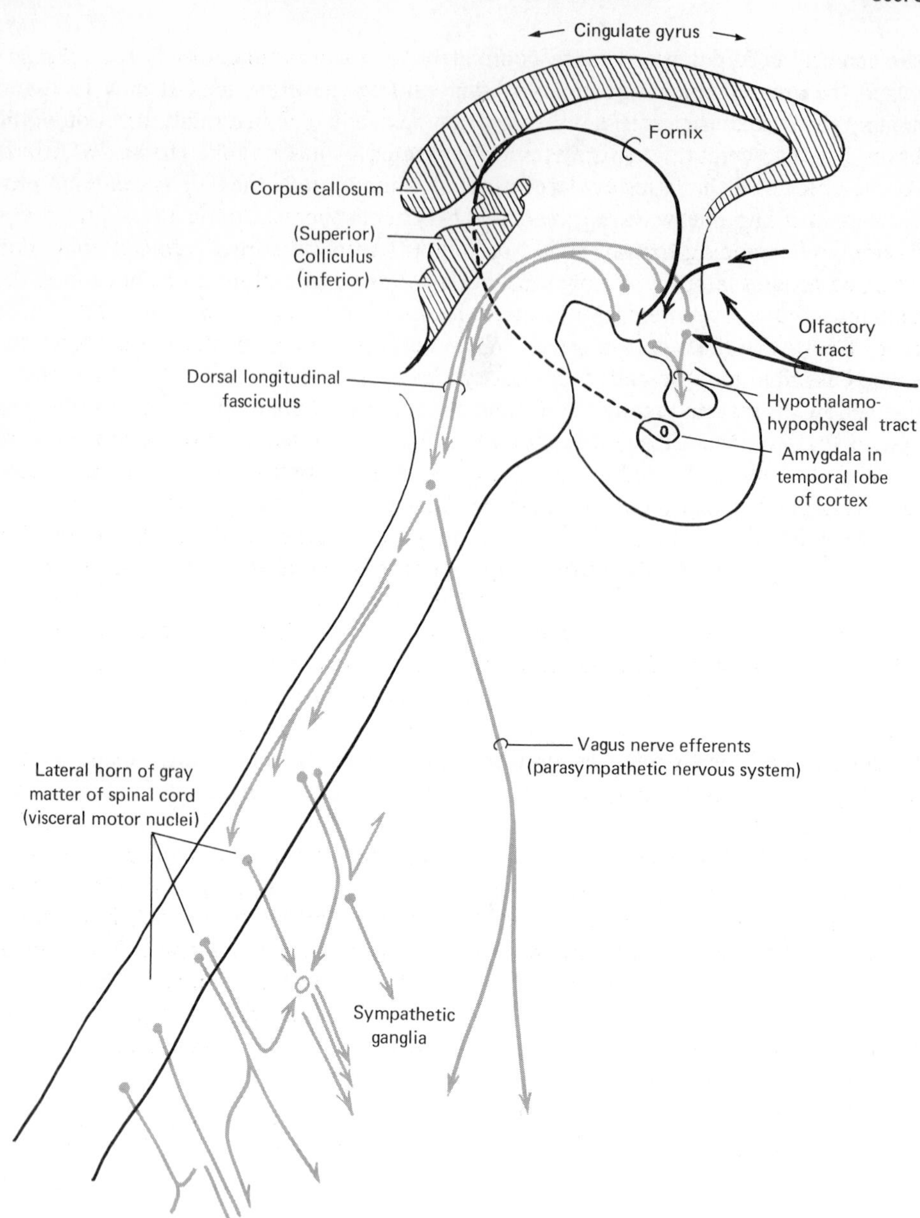

Fig. 9-6. Hypothalamic control of autonomic nervous system. Black arrows show input to the hypothalamus from higher levels. Note relation to olfactory input and to centers in the cerebral cortex. Gray arrows show outflow pathways from the hypothalamus, particularly to the neural lobe of the pituitary and, along the dorsal longitudinal fasciculus, to the brainstem visceral motor nuclei of the vagus parasympathetic and to the visceral motor nuclei of the thoracic region of the spinal cord, governing the sympathetic system.

penetrates the hypothalamus, and, consequently, the hypothalamic neurons are well placed to respond to changes in the characteristics of the blood. Blood temperature and osmotic pressure, for example, affect the rate of firing of these neurons. The outflow of impulses from the hypothalamus occurs particularly along the *dorsal longitudinal fasciculus*, which distributes to the midbrain and the medulla oblongata. From these regions, in the reticular system, connections are further made with the central neurons that give rise to the preganglionic fibers of the autonomic nervous system. Some of the nuclei of the hypothalamus are particularly outstanding, for their axon terminals end in the neural or posterior lobe of the pituitary and are the source of the hormones of that lobe. In addition, certain nuclei of the hypothalamus secrete hormones that control the secretory function of the anterior pituitary lobe and hence nearly the entire *endocrine system*. The endocrine system is a group of glands that pour into the bloodstream secretions that exert profound effects on nearly all the cells of the body. The effects are related especially to the maintenance of a suitable environment for the adequate functioning of the body cells.

9.2 The Endocrine System

Through somatic and visceral reflexes, we make moderately rapid adjustments of the body to environmental changes. Part of this adjustment is toward maintaining the proper balance of conditions in the internal environment bathing the cells of the body. This internal environment is controlled finally by the blood. The vessels through which the blood flows serve as a distribution system for a tremendous variety of chemical substances forming part of the environment of the cells. How well the cells carry out their normal functions (indeed, the stimulus to carry out these functions at all) is due not only to the proper patterns of nerve impulses from the central nervous system, but also to the proper balance of chemical agents coming in contact with, and often penetrating, the membranes of the cells. For example, in a smooth muscle the degree of contraction may depend on a suitable level of adrenaline-like compounds produced by post-ganglionic sympathetic neurons or by the adrenal medulla. Like nerve impulses, these compounds serve as commands to "call forth" responses in susceptible cells. The substances may be delivered to the responding cells by nerve endings or by way of the bloodstream. Special organic molecules secreted by some cells and capable of evoking changes in cellular metabolism in other cells are called *hormones.* The glands that secrete such substances are called *endocrine*, or ductless, glands. *Endo*crine is thus to be distinguished from *exo*crine. The latter term refers to glands that secrete to epithelial surfaces other than into blood vessels—for example, into the digestive tract or onto the skin. The endocrine glands, on the other hand, pour their secretions into the bloodstream. The hormones serve as messengers, carrying information to the cells that "understand" the message, just as the nervous system provides, more rapidly by way of nerve impulses, commands to the special cells to which they come. Hormones generally act more slowly and bring about long-lasting effects, for example, involving metabolism, cell division, and cell growth.

There are several endocrine glands in the body, collectively forming the *endocrine system*. Each of the body systems is regulated in some way by some part of the endocrine system, just as each is controlled in some way by the nervous system.

Hormones seem generally to regulate cellular function by stimulation or inhibition of some specific normal function of the cells.

Pituitary hormones

Some of the endocrine organs are interrelated through the control exerted over them by the *pituitary*, itself an endocrine gland (Fig. 9-7). This organ is situated at the base of the brain in close association with the hypothalamus. The hypothalamus controls not only the autonomic nervous system, but to a considerable extent it governs the endocrine system as well.

The pituitary has a double embryonic origin. An *anterior lobe* arises as an evagination from the roof of the mouth, whereas the *posterior lobe* develops as an outpocketing from the base of the brain below the hypothalamus. The posterior lobe consists mainly of modified nerve fiber terminals whose cell bodies lie in the hypothalamus. These cells synthesize hormones that are stored in the cell terminals and are released into the circulation when the neurons are suitably stimulated by impulses or by changes in the chemistry of the fluids bathing the cells. *Antidiuretic hormone* or *vasopressin* from the posterior lobe regulates urine flow and the tone of smooth muscle in blood vessels. *Oxytocin* from the posterior lobe affects the tone of smooth muscle in the reproductive system.

Since the source of these hormones is actually in the hypothalamus, a deficiency of them may result from a lesion or local injury to that part of the brain. *Diabetes insipidus* may result from such a lesion. This defect involves a copious urine flow, without loss of sugar from the blood into the urine, in contrast to *diabetes mellitus*, involving a pancreatic deficiency. There are neuroglial cells in the *posterior lobe* of the pituitary as there are elsewhere in the nervous system. Those in the pituitary are called *pituicytes*. They are not known to secrete any hormones.

The cells of the *anterior lobe* secrete several hormones, some of which regulate cellular metabolism directly, while others exert an indirect effect by controlling the output of hormones from some of the other endocrine glands. The secretory cells of the anterior lobe are themselves subject to the influence of specific substances in the blood. For example, hormones from some of the other endocrine glands partially control the output of pituitary hormones. (see Chap. 9.2) Control by the nervous system is indirect. There is a rich capillary network between the hypothalamus and the anterior lobe of the pituitary (*the hypothalamic-hypophyseal portal system*). Neurosecretions from cells of the hypothalamus may be transported to the anterior lobe

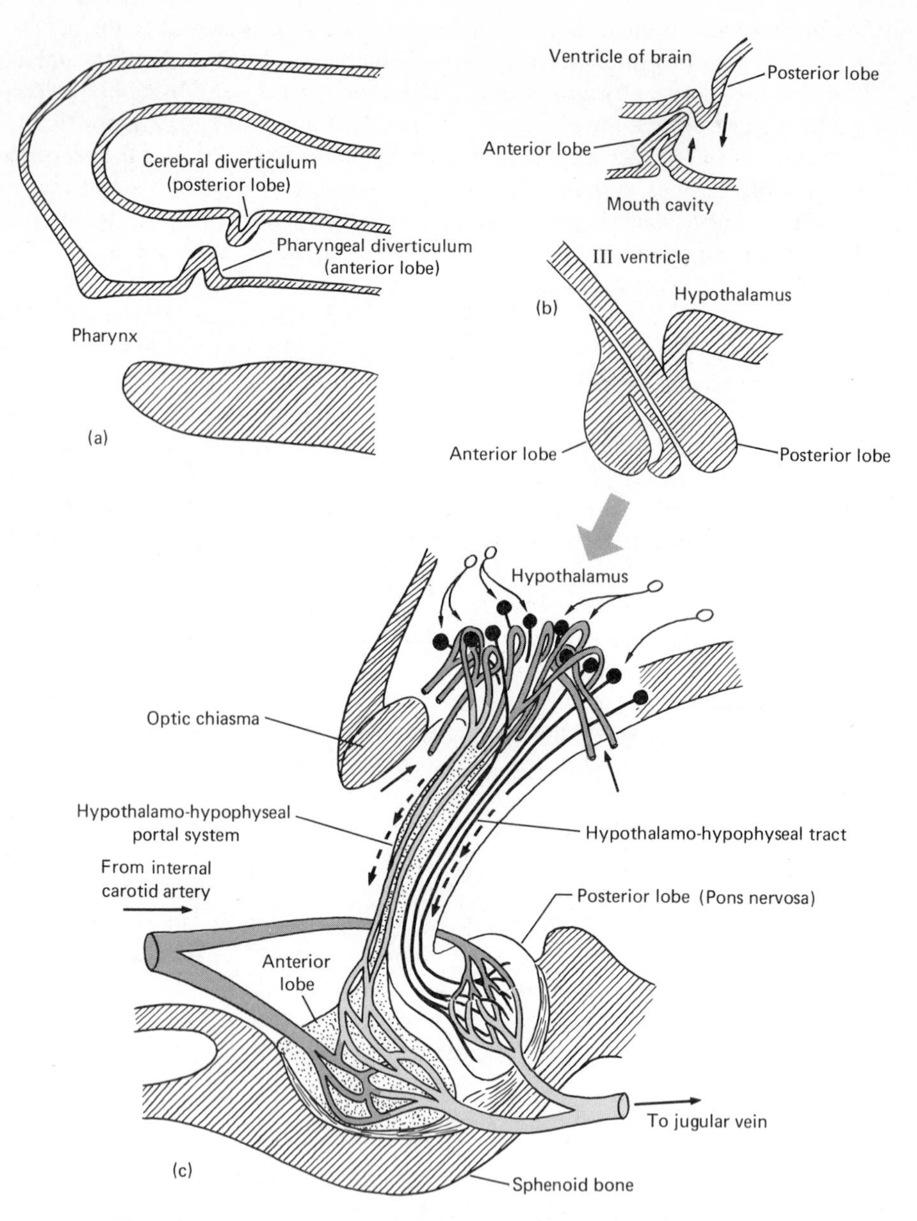

Fig. 9-7. Pituitary lobe. (a, b) At early stage in embryo, the pituitary begins to form from outpocketings from top of pharynx and base of brain, which become respectively anterior and posterior lobes. (c) In the fully developed pituitary, the posterior lobe consists of ends of fibers whose cell bodies are in the hypothalamus. Communication between hypothalamus and anterior lobe is via blood vessels between hypothalamus and pituitary—the hypothalamo-hypophyseal portal system.

cells along this route. Secretion of hormones by the cell of particular hypothalamic nuclei is stimulated by nerve impulses or by changes in the composition of the blood.

Each of several anterior lobe hormones acts specifically on each of several body structures, which are said to be the *targets* of the hormone. These hormones from the anterior lobe of the pituitary are named according to the target organ, to which the suffix "tropic," meaning "directed toward," is added. The *growth hormone*, for example, influences the entire body or *soma* and is therefore called the *somatotropic hormone* or *somatotropin.*

Somatotropin favors retention of protein by the body. There is a decline of nitrogen products in the urine when the concentration of this hormone in the blood rises. Anatomically, the effect of excessive growth hormone is seen in continued growth of the bones and increased mass of the body beyond the normal growth level.

Besides somatotropin, the anterior lobe secretes hormones that in their actions are directed toward the other endocrine glands: *thyrotropin* toward the *thyroid gland*, *adrenocorticotropin* toward the outer layer or *cortex* of the *adrenal gland*, *gonadotropin* toward the *gonads* (*ovaries* and *testes*) (see Table 9-1).

Each hormone of the anterior pituitary lobe seems to be secreted by cells of a characteristic type. A simple distinction can be made on the basis of the affinity of the granules in the cells for acidic or basic dyes. The cells are classified as *acidophilic* if

Table 9-1 Pituitary hormones

Source	Anterior Pituitary Lobe (Adenohypophysis of Distal Lobe)	Posterior Pituitary Lobe (Neurohypophysis of Neural Lobe)
Hormones	1. Growth hormone (somatotropin) Affects body growth	1. Antidiuretic hormone ADH (directed toward kidney)
	2. Thyrotropin Controls thyroid	2. Oxytocin Directed toward uterus and some other smooth muscle
	3. Adrenocorticotropin ACTH Affects adrenal cortex	
	4. Gonadotropin Governs maturation of eggs and sperm (Follicle stimulating hormone, FHS) (Luteinizing hormone, LH)	
	5. Prolactin Stimulates milk ejection	
Stimuli	"Releasing" hormones from hypothalamus	1. Osmotic pressure of blood 2. Nerve impulses, blood chemistry

they stain with acid dyes, *basophilic* if with basic dyes, and *chromophobic* if neither stains. The functions of the cells can be determined if histological changes are studied under conditions of excess or deficiency of specific hormones. The chromophobes, which lack granules, constitute about half the cells and may be acidophil or basophil cells that are exhausted of their secretions. Three kinds of basophils, differing in shape and distribution, separately produce thyrotropin (TSH or thyroid-stimulating hormone), and two gonadotropins—follicle-stimulating hormone (FSH) and luteinizing hormone (LH). Growth hormone may be in the acidophilic cells. The tropic hormones released by these cells of the anterior pituitary lobe are all small protein molecules. TSH, LH, and FSH all have molecular weights (MW) of about 25,000, while another hormone from the anterior lobe, adrenocorticotropic hormone, or ACTH has a molecular weight of less than 5000.

The hypothalamico-hypophyseal portal system provides a direct line of communication between cells of the hypothalamus and the pituitary. Chemical messengers are sent along this route, from hypothalamic cells to cells of the anterior lobe of the pituitary. These substances are *releasing factors* that command the granule-laden cells of the adenohypophysis to release their burden of hormones into the bloodstream. Different regions of the hypothalamus produce different releasing factors. Anterior, middle, and posterior hypothalamus produce, respectively, releasing factors for TSH, LH, and ACTH.

Several of the endocrine glands do not appear to be controlled by the pituitary: the *parathyroid gland*, the *islets of Langerhans* in the pancreas, and the lining of the digestive tract which secretes hormones that regulate the secretion of digestive juices.

Thyroid gland

The *thyroid gland* is a bilobed gland in the neck (Fig. 9-8). The substance of the gland consists of a large number of spheres or *follicles*. The wall of each follicle is a layer of epithelium. The amorphous material, or *colloid*, within each follicle, is *thyroglobulin*, the thyroid hormone in combination with a protein.

When thyroid hormone is required in the circulation, a proteolytic ("protein splitting") enzyme is secreted by the follicle epithelium and separates the hormone from its protein complex. The free hormone can then diffuse through the wall of the follicle and into the blood in the surrounding rich network of capillaries.

The general effect of thyroid hormone is to increase the rate of metabolism of many tissues of the body. Its effect is particularly striking on the nervous system. In an unknown way, it has an effect on the development of the higher nervous centers. When insufficient thyroid hormone is present early in life, general body growth is poor—in the extreme case, the individual may be a *cretin*, deformed physically and retarded mentally.

All the hormones are manufactured by the cells of the endocrine glands that secrete them. If the raw materials are not available, adequate synthesis cannot be carried out. In the thyroid hormone, *iodine* is an essential constituent. If iodine is not

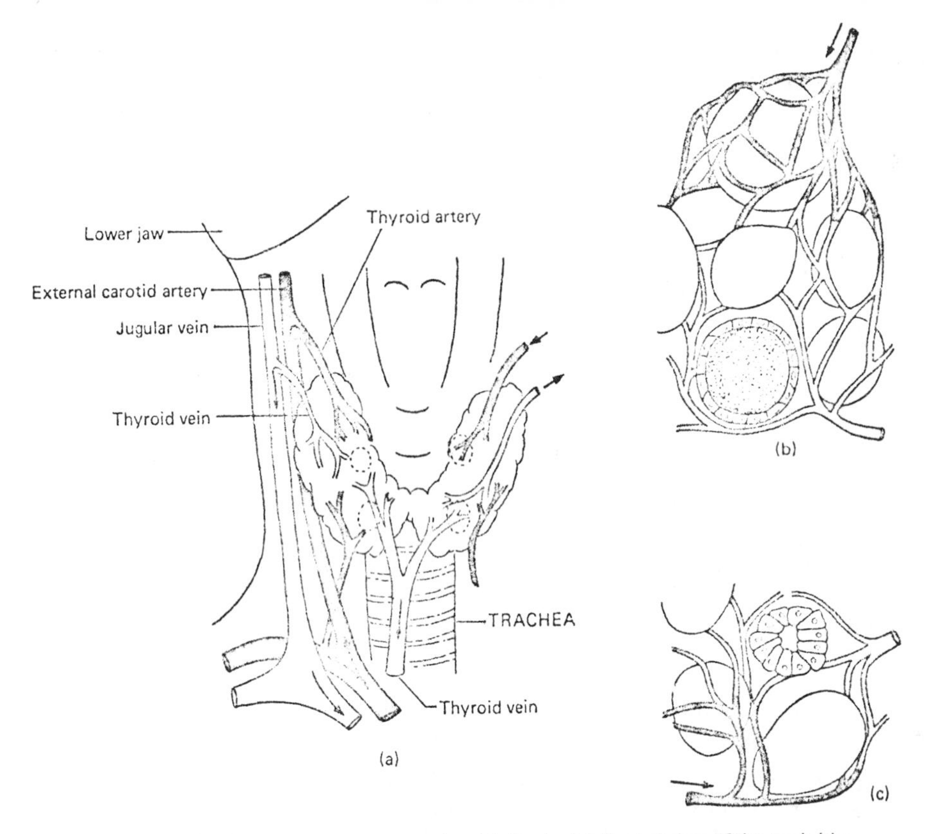

Fig. 9-8. The thyroid and parathyroid glands. (a) Ventral view of the neck (the head is thrown back) showing thyroid gland just under the larynx. The four parathyroid glands on the dorsal surface of the thyroid glands are shown as dotted circles. (b) Appearance of quiescent thyroid follicles. Thin epithelium, stored colloid. (c) Appearance of follicles stimulated by thyrotropic hormone from anterior pituitary. Tall cells, little colloid storage.

adequately available in the diet, sufficient amounts of thyroid hormone cannot be synthesized. Today, the iodine commonly included in table salt makes such deficiency generally unlikely. If the thyroid gland works "overtime," attempting unsuccessfully to manufacture thyroid hormone when iodine is insufficient, the gland may increase in size and produce a swelling in the neck called a *goiter*. The individual, in this instance, has *hypothyroidism*, a thyroid hormone deficiency.

The thyroid, like any other endocrine gland, must produce the right amount of hormone if the body cells are to function properly. If the thyroid is, for some reason, stimulated to produce an unsuitably high concentration of thyroid hormone into the bloodstream, body metabolism will be raised above normal levels, and heat production will be excessive. The nervous system will be more than usually excitable, and a person suffering thus from *hyperthyroidism* will appear nervous and uneasy. A goiter may also be present in hyperthyroidism; that is, there may be more than a normal amount of active thyroid tissue. Goiter refers to the enlarged condition and not to the functional state of the thyroid.

Thyroid hormone is an organic, iodinated ring compound, iodothyronine (Fig. 9-9). In the bloodstream, the form with four I atoms is present in highest concentration and is called *thyroxine* or tetraiodothyronine. A form with only three I atoms, and

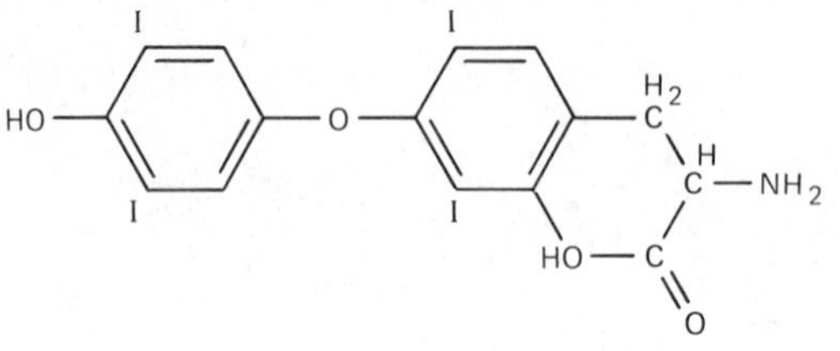

Fig. 9-9. Iodothyronine. With the encircled I present, this molecule is thyroxine or tetraiodothyronine. Without that I, it is triiodothyronine.

called triiodothyronine, is present in much lower concentration, but it acts at lower concentration than does thyroxine. Iodothyronine is synthesized by the thyroid from the amino acid tyrosine plus iodide. Both substances are therefore essential for good nutrition.

Thyroxine circulating in the bloodstream is bound to serum globulin, and the level of hormone may ordinarily be measured by the amount of protein-bound iodine in the blood. Thyroid deficiency may be compensated for by adding iodine or thyroxine to the diet, whereas thyroid excess may be combated by means of specific substances that cut down the effectiveness of the thyroid hormone. Alternatively, some of the thyroid tissue may be inactivated or removed. Sodium thiocyanate (NaSCN) and thiouracil are well known as antithyroid drugs that inhibit the action of thyroid hormone.

Feedback control of thyroid activity

The secretion of thyroid hormone depends on the concentration of thyrotropin put into the blood by the anterior pituitary lobe. On the other hand, the secretion of thyrotropic hormone is regulated by the concentration of thyroid hormone in the bloodstream. When the thyroid hormone level is low, much thyrotropic hormone is secreted; when it is high, the thyrotropic hormone declines. Thus the call to secrete thyrotropic hormone depends on information fed back to the pituitary from the thyroid gland in the form of the concentration of thyroid hormone present in the bloodstream. A suitable level of thyroid hormone is, therefore, maintained in the normal bloodstream according to the needs of the cells that consume the hormone. All of those endocrine glands that are regulated by the pituitary are controlled by *feedback* arrangements of this sort, which tend to maintain a constant level of the hormones and thereby contribute to the homeostasis or constancy of the internal environment.

Another level of control over TSH exists in cells of the hypothalamus, which secrete *TSH-releasing* factor. It is probable that the amount of this factor that is released is controlled by the neurons of the temperature–regulating centers in the hypothalamus. The secretion passes to the anterior pituitary lobe by way of the hypothalamo-hypophyseal portal system along with other releasing factors.

The parathyroid gland

Imbedded on the dorsal surface of the thyroid are the *parathyroid glands* (Fig. 9-8). These tiny organs secrete *parathormone*, which regulates the concentration of calcium in the blood and thus indirectly controls the excitability of the nervous system, which is very much dependent on calcium. If the parathyroid glands are removed in an experimental animal, a small stimulus to the animal may result in uncontrolled muscle contractions, which being sustained, are called *tetani.* This condition is temporarily relieved by injection of calcium salts or parathormone into the bloodstream. Parathormone is a small protein molecule with a molecular weight of about 10,000.

The concentration of calcium in the blood is determined not only by the amount taken in food but also by the amount taken up by the bones and the amount lost through the kidneys. Parathormone increases the Ca^{++} level in the bloodstream by increasing the action of osteoclasts that resorb Ca^{++} from bone. In addition, the kidney is stimulated to resorb Ca^{++} from the renal tubules. The increased resorption of Ca^{++} is accompanied by increased excretion of phosphate. The decrease of phosphate level in the bloodstream allows a higher concentration of Ca^{++} to be present; although calcium phosphate is only slightly soluble, it will not precipitate until the product $[Ca^{++}][PO_4^{=}]$ exceeds a particular quantity, the *solubility product.* The lower the $[PO_4^{=}]$, the higher the $[Ca^{++}]$ can be before precipitation occurs.

There is a second hormone that tends to *decrease* the blood Ca^{++} concentration. This *hypocalcemic factor* is *calcitonin*, secreted by *parafollicular cells* that are dispersed among the thyroid follicles. The action of calcitonin is not only opposite to the effect of parathormone; it is also of short duration. The calcitonin molecule is only about one-third the molecular weight of the parathormone.

The adrenal gland—adrenal medulla

The adrenal or suprarenal gland derives its name from its location on top of the kidney ("renal" = kidney). It is an endocrine gland of double origin and double function (Fig. 9-10). The outside layer (cortex) arises from mesodermal tissue closely related to the gonads. The cells of the central portion (*medulla*) have the same embryonic origin (ectodermal) as, and are homologous to, the postganglionic neurons of the sympathetic nervous system. They are, however, entirely secretory and do not produce impulses. Instead they secrete two hormones, the catecholamines *adrenaline*

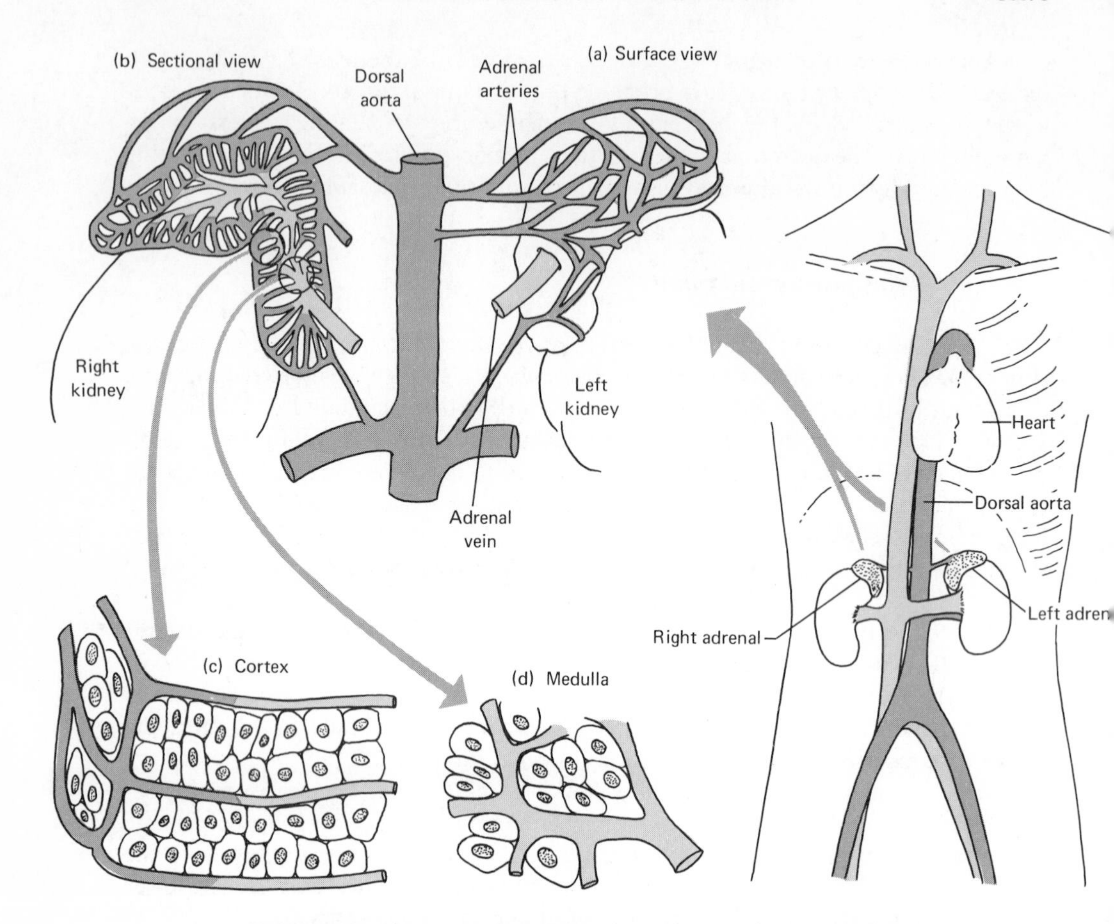

Fig. 9-10. The adrenal gland. (a) Surface view of adrenal gland. Arterial blood supply from numerous small anastomosing branches. (b) Sectional view of adrenal gland. Extensive capillaries in parallel array pass through the cortex—see (c)—and collect together to provide the venous drainage from the medulla—see (d). The cells in the cortex are arranged in regular rows between the capillaries (c), but the cells of the medulla are quite irregular in arrangement (d).

(epinephrine) and the similar, but not identical, *noradrenaline* (norepinephrine) (Fig. 9-11). Noradrenaline is also released from the sympathetic postganglionic terminals. Release of adrenaline into the bloodstream is stimulated by impulses in sympathetic preganglionic neurons. The hormone produces bodily effects approximating generalized sympathetic nervous system activity. Smooth muscle and heart muscle are the special targets of both adrenaline and of noradrenaline. Generally adrenaline has the more intense effect on these targets. In the arterioles of skeletal muscle, the two hormones have opposite actions. The adrenaline provokes relaxation of the smooth

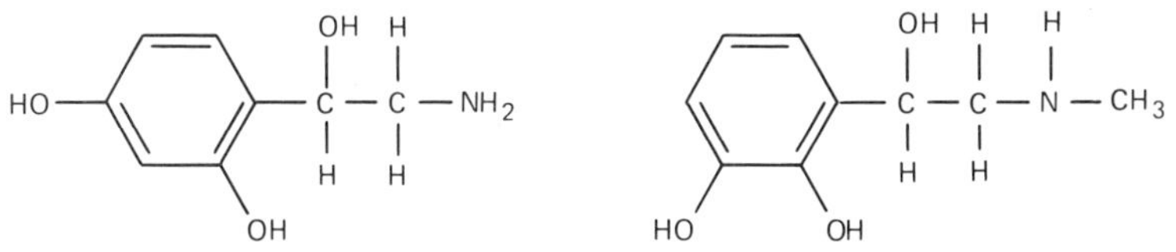

Norepinephrine
(noradrenalin)

Epinephrine
(adrenalin)

Fig. 9-11. Catecholamines.

muscle of these small blood vessels, whereas the noradrenaline causes contraction. Both hormones, and especially adrenaline, influence the rate of production and use of sugar by the body cells. One effect is to increase the splitting of glycogen to glucose. The glucose released in liver increases the blood sugar level. The glucose made available in muscle is oxidized to lactic acid, which enters the bloodstream and is converted to glycogen on reaching the liver.

The adrenaline and the noradrenaline are apparently secreted by different cells of the adrenal medulla and effects of these two substances are not identical. Different kinds of stimuli perhaps activate the two sets of cells to different extents because the amounts of each of the two catecholamines that the adrenal medulla secretes depend on the kinds of stimuli.

The adrenal cortex

Three histologically distinguishable layers of cells form the adrenal cortex and secrete corticosteroid hormones that are structurally similar to the sex hormones. These adrenocortical hormones control salt balance and carbohydrate metabolism. The cells of the outermost layer, the *zona glomerulosa*, secrete *aldosterone*, which is produced relatively independently of ACTH. Because of its effect on mineral salt distribution, aldosterone is called a *mineralcorticoid.* The formula for aldosterone is shown in Fig. 9-12. The secretion of aldosterone and other mineral corticoids is stimulated to occur when the extracellular fluid volume or the Na^+ level is lower than normal, or when the K^+ level of the body fluids is above normal. Neither the receptors for these changes nor the means by which they stimulate the adrenal cortex are known. It is clear, however, that the secretion of the aldosterone will help restore the Na^+ and increase the extracellular fluid volume and eliminate the surplus K^+. These effects will be mainly on the salt and water-balance control systems of the kidneys.

The inner two layers of the adrenal cortex also secrete steroids that have mineralcorticoid effects. The cells of these layers are very much dependent on the presence of ACTH secreted by the anterior pituitary lobe. The effect of ACTH on the adrenal glands appears to be to stimulate the enzyme systems responsible for the synthesis of corticosteroids.

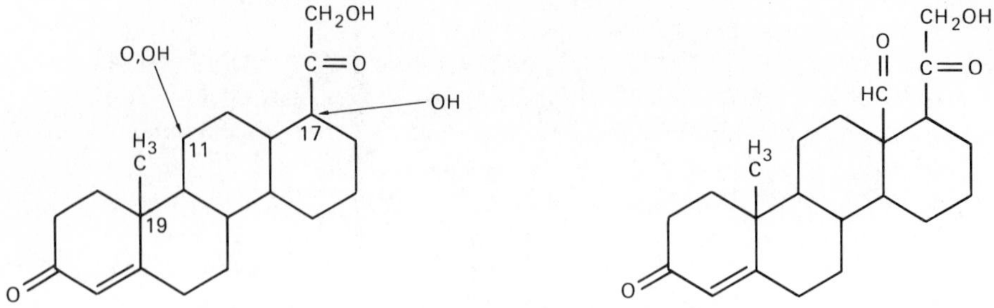

Corticosterone
(mainly glucocorticoid)

Aldosterone
(mainly mineralcorticoid)

Fig. 9-12. Adrenocortical steroids. The relative effects of these substances on mineral or on glucose metabolism may be altered by the presence or absence of CH_3 at position 19, O or OH at 11, OH at 17.

The steroids from these layers influence the metabolic transformation of amino acids to sugar and glycogen, especially in the liver. This action is called a *glucocorticoid* effect of the adrenal steroid hormones. The glucocorticoid action seems to depend on the presence of O or OH at position 11 in the steroid ring (Fig. 9-12). Adrenal hormones are secreted out of the blood and into the urine after being slightly altered in structure by metabolic action of the liver.

The corticosteroids are often administered to help heal inflammations and to rectify deficiencies in salt balance. According to the general principles of feedback control in the endocrine system, added adrenocorticosteroid hormone may act on the hypothalamus and pituitary to depress the secretion of ACTH. No longer stimulated by ACTH, the adrenal cortex may atrophy unless the level of injected corticosteroids is reduced, preferably gradually.

If the adrenal cortex is removed or is defective, body cells cannot properly utilize glucose, and the concentration of sodium in the blood falls, while that of potassium rises. Both changes have particularly profound effects on the nervous system, and the adrenal cortex is quite necessary for continued life.

The adrenal steroids are much like the sex hormones in structure, and, in fact, the adrenal cortex may secrete considerable androgen, or male sex hormone, and estrogen, a female sex hormone. In certain diseases of the adrenal cortex, excessive amounts of male sex hormone may be produced. In a female, the result is a change of body features toward a masculine appearance.

The pancreas

Several parts of the digestive tract serve as endocrine glands and secrete hormones independently of the pituitary or any other endocrine glands. These hormones regulate the flow of digestive juices, for example, from the pancreas into the small intestine. They are described in detail in Chap. 13.3. The *pancreas* is a double gland (Fig. 9-13). One part is exocrine—that is, it pours secretions to the outside of the body rather than into the circulatory system. Outside, in this instance, is the lumen of the small intestine, and the secretions are digestive juices. The secretory activity of this part is regulated by the vagus parasympathetic nerve and by hormones produced by the wall of the small intestine. The other part of the pancreas consists of patches of endocrine gland tissue distributed throughout the organ. These *islets of Langerhans* are also not controlled by the pituitary.

Two kinds of cells are distinguishable in the islets of Langerhans. One, the *beta cells*, secrete *insulin*, the better known of the pancreatic hormones. Insulin is thought to increase the permeability of cells to glucose, and thus more glucose is made available to be stored as glycogen in liver and muscle. At the same time, the increased availability of glucose permits an increase in utilization of glucose by the cells.

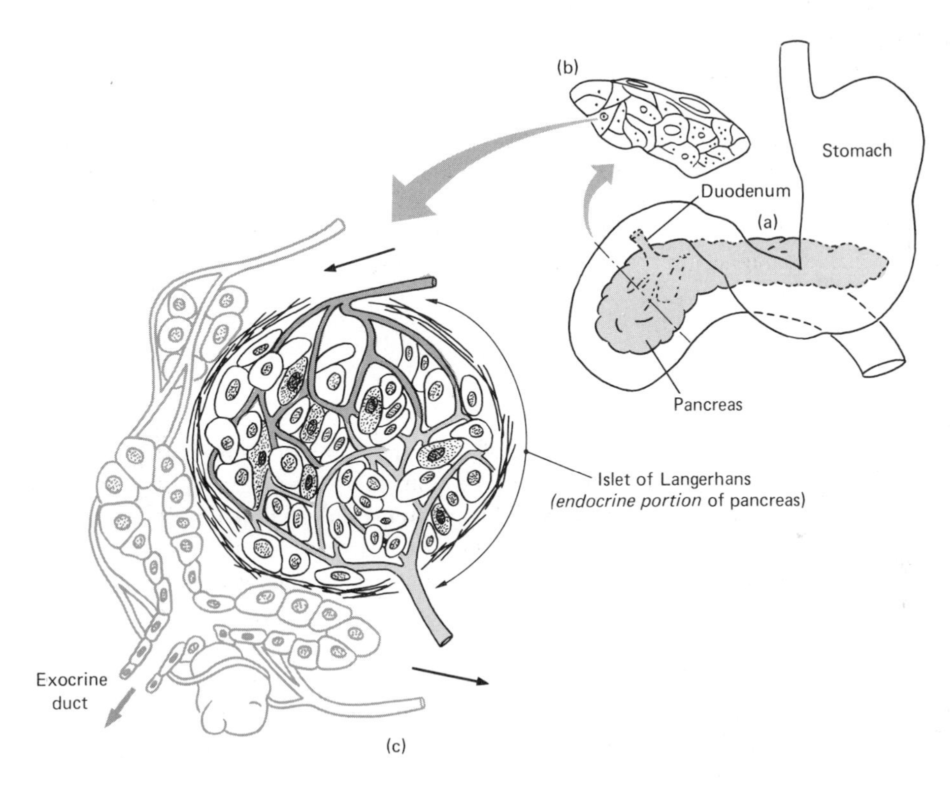

Fig. 9-13. The Islets of Langerhans of the pancreas. (a) View of location of pancreas. (b) Cross section of pancreas—most of the substance is exocrine, secreting digestive enzymes. The small dots are endocrine tissue, the Islets of Langerhans. (c) Enlargement of an Islet of Langerhans. Cells of this structure manufacture glucogon and insulin. Below are diagrammed cells of exocrine portion of pancreas.

The most important stimulus to the islets of Langerhans is the concentration of glucose in the bloodstream. If glucose concentration gets above normal, the islets are stimulated to secrete insulin into the blood. When the glucose level drops below normal, the secretion is decreased.

When the insulin concentration in the bloodstream is chronically low, sugar is inadequately utilized by the tissues and is lost through the kidneys, in the urine. This is the disease *diabetes mellitus*, characterized by a large flow of urine associated with the loss of glucose.

In this disease, glucose is not adequately utilized, but fatty acids are consumed at a greater than normal rate. There is an accumulation of molecules with a ketone structure. Acetone is a volatile "ketone body" having a characteristic odor, which when detected in the breath is diagnostic of diabetes mellitus. Fortunately, this disease can be kept under control by the regular administration of insulin by injection or by other substances taken by mouth. Insulin is a polypeptide with a molecular weight of about 6000. The action of an injection of insulin may be prolonged if the hormone is first joined chemically with a larger molecule, which will not be so rapidly taken into the bloodstream and destroyed.

The storage of glycogen is opposed by the hormone *glucagon*, a polypeptide secreted by the *alpha cells* of the pancreas. Glucagon makes more sugar available by stimulating the conversion of glycogen to glucose. Its outstanding effect is to increase the secretion of insulin by the β cells, into the bloodstream.

When blood glucose declines, the *alpha cells* are stimulated to increase their secretion of glucagon, and more glucose (and more insulin) is mobilized into the bloodstream.

The sex hormones

The *gonads* produce the sex cells—sperm from the *testes* in the male, eggs (ova) from the *ovaries* of the female. In addition, the gonads are endocrine glands whose secretions into the bloodstream have important general effects relating to sexual activity and reproduction. Representations of these glands may be seen in Figs. 16-4 and 16-5. Secretion of hormones from the gonads is controlled by two *gonadotropins* from the anterior lobe of the pituitary. One of these (*follicle-stimulating hormone*) is directed toward the ovarian follicles that produce the eggs in the female; in the male it is directed toward the testicular tubules where the sperm are produced. The ovarian follicles also secrete *estrogen*, a hormone that affects the excitability of the nervous system and affects the female reproductive tract in a way that favors sexual activity.

The second gonadotropin is *luteinizing hormone*, which has as its target organ the transformed follicular cells that remain in the ovary after the egg has been ejected from the follicle. These cells form the *corpus luteum*, an organ named for its yellow color. Secretion of *luteotropin* from the anterior lobe of the pituitary calls forth the hormone *progesterone* from the corpus luteum. *Progesterone* is concerned with maintaining pregnancy. During pregnancy the *placenta*, the organ by means of which the embryo is attached to the mother's body, also is an endocrine organ, secreting *placental gonadotropins* that are also directed toward the corpora lutea, stimulating those organs to secrete the progesterone, which assists in keeping the placenta firmly attached.

In the male, the anterior lobe of the pituitary secretes *interstitial cell-stimulating hormone*, a name that indicates the cells of the testes on which this hormone acts. Interstitial cell-stimulating hormone corresponds to the luteinizing hormone of the female. These cells of the testes secrete testosterone, which has effects on the body generally and especially on the reproductive tract. Testosterone favors development of sexual activity in the male. The sex hormones are considered more completely in the chapter on reproduction (Chap. 16.1).

General functions of hormones

The hormones form part of the chemical environment in which the cells of the body are bathed. The effectiveness with which cells carry out their functions depends on the proportion of various substances available to them in their surroundings. The hormones are all organic molecules that probably participate in cellular biochemical reactions. Perhaps they alter membrane permeability and thereby permit certain necessary materials to pass more easily into the cell. In that way certain chemical reactions in the cell may be carried out more effectively. The fact that the hormones are relatively specific in their action reflects the existence of subtle differences in molecular structure. Such differences explain why particular hormones act mainly on particular cells. Part of the specificity of hormone action is perhaps deceptive. Most of the hormones, as a matter of fact, have a rather general effect on all cells. This is probably true of the somatotropic hormones (protein utilization), of thyroid hormone (cellular metabolism and heat production), of insulin (carbohydrate utilization and storage), of adrenocorticosteroid hormone (membrane permeability, salt regulation), and of parathormone (regulation of calcium level). The sex hormones stimulate cell division, particularly along the reproductive tract. They probably influence cell permeability, especially in muscle and nerve, since muscle contractility and the excitability of the nervous system are both influenced by the sex hormones.

The actions of some hormones can be understood in terms of a "second messenger" concept. The first messenger is the particular homone that has affinity for cells of a particular target organ. The second messenger is a substance produced within the cell as a result of the interaction of the hormone with an enzyme in the cell membrane. Although there are several first messengers, there may be only one second messenger that has an effect on the biochemical machinery inside the cell. The basic biochemical process that is influenced is phosphorylation. The result of a stimulus to phosphorylation will depend on the kind of specialized activity that the energy is utilized for in the cell and whether one is looking at, say, protein synthesis, carbohydrate metabolism, secretion by the cells, membrane permeability, or cell movements.

The second messenger comes from ATP by the loss of two phosphates and the formation of a ring structure. The result is *cyclic AMP—cyclic adenosine monophosphate*. The effect of many hormones seems to be to alter the activity of the enzyme *adenylcyclase*, which catalyzes the reaction, ATP ⟶ cyclic AMP. For example, glucagon, catecholamine, ACTH, and LH increase the production of cyclic AMP, whereas insulin decreases it. An action of the cAMP is to regulate the amount of active phosphorylase in the cell. These enzymes, in turn, catalyze the conversion of glycogen to glucose 1–phosphate, which can then undergo breakdown in order to provide energy for the cells' activities.

Section 9: *Questions, Problems, and Projects*

1. Define the term hormone and give examples.
2. Where is the adenohypophysis (anterior lobe of the pituitary)?
3. Where are the following hormones produced: somatotropic, thyrotropic, adrenocorticotropic, gonadotropic?
4. What are the general effects of secretion from the thyroid? From the adrenal medulla? What are the names of these hormones?
5. What is the role of adrenal cortical hormones in mineral and carbohydrate metabolism?
6. What hormone is secreted from the pancreas? What is its action?
7. What are the effects of sex hormones? Of parathormone?
8. Which of these are endocrine, which exocrine glands: pituitary, pancreas, sweat glands?
9. Couple properly: neurohypophysis, adenohypophysis, anterior pituitary, posterior pituitary. Which structure is connected functionally to the hypothalamus?
10. An extract of this gland, injected into a young individual, causes an increase in length and thickness of bones, thickness of skin, mass of muscle, increase in metabolic rate. Which is the hormone responsible, and what is its source?
11. Diagnose the difficulty: The individual is 20 years old but has the appearance and general development of a young child.
12. Contrast giantism and acromegaly as to origin and characteristics.
13. Acromegaly involves excess secretion of growth hormone, yet the person is of normal height. Explain.
14. To what extent does the thyroid gland develop in the absence of thyrotropin?

15. What is this molecular structure?

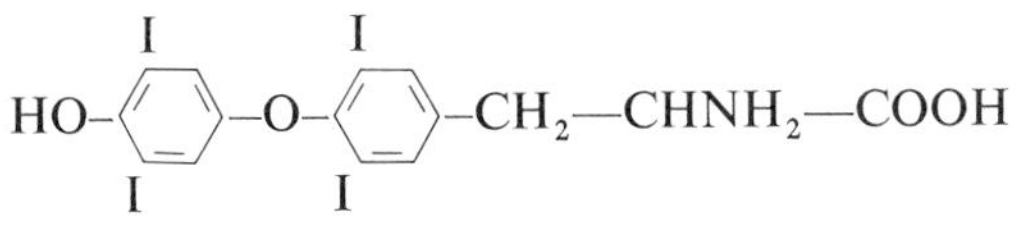

$$HO-\underset{I}{\overset{I}{\bigcirc}}-O-\underset{I}{\overset{I}{\bigcirc}}-CH_2-CHNH_2-COOH$$

16. Where is thyroglobulin stored?
17. What fills the follicles of the thyroid gland?
18. What happens to thyroxine poured into the bloodstream?
19. What is the source of antidiuretic and of oxytocic hormone? Where are they stored and released from?
20. Explain how the osmoreceptors of the hypothalamus respond to osmotic changes of the blood and how they regulate the flow of urine.
21. To what cortex is the corticotropin directed?
22. Where is 99 percent of the body Ca^{++}? The remainder is partly bound, partly "free"; where is it?
23. What is the role of Ca^{++} in cell membranes? What effect does it have on nerve and muscle? In particular, what is the effect of a deficiency?
24. What is the role of Ca^{++} in blood clotting?
25. $$Ca^{++} + HPO_4^{=} \rightleftharpoons CaHPO_4;\ k = \frac{(Ca^{++})(HPO_4^{=})}{(CaHPO_4)}$$

    In these equations, the ratio determined by k must remain constant. If either Ca^{++} or $HPO_4^{=}$ increases in concentration, what happens to the concentration of $CaHPO_4$, a precipitate?
26. Parathormone causes bone absorption. What then happens to the level of Ca^{++} and $HPO_4^{=}$ in the blood?
27. If the level of Ca^{++} and $HPO_4^{=}$ in the blood is changed because of parathormone, which of these ions is excreted most effectively in the urine?
28. Prolonged Vitamin D deficiency results in inadequate absorption of Ca^{++} from the gut. What is the consequence for bone and muscle? Explain.
29. What effect might parathormone have on the number of osteoclasts in bone? What therefore, is the mechanism by which parathormone alters the Ca^{++} and $HPO_4^{=}$ concentration in blood?
30. Describe the relation of parathyroid activity (what is parathyroid activity?) to the Ca^{++} level in the blood.
31. What is the principal value of parathyroid hormone?
32. The parathyroid is one of many homeostatic devices. Explain.
33. What are the consequences of hyperparathyroidism?

34. In the presence of thyroxine, the actions of several intracellular enzymes is increased. How, then, does thyroxine exert its action?
35. Explain in terms of vasodilation and cardiac activity the rise of systolic blood pressure as a consequence of thyroxine secretion.
36. What effect do you predict thyroxine has on the nervous system? On the gastro-intestinal tract?
37. Hypothalamus, hypophysis, the thyroid gland, and the general body cells form an interacting feedback system. Explain.
38. Would you expect thyroid secretion to increase if you are exposed to a cold or a warm climate? How, then, might the basal metabolic rate of inhabitants of the Tropics compare with inhabitants of the Arctic?
39. What are the consequences of increased secretion of thyrotropin?
40. How do you recognize hyperthyroidism?
41. Why is radioactive iodine a useful treatment for hyperthyroidism?
42. If a person is lethargic, sleeps 12 to 15 hours a day, is obese, and his skin has an edematous appearance, what would be your diagnosis?
43. A goiter may exist in hyper- or hypothyroidism. Explain.
44. What is the consequence for thyroid production of a deficiency of iodine? What are the results in the thyroid?
45. What is endemic goiter? What is the cure?
46. While not functionally connected to the hypothalamus the way the neurohypophysis (posterior lobe) is, the adenohypophysis (anterior lobe) is in communication with the hypothalamus via the vascular system. Name the pathway.
47. Between what regions does the hypothalamo–hypophyseal portal system run?
48. What is the significance of the extensive vascular communication between hypothalamus and adenohypophysis?
49. What are the origin and the target organ of the thyrotropin–releasing factor? The corticotropin–releasing factor?
50. How does the structure of the endocrine machinery of the neurohypophysis differ from that of the adenohypophysis?
51. The pancreas is a double organ. What are its two parts?
52. The pancreas serves both endocrine and an exocrine function. Elucidate.
53. If the islets of Langerhans are destroyed, what is the result for body function?
54. What are the acini of the pancreas?
55. The beta cells of the pancreatic islets secrete the best-known pancreatic hormone. Name it. What do the alpha cells produce?
56. What is the specific action of insulin on glucose metabolism?
57. How could you demonstrate that insulin increases the rate of transport of glucose into cells? That insulin is concentrated in the membrane?

58. What is the mechanism of reduction of concentration of blood glucose by insulin?
59. Since glucose is inadequately utilized when insulin is deficient, what happens to fat metabolism? What is the source of the keto-acids characteristic of insulin deficiency?
60. What is the normal effect of glucose level on the secretion of insulin?
61. A large volume of urine is produced in diabetes. Explain.
62. Blood glucose is high in diabetes, but the diabetic is likely to be very hungry. Explain.
63. What is the effect that diabetes has on ventilation? Explain.
64. The neurons depend on the blood glucose level. What, therefore, is the neural effect of hyperinsulinism?
65. What is glycogenolysis? Where does it mainly occur?
66. Where is glucagon produced? What is its action?
67. Over what general functions does the autonomic nervous system have control?
68. What are the two divisions of the autonomic nervous system, and why is this separation made?
69. Along which spinal roots do the sympathetic fibers leave the spinal cord?
70. What is the white ramus of the sympathetic chain?
71. What gives the color to the white ramus, and where do the fibers of the white ramus go?
72. Where do the fibers of the gray ramus arise? Why are they gray? Where do they go?
73. From what region of the CNS do preganglionic sympathetic fibers arise?
74. The autonomic nervous system is a motor system. Where are the sensory fibers that are involved in visceral reflexes?
75. What route do visceral sensory fibers follow to get into the CNS?
76. What is a synaptic ganglion? Distinguish it from a sensory ganglion.
77. Where do the preganglionic sympathetic fibers terminate?
78. How many synaptic levels are between CNS and end organ innervated by a sympathetic fiber? Contrast with arrangement of skeletal muscle innervation.
79. In which cranial nerve do most of the parasympathetic nerve fibers travel?
80. Where do the vagus parasympathetic fibers terminate?
81. What autonomic nerve fibers travel in III, V, VII cranial nerves and in sacral roots?
82. Compare the general features of anatomy of parasympathetic and sympathetic nervous systems.
83. Describe, in terms of number and location of synaptic levels, the vagus parasympathetic pathway to the viscera.

84. Where are the postganglionic soma of the parasympathetic nervous system?
85. What neurohumor is secreted by postganglionic parasympathetic fibers?
86. What is a cholinergic neuron?
87. Where do you find adrenergic neurons?
88. Considering that the parasympathetic is a protective, conservative system, what action would you expect the autonomic fibers of the III cranial nerve would have on the pupillary sphincter muscle? What would you predict to be the effect of the sympathetic fibers?
89. What is the main neural control of the digestive glands?
90. Stimulation of parasympathetic centers in the brain will cause the sweat glands to secrete. Would you therefore predict the sweat gland innervation to be cholinergic or adrenergic? Why?
91. What nerve to the heart would you stimulate to increase the beat frequency? Considering the added nutritional requirements, in this circumstance, what effect do you suppose the same innervation has on the coronary arteries?
92. Extreme muscle activity may be necessary to meet the sort of stresses that mobilize the sympathetic nervous system. What then would you expect to be the effect of sympathetic impulses to the arterioles of skeletal muscle?
93. What is the effect of parasympathetic innervation on the heart?
94. Bronchiolar dilation increases access of air to the lungs. What would you predict to be the effect of sympathetic innervation on the bronchiolar smooth muscle?
95. Distinguish between the adrenal cortex and medulla.
96. What does the adrenal medulla secrete?
97. Compare the effects of adrenal medulla secretion and discharge of the sympathetic nervous system.
98. What is autonomic tone?
99. Explain how existence of autonomic tone places the innervated organs under more delicate control than would otherwise be possible.
100. Is respiration controlled by the autonomic nervous system? Explain.
101. What effect does sympathetic stimulation have on the metabolic rate of body cells generally?
102. The anterior lobe of the pituitary gland, unlike the posterior lobe, does not receive axons from the hypothalamus. It does, however, receive sympathetic fibers (where do they come from?), which stimulate it to secrete a hormone that affects the adrenal gland. What is the hormone and what is its effect?
103. Under what aspect of autonomic excitation do you suppose you are more mentally alert?
104. What is the general effect of a massive sympathetic discharge?
105. Of what significance is vasodilation of skin arterioles? What effect does heat

applied to the skin have on these arterioles? What is the reflex mechanism of the effect?

106. Vasodilation of muscle arterioles occurs in sympathetic excitation associated with exercise. What happens to the arterioles elsewhere in this circumstance? What useful purpose is served thereby?

10

BLOOD AND BODY FLUIDS

10.1 Body Fluid Compartments

All the living cells of the body are bathed in a fluid environment. The volume of this fluid in the immediate neighborhood of most of the cells is very small indeed. Cells touch their neighbors closely in many places, but where they are not in contact, the separation is generally of the order of only a few hundred Ångstroms. The fluid in these very small spaces is ordinarily called *tissue fluid*, and its composition is regulated with great precision. This fluid is the *milieu interiere*, or the *internal environment* of the body. The processes tending to keep constant the characteristics of this fluid are *homeostatic mechanisms*, and *homeostasis* is the condition of constancy of the internal environment toward which all the body's functions seem to strive.

The *blood*, the *lymph*, the *coelomic fluid*, and the *cerebrospinal fluid* are very much like the tissue fluids. These fluids are within vessels coursing among the cells generally, in the case of the blood and lymph, and among the cells of the nervous system, in the case of the *cerebrospinal fluid*. The *coelomic fluid* fills the small spaces between the organs of the abdominal and thoracic cavities. It lubricates these organs and the walls of the cavities in which they are contained. This fluid is *serosal fluid*, the *pleural fluid* in the thoracic cavity, *peritoneal fluid* in the abdominal cavity. All these fluids are *within* the body in the sense that they are closed off from and do not have direct access to the exterior. Each fluid is sufficiently distinct from the others that we may speak of separate fluid *compartments*, and this separation extends to the other classes of fluids to be mentioned (Figs. 10-1 and 10-2).

The constancy of the internal body fluids is made possible by other body fluids that *do* have direct access to the exterior. These fluids are also separated from one another by layers of cells through which solvent water and solutes may pass only by going between or through the cells.

The *input* to the body fluids is mainly via fluid taken into the gastrointestinal tract. To this volume are also added water, salts, and digestive juices that are secreted by the glands of that tract. These fluids form, on the whole, a rather unnatural environment for body cells. However, the gastrointestinal epithelium is protected against deleterious effects of the fluids by the layer of mucus that covers the epithelium.

The *gastrointestinal juices* serve digestive functions and at the same time form part of the *output* or excretory side of the fluid regulatory system. In certain special circumstances water and several dissolved substances may be absorbed by the skin, but in general, the skin, particularly the sweat glands, is a route for *loss* of fluid.

The *lining of the lungs* is another epithelium through which gain or loss of fluid and dissolved substances may occur. However, the largest part of the output side of the fluid regulatory system is in the kidneys, which control both fluid volumes and the chemical composition of the blood (see Chap. 15.1).

It is clear that there are several routes for fluid intake—the epithelium of the digestive tract, the respiratory epithelium, and the skin—and several avenues for fluid loss: the kidneys, the skin, the digestive tract, the respiratory epithelium. Income and outgo of water and solutes at these surfaces must be properly balanced in order to maintain the constancy of composition and volumes of the internal fluid compartments. The fluids in the external compartments are prevented from doing damage to the epithelium with which they come in contact, usually because of a layer of mucus, or of keratinized cells.

Relations among fluid compartments

The continually circulating blood provides the connection between the external and the internal fluid compartments of the body. Intake into the blood is mainly via the digestive tract (Fig. 10-1). The blood distributes its load to the internal fluid compartments (Fig. 10-2), from which the fluid also returns to the blood. Outgo from the blood is via the external fluid compartments (Fig. 10-1). The internal fluid compartments constitute the internal environment of the body—the external environment of the body cells (Fig. 10-2). The external fluid compartments maintain the constancy of that environment by removing surplus constituents in the face of a continual influx. The fluid compartments are regulated as to volume, constituents, osmotic pressure, and *p*H.

In an average individual, about 45 percent of the total weight will be of extracellular water or water outside of cells in the fluid compartment. In a person weighing 70 kg, there will be perhaps 25 liters of interstitial (tissue) fluid and lymph, of which about 10 liters will be in spaces between cells of bone and cartilage. The other 15 liters will be distributed among the cells of looser tissue and will therefore participate easily in the exchanges among the body fluids. Outside the viscera, in the abdominal

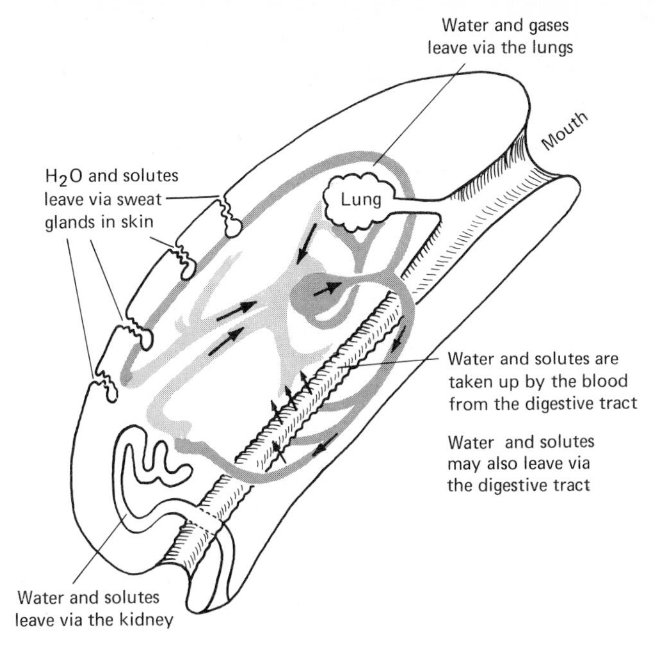

Fig. 10-1. Relation among external fluid compartments and blood. Fluid filtered, or secreted, from blood moves into the external fluid compartment. Conversely, certain constituents of these compartments move into the blood.

cavities, there is also a small amount of fluid in which the organs move and slide over one another; that is, the fluid has a lubricating action. About 5 liters of extracellular fluid will be plasma. The remaining 2 liters include the external compartments through which large quantities of fluid may be lost or gained, but which contain at any moment only a small proportion of the total body water.

The extent to which the various compartments are filled varies with circumstances. Gastrointestinal and urinary volumes undergo daily large variations, but other volumes change radically only in extreme or pathological conditions. If the blood does not move adequately through the blood vessels, interstitial fluid may accumulate locally, as it also will if there is local damage to small blood vessels. The swelling that results from such accumulation is called *edema*.

Measurement of the volumes of the various compartments is beset with difficulties, and the dilution method is the only practical one for measuring fluid volumes in the intact body. According to this method, if a known amount of material is diluted in an unknown volume, that volume may be calculated if the concentration of the substance in the volume can be determined. The method is valid if the material stays in the volume that is to be measured and if it becomes uniformly diluted throughout it. Unfortunately, because of the intercommunication among the various compartments, the measurement of the volume in which a particular substance exists may be subject to considerable error. A substance of sufficiently large molecular size will be limited to the internal fluid compartments at least for a short time, until it is broken down and/or excreted. Properly, one speaks of a *chloride*, or *inulin*, or *sucrose space*, to indicate the particular volume throughout which the tracer molecules are distributed. The relations of these functional volumes to the anatomical volumes may not always

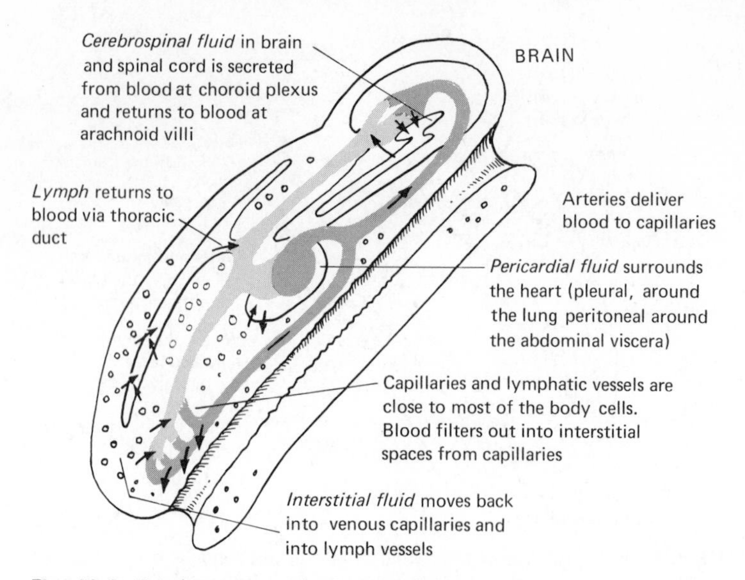

Fig. 10-2. Relations among the internal fluid compartments of the body. Blood is forced out along the arteries and into the small capillaries where it is filtered out into the interstitial spaces. Cerebrospinal fluid is secreted from the blood into the cerebrospinal canal. Interstitial fluid drains back into the veins and into the lymph vessels. Cerebrospinal fluid and lymph return to the veins. From the heart, follow the large arrows showing the pathway of blood from arteries to capillaries to veins and back to the heart. Note short arrows showing where fluid leaves or returns to the blood or other compartment.

be entirely clear. An example showing the measurement of blood volume by the dilution method will be found in the next chapter.

The various fluid compartments of the body are separated from one another by layers of flattened epithelial cells. These cells are in such close contact with one another that diffusion of many small molecules from one side to the other is ordinarily more likely to occur through the membranes of the cells rather than in the spaces between the cells. The layer of cells, therefore, acts as a continuous double membrane and a barrier to free diffusion.

An extracellular compartment that is different from the others in its relation to the blood is the *cerebrospinal fluid.* A number of substances will not pass from capillaries into interstital space in the brain, but if introduced into the cerebrospinal fluid, they will become distributed into that volume. Thus there appears to be a *blood-brain barrier* that diminishes access of blood to the neurons. The cerebrospinal fluid is secreted from blood constituents by cells of the thin roof of the cerebral ventricles. These cells are in a very vascular membrane called the *choroid plexus,* a network of blood vessels. During development the cerebral lobes grow up and over, leaving the original roof of the cerebral canal buried deep in the brain.

The fluid that is secreted into the cerebral ventricles leaves the cerebrospinal canal via passageways in the roof of the medulla oblongata. If there is obstruction to the flow of cerebrospinal fluid, pressure may build up within the canal and cause injury to the brain. Ordinarily the fluid moves out into the spaces surrounding the central nervous system. These spaces are formed by the *meninges* that enclose the entire central nervous system.

The meninges are of several layers. Close to the surface of the brain is the *pia mater*, whereas the tough *dura mater* is next to the bone. Between these two layers is the delicate *arachnoid layer*—like a filmy spider web. The fluid in the space between the layers moves out into the venous sinuses of the skull. The passageway is through tiny one-way valves that project out in the arachnoid layer over the surface of the brain. Thus the cerebrospinal fluid is finally drained into the bloodstream. The relationship of neurons to the tissue fluids is not the same as in other tissues, where fluid moves out of capillaries directly into the spaces surrounding the cells. In the nervous system, the glial cells intervene between capillaries and neurons.

Materials in the bloodstream cannot pass directly into the spaces around the neurons. Instead the substances must pass through the glial cells, which are in contact with the neurons. The extraneuronal fluid space of the nervous system appears to be largely occupied by glia. The failure of certain substances to penetrate from the blood into neurons is due to the glia. The glia and the choroid plexus provide the blood-brain barrier. The fluid surrounding the central nervous system not only plays a nutritive role but also has an important function of mechanical support. It acts as a cushion to absorb the shocks of movement and vibration to which the skull and spinal column may be subject.

Figure 10-3 compares the composition of the cerebrospinal fluid and of plasma,

Ion	Cerebrospinal Fluid (mM/l)	Blood Plasma (mM/l)
Na^+	142	152
K^+	2.4	4.3
Mg^{++}	2.8	2.2
Ca^{++}	2.7	5.4
HCO_3^-	25	28
Cl^-	133	110

Fig. 10-3. Comparison of major ionic constituents of cerebrospinal fluid and blood.

the fluid part of the blood. Among those ions present in greatest concentration, the most striking difference is in the lower concentration of negative charge in the blood compared to CSF. This seeming deficit is made up by negative charges on protein, which is present in higher concentration in the blood than in the other extracellular fluids.

The composition of each compartment differs from that of any other. The differences arise from the various metabolic activities of cells associated with each compartment. Continuous synthesis of material is one factor affecting the movement of material among the various compartments.

The kinds of differences that may be found between extracellular and intracellular fluids are illustrated in Fig. 10-4. Although the proportions of the different kinds of ions are quite different, the total number of ions per unit volume on the inside is nearly the same as on the outside. The difference, although small, is important, as we shall see.

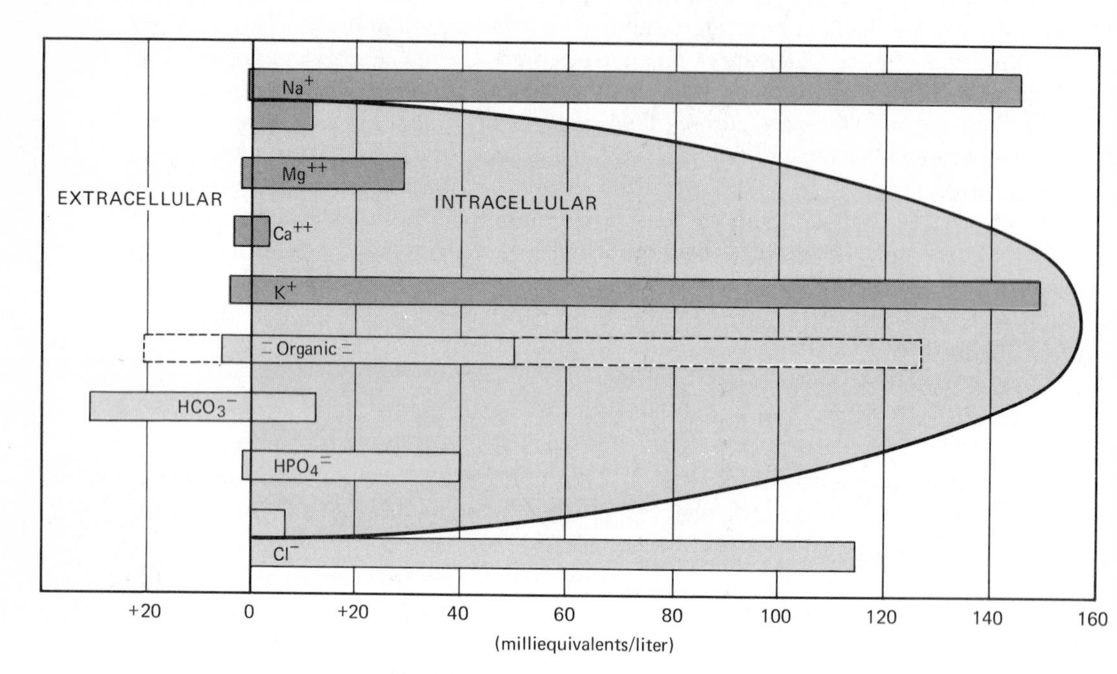

Fig. 10-4. Extracellular and intracellular ions. *Horizontal scale,* concentration of ions in meq/liter. The relative numbers of particles of each kind, in the intracellular volume (shaded) and in the extracellular volume (clear) are directly related to the areas of the bars. Note: Na^+, Cl^+, HCO_3^- are relatively high outside; K^+, Mg^{++}, HPO_4^{--} and organic anions are higher inside the cell.

The total number of positive charges on the inside appears to be about the same as the total number of negative charges on the outside in an equal volume: but, inside, the number of negative charges seems to be insufficient to balance the positive charges. Actually, the apparent deficit is made up by organic ions, the nondiffusible ions, which carry several negative charges on each molecule. These nondiffusible ions are the source of the electric potential and osmotic differences that can be measured across the membrane separating two compartments—for example, between the inside and the outside of a cell (see Chap. 6.1). In order to account for these differences, we use principles of diffusion that apply to the relation of a cell to its environment as well as to the relation between two compartments separated by epithelia.

Electric potential and osmotic pressure differences between fluid compartments

There is a higher concentration of protein in the blood than in the various tissue fluids. This protein has a net negative charge, which in the bulk volume of the blood is compensated by ionic charges such as Na^+. In other words, the solution as a whole is electrically neutral. The smaller ions like Na^+ will tend to diffuse through the membrane separating the compartments and thereby obliterate any difference that may exist. However, this diffusion tendency will be limited by the fact that the electrical attraction prevents the diffusible ions from moving far from the nondiffusible, charged protein molecules with which they are matched. The large, negatively charged protein molecules stay on one side of the membrane, and the diffusible ions will be able to move only a very short distance toward the other side. The effectiveness of the restraint that the nonmobile ions impose on the diffusion tendency of the mobile ions will be reflected in the magnitude of the membrane potential that may be measured between one compartment and the other.

The situation that gives rise to this potential difference can be explained by a physical model called the Gibbs-Donnan Equilibrium, which is illustrated in simple form in Fig. 10-5. In the model, K^+ and Cl^- are used as the diffusible ions because they appear to be most relevant in explaining the resting membrane potential of a cell (see also Chap. 6.1). $P^{\equiv}$ represents a nondiffusible, charged protein molecule.

Figure 10-5 also shows another consequence of the presence of nondiffusible ions: the concentration of diffusible ions will not be the same in the solution on one side as on the other side of the membrane, and a difference in osmotic pressure will therefore exist.

Osmotic pressure can be measured in terms of the concentration of solute particles to which a particular membrane is *not* permeable. The number of protein molecules may not be enough to make a significant contribution to the osmotic pressure by themselves; however, the diffusible ions that are restrained from leaving the compartment become, because of that restraint, limited to that compartment and thereby contribute to the osmotic pressure. The osmotic pressure that may be attributed to the protein plus the attracted smaller ions is called the *colloid osmotic pressure.* Since the colloid osmotic pressure on one side is greater than on the other, water would tend to flow from the latter to the former.

Figure 10-5(e) illustrates the situation. Under the conditions described, the cell or the compartment will swell unless the membrane is strong enough to resist the inward flow of water, or is extremely leaky; in the latter case, the differences we have postulated do not exist in the first place. Actually, the influx of water seems to be taken

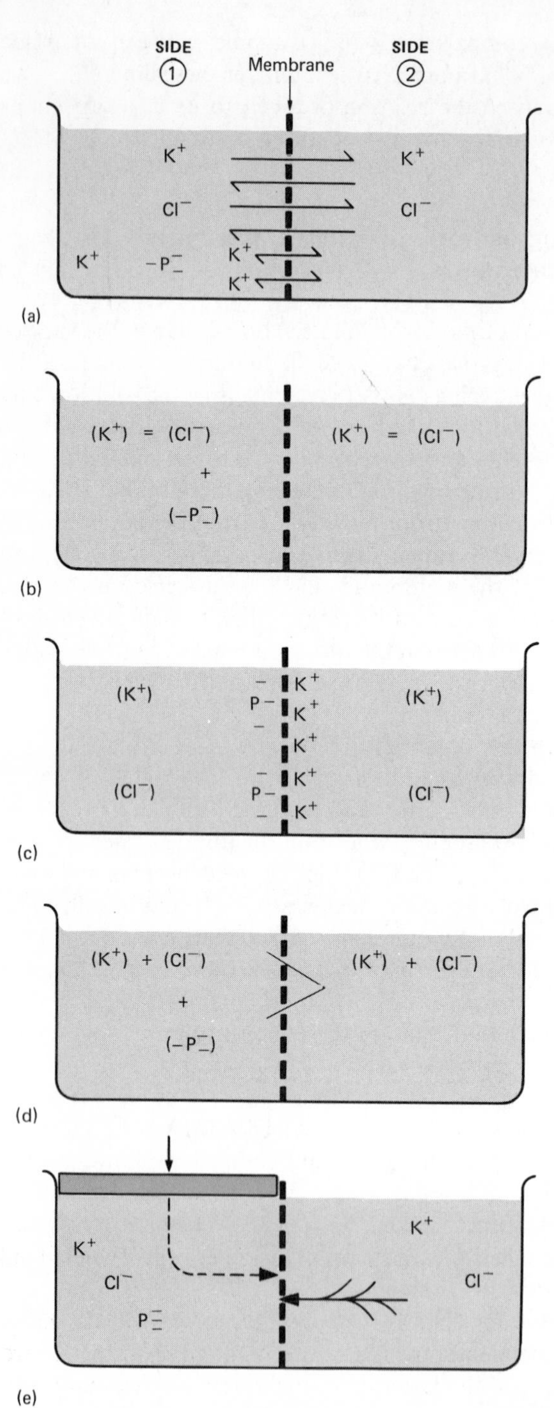

Fig. 10-5. Gibbs-Donnan equilibrium. Electric potential and osmotic pressure differences exist between one side and another of a membrane impermeable to large charged protein molecule (P^-). The membrane in the example is assumed to be permeable to the only other ions present, K^+ and Cl^- (a). In the bulk solution of each compartment, there must exist electrical neutrality—i.e., the sum of all the negative charges must equal the sum of all the positive charges (b). *But,* $(K^+)_1$, neutralizing $(Cl^-)_1 + (P^-)$, is greater in concentration than is $(K^+)_2$, so there is a net tendency for $K_1{}^+$ ions to diffuse toward side 2, and an electrical double layer is set up, such that *side 2 is positive to side 1* (c). Because of the presence of the nondiffusible ions, the sum of all the charges on side 1 exceeds that on side 2 (d). Therefore there is a *tendency for water to flow from side 2 to side 1* (e).

care of by a *metabolic pump* that ejects through the membrane water entering the cell [illustrated in Fig. 10-5(e)]. The pump probably works by ejecting Na^+ ions, and water moves along with greatly hydrated Na^+ ions. The activities of the cells determine the composition of the extracellular fluids. The biochemical machinery seems to select the K^+ ion, while excluding Na^+ and Cl^-. A good deal of evidence seems to suggest that the critical event is the exclusion of Na^+. Almost all cells seem to have a Na^+ pump that ejects Na^+ when this ion diffuses into the cell. This universal condition is probably related to the utilization of Na^+ by cells during the synthesis of protein.

Acid and base balance of fluid compartments

The relative acidic or basic quality, the acid-base balance, is an important characteristic that is maintained quite constant in the internal body fluids. The constancy is particularly required in order to keep the integrity of the molecular structure of cell membranes and to allow cellular enzymes to function properly.

The internal body fluids are only slightly more alkaline than pure water, being maintained at a *p*H about 7.4 compared to 7.0 for water. In contrast, the *p*H of the external body fluids is different and variable. The stomach contents may be as low as *p*H 2; the urine may be as acidic as *p*H 4 and as alkaline as *p*H 8, depending on what is excreted.

Most materials taken into the stomach have little opportunity to affect the *p*H of the internal body fluids, for the process of digestion and absorption intervenes. Acid products of metabolism, however, continually threaten to upset the *p*H balance. Substances in the blood and tissue fluids serve as *p*H buffers (see Chap. 3.1) against this threat.

One especially important buffering substance in body fluids is *sodium bicarbonate*. In dilute solution, the sodium bicarbonate molecules are *completely dissociated*, as indicated by the single arrow below.

$$NaHCO_3 \longrightarrow Na^+ + HCO_3^-$$

If an acid is added to the sodium bicarbonate, the HCO_3^- ion will tend to become associated with the added H^+ ion.

$$H^+ + HCO_3^- \rightleftharpoons H_2CO_3$$

The small arrow means that in a mixture of this sort, some of the ions will remain associated. The carbonic acid (H_2CO_3) is called a *weak acid* because only a relatively small fraction of it remains dissociated in solution. It is precisely this characteristic of being able to soak up H^+ ion that makes bicarbonate ion a good buffer against acid. CO_2 produced by cells participates in the above reactions in a manner more fully described in Chap. 12.2.

10.2 Anatomy and Function of the Blood

The cells of the body are provided with oxygen, nutrients, and other stimulating (or depressing) substances by the *blood*, which also removes waste products and secretions from the regions surrounding the cells. The blood is contained within the *vascular system*, which includes the heart and blood vessels. The vascular system, and the blood moving within it, constitutes the *circulatory system*. Exchange of materials between the blood and the tissue fluids occurs through the walls of the capillaries, the smallest subdivisions of the vascular system.

Functions and composition of the blood

Most cells of the body can function properly only if their immediate surroundings—the body's *internal environment*—remain relatively unchanged. The fluid surrounding the body cells continually exchanges with the blood through the capillary walls, and the constancy of the internal environment depends on the constancy of the properties of the blood. Into the bloodstream are poured waste products of metabolism that are removed by the excretory system. Oxygen comes to the bloodstream from the air in the lungs and is caught up by the red blood cells. Dissolved food materials come into the bloodstream through the walls of the digestive tract. Cells that consume agents of disease are suspended in it. Dissolved in it are proteins which, by immunity reaction, inactivate destructive materials that may gain access to the circulation; other proteins dissolved in the blood can coagulate at appropriate times and prevent blood loss through broken vessels. Like the nerves, the blood provides a system for communication among all tissues of the body. The information it carries is contained in certain chemical substances, particularly hormones from the endocrine glands. These hormones are distributed throughout the body by the circulation and affect the functioning of particular cells.

The fluid part of the blood, the *plasma*, is rather similar to diluted seawater, with the addition of dissolved organic compounds—of which the proteins are especially important. The proportions of the dissolved salts are about the same, but there is relatively more water in a given volume of human blood plasma than there is in an equal volume of seawater. The concentration of salts in blood is similar to that in interstitial fluid. The major difference is that blood throughout the body contains a higher concentration of protein. The capillary walls are relatively impermeable to the protein molecules, which, because of their charge, set up a Gibbs-Donnan situation in which the (Na^+)—that is, the *concentration* of Na^+—is about $\frac{1}{20}$ higher, and the (Cl^-) about $\frac{1}{20}$ lower in the blood than in the interstitial fluid.

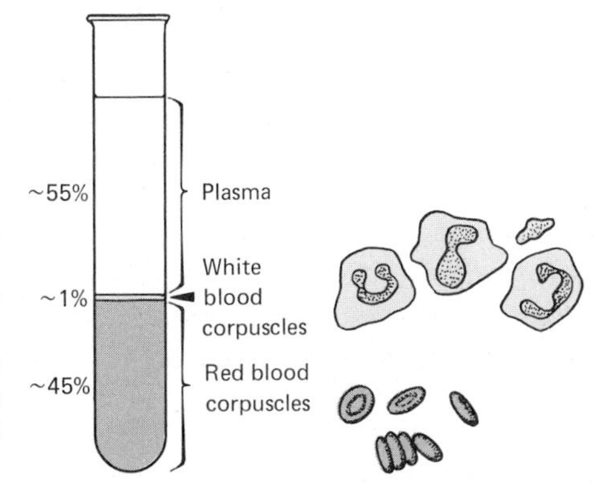

Fig. 10-6. Plasma and cells of blood. The plasma and cells are separated by centrifugation. The heavier red cells sink to the bottom, while the white cells (which have a lower concentration of dense material and more water) accumulate between the plasma layer and the red cell layer.

Nearly half the volume of the blood consists of blood cells (Fig. 10-6), and one commonly speaks of the blood as a tissue. It differs from other tissues chiefly in the fact that the cells are not attached but move with the flow of blood. The red blood cells, or *erythrocytes*, contain the red pigment, *hemoglobin*, which combines with oxygen in the circulation of the lungs and gives up oxygen to the tissues. In the adult, the erythrocytes arise through cell division in the red marrow of the bones, but in the middle period of embryonic development they are manufactured in the liver and spleen as well. A mature, circulating erythrocyte is remarkable for having lost its nucleus. In spite of this shortcoming, under normal circumstances the red cells remain alive in the circulation for about 4 months. That they do not survive longer may be due to the continual abrasion they receive in tumbling about in the circulation, particularly when squeezed through some of the smaller branches, the capillaries. The shape of the red blood cell, a disc flattened more in the middle than at the rim, provides a very large exposed surface area, relative to the volume of the cell. This feature favors rapid passage of oxygen and carbon dioxide through the cell surface.

The *white blood cells* (*leucocytes*) are not really white but colorless, and they constitute about 0.5 to 1 percent of the blood volume. When blood is centrifuged, the red blood cells collect at the bottom of the centrifuge tube because they are heavier than the plasma and are forced down by the centrifugal force. The white blood cells are lighter than the red cells and appear at the boundary between the erythrocytes and the overlying fluid plasma. Like froth in beer, the aggregation of these millions of tiny, colorless cells gives a whitish appearance. Most of the leucocytes are not in the bloodstream but are distributed among the tissues of the organs of the body and only enter the circulation transiently.

The nucleus of a white cell is, like that of most cells, retained in the mature cell. The *granular leucocytes* appear to have coarse particles in the cytoplasm, and the nucleus is irregular in shape or lobulated. The cells are therefore described as *polymorphonuclear*. They, together with the *monocytes*, smaller cells with little cytoplasm and a simple nucleus, arise in the bone marrow. *Nongranular leucocytes*, such as the *lymphocytes*, originate in lymph nodes and increase in number in the bloodstream during recovery from infection. The monocytes accumulate near injuries and assist in the repair of wounds by *phagocytosis*, a process of engulfing particles.

The polymorphonuclear leucocytes are distinguished according to the staining characteristics of their granules. The granules in the *basophils* are stained by basic dyes (e.g., blue by hemotoxylin), those in the *eosinophils* are stained red by the acid dye eosin, while the *neutrophils* are stained by both dyes.

The neutrophils are phagocytes; that is, they ingest debris of various kinds, including bacteria that may have invaded the bloodstream. They are especially numerous during infections. The eosinophils may consume dissolved foreign protein. Allergic diseases and parasitic infection are characterized by *eosinophilia*, an increase in concentration of eosinophils in the bloodstream.

Blood *platelets* found with the white cell layer after centrifugation of the blood are not really cells but parts of cells—*megakaryocytes*, giant cells in the bone marrow, which break up and become these tiny fragments. The fragments remain intact until they touch a foreign surface, such as the skin, when they literally explode and release substances that aid blood clotting.

Calculation of blood volume and cell concentration

The fraction of the blood volume that is occupied by cells may be determined by centrifuging a sample of blood in a test tube (Fig. 10-6). The measure of the fraction of the blood that consists of cells is called the *hematocrit*. Usually it is about

0.45, but it may be as high as 0.8 for a sample of blood taken from the spleen or as low as 0.2 in blood from the circulation of the kidney.

It is sometimes of practical importance to know whether the total blood volume is greater or less than normal. The average amount of blood in an average person is about 5 liters. Direct measurement of this volume by collecting all the blood from the body is neither convenient nor comfortable for the subject. The volume may, however, be measured *indirectly* by means of the *dilution principle*. According to this principle, when a solution is diluted, the concentration of the dissolved material varies inversely with the volume of the solution; that is, volume and concentration change in opposite directions to one another. When the volume is made greater, the concentration is proportionately decreased. Put another way, the product of concentration and volume remains constant when the concentration of a material in solution is changed by the addition or removal of solvent. *This simple physical rule has wide application in the study of the kidney and the lungs as well as the circulation.*

Suppose that a small volume and known concentration of a conveniently measurable substance, such as a dye, is injected into the bloodstream. When the material becomes distributed throughout the blood, a sample of blood may be taken and the concentration of the diluted material may be measured (Fig. 10-7).

Let us define *concentration* as *amount per unit volume*. That is, in general,

$$C = \frac{a}{V} \quad \text{and} \quad a = CV$$

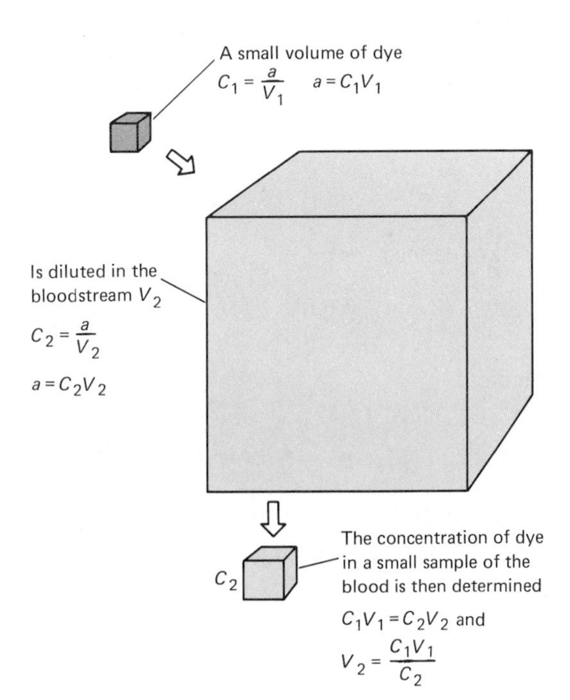

Fig. 10-7. Calculation of blood volume by means of the dilution principle (see text).

If the amount of the substance injected is in a volume V_1 and at a concentration C_1, then $a = C_1V_1$. The amount a becomes diluted to volume V_2 in the bloodstream, where its concentration is C_2. Then $a = C_2V_2$. Two quantities equal to the same quantity are equal to each other. Therefore

$$C_2V_2 = C_1V_1 \quad \text{and} \quad V_2 = \frac{C_1V_1}{C_2}$$

If $C_1 = 1$ mg/cc and $V_1 = 5$ cc, and if C_2 is measured to be 0.001 mg/cc, then

$$V_2 = \frac{1 \text{ mg/cc} \times 5 \text{ cc}}{0.001 \text{ mg/cc}} = 5000 \text{ cc} = 5 \text{ liters of blood}$$

This calculation holds, of course, only if the material is not lost from the blood, either through the kidney or in the tissue fluids. For short intervals of time, some substances may be limited to the bloodstream sufficiently to permit a valid calculation of blood volume to be made.

The concentration of blood cells in the human bloodstream may also be determined by means of the dilution principle (Fig. 10-8). An ordinary blood-counting slide has numerous lines that form tiny squares $\frac{1}{20}$ mm on a side in a depression $\frac{1}{10}$ mm deep. The red cells in a drop of whole blood placed on this slide will be so numerous that they cannot be counted directly even in one of the smallest squares. Therefore the blood is first diluted (usually 200 times). Then we might find, in a particular instance, an average of six red blood cells in each small square.

Problem: What was the concentration of red cells in the original blood (Fig. 10-8)?

We know that the concentration of cells in the dilute sample (C_2) is only $\frac{1}{200}$ of the concentration in the original blood (C_1). That is,

$$C_2 = \frac{1}{200} \times C_1 \quad \text{or} \quad C_1 = 200 \times C_2$$

We have defined

$$C_2 = \frac{a_2}{V_2}$$

Therefore

$$C_1 = 200 \times \frac{a_2}{V_2}$$

where a_2 is the number of cells counted in a small square of the counting slide and V_2 is the volume within the small square. From the given dimensions,

$$V_2 = \frac{1}{20} \text{ mm} \times \frac{1}{20} \text{ mm} \times \frac{1}{10} \text{ mm} = \frac{1}{4000} \text{ mm}^3$$

and therefore

$$C_1 = 200 \times \frac{6}{1/4000} = 1200 \times 4000 = 4{,}800{,}000 \text{ cells/mm}^3$$

1 liter = 1000 cc; that is, a cube 10 cm × 10 cm × 10 cm or 100 mm × 100 mm × 100 mm, or 1,000,000 mm^3. In 5 liters of blood there are 5,000,000 × 4,800,000 = 24,000,000,000,000 cells, the total number of red blood cells in the entire body. Such large numbers are awkward to handle. It is easier to write 2.4×10^{13}; that is, 2.4 × the quantity (10 multiplied by itself 13 times).

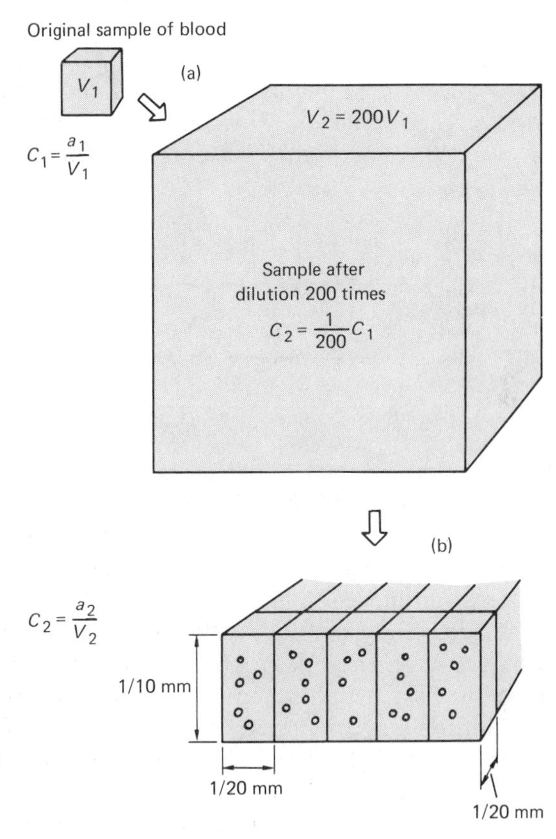

Fig. 10-8. Calculation of concentration of cells in blood by means of the dilution principle.

Organic solutes in blood: Clotting substances

Besides mineral salts, organic materials of considerable importance are dissolved in blood. Some are absorbed from the digestive tract and are merely transported by the blood to cells where they will be used. Of major importance for the function of the blood are the proteins, which are manufactured mainly in the liver.

Special clotting proteins confer on the blood an ability to prevent the loss of itself. When blood vessels are cut, they retract and constrict, reducing the blood flow at the region of the cut. In addition, the blood *coagulates*—that is, a blood clot forms and serves as a plug. The process of clot formation is called *coagulation*. The following description is a very much simplified scheme of the actual sequence of events occurring during the coagulation process.

The clot is a mass of fibers of a protein material called *fibrin*, which is insoluble in blood. It comes from the blood but is not present as fibrin in the circulation. It arises under special circumstances from *fibrinogen*, a soluble blood protein. The transformation of fibrinogen to fibrin occurs in the presence of an activator, *thrombin* ("clot-former") (Fig. 10-9), a proteolytic enzyme that catalyzes a chemical change in the fibrinogen to smaller fibrin molecules, which aggregate to form the clot.

Obviously this substance is also not present as such in circulating blood. It arises from a precursor, *prothrombin*, in the blood. This transformation also requires an activator. The activator, *thromboplastin*, is released from injured cells and from the tiny platelets. The thromboplastin is a phospholipid, a constituent of cell membranes.

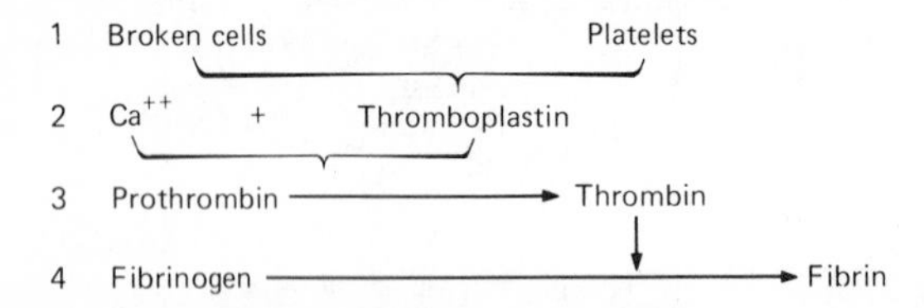

Fig. 10-9. Sequence of events in blood coagulation. These steps, numbered in sequence, occur during blood coagulation.

The platelets remain intact when the blood is inside the blood vessels but are easily broken when the blood wets a foreign surface, as when it flows out of a cut on to the skin. The calcium ion (Ca^{++}), which is always present in the blood, is also required if the change of prothrombin to thrombin is to occur, and it hastens the conversion of fibrinogen to fibrin.

A complete picture of clotting would involve half a dozen more steps of successively activated enzymes prior to the prothrombin and fibrinogen transformations before the final step, fibrinogen to fibrin, is reached.

Clotting may be slowed or stopped by interference with any of these steps. If the blood is collected in a wax-lined or silicone-coated vessel that is not wet by the blood, the platelets will not be broken and thromboplastin will not be released. The Ca^{++} must be available as a free ion. If a chemical with which the Ca^{++} will readily combine (such as citrate or oxalate) is added, then the Ca^{++} will not be available to help in the formation of thrombin. The action of thrombin may also be interfered with by an anticoagulant like *heparin*, found in the liver, or *hirudin*, produced by leeches. Vitamin K is required for prothrombin formation by the liver; an insufficiency of this vitamin will increase clotting time. *Hemophilia* is a hereditary deficiency of another clotting factor. It is characterized by extensive bleeding from even small cuts. It appears only in males because affected females do not even survive long enough to be born. The hereditary factor for the deficiency may be carried by a woman who does not, however, have the disease. She may pass this factor on to her sons, who will then be afflicted.

When a clot forms in a blood vessel, as may happen occasionally under conditions of slow blood flow or injury to a vessel, it is called a *thrombus*. Thrombus formation may be inhibited by heparin, or by *Dicumarol*, a compound first found in spoiled clover.

Several days after a clot has formed, regrowth of capillaries into the region may start, and the clot will begin to dissolve. A proteolytic enzyme, *plasmin*, is responsible for the destruction of the clot.

Immunity reactions

The detailed shape and the chemical composition of protein are quite specific for each kind of animal, and even different individuals of a particular species are constructed of slightly different proteins. Proteins that get into the digestive tract are ordinarily broken into their constituent amino acids before being absorbed into the bloodstream. The molecular structure of protein of the particular individual can then be built from these smaller molecules.

A dramatically different situation exists when intact foreign protein or organisms enter the bloodstream. Certain cells of the circulatory system line the small blood vessels of the liver, bone marrow, and spleen; collectively they comprise the *reticulo-endothelial system*, (the RES). These cells are stimulated by the foreign material to manufacture special protein molecules. The molecules are *antibodies* that will react against the strange substances (see Fig. 10-10). Any material that stimulates the production of antibodies is called an *antigen*.

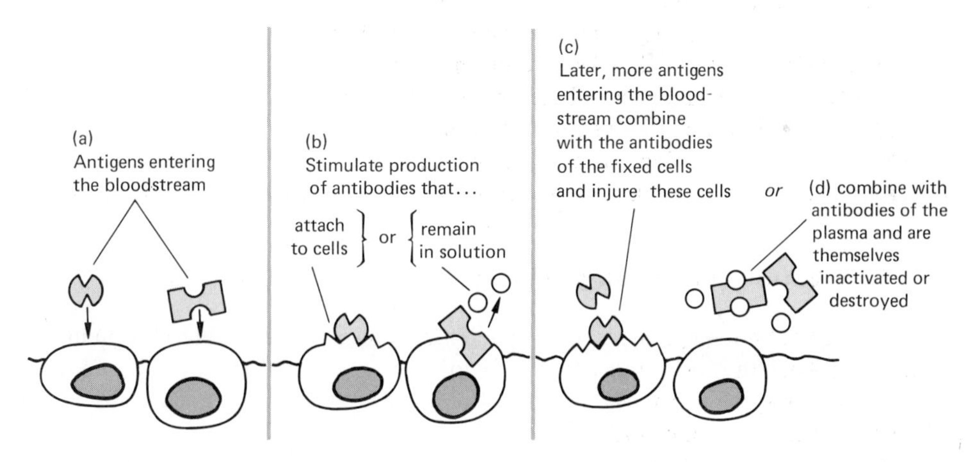

Fig. 10-10. Scheme for the immunity reaction. The shapes of the shaded figures represent specific foreign protein material antigen. The corresponding antibody, fixed or free, is unshaded.

Following the injection of a foreign protein, an excess of antibody is produced, which can combine with more antigen that may appear subsequently. Each time antigen appears, the body synthesizes a larger quantity of antibody. The antibodies are formed mainly by certain white cells, the *plasma cells*, and to some extent by the lymphocytes.

The *thymus* is an organ that has an important bearing on antibody formation, even though no significant number of antibody-forming cells are found in this organ in the adult. If the thymus is removed from a young individual, that individual cannot produce antibodies when he is an adult. The thymus is perhaps the place where plasma cells arise, move out into the circulation, and become established in other organs. Alternatively, the thymus may be the source of a hormone that regulates the immunological response of plasma and lymph cells.

Molecular anatomy of antigen: Antibody reaction

The antibodies in the bloodstream are a particular type of protein known as *gamma globulin* or *γG*. The γG molecule is of a form shown in Fig. 10-11. Two long peptide chains of amino acids are connected to one another by S—S bridges. A shorter chain is also connected to each long chain by S—S bridges. The large chains are called *heavy* because they have the larger molecular weight, whereas the short chains are called *light* because they have the smaller molecular weight. The long, heavy chains have the greater density and therefore move more rapidly toward the bottom of a tube during centrifugation.

The short (light) chains and the associated parts of the long (heavy) chains are the regions of the antigenetic specificity. The sequences of amino acids in these other parts depend on the nature of the antigens that stimulate their synthesis. These parts of the antibody molecule, in fact, combine with antigens, and this region of the molecule is called the *binding fragment*. The other ends of the heavy chains can be removed, and since they are all alike, a solution of these molecular fragments will crystallize. Therefore this end is called the *crystallizing fragment*.

The heavy chains can move apart by motion at the hinge points (shown by arrows), and the combinations of antigens with antibodies can therefore exist as triangles and squares, as can be seen in electron microscopic pictures.

Various immunological responses

Antibodies against foreign protein may be found in the blood protein and in the cells, both in those circulating freely in the blood and in those lining the walls of the vascular system. Antibodies against *organisms* are found mainly in the blood plasma, particularly in the globulin protein manufactured by cells in the bone marrow and lymphatic glands. An organism against which immunity has been developed will be coated with antibody if it enters the bloodstream, and cells will be destroyed. In the case of bacteria, if the body is able to withstand the poisons of a first infection,

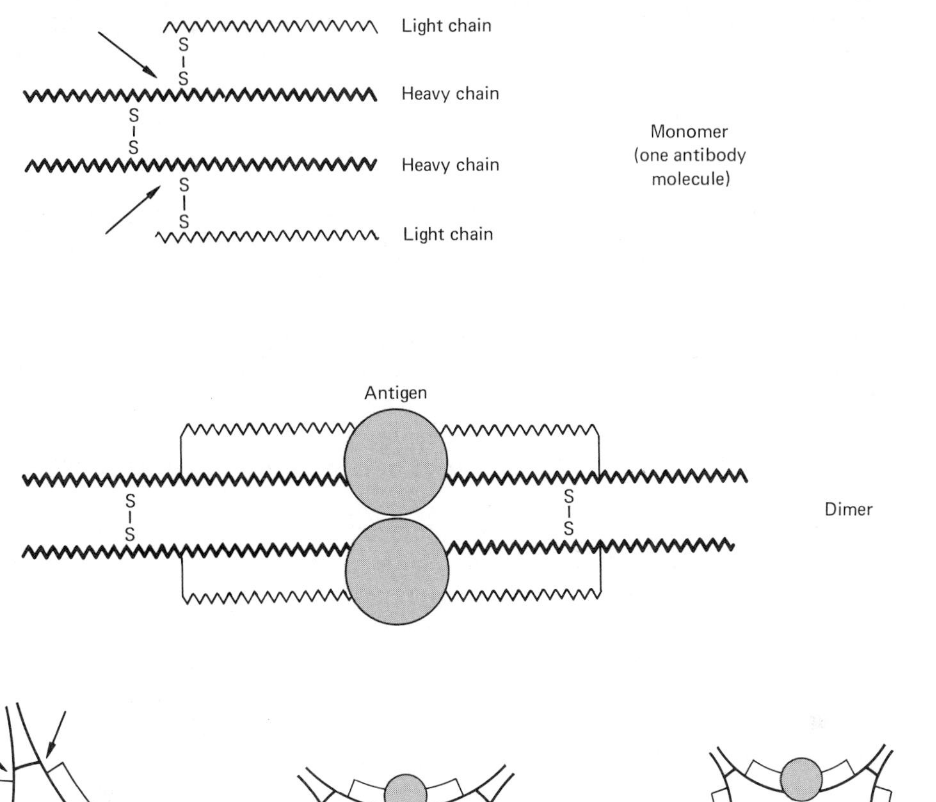

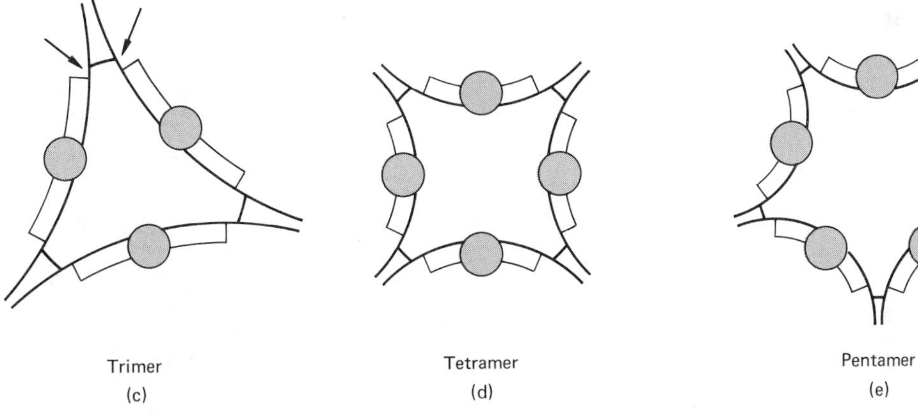

Fig. 10-11. Gamma globulin antibody-antigen combinations. (a) Scheme of antibody molecule structure: Two heavy (long) polypeptide chains (∧∧∧) are held together by S—S bond. A light (short) polypeptide chain is held to each heavy chain by an S—S bond. (b) Two antibody molecules are joined to an antigen molecule (gray). (c), (d), (e) Three, four, and five antibody molecules are joined to antigen molecules.

The chains bend at the arrows to allows formation of structures shown in (c), (d), (e).

sufficient antibodies may be synthesized to destroy the members of a second invasion. Immunity may also be conferred by injection of weakened but living (attenuated) organisms to stimulate the synthesis of antibodies.

Occasionally intact protein of foreign origin will be absorbed through the wall of the digestive tract or the respiratory system. These protein molecules stimulate the formation of antibodies mainly in cells, particularly in those lining the vascular system. When such sensitized cells are subsequently presented with the foreign protein a second time, the antibody-antigen reaction that takes place on their surfaces has an adverse effect on the whole body. Cell membrane permeability increases greatly, either as a direct effect of the antibody-antigen reaction in the cells or as the result of secretion of *histamine* by the cells following such reaction. Skin rash, puffiness of the skin, and other unpleasant effects that we recognize as allergic reactions may result. Such effects may occur if egg albumen or fish protein is absorbed undigested. Young children are sometimes particularly susceptible, but often they outgrow their sensitivity. Foreign proteins inhaled as dust or pollen may have a localized effect on the respiratory tract and may produce asthma or hay fever.

In transfusion of blood from one animal to another, foreign protein, as red blood cells and as soluble protein in the plasma, is introduced into the bloodstream. No difficulty may arise from a first transfusion if, for example, blood of a rabbit is injected into a guinea pig. However, a guinea pig receiving a second injection of rabbit blood may die from the intense reaction that follows. The reaction is called *anaphylactic shock* and occurs when the plasma antigens of the rabbit blood react with the antibodies of the fixed cells of the guinea pig vascular system. In anaphylactic shock, the cells of capillaries and other structures become excessively permeable, nerves become hyperexcitable, and fluid leaks through capillary walls. A foreign protein, such as egg white, injected into the bloodstream a second time several weeks after an initial dose will also provoke an anaphylactic shock. In the case of the injection of blood, the red cells of the transfused blood will also have been destroyed by the combination of their antigens with the antibodies that were synthesized in the guinea pig plasma.

Immunology and transplants

Another circumstance in which the antigen-antibody reaction is important is in the case of tissue and organ transplants. A piece of skin transplanted from one place to another on a person's body will grow perfectly well, as you might expect. So, too, will a similar transplant made from one identical twin to another. A transplant from a donor not genetically identical with the recipient is likely to run into trouble. The proteins of the donor cells are foreign to the recipient. They are antigens that stimulate the production of antibodies in the recipient. Antibodies combine with antigens and destroy the donor cells. Fortunately, as the many successful transplants of skin, kidneys, and hearts certainly demonstrate, there are ways of preventing an immunity reaction between the graft and the host. An effective method is the removal of the thymus at birth. The adult will then accept any graft. This procedure has been carried out in experimental animals. Sometime before birth the apparatus for producing antibodies is not yet developed, so that foreign substances are not rejected. On the contrary, the fetus develops a tolerance for otherwise-strange protein structures, which are therefore treated instead as if they were native to the fetus' own molecular form. This acceptance persists into the adult. If cells of a potential donor are made available to a fetus, that fetus will develop an immunological tolerance to the donor. The fetus grown to an adult will then accept a transplant from the donor to which it was conditioned. Of more immediate practical use is the reduction of immunological response by treatment of the host with X-irradiation, or, better, with particular chemicals. Some compounds seem to poison the immunological system. In certain cases, massive doses of antigen will reduce the immunological response and protect the transplanted tissue or organ from destruction.

A, B, O blood types

Naturally occurring antibody-antigen combinations provide the basis for the human blood groups. Antigens corresponding to a person's own antibodies do not, reasonably enough, occur in his own bloodstream. In transfusions, however, care must be taken that the red blood cells of the *donor* (the person donating the blood) do not have antigens that would react with the *recipient's* antibodies. When the cells are in the presence of antibodies with which the cell antigens react, the cells become clumped together (agglutinate), their membranes become leaky, and the hemoglobin diffuses out. Cells thus destroyed are said to be *hemolyzed.* The hemolobin, freed from the cells, filters into the kidney tubules, where it may injure the cells of the tubules as it crystallizes in the concentrated urine.

In order to determine the common blood types accurately, the cells in several blood samples are separated from the plasma. The cells from each sample are mixed with each plasma sample. In those combinations in which agglutination of the cells occurs, the antigen of the red cells must have met its corresponding antibody in the plasma to which the cells were added.

Many potential antigen-antibody combinations exist in blood, but most are too weak to be of any practical importance, or are developed only under exceptional experimental conditions. Two particular antigens are of most general importance. They are designated A and B, and they may occur in cells in the combinations: *separately*, *together*, or *not at all.* Thus there are the blood types A, B, AB, O (Fig. 10-12). The *antibodies* in plasma corresponding to the A and B antigens may be designated *anti-A* and *anti-B*. Since clumping of the red cells would occur if the antibody were present in the blood with its corresponding antigen, type A blood does *not* have anti-A plasma but rather contains anti-B, and type B blood has anti-A. Type AB has neither, while type O blood has both antibodies. When a transfusion is to be made, bloods that have corresponding antigen-antibody combinations must not be mixed, if agglutination is to be avoided. Because they are involved in agglutination, the antigens are called *agglutinogens*, and the antibodies are called *agglutinins*.

RECIPIENT (Antigens, Antibodies) \ DONOR (Antigens, Antibodies)	O αβ	A β	B α	AB —
O αβ	—	Aα	Bβ	AαBβ
A β	Aα	—	AαBβ	Bβ
B α	Bβ	AαBβ	—	Aα
AB —	AαBβ	Bβ	Aα	—

Fig. 10-12. The ABO Blood Groups.

Columns: Donor blood types (antigens) are in capital letters; plasma antibodies in Greek letters.

Rows: Recipient antigens and antibodies. The chart shows the antigen-antibody combinations that result in clumping of cells when mixture is made on microscope slide.

Donor's cells clumped. These combinations are not allowable for transfusion (shown in gray).

Recipient's cells clumped. These combinations may be allowed for transfusion (shown in black on white).

In transfusions of whole blood, it is particularly important that the red cell antigen of the *donor* not be agglutinated by the plasma antibodies of the recipient. To understand why clumping of the recipient's cells by antibodies of the donor's plasma is likely to be less dangerous and less likely to occur, imagine one liter of donor blood added to 4 liters of recipient blood volume. The donor cells enter a heavy concentration of recipient plasma antibodies and will readily be clumped if the appropriate antibody is present. The donor plasma, on the other hand, becomes diluted in the large volume of recipient plasma. The donor antibodies are therefore less likely to combine with the recipient red cell antigens, even if the appropriate ones should be present. The cells of type O blood have no antigens (of the AB group), and an individual with type O blood is therefore called a *universal donor*. A type AB individual is called a *universal recipient* because his plasma has no antibodies to clump any donor cells (of the AB group) (Fig. 10-13).

Before making a transfusion, compatibility of blood types may be determined by mixing a suspension of donor cells with a drop of recipient plasma. If clumping occurs, the donor blood cannot be used. If clumping *does not* occur, the blood of the donor and of the recipient are not necessarily of the same type, although a transfusion in this instance would be moderately safe, as noted above. If clumping still does not occur when a drop of donor plasma is mixed with a suspension of recipient cells, then the two bloods are of the same type, and the transfusion will be completely safe.

The Rh factor

The *Rh factor* or antigen is another red cell antigen that is of importance in transfusions, but it is also important in another way. The *Rh antibody* is not normally

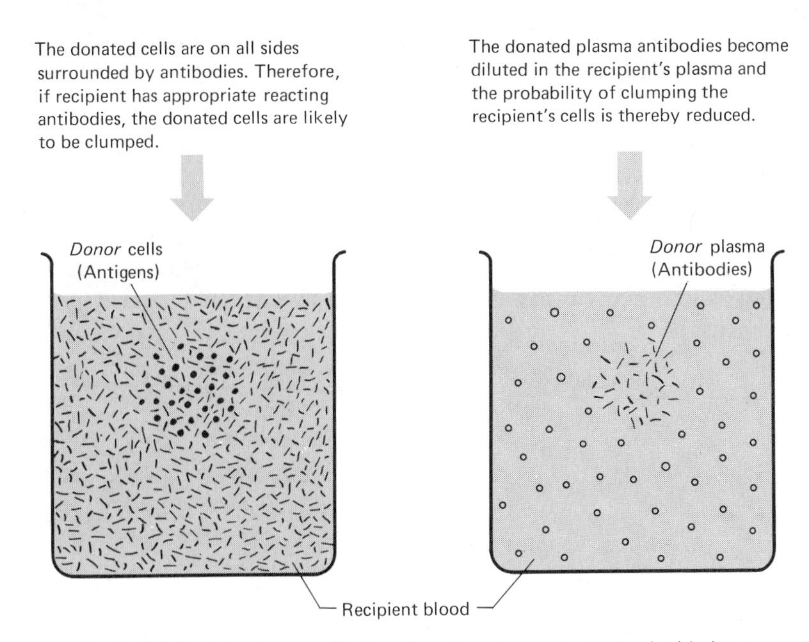

Fig. 10-13. Transfusion and dilution. Consider *donor* whole blood added to a larger volume of *recipient* whole blood.

present in the plasma but is stimulated to be produced under special circumstances. The *Rh antigen* is present in the red cells of about 85 percent of the population. Those who carry the factor in their red cells are designated Rh^+, those who do not are Rh^-. A father who is Rh^+ may have an Rh^+ child, even if the mother is Rh^-. In this instance, a small amount of Rh antigen of the child's red cells manages to pass across the placental barrier from the fetal circulation into the mother's blood. The mother's reticulo-endothelial system is stimulated to produce in the plasma, the anti-Rh antibodies which will react with the antigens on the Rh^+ cells. Agglutination of the child's blood will result if some of these antibodies pass from the maternal circulation into the blood of the child. During successive pregnancies, the amount of anti-Rh antibodies may increase sufficiently to cause real damage to the child's red blood cells. When red cells are destroyed, inadequate oxygen may be carried to the tissues.

Red blood cells go through stages of growth (maturation) in the bone marrow and are ordinarily released only when mature. When large numbers of mature cells are destroyed, many immature cells (*erythroblasts*) are released into the circulation. In the fetus this condition is called *erythroblastosis fetalis*. The first Rh^+ child of an Rh^- mother will generally not be affected, and the deficiency of later ones may often be alleviated by a complete transfusion that provides the child with a new volume of Rh^- blood. Because the child has a large supply of Rh^+ antibody, Rh^+ blood cells bearing Rh^+ antigens would be inappropriate because clumping of the cells could occur. A complete transfusion, replacing the child's blood with Rh^- blood of the mother, will be satisfactory, however. The problem can be circumvented much less traumatically by injecting into the mother's bloodstream, at some time early during the pregnancy, a small amount of Rh^+ antibody that will react with and destroy any Rh^+ antigens that may have moved from the child's circulation into her own. By this means the mother is immunized against the Rh^+ antigen, and no Rh^+ antigen will be available to stimulate production of more Rh^+ antibody. The difficulty is best prevented by avoiding pregnancy when there is a high probability that the child of an Rh^- mother will be Rh^+. By examining family blood types, it is now possible to predict whether there may be, for example, a 50 percent chance for that to occur. Erythroblastosis fetalis and many other diseases may be avoided if intended parents seek genetic counseling.

Section 10: *Questions, Problems, and Projects*

1. If, in a 70-kg person, the volume of blood in the systemic veins is about 3 liters, in the lung capillaries about 1 liter, and in all the rest of the circulation about 1 liter, what percent of the total blood volume is in each of these compartments?
2. Write, as a mathematical equation, a definition of hematocrit (Hct) in terms of total V_t and cellular V_c volumes of the blood.
   (a) If the hematocrit is 45 percent, what volume (of the entire blood) is plasma?
   (b) How many cubic millimeters are in 1 liter of blood?
   (c) If there are 5×10^6 RBC/mm^3, how many RBC are there in the average 70-kg man?
   (d) Each gram of the 14.5 g of hemoglobin in each 100 ml of blood combines with 1.34 ml O_2. How many milliliters of O_2 can combine with the entire blood volume of 5 liters?
   (e) If one gram of Hb combines with 1.34 ml of O_2, how many grams Hb are required to combine with 1 mole of O_2? (There are 22.4 liters in one mole of a gas.)
   (f) If one molecule of O_2 combines with 1 molecule of Hb, what appears to be the molecular weight of Hb?
3. What is the relation between hemoglobin and jaundice?
4. Distinguish: serum, plasma.
5. Identify, in the blood coagulation scheme, the precursors, the activators, and inhibitors at various stages of the process.
6. Construct a chart showing blood types (of the ABO system) of donors in vertical columns and blood types of recipients in horizontal rows. Then
   (a) in the appropriate column, indicate by anti-A or anti-B the antibodies that are present in the plasma of the blood types of the recipients.

(b) draw lines to the appropriate boxes and mark with an X to show the combinations in which the donor's cells will be agglutinated.
(c) encircle and label to show the "universal donor" and the "universal recipient."
(d) show which, if any, of the donors carry antibodies that will react with antigens of the recipients. How then can the recipient be universal?

7. How would you define: hematopoiesis? Leucopoiesis?

11

TRANSPORT OF BODY FLUIDS

11.1 Functional Anatomy of the Vascular System

In its circuit through the body, the blood moves from systemic capillaries to veins and back to the right side of the double pump that is the heart. The blood then travels to the lungs, where carbon dioxide (CO_2) is lost and oxygen (O_2) is gained by the blood. From the lungs, the blood returns to the left side of the heart and is then forced out to the body capillaries via the arteries. Flowing through the capillaries, it again reaches the veins and returns to the heart (Fig. 11-1). Vessels that carry blood away from the heart to the tissues are called *arteries*, while those that carry blood from the tissues to the heart are called *veins*. All arteries except the pulmonary are *systemic arteries;* all veins except the pulmonary are *systemic veins*. Blood in the systemic arteries (commonly called arterial blood) is *oxygenated*, rich in oxygen, while blood in the systemic veins (commonly called venous blood) is lower in oxygen content or is *deoxygenated* blood. Oxygenated blood flows in the pulmonary veins, while deoxygenated blood flows in the pulmonary arteries on its way to the lungs.

The simplest functional representation of the vascular system is shown in Fig. 11-1. The heart is represented as two pumps. The right-side heart pump receives deoxygenated blood from the general body circulation and delivers it to the pulmonary circuit of the lungs. The left side receives reoxygenated blood from the pulmonary circuit and delivers it to the general body tissues (systemic circulation). The continuous circuit can be visualized as a figure-8 pattern.

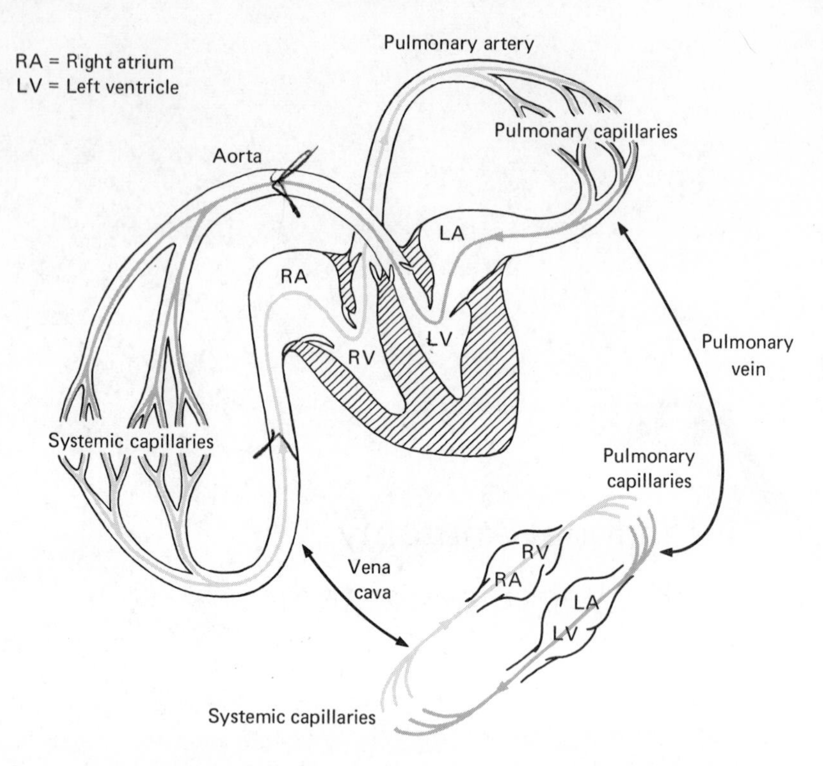

Fig. 11-1. Simplified diagram to show the double nature of the circulation. Gray: Oxygenated blood from pulmonary capillaries to pulmonary veins, into left heart, and systemic arteries. Light gray: Deoxygenated blood from systemic capillaries to systemic veins, into right heart and pulmonary arteries.

The heart

The movement of blood through the vascular system depends primarily on the force of contraction of the heart muscle during the heartbeat. The muscle fibers enclose the cavity of the heart; when they shorten, the volume of this cavity decreases, and the blood is forced out of the heart into the blood vessels. The direction in which the blood moves is determined by the *valves*. Like a water pump, the heart has valves which ensure that the movement of the fluid will be in only one direction. Each pump of the mammalian heart has a receiving chamber, the *atrium*, and a pushing chamber, the *ventricle*. The passageway between atrium and ventricle is guarded by the *atrioventricular valves*, which prevent backward flow of blood from ventricle into atrium. Blood comes from the systemic circulation to the *right atrium*. It then flows into the *right ventricle*, which forces the blood into the *pulmonary artery*. Blood that has passed through the pulmonary capillaries returns to the *left atrium* and is forced into the aorta by the *left ventricle* (see Figs. 11-1, 11-2, and 11-3).

The arterial system

In some animals the blood comes in direct contact with the tissue cells, but in the vertebrates the vascular system is an entirely closed set of tubes. Food materials and oxygen are supplied to the tissues by diffusion and filtration of molecules into the tissue spaces through the walls of the *capillaries*, the smallest subdivisions of the vascular system. Waste products are removed by passage of molecules in the opposite direction.

The *aorta* is the largest artery (Fig. 11-2). In it the blood moves toward the systemic circulation. Its first branch is the *coronary artery*, which supplies the muscle of the heart itself. The heart requires such a vessel because no nourishment is provided to the heart muscle during the passage of blood through the chambers of the heart. From the arch of the aorta, branches lead to the head, neck, and upper limbs. The *carotid* artery is especially important because it supplies blood to the brain. The name carotid has reference to sleep, and indeed, pressure on the carotid artery may deprive the brain of blood sufficiently to render a person unconscious.

The *carotid* on the right side of the body arises from a common junction with the *subclavian artery*, which supplies the right limb. The name subclavian refers to the location of the vessel beneath the clavicle. On the left side, the two vessels leave the aortic arch at separate points. As it runs into the arm, the subclavian becomes successively the *axillary* and the *brachial*. The *ulnar* and *radial* are branches of the brachial into the lower arm where they run in relation to the bones of the same name. The radial is close to the surface at the wrist, where the impact of the heartbeat can be felt as the *pulse*.

Beyond the arch, the aorta is a straight tube (the *dorsal aorta*) located ventral to the vertebral column. In the thoracic region it sends small branches onto the surface of the pericardium (*pericardial arteries*) and to the esophagus. If the coronary circulation is occluded, as may occur during *coronary thrombosis*, the pericardial vessels may provide some of the supply of blood to the affected region of heart muscle. The dorsal aorta also gives rise to nine pairs of *intercostal arteries*, supplying the muscles and skin of the back.

Below the diaphragm, the aorta supplies branches to the abdominal viscera. The *splenic* artery supplies the spleen, an organ that stores large quantities of red blood cells that can be poured into the circulation on demand. The splenic artery arises from the *coeliac* in common with the *gastric* supplying the stomach, and the *hepatic*, which supplies the arterial blood to the liver. Most of the intestine is supplied by the *superior mesenteric artery*, which receives its name from its course through the mesentery of the digestive tract. More caudally, the small *inferior mesenteric* supplies the distal part of the colon. Between the two mesenterics, the dorsal aorta sends branches

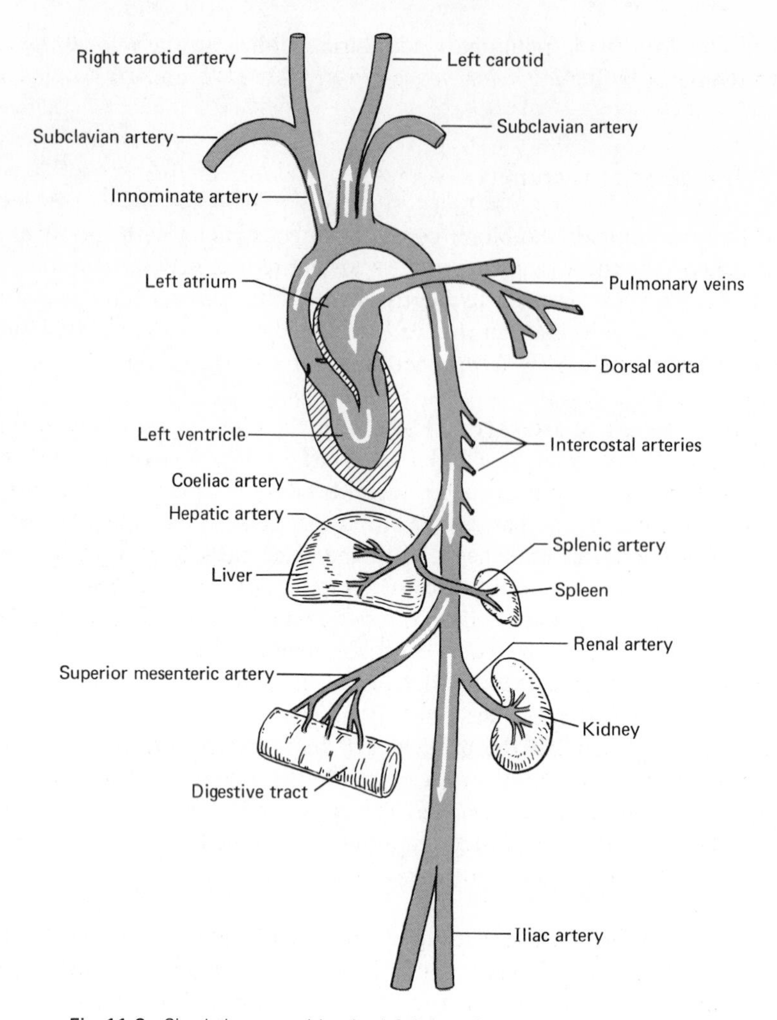

Fig. 11-2. Circulation served by the left heart. Gray indicates oxygenated blood coming to the heart from the lungs and being distributed throughout the body by the systemic arteries. The major vessels carry blood away from the left heart (arrows).

to the kidneys by way of the *renal arteries* and to the gonads by way of the *spermatics* or the *ovarians*. Beyond the mesenteric, the aorta divides into the two *iliac* arteries. The main branch of each iliac artery is the *femoral*, supplying the lower limb.

The venous system

The blood that comes to the organs must also leave. After flowing through the capillaries, the blood ultimately reaches the veins that finally return blood to the right side of the heart. Blood from the heart muscle returns to the cavity of the right atrium through the *coronary veins*. Blood from the head, neck, and arms enters the right atrium via the *superior vena cava* (see Fig. 11-3). The *internal jugular vein* drains the capillaries of the brain, while the *external jugular* collects from the more superficial parts of the head. Capillaries in the arms are drained by the *brachial* veins, which empty into the *axillary*. *Axillary*, *subclavian*, and *innominate* are different regions of the vessels that pour blood into the superior vena cava on each side.

The *inferior vena cava* runs along with the dorsal aorta, returning blood to the right atrium from the lower limbs and from all the abdominal viscera except the diges-

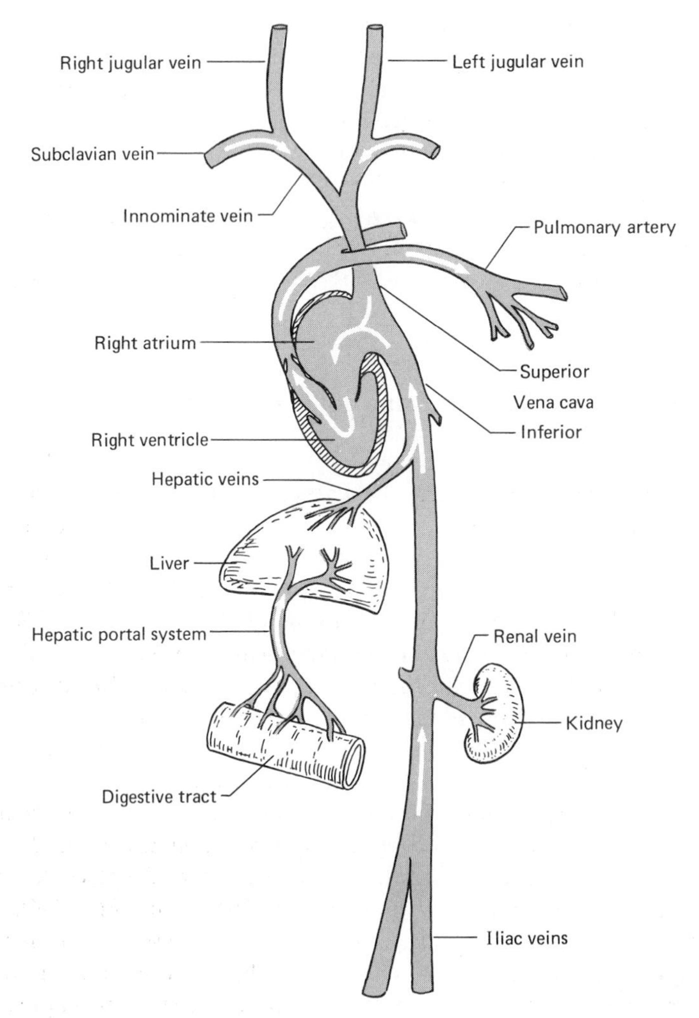

Fig. 11-3. Circulation served by the right heart. Complementary figure to Fig. 11-2. Light gray indicates blood low in oxygen (deoxygenated). The major vessels carrying blood *to the heart* are associated with the *right heart* (arrows).

tive tract. Blood from the gonads drains into the inferior vena cava by way of the *spermatic* or *ovarian* veins, and from the kidneys by way of the *renal* veins. Each *femoral* vein drains the deeper portions of one of the lower limbs. The *saphenous* is a much-branched system of veins draining the superficial parts of the limb. Its branches are complex and tortuous, and in some individuals the blood may stagnate and not be able to return rapidly enough to the heart. In such instances, the vessel walls may be excessively stretched from the pressure of blood. Regions of such swelling in any vein are called *varicosities*, and the vessels are said to be *varicose*. The varicosities occur especially in the regions of the *valves* that prevent backward flow of blood in the veins.

Blood from the digestive tract has a special pathway all its own back to the heart. As the arterial blood from the gastric and superior mesenteric arteries enters the capillaries of the gastric and intestinal wall, digested food materials are absorbed into the blood. The capillaries of the digestive tract then empty into the *hepatic portal* system. A *portal system* is a blood vessel that has capillaries at both ends. One end of the hepatic portal vein collects from the capillaries of the stomach and intestine. The other end delivers blood to the capillaries of the liver, where the transported food materials are utilized by the liver cells. Blood from both the hepatic portal vein and from the hepatic artery flows from the *hepatic veins* that collect together to enter the inferior vena cava.

Pathway of blood in fetus compared with adult

In the heart and lungs, the pathways through which the blood flows are not the same in a child before as after he has been born (see Fig. 11-4). The fetus secures O_2 and food from the *placenta* rather than from its own pulmonary and digestive system. Following birth, there are alterations in the blood flow corresponding to the sudden need to use the lungs and gastrointestinal tract. Before birth, venous blood from the aorta of the fetus is brought to the placenta via the *umbilical arteries* that branch from the iliac arteries of the fetus. After oxygenation at the placenta, the blood, now arterial, returns via the *umbilical vein* and the *ductus venosus* to the inferior vena cava of the fetus. It then enters the right atrium along with venous blood from the systemic capillaries of the fetus. The blood from the vena cavae mainly follows the route from the right atrium into the right ventricle. As the lungs are collapsed and of small volume, little blood enters the pulmonary circuit. Instead it gets into the aorta primarily via a bypass, the *ductus arteriosus*, and is delivered mostly to the placenta. The blood coming to the heart from the placenta goes principally through a shortcut, the *foramen ovale*, into the left atrium (Fig. 11-4). This route relieves the load on the right ventricle, which would otherwise have to handle all the blood passing through the heart at any moment. The blood is forced into the aorta by the left ventricle and is then distributed mainly to the systemic capillaries of the fetus. Note that the oxygenated blood from the placenta and the deoxygenated blood from the fetus' systemic capillaries have parts of their routes in common, and some

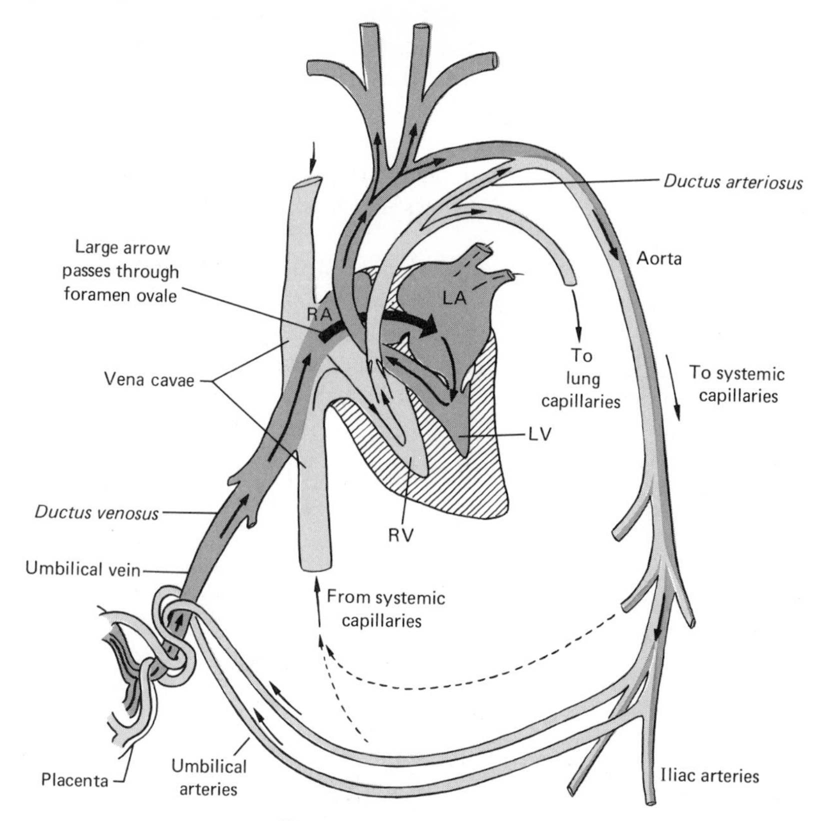

Fig. 11-4. Fetal circulation

mixing must therefore occur. Apparently mixing is not complete, however, and the two streams may be separate within the same vessels.

Before birth, left and right ventricles make approximately equal contributions to the systemic blood supply. Following birth, the lungs fill with air and blood flows into the large pulmonary capillary bed from the right ventricle. Blood returning from the pulmonary circulation to the left atrium tends to close the foramen ovale. The blood entering the right ventricle finally goes exclusively to the pulmonary circulation, while the systemic circulation is provided entirely by the left ventricle.

The pressure required to force the blood adequately around the systemic circulation is higher than that necessary for the smaller pulmonary circuit. The pulmonary artery pressure is about 75 mm Hg, whereas the aortic pressure is about 120 mm Hg. Because of this pressure difference, the direction of blood flow in the ductus arteriosus before birth is not maintained following birth, but may even be reversed, and the vessel usually becomes obliterated. If the ductus arteriosus does not become sealed, the load in the left side of the heart may be excessive and result in heart failure. If the foramen ovale remains partly open, the pulmonary circulation is reduced, blood oxygenation is inadequate, and the baby may be blue from the color that the insufficiently oxygenated blood in the capillaries gives to the skin.

When the placenta is lost, the blood that had entered the umbilical artery must find another route. Circulation to the digestive system is increased, and this load of blood enters the hepatic portal system and finally returns to the right atrium via the hepatic veins. The *ductus venosus*, no longer carrying any significant volume of blood, becomes obliterated. The umbilical arteries no longer carry blood and persist only as an umbilical ligament.

In summary, the changes in circulation at birth are: the *umbilical arteries* are tied off and blood can no longer go to the placenta. The *umbilical vein* is tied off and the *ductus venous* receives the blood from the gut instead. When air fills the lungs, the pulmonary capillaries are opened up, and the *ductus arteriosus* becomes nonfunctional as blood flows preferentially into the lungs instead of into the aorta. As pressure of blood from the lungs builds up in the left atrium, the *foramen ovale* is sealed off and blood is no longer diverted from right to left atrium.

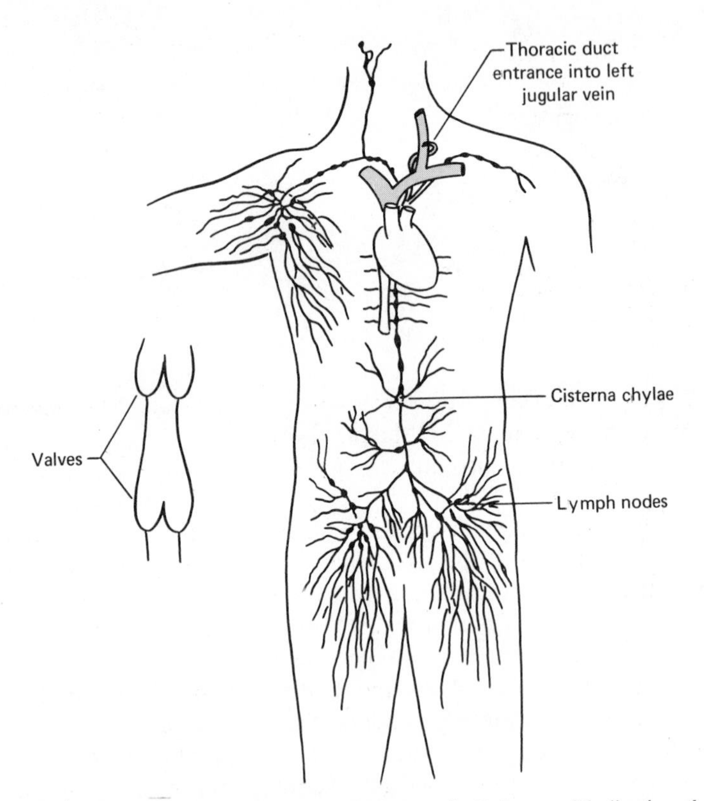

Fig. 11-5. The distribution of the lymphatic vessels. Only a small indication of the distribution of the lymphatic vessels over the entire body is shown. Note that the thin vessels swell and have a beaded appearance at the frequent valves. The nodes are aggregations of special tissue where lymph cells arise.

The lymphatic system

Another set of vessels, working along with the arteries and veins, is the lymphatic system (Fig. 11-5). Its smallest branches, which are about the size of capillaries, are, in contrast to capillaries, blind-ended tubes. Fluid from the blood is forced out of the capillaries into the tissue spaces. Some of this fluid returns to the circulation on the venous side of the capillaries, but some of it passes into the blind-ended *lymphatic vessels*. Most of the tissues of the body, with the notable exception of the central nervous system, are drained by the lymphatics. The fluid part of the lymph consists mainly of a filtrate from the blood, but most of the cells, particularly the *lymphocytes*, arise in special aggregations of tissue, the *lymph nodes*, located at strategic points in the system.

The lymph is moved from the tissue spaces back into the veins, by way of the lymphatic vessels. Much of the abdominal and thoracic regions are drained into the *cisterna chyli*. This dilated vessel is the beginning of the *thoracic* duct that carries the collected lymph into the left subclavian vein. The lymphatic vessels are particularly obvious after a meal, when they may be seen in the mesentery along the blood vessels because of the milky suspension of fat globules absorbed from the digestive tract.

Movement of the lymph is due mainly to the force of muscle contraction pushing on the thin-walled vessels. Unidirectional flow of the lymph is ensured by the numerous valves in these vessels.

11.2 The Heart and the Blood Pressure

The heart begins, both embryonically and phylogenetically, as a pulsating blood vessel coiled upon itself. In fish it consists mainly of three cavities: the *sinus venosus*, which receives blood from the veins, the *ventricle*, which forces blood into the arteries, and the *atrium* between these two chambers (Fig. 11-6). In the mammalian heart, a *septum* divides both the atrium and ventricle into right and left halves. The sinus disappears as a chamber and remains only as a knot of tissue in the right atrium.

In the adult mammal, the right side of the heart carries blood from the body tissues to the lungs, whereas the left side carries blood from the lungs to the tissues [Fig. 11-7 (a), (b), (c)]. Figure 11-7 shows an external view of the human heart from the ventral aspect. The apex of the heart is pointed toward the lower left of the sub-

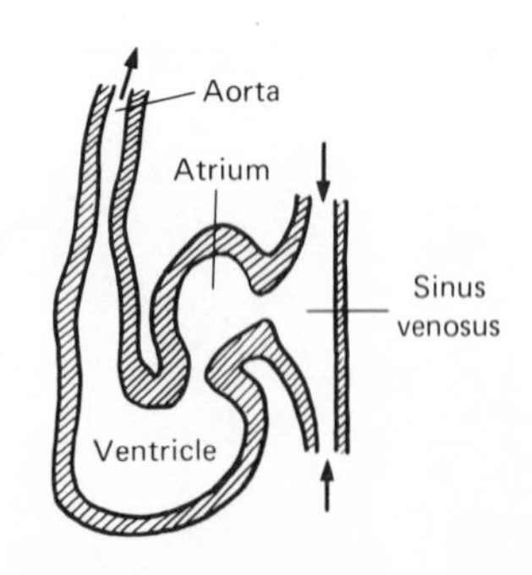

Fig. 11-6. Longitudinal section of embryonic human heart. Arrows show direction of blood flow through the heart. At this stage, there are three chambers. During development, the sinus venosus disappears as a cavity, while the atrium and ventricle are partitioned into pairs.

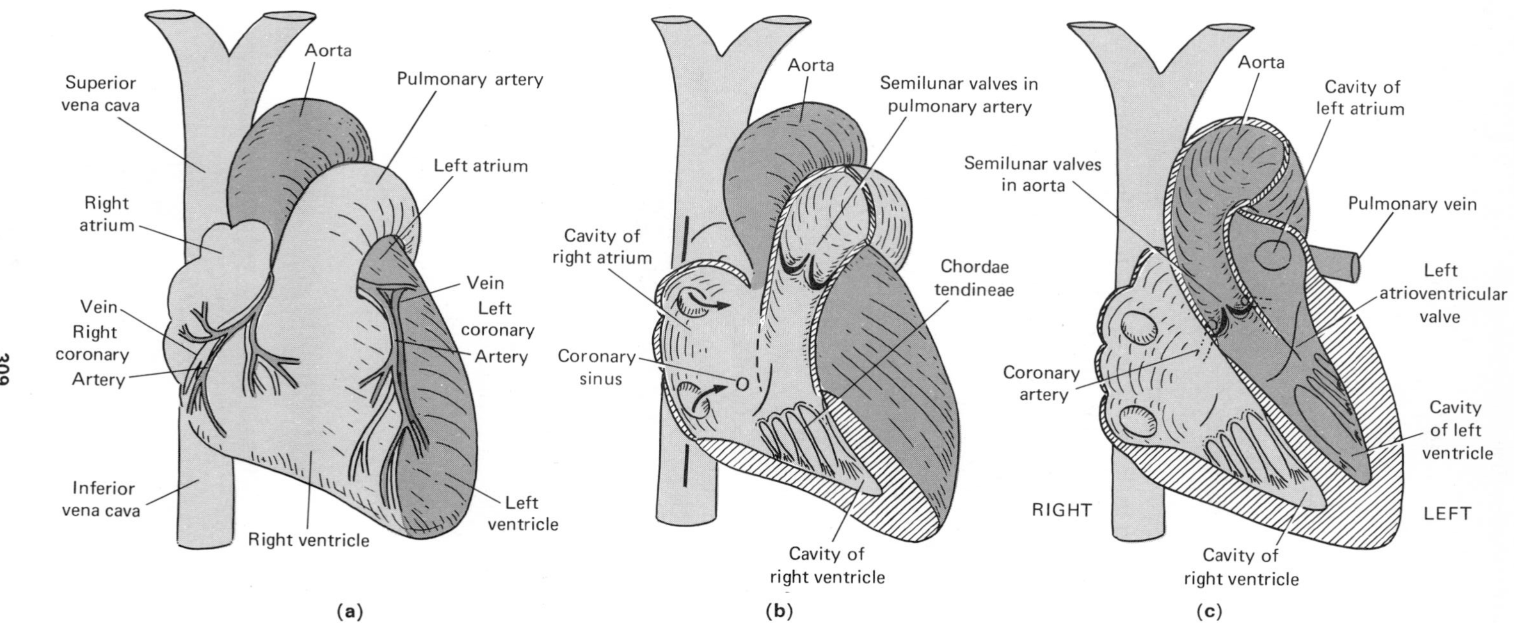

Fig. 11-7. Human heart. Gray: Oxygenated blood in these cavities; Light gray: deoxygenated blood in these cavities. (a) External surface view. (b) Chambers of heart exposed on right-hand side. (c) Chambers of heart exposed on right and left sides.

ject (your lower right as you face the illustration). Valves between atria and ventricles ensure one-way flow of the blood [Fig. 11-7 (b), (c); see also Figs. 11-15 and 11-16]. The motive force is provided mainly by the thick muscular walls of the ventricles. Note in Fig. 11-7(a) the coronary arteries and veins that supply the circulation to that heavy ventricular muscle.

Origin of the heartbeat

The time course of the heartbeat may be recorded by means of a lever attached to the tip of the ventricle. In the record from the heart of a fish or frog, three peaks may occur during each heartbeat. These peaks show the contraction sequence: sinus venosus, atrium, ventricle. The heart will continue to beat even if cut free and removed from the body. If the heart is cut into pieces, each piece may continue to contract rhythmically. If the sinus venosus, atrium, and ventricle are separated from one another, the sinus venosus will beat at the rate of the intact heart, the atrium will beat less frequently, and the ventricle will beat slowest of all. The sinus venosus, which drives the other heart chambers at its rate, is known as the *pacemaker* of the heart. The fibers of the heart muscle act as if they were continuous with one another (i.e., a *syncytium*), and the muscle impulse that starts in the sinus venosus is conducted throughout the entire heart. The contraction of each fiber follows in rapid sucession.

In mammalian hearts the sinus venosus, although it remains only as a knot of tissue imbedded in the right atrium, still initiates the heartbeat. This tissue is rather different in appearance from the ordinary heart muscle and is called the *sinoatrial* (SA) *node* (Fig. 11-8). The cells contain very few striated fibrils, and the cytoplasm is clear, similar to nerve tissue. These modified muscle cells have no significant ability to contract; rather they are specialized for the production of impulses that propagate over and stimulate the atrial muscle fibers. In mammals the atria are separated from the ventricles by a connective tissue band, but continuity between the structures is maintained by means of a special conduction pathway, the *atrioventricular bundle*, which extends from the atrioventricular node to the fibers of the ventricles. The *atrioventricular* (AV) *node* is similar in appearance to the sinoatrial node and is specialized for conduction of the impulse. It is excited by impulses of the atrial muscle fibers.

Electrical activity of the heart

Because the heart is a very large muscle, and many of its cells become active almost simultaneously, the electrical effect of its activity is considerable and spreads widely throughout the body (see Fig. 11-9). An *electrocardiogram* (ECG) is a record of currents that flow from the heart to those regions of the body where electrodes are attached. An electrode placed on the chest near the heart provides a relatively simple record when the other electrode is connected in such a way as to be practically unaffected by the action currents of the heart. As Fig. 11-9 shows, when excitation begins,

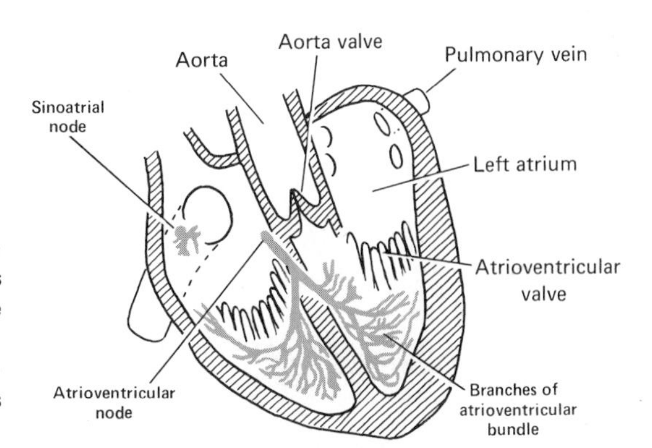

Fig. 11-8. The conduction system of the human heart. Branches from both nodes are effectively continuous with the ordinary cells of the atrium (S-A node) and of the ventricle (A-V node). The pulmonary artery is omitted from this figure.

the depolarized region is small, and the currents flowing through the body produce only a small change at the electrode on the chest (1, 2). As the depolarized area comes to involve more of the heart, the currents increase, and the record is larger (3). When the depolarized region is very large, however, the record decreases because the unpolarized area supplying the current is small and supplies less current to the depolarized zone (4). Finally, when the entire heart is depolarized, there are essentially no external currents, and the record returns to zero (5). During recovery of the membrane potential, the currents flow in the opposite direction (6). The actual electrocardiogram is somewhat more complicated because it involves currents of both repolarization and depolarization, in both atria and ventricles. In addition, the connections to the body are commonly made so that the currents of the heart action will affect both electrodes, and the resulting ECG is the *voltage difference* between the two electrodes.

In an ordinary ECG, the *P wave* indicates *external current* flow during the depolarization of the atria, while the *QRS wave* represents the depolarization of the ventricles (Fig. 11-10). During the time that the ventricles are completely depolarized, virtually no external potential difference exists between parts of the heart, and, consequently, there is essentially no external current flow. This interval is the *S-T interval*. The *T wave* terminating this interval represents the external currents flowing during repolarization of the ventricle. There are also practically no external currents during the *T-P interval*, when the heart is relaxing. These ideas are shown graphically in the simultaneous records of an intracellular ventricular action potential and the conventional ECG in Fig. 11-10. Note that the actual membrane potential change occurring in a cardiac muscle fiber is quite prolonged compared with the action potential of a skeletal muscle fiber or of a nerve.

The usual method of obtaining the ECG involves attachment of electrodes as shown in Fig. 11-11. The conventional limb leads are designated I, II, III. In lead I, the electrical activity of the heart is detected by one electrode attached on the right arm and another electrode on the left arm. The excitation wave sweeps over the heart approximately in the direction shown by the arrow. The region of initial depolarization is closer to the right arm than to the left, and, for this reason, the electrical

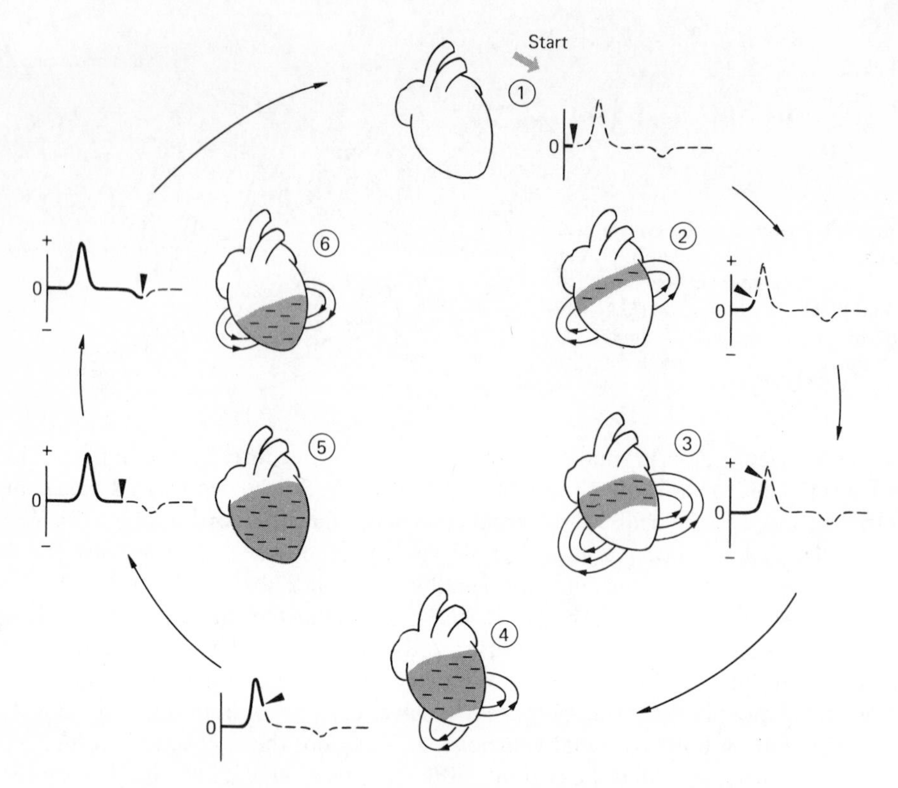

Fig. 11-9. Monopolar electrocardiogram of ventricle. Starting at the top (1), move clockwise to see how the ECG comes about from the currents flowing with respect to the depolarized zone of the heart during the heart action potential. 0 in each case shows the position of one electrode, the other being so arranged that it is unaffected by the currents. The heart is to the left of each record. The small marker on the record shows the extent of the ECG traced out by the time the depolarization has gone as far as shown in each instance. 0 also indicates zero potential on the ECG record. At (5), the ventricle is uniformly depolarized and no current flows. Normally, depolarization follows a path such that its signal is in the same direction as the Q wave. Here that T wave is shown inverted (6).

change seen at that moment and each moment thereafter will not be the same at these two electrodes. Lead I reports the difference between the time course of the events at the two electrodes. The events seen by Lead II are somewhat similar when they are recorded in both cases, so that if a negative signal is seen at the right arm, there will be a downward deflection of the record. The depolarization wave sweeps more or less toward both electrodes used in Lead III but somewhat more toward the leg electrode. In this instance, the signal is positive upward if a positive signal is applied to the left leg electrode.

A number of malfunctions due to injury of the heart can be interpreted from the

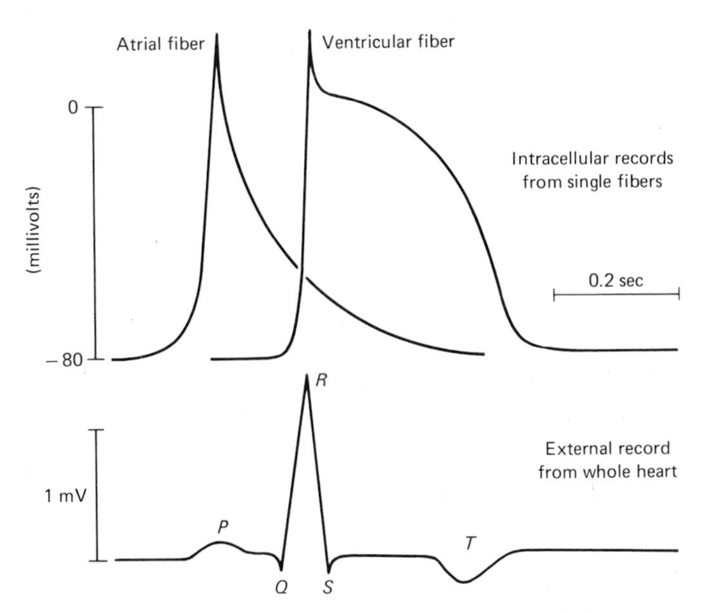

Fig. 11-10. ECG and intracellular membrane potential changes in heart muscle. The QRS part of the conventional ECG record represents the currents flowing outside the heart muscle when the cells become depolarized. The ECG is flat during the time the cells remain depolarized (S-T interval). There is a small deflection when the cells repolarize (T wave). The small P wave reflects depolarization of the atrium.

form of the ECG. The atria may beat very rapidly, and excitation may circulate several times in the atria before the ventricles are stimulated. The QRS waves may then appear at the usual frequency, but the base line of the record will be irregular because of repeated atrial excitation (*atrial fibrillation*). In another situation, some of the impulses from the atrium may not be effective in exciting the ventricle. This is *atrioventricular* block. In this instance, the P wave appears at its normal frequency, but the QRS wave follows only every other, or every third (or more), P wave.

If there is an area of killed tissue in the heart, there will be a *current of injury* that flows when the heart is relaxing, during diastole (Fig. 11-12). This current provides a steady electrical record, and the ECG changes will be superimposed on it. The *S-T* segment appears to be lifted above the usual base line. This *elevated S-T segment* disappears as the injury current is reduced during healing.

Nervous control of the heart rate

Although the heartbeat is started by the sinoatrial pacemaker, the frequency of the beat may vary considerably, especially under the influence of autonomic nerves (Fig. 11-13). The *vagus parasympathetic preganglionic fibers* (*nerve X*) end among

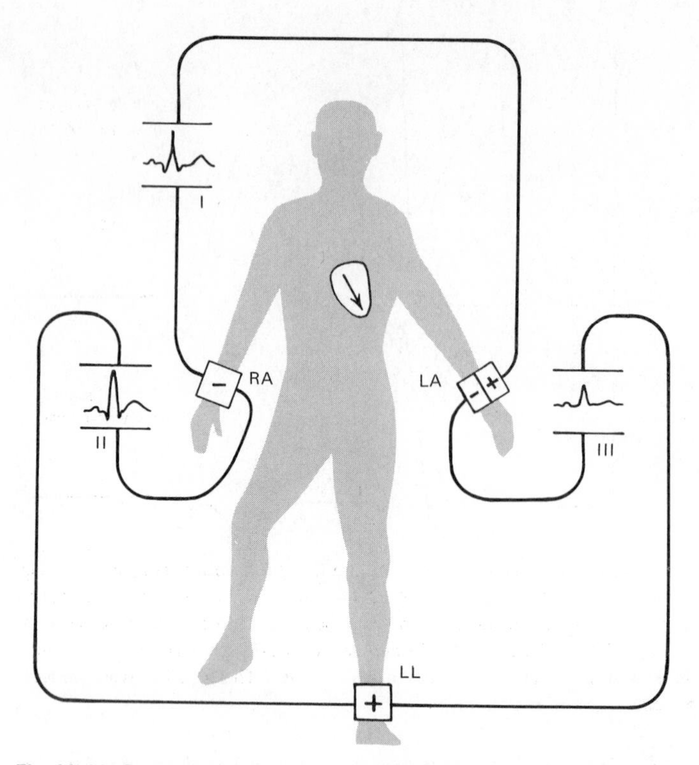

Fig. 11-11. Conventional electrocardiographic leads. Positive signal applied to electrode designated + gives upward deflection in each instance. Each record may be considered to be the difference between two monopolar leads.

nerve cell bodies of ganglia buried amid the special muscle fibers of the nodes. Fibers from these cell bodies innervate the nodes. The right vagus mostly supplies the SA node, while the left innervates the AV node. The arrival of vagus nerve impulses in these regions may be followed by an increase in resting membrane potential of the nodal cells. The cells become less excitable, and the frequency of the heartbeat decreases. Contrariwise, the fibers of the sympathetic nervous system stimulate the heart to an increased frequency of contraction. The vagus action is accompanied by secretion of acetylcholine from the postganglionic terminals, while the sympathetic postganglionic fibers produce noradrenaline.

The heart rate at any moment is determined by the balance of sympathetic and parasympathetic impulses reaching the heart. The impulse traffic in these nerves is determined by many factors, particularly by reflex adjustments involving chemical and pressure receptors in the vascular system. The impulse traffic in the autonomic nerves is referred to as sympathetic and vagal *tone*. The existence of vagal tone is demonstrated by the fact that there is an increase in heart rate (tachycardia) when the

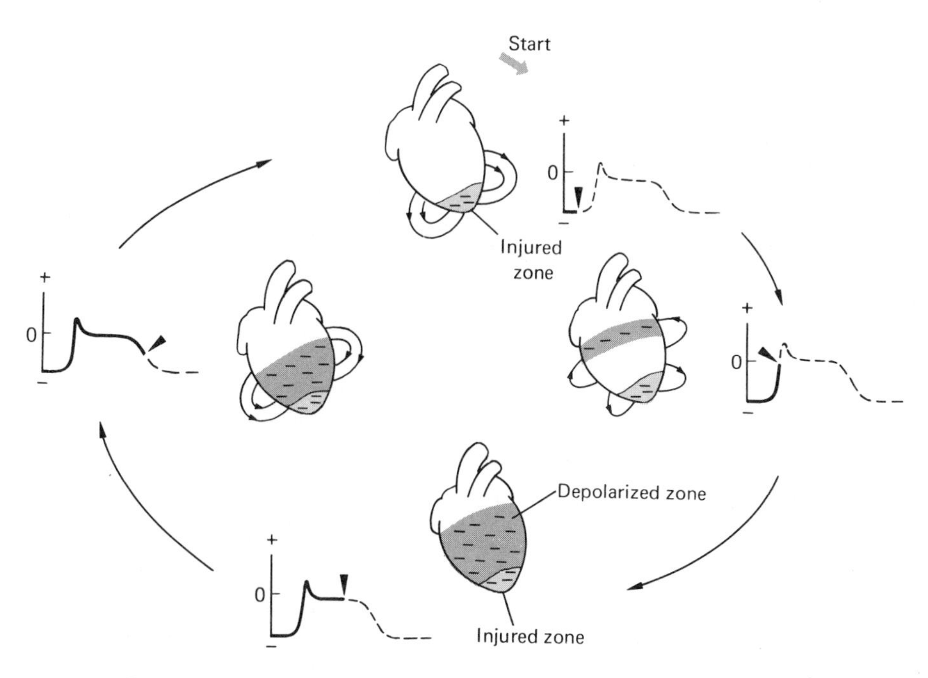

Fig. 11-12. Monopolar electrocardiogram of injured heart. As in Fig. 11-9, but assuming the ventricle to be injured (gray shading). Thus the record starts out negative, and moves toward zero as the ventricle becomes completely depolarized during the action potential.

vagus nerves to the heart are cut. Conversely, cutting the sympathetic supply results in a slowing of heart rate (bradycardia), implying the normal presence of sympathetic tone tending to increase the rate.

The force of the heartbeat—blood pressure

The effectiveness with which a pump moves its load depends on the force that the pump can provide. The rate of delivery of blood to the tissues depends primarily on the force of the heartbeat. Like any other force, the force of the contraction of the heart can be measured in terms of a weight that the force can balance. The force within the heart is distributed evenly over the entire inside wall. If one end of a long tube is inserted in the aorta, the other end being open to the atmosphere, the blood will rise vertically in the tube until the force exerted by the weight of the column of fluid equals the force exerted by the heart on that part of its wall represented by the cross section of the tube. The height of the column of fluid will not depend on the tube diameter. This follows from the fact that the *force per unit area* is, of course, the same regardless of the diameter of the tube used to measure the force.

Force per unit area is defined as the *pressure;* that is, $P = F/A$. The walls of the heart exert a force on the contained blood even between contractions when the heart is relaxed, in *diastole*. This force is a result of the fact that the vascular system is in effect *overfilled* and the walls are stretched by the contained blood. The height of the column of blood in the tube is a measure of the pressure of the walls of the vessels on the blood (Fig. 11-13).

Suppose that the force within the heart during diastole balances a column of blood 136 cm high when the cross-sectional area of the tube at the heart is 1 cm^2. On each square centimeter of area at the bottom of the tube there rests 136 cm^3 or 136 g

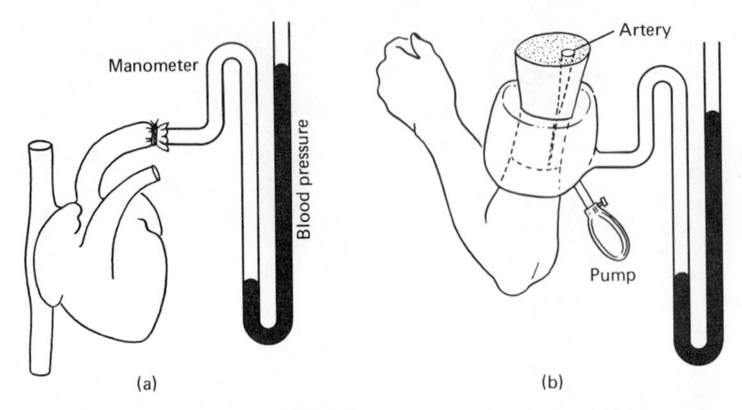

Fig. 11-13. Direct and indirect measurement of blood pressure. (a) The pressure of the blood supports a column of fluid and the blood pressure is measured directly. (b) The rubber bag wrapped around the limb is inflated until the pressure in the bag is just sufficient to stop the flow of blood through the artery. The air pressure, measured by means of the manometer filled with mercury and attached to the bag, is then equal to the blood pressure.

of blood (blood weighs about 1 g/cm^3). The pressure is therefore 136 g/cm^2, expressed as force per unit area. Evidently the height of the column supported is a direct measure of the pressure. If the liquid in the tube is mercury (Hg), the column will be only 100 mm high, since mercury is 13.6 times as heavy as water. Mercury is ordinarily used for measurements of this sort, since it occupies only 1/13.6 the volume of an equal weight of water. In an average individual, 100 mm Hg is the average pressure in the left ventricle between contractions, during *diastole*. During *systole*, the contraction of the heart muscle provides an extra force on the contained blood. On the average, the pressure may then rise to 120 mm Hg.

The elastic force of the walls of the blood vessels acting against the force produced by the contraction of the heart on the contained blood produces the *blood pressure*. In general, the blood pressure decreases with distance from the heart. It may be measured *directly* by means of a tube placed in a vessel at a desired place, in which instance the blood (or the mercury) pushes against the pressure of the atmosphere (Fig. 11-13). The rest of the body is also at atmospheric pressure, so the height of the column is a measure of the magnitude of the blood pressure, above atmospheric pressure.

Indirect measurement of blood pressure

In human subjects, an *indirect method* of measurement is more convenient (Fig. 11-13) Pressure in a blood vessel is measured indirectly by noting the external pressure that is necessary to compress the vessel sufficiently to stop the flow of blood through it.

In the indirect measurement of blood pressure, air is pumped into a rubber bag wound around the arm, and the pressure in the bag at this level is ordinarily taken as the standard for comparison of blood pressures. The pressure in the bag is measured by a manometer, a device involving a column of fluid or a thin membrane. As the vessel is occluded by the externally applied pressure, the *turbulence* of the blood flow provides a sound that may be heard by means of a stethoscope bell placed on the arm, just below the cuff. When the flow is completely occluded, no sound can be heard, nor can the pulse be detected on the radial artery at the wrist (Fig. 11-14).

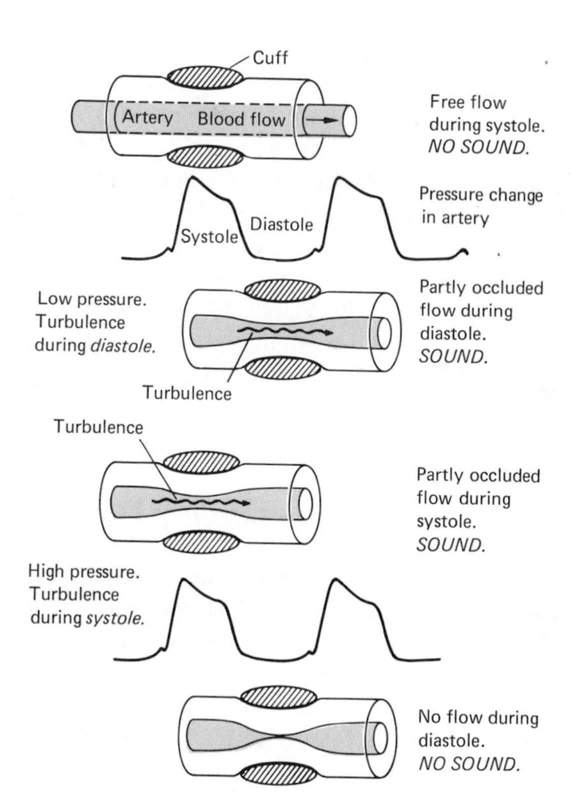

Fig. 11-14. Indirect measurement of systolic and diastolic blood pressure.

If the pressure in the bag is gradually decreased, a sound will be heard when the vessel opens momentarily during the maximum force of the ventricular systole. With further decrease of external pressure, the vessel remains closed only during diastole. At sufficiently low external pressure, no sound is heard because the blood flows without obstruction during both systole and diastole.

Pressure cycle within the heart

The pressure within the various chambers of the heart may be measured directly by means of a pressure-sensing device that can be inserted within the heart. The direction of the flow of blood is determined by the pressure gradient; that is, the blood moves toward the lower pressure. The figure shows the pressure at various places in the heart during the cycle of the heartbeat (Fig. 11-15).

During ventricular *diastole*, the ventricles are relaxed, and blood flows through the atrioventricular valves because the pressure in the veins and atrium is greater than in the ventricle. The weak atrial systole also helps slightly to move the blood. When ventricular systole begins, the atrioventricular valves snap shut as blood is forced back against them. The contraction of the ventricular muscle, the click of the valves, and the turbulence of the blood flow during the sudden stop in the movement of blood through the valves—all contribute to the vibrations of the *first heart sound* heard by means of a stethoscope placed on the chest. At the moment of the first heart sound, the aorta is still filled with blood from the previous ventricular systole. The *semilunar valves* prevent the blood from flowing backward from the aorta into the ventricle, and the forward motion of the blood through the arterioles is maintained by the elastic forces in the wall of the slightly overfilled aorta. The ventricular systolic pressure increases as the muscle fibers contract, and when it exceeds the aortic blood pressure, the blood moves out of the left ventricle through the semilunar valves and into the aorta, as well as out of the right ventricle and into the pulmonary artery. As the contraction force of the ventricle declines, the blood, being under high pressure inside the stretched aorta, begins to move backward. The semilunar valves immediately snap shut, and the *second heart sound* may be heard.

Pressure in each side of the heart

The blood pressure in the left ventricle is greater than in the right. The capillary network of most of the body into which the left heart forces blood is much greater in extent than the network of the lungs. It follows that more force is necessary to overcome the friction of the enormous surface with which the blood in the tissues is in contact. The pressure in the pulmonary circuit is much lower. The differences in pressure in the systemic and pulmonary circuits are shown in Fig. 11-16. Note that

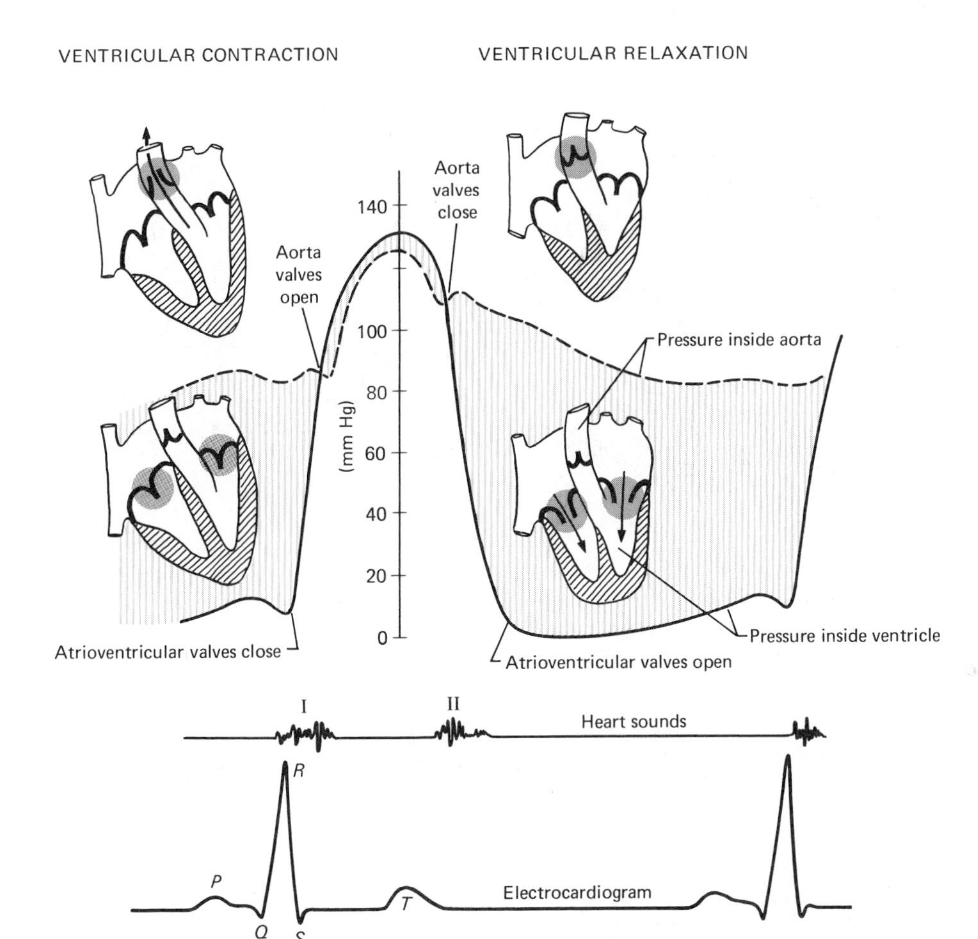

Fig. 11-15. Events during the cycle of heart contraction. The pressures in the aorta (dashed line) and in the ventricle (solid line) are shown during a heartbeat. During diastole (most of the large shaded areas) the pressure in the aorta is higher than the pressure in the left ventricle. During systole, the pressure in the left ventricle rises quickly above the pressure in the aorta (small shaded area), and the blood is therefore moved into the aorta.

pressure in the left ventricle has a range of about 0 to 120 mm Hg, whereas in the right ventricle the pressure rises only to about 25 mm Hg. Since right and left sides of the circulation are ultimately continuous, as shown in Fig. 11-16, the average rate of volume flow of blood through the pulmonary circuit under low pressure is the same as the flow through the systemic circuit under high pressure.

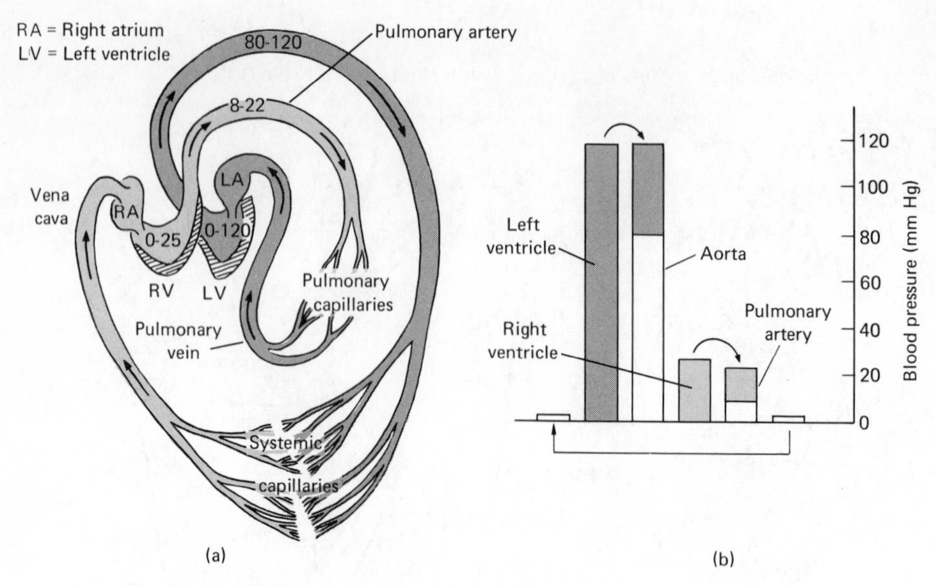

Fig. 11-16. (a) Direction of blood flow in the vessels. (b) Comparison of pressures in the heart, pulmonary artery, and aorta. Arrows indicate direction of decreasing pressure and range over which pressures change, allowing, with the help of valves, directional flow of blood.

11.3 Regulation of Blood Distribution

Many tissues require blood, not at a constant level of supply but in varying degree as body activities change. The greatest changes are along the digestive tract, where large supplies of blood are needed during digestion, in the sex organs, where an increased blood flow occurs during sexual excitement, and in skeletal muscle and heart muscle, which demand more blood during strenuous exercise. The heart is of primary importance in moving the blood through the vascular system. Within limits and under ordinary conditions, the faster the beat, the more rapidly will blood be supplied to the tissues and the higher will be the pressure of blood as it is forced into the arteries. But the entry of blood into the organs is also determined by the degree to which the blood has free access to the capillaries. Entrance to the capillaries is via arterioles that act as valves controlling the flow of blood into the capillaries beyond. Flow of blood to particular regions is controlled by the degree to which the arterioles are dilated or constricted. The volume of blood moved per unit time generally depends on the force, frequency, and volume of the heartbeat, the massaging effect of muscles, and the degree of arteriolar tone. Mechanical, chemical, and nervous factors control the effectiveness of the heartbeat and the tone of the arterioles.

Factors governing heart output

The rate at which blood is provided to the capillaries depends partly on how fast it is forced out of the heart—that is, the volume ejected per minute, the *minute volume* or cardiac output.

The minute volume output of the heart into the aorta and into the pulmonary artery is determined by the product of beat frequency and volume per beat. On the average, each ventricle puts out about 60 ml of blood per beat. At 60 beats per minute, 3.6 liters are ejected per minute, and the entire blood volume therefore passes through the heart in a minute and a half or less. Under the stress of exercise, the demand of the tissues for arterial blood increases, and the heart output is adjusted accordingly. How rapidly blood returns to the heart from the capillaries affects the volume per beat. Within limits, the more that returns to the heart during diastole, the more that can be ejected during systole. In this respect, heart muscle is similar to skeletal muscle in that the more it is stretched, the more force it can exert, at least for part of its functional range.

The return of blood to the heart is determined partly by the *force of the heartbeat* that sends the blood towards the capillaries. It is determined partly by the resistance the blood meets in passing from the arterial to the venous side of the circulation. This resistance is mainly in the arterioles and is called the *peripheral resistance*. It is also determined partly by *contractions of skeletal muscles*, which provide pressure external to the veins, helping to move the contained blood along toward the heart. *Valves in the veins* ensure unidirectional flow. *Respiratory movements*, by influencing heart rate and blood pressure, also affect the return of blood to the heart.

Pressure, volume, and velocity of blood flow

The blood hydraulic pressure reaches its highest levels within the left ventricle. The peak systolic pressure in the aorta is somewhat lower, but the minimum (diastolic) pressure in the aorta does not fall as low as it does in the left ventricle. The very large fluctuations between the systolic and diastolic pressures in the ventricle are damped as the blood is forced into the aorta.

During ventricular *systole*, blood forced into the aorta stretches the walls of that vessel and of the arteries branching from it. During ventricular *diastole*, the stretched elastic walls of the vessel squeeze on the contained blood and keep the pressure high. *Fluctuations* in pressure in the aorta are therefore less than in the ventricle. If we move farther from the ventricle along the arteries into the arterioles, we find that the fluctuations (the pulse pressure) and the average pressure both decrease dramatically (Fig. 11-17). The pressure is lower still in the capillaries and decreases even further as we follow the blood into the veins.

Pressure distribution in the vessels

If a fluid such as blood is contained within a firm–walled pipe, the same pressure that is applied at one end will be measurable at the other end because the fluid is incompressible. Although the blood is incompressible, the walls of the blood vessels are elastic. Therefore much of the energy available in the force applied to the blood by each contraction of the heart is used in stretching the arterial walls, and thus the pressure decreases with distance from the heart. In a closed system a steady pressure applied at one end will finally be distributed uniformly throughout, and if the pressure oscillates, the average pressure will be the same everywhere. However, although it is in a closed system, the blood flows continuously in one direction. Pressure remains low in the smaller vessels where it is not allowed to build up because the blood is continually running out of those vessels and into the veins. The blood in the arteries is under high pressure, like the water in a pipe. The blood entering the capillaries undergoes a great fall in pressure just as does pressure of the water coming out of the tap and into the sink. Figure 11-17 compares the cumulative cross-sectional areas at various parts of the circulation and shows the corresponding pressures.

Control of volume rate of flow

The volume rate of flow through the vascular system is important because it affects the efficiency of delivery of O_2 to and removal of CO_2 from the tissues (as well, of course, as other important constituents in the blood). Since the vascular system is a continuous system of tubes, at any moment the total volume flow at one level must equal the total volume flow at any other level farther along in the system. The flow through a capillary network per unit time must be the same as the flow through the large artery that serves it. Figure 11-18 clarifies this idea. The volume moving per unit time ($\dot{Q}$) in the smaller section must be equal to the volume moving per unit time ($\dot{Q}$) in the larger section. This situation can be diagrammed as the time to move the volume from A to B in the one instance and from A' to B' in the other instance. The times for these two distances, ΔL_1 and ΔL_2, are equal.

What about the *velocity* of the flow through the two vessels? Since $v = s/t$, evidently the velocity must be greater in the smaller vessel. Each individual capillary has a very small cross section, but the volume of blood is carried by a very large number of these small vessels arranged in parallel with one another. Figure 11-17 shows several sets of capillaries arranged in parallel with one another. If the arterioles are clamped down, at the entrance of a capillary bed, the blood is diverted preferentially to other channels. If the total cross section of the other channels is less, the velocity of flow through the channels will increase. The velocity of the blood at various regions is also shown in Fig. 11-17, where it may be compared with the volumes (proportional to cross-sectional areas) and pressures.

Force of heartbeat, "thickness" or viscosity of blood, length and cross-sectional

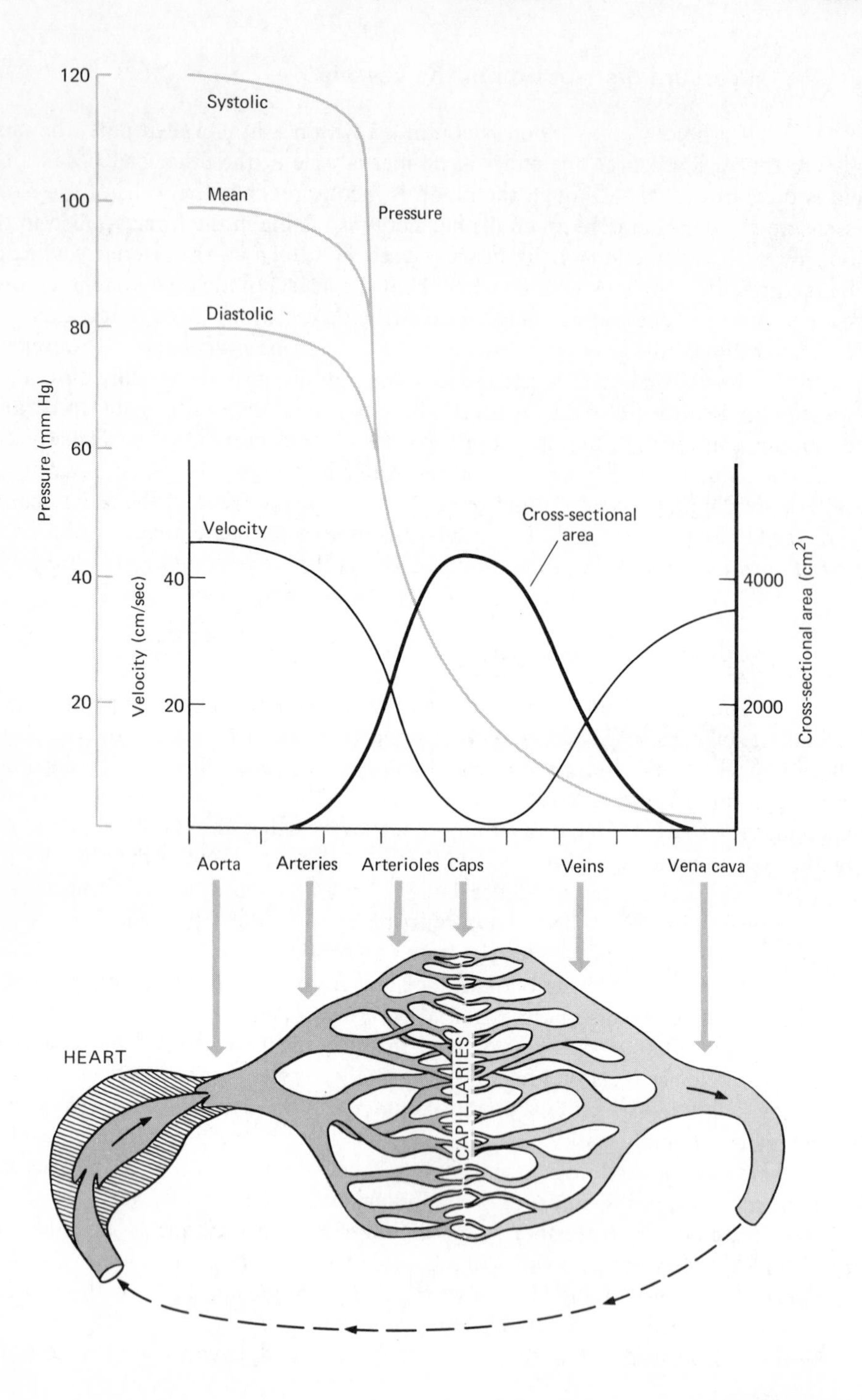

120
100
80
60
40
20
Pressure (mm Hg)
Systolic
Mean
Diastolic
Pressure
Velocity
Cross-sectional area
40
20
Velocity (cm/sec)
4000
2000
Cross-sectional area (cm^2)
Aorta
Arteries
Arterioles
Caps
Veins
Vena cava
HEART
CAPILLARIES

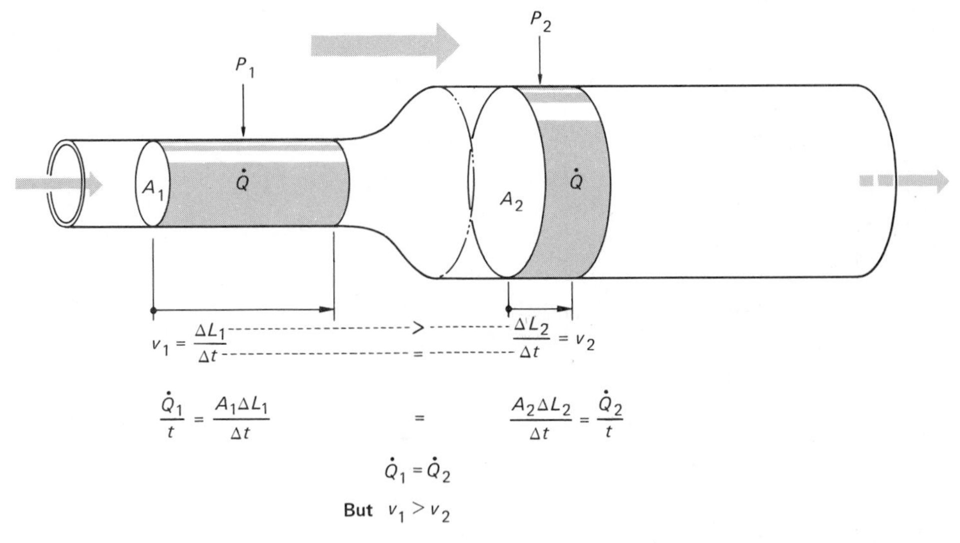

Fig. 11-18. Comparing velocity and volume rate of flow. The shaded areas represent equal volumes of blood in (1), an artery, and in (2), a large bed of capillaries. The cross-sectional area A_2 is much greater than A_1. The length ΔL_1 must therefore be greater than ΔL_2, because the volume $A_1 \cdot \Delta L_1 = A_2 \cdot \Delta L_2$. Consider equal volume rates of flow—you can see that the blood in the artery must move faster than the blood in the capillaries in order to get the shaded volume past a point in a given time.

diameter of the vessels, all can affect the volume rate of flow (and hence the velocity of flow) of blood in the vessels. We will consider each factor in turn in a simple system of pipes (the vessels) through which a liquid flows because it is driven by a head of pressure (mainly the force of the heartbeat). We will take each of the parameters in turn, and we will arrive at a simple mathematical description that will include all of them.

The blood flow equation

DERIVATION OF THE EQUATION. The volume rate of flow ($\dot{Q}$) depends on the pumping force provided by the heart. We can represent that force by a pressure head in a column of water: in Fig. 11-19(a) the volume rate of flow $\dot{Q}$ is greater for P_2 than for P_1.

Fig. 11-17. *Pressure, velocity and cross section in the circulation.*

1. *Gray line and scale.* Blood pressure in various parts of the system.
2. *Heavy line and scale.* Summated cross-sectional areas at each level.
3. *Thin black line and scale.* Velocity of blood flow in vessels.

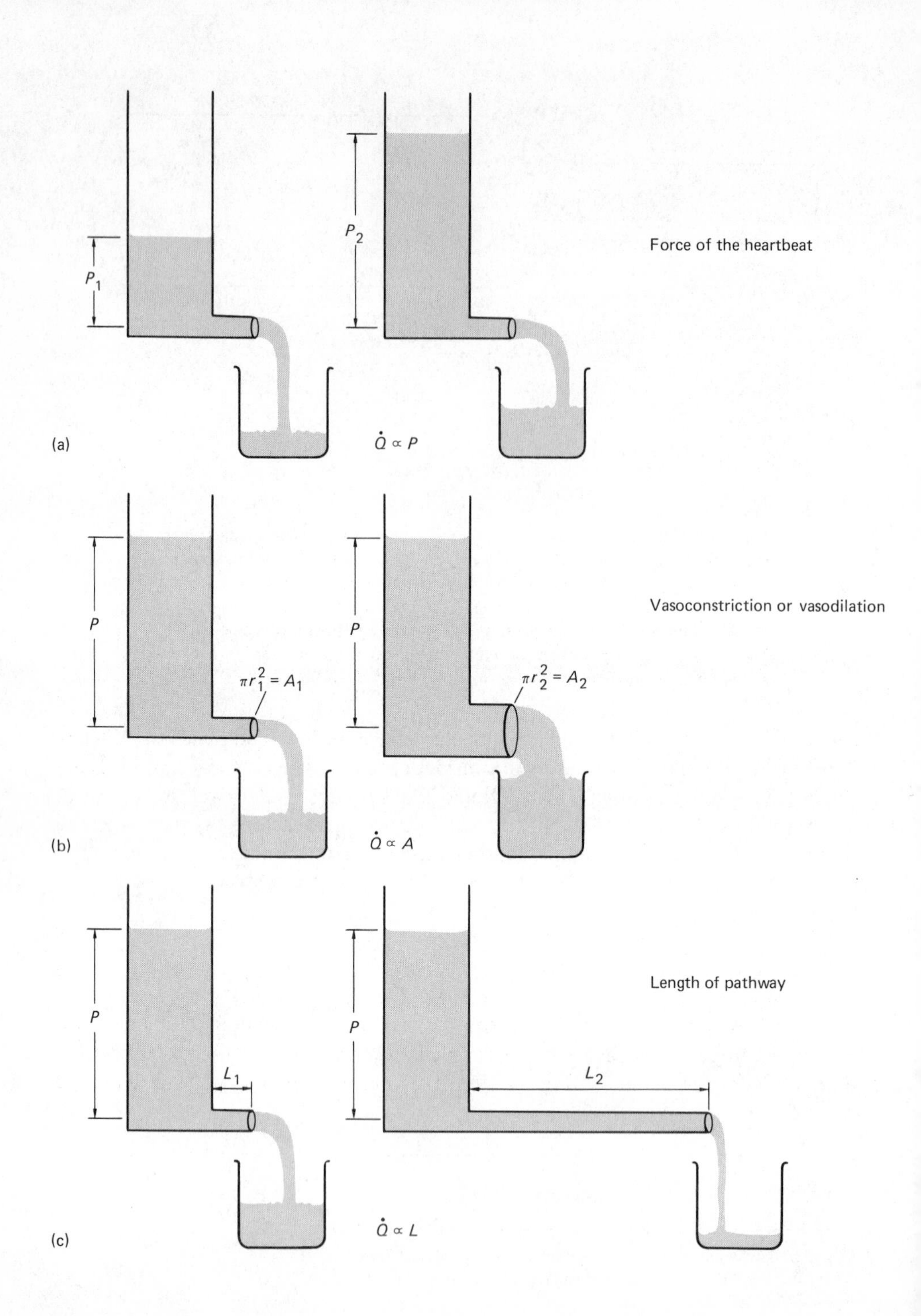
P_1
P_2
Force of the heartbeat
(a)
$\dot{Q} \propto P$
P
P
$\pi r_1^2 = A_1$
$\pi r_2^2 = A_2$
Vasoconstriction or vasodilation
(b)
$\dot{Q} \propto A$
P
P
L_1
L_2
Length of pathway
(c)
$\dot{Q} \propto L$

For a particular pressure, $\dot{Q}$ also depends on the size of the delivery pipe. In Fig. 11-19(b), $r' > r$; therefore $\pi r'^2 > \pi r^2$, the cross-sectional area A' is greater, and the volume per unit length is greater. It follows that the volume delivered per unit time, $\dot{Q}'$, is greater than $\dot{Q}$, since $\dot{Q} \propto A$ and therefore $Q \propto \pi r^2$.

In a long pipe, energy is used in overcoming *frictional resistance* (Fig. 11-19) of the fluid in relation to the container. Therefore, for a particular pressure, $\dot{Q}$ depends on the length of pipe and $\dot{Q} \propto 1/l$.

The frictional resistance to flow also depends on the cross-sectional diameter. Not only is $\dot{Q}$ greater for a large pipe simply because the larger pipe carries a large volume, but the flow is also easier in a large pipe compared to a small one because the frictional resistance to flow is less in the former. It can be shown that the proportionality is such that $\dot{Q} \propto r^2/8$ because of frictional resistance as affected by diameter. Finally, the volume flow depends on the internal frictional resistance of the fluid. This is called *viscosity*. Blood can change in its viscosity. The viscosity of blood is increased when the red cell concentration rises. Therefore the volume rate of flow decreases. So we can write $\dot{Q} \propto (1/\eta)$, where η is the viscosity.

If we combine all these proportionalities, we get

$$\dot{Q} = P \times \pi r^2 \times \frac{1}{l} \times \frac{r^2}{8} \times \frac{1}{\eta} = \dot{Q} = \frac{P\pi r^4}{8l\eta}$$

Having obtained this equation, and comprehending somewhat how we derived it, we can use it to help understand how flow through the capillaries is influenced by changes in blood pressure, vessel diameter, and blood viscosity. Vessel length generally remains constant.

PREDICTIONS FROM THE BLOOD FLOW EQUATION. Suppose that we consider how the flow through a vessel is affected when we change the *cross-sectional area* of the vessel. Assume that the radius is changed from 10 to 20 μm, for example,

---

Fig. 11-19. Conditions governing volume rate of blood flow.

(a) Volume rate of flow is directly proportional to the driving pressure: $\dot{Q} = K_1 \Delta P$.

(b) Volume rate of flow is directly proportional to the cross-sectional area of the outflow vessel: $A = \pi r^2$, $\dot{Q} = K_2 \pi r^2$. The effect of internal friction of flow is reduced, and, therefore, flow itself is increased, proportional to the factor $r^2/8$ (see text): $\dot{Q} = K_3 r^2/8$.

(c) Volume rate of flow is inversely related to the length of the outflow vessel: $\dot{Q} = K_4/L$. $\dot{Q} \propto 1/l$.

(d) (Not illustrated). Volume rate of flow is inversely related to the viscosity (internal friction of fluid): $\dot{Q} = K_5/r$.

$$\therefore \quad \dot{Q} = K_1 K_2 K_3 K_4 K_5 \frac{\Delta P \pi r^4}{8\eta L} = \frac{k \Delta P \pi r^4}{8\eta L}$$

If units are properly chosen,

$$\dot{Q} = \frac{\Delta P \pi r^4}{8\eta L}$$

and that all other terms in the equation remain constant: Then

$$\dot{Q} = \left(\frac{P\pi}{8l\eta}\right) r^4 = Kr^4$$

$$\dot{Q}_1 = K \times r_1^4 = K \times 10^4 = K \times 10{,}000$$

$$\dot{Q}_2 = K \times r_2^4 = K \times 20^4 = K \times 160{,}000$$

For a doubling of diameter, there has been a 16 times increase in the flow through the vessel. The vessel diameter may easily alter under the influence of vasomotor changes. Nerve impulses and/or chemical substances can affect the tone in the vascular smooth muscle, and hence the vessel diameter.

How will the flow be affected by a change in blood pressure? In that instance, let the diameter and other terms except pressure remain constant: Then

$$\dot{Q} = \left(\frac{\pi r^4}{81\eta}\right) P \quad \text{and} \quad \dot{Q} = KP$$

The relevant driving force for blood through the vessels is the difference between the pressure on the arterial and the venous side. Venous pressure is near zero and small compared to arterial pressure. So, as an approximation, we may let the blood arterial pressure equal P. Let the blood pressure change from an average of 100 mm Hg to an average of 200 mm Hg. Then

$$\dot{Q}_1 = KP_1 = K \cdot 100$$

$$\dot{Q}_2 = KP_2 = K \cdot 200$$

A doubling of the pressure with no change in other factors has only doubled the flow. A pressure change of this magnitude may easily occur during exercise. Ordinarily, however, it is also accompanied by an increase in vessel diameter, a change that has a more profound effect on flow as noted above. When the increase in pressure is a result of arteriosclerosis, the vessel walls are not flexible and do not adequately change in diameter.

Arteriolar control of blood distribution

Although the vascular system usually behaves as if it were overfilled, it is easily capable of holding a great deal more blood than it normally contains. For example, a sudden increase in vascular capacity occurs during the generalized vasodilation that precedes a faint. In this instance, the available blood fills the capacious capillary beds of viscera and muscle, and none is left for the head. The sudden drop in blood pressure prevents ascent of blood to the brain, and the individual collapses. This reaction is quite appropriate, since all parts of the body will be equally well supplied with blood when the body is horizontal.

When the circumferentially arranged arteriolar smooth muscle contracts, the arteriole decreases in diameter; that is, it is in a state of *vasoconstriction*. Conversely, when the smooth muscle is relaxed, the vessel is in a state of *vasodilation*. The degree of *tone* in arteriolar smooth muscle of a particular region determines the total flow of blood through that region. The distribution of blood within a capillary bed depends on the preferential channels (*metarterioles*) and the *precapillary sphincters*. The arrangement and function of these pathways are shown in Fig. 11-20. The metarterioles have walls with somewhat less smooth muscle than the true arterioles. A precapillary sphincter is a small ring of smooth muscle at the entrance to a small set of capillaries. When it is constricted, blood is excluded from that region and flows instead through the preferential metarteriolar channels and through less-constricted precapillary sphincters.

Factors affecting blood distribution

The changes in arteriolar tone are determined by chemical factors, particularly by acid metabolites and hormones, as well as by nerve impulses to the arteriolar smooth muscle. The cell requirements change according to the cell activity, and the amount of blood moved through the capillaries per unit time changes correspondingly. In resting muscle, only a fraction of the capillaries are open, but actively working muscle requires a tremendously augmented blood flow. CO_2 and lactic acid, produced during muscle contraction, act directly on the arterioles and bring about local vasodilation. Thus more capillaries are made available to supply blood to the muscle cells.

In the viscera, considerable blood flow, and hence vasodilation, is required during digestion. However, during stress, adrenaline provokes vasoconstriction in the visceral arterioles and in the skin, although it brings about vasodilation of the arterioles of muscle.

Nervous control of vascular tone

The two most important factors determining blood pressure are the contractions of the heart and the peripheral resistance of the vascular system. The peripheral resistance is the resistance to flow that the blood meets in its movement away from the heart. This resistance depends, of course, on the extent to which the vessels are constricted. The greatest capacity for control over the vessel diameter is in the veins and in the arterioles. Smooth muscle cells oriented mainly circumferentially in these vessels are innervated by sympathetic postganglionic fibers that produce vasoconstriction. The most effective control of the vessel diameter is through these vasoconstrictor fibers. Although *vasoconstrictor* innervation is entirely postganglionic sympathetic in nature, *vasodilation*, resulting in decrease of peripheral resistance, has different origins in different places (Fig. 11-21). Generalized vasodilation occurs in relation to vegetative functions of the body, when, for example, the blood supply is

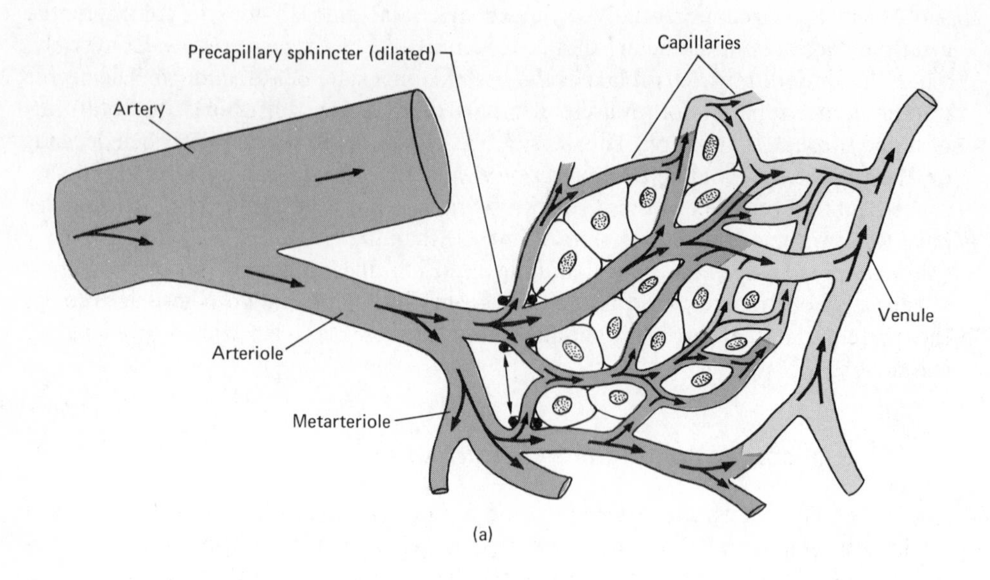

(a)

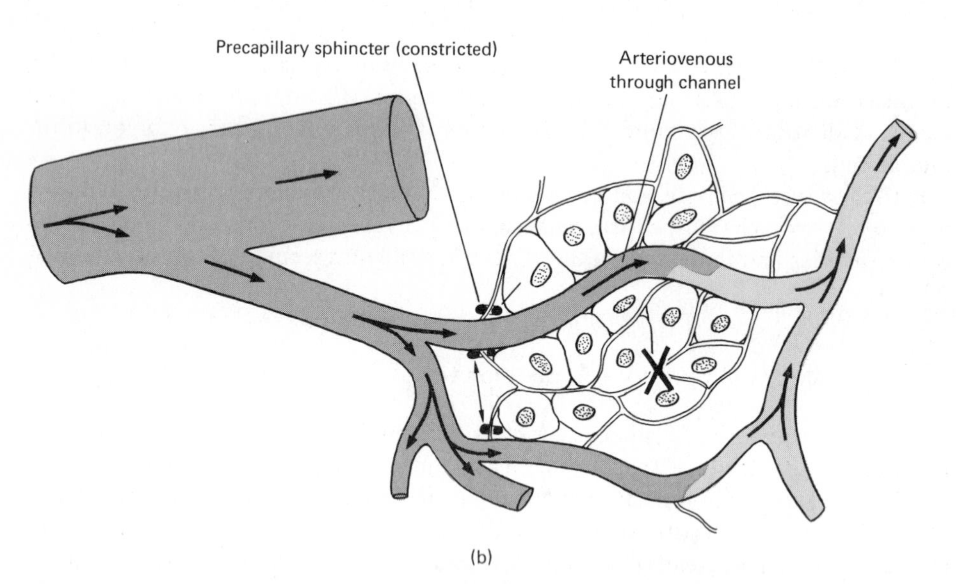

(b)

Fig. 11-20. *Capillary network.* (a) Blood flows through all the capillaries of the network. *Arteriole* has complete investment of smooth muscle. *Metarteriole* has discontinuous rings of smooth muscle. *Precapillary sphincter* is ring of smooth muscle at entrance to capillary. (b) Precapillary sphincters are constricted, blood is excluded from capillaries and flows preferentially in arterio-venous through channels. The cells at X have a particularly reduced blood supply.

particularly ample to the digestive system. Generalized vasoconstriction is more evident in circumstances of fear, fight, or flight, when blood is shunted away from the viscera. Then, vasodilation in arterioles of skeletal muscle may also occur, because of local action of substances produced by active muscles, plus the action of adrenergic vasodilator fibers. One effective vasodilation mechanism is simply the decrease in vasoconstrictor tone.

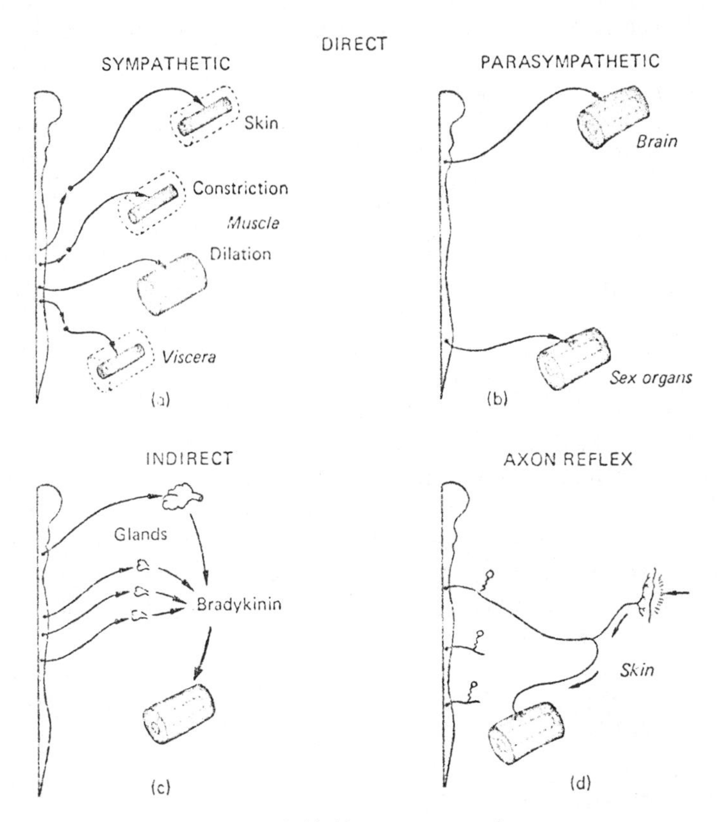

Fig. 11-21. Vasomotor control.

Active vasodilation may be induced indirectly through the action of *bradykinin*. Bradykinin is formed from a specific protein of blood and tissue fluid, in the presence of a suitable enzyme or activating substance, and causes contraction of some smooth muscle. In arterioles it evokes vasodilation, particularly in certain parts of the skin, where cholinergic (preganglionic) sympathetic fibers activate the sweat glands.

There are some places in the body where active vasodilation is caused by vasodilator parasympathetic fibers acting directly on arterioles [Fig. 11-21(b)]. The blood flow to the external sex organs depends on parasympathetic vasodilator fibers. Other parasympathetic vasodilator fibers innervate arterioles of the cerebral cortex. Unlike other organs, the brain cannot change volume when vasodilation or vasoconstriction occurs. The unyielding cranium prevents any overall volume change, and vasodilation in one region can occur *only* if there is vasoconstriction elsewhere in the brain.

The arterioles of skeletal muscle are evidently very important, for the blood flow in muscle may increase as much as 100 times during exercise. The necessary vasodilation is produced by impulses in cholinergic sympathetic fibers [Fig. 11-21(a)]. Finally, vasodilation can be brought about by the peripheral branches of certain dorsal root nerve fibers [Fig. 11-21(d)]. This does not mean that impulses normally travel from the central nervous system over dorsal root fibers and cause vasodilation. Apparently the normal action is entirely a peripheral one depending on what is called an *axon reflex*. A sensory fiber apparently may have a branch to the skin *and* to blood vessels. When the skin branch is stimulated, impulses go not only into the central nervous system but also go via the other branch to the blood vessels, where the impulses bring about vasodilation.

Some effects of changes in blood flow

The vasomotor tone of an organ can be judged without direct examination of the arterioles. During vasoconstriction, tissues that are not highly pigmented become pale, whereas during vasodilation they become flushed. An exception exists in the effect of extreme cold on the skin. Mild cold causes vasoconstriction, and the skin may appear pale. In more extreme cold, vasodilation ensues, but the blood flows slowly, is anoxic, and imparts a blue color to the skin.

An increase in temperature accompanies vasodilation in an appendage. The reason is that the augmented blood flow comes from the central part (the *core*) of the body, where the temperature is higher than at the periphery. Conversely, a decrease in temperature is characteristic of vasoconstriction. During vasodilation, the volume of an organ will increase; it will decrease during vasoconstriction. Blood flow out of the veins is rapid during vasodilation and relatively slow during vasoconstriction.

The capillaries

So far as the nourishment of the tissues of the body is concerned, the capillaries are the most important parts of the vascular system. It is through the thin walls of these vessels that the cells receive their energy supplies and get rid of their wastes.

Most capillary walls are permeable only to water and to small dissolved particles and generally do not allow the exit of proteins and cells from the blood. The capillaries of the liver are an exception and normally allow the passage of protein through their walls. Other capillaries may become excessively permeable when subject to extreme stress or disease. Materials do not merely diffuse across the capillary wall. They are actually forced out of the capillaries by the blood pressure.

There are two means by which water and salts forced out of the capillaries into the tissue spaces are returned to the blood. Some fluid is collected and returned to the blood by the lymphatic system. In addition, water and small molecules that leave the capillaries especially on the arterial side may return especially on the venous side of the capillary bed.

OSMOTIC AND HYDROSTATIC PRESSURES IN CAPILLARIES. The movement of water and solutes between blood and tissue fluids at any particular place in the capillary network is determined by the net pressure difference between the inside and outside of the capillary.

The hydrostatic head of blood pressure tends to force blood plasma out of the capillaries, but the osmotic pressure of the blood tends to draw the fluid back into the capillaries. Correspondingly, hydrostatic pressure and osmotic pressure of the tissue fluids will also influence the direction of fluid movement. It is easy to see the role of osmotic pressure in this situation. The osmotic pressure of blood and of tissue fluids depends on the concentration of particles in those two compartments, and the water will tend to flow in the direction toward the highest salt concentration. However, the capillary wall is permeable not only to water but also to most of the solutes in the plasma and lymph. It is relatively impermeable to only larger protein molecules. The relevant osmotic pressure difference therefore depends on the concentration of protein molecules having a molecular weight of about 70,000 or more. The concentration of such molecules in the plasma exceeds that in the tissue fluid surrounding most capillaries, except in the liver, where the blood proteins are manufactured and enter the blood from the intercellular spaces.

Although the capillary membranes are permeable to small solute particles, such as Na^+ and Cl^-, these ions do make a contribution to the effective osmotic concentration for a reason that depends on properties of the protein molecules. The protein molecules are electrically charged and will therefore tend to hold in their vicinity other particles of opposite sign. In another context, we have seen how this situation gives rise to an electric potential difference across the membrane (see Chaps. 3.1 and 10.1). In addition, it gives rise to an osmotic pressure difference across the membrane. The net *hydrostatic pressure* outward may be described by the equation

$$P_{\text{bp}} - P_{\text{tp}} = \Delta P_{\text{hp}}$$

(i.e., blood hydrostatic pressure minus tissue hydrostatic pressure equals net hydrostatic pressure).

Tissue pressure working in opposition to blood pressure tends to collapse the capillaries as well as drive fluid inward. Pressure on the surface of the skin or pressure exerted on vessels during contraction of skeletal muscles has this effect. We are immersed in air at atmospheric pressure. The pressures we have been talking about are usually in excess of atmospheric pressure and are therefore called *positive pressures.* If the tissue pressure were less than atmospheric, the capillaries would tend to balloon outward, and outward flow from the blood plasma would be greater to the extent that the tissue pressure is negative—that is, less than atmospheric. Apparently the tissue hydrostatic pressure is occasionally negative, tending in effect to pull fluid through the capillary wall and into the tissue spaces. This effect is generally small, compared to the positive pressures encountered.

Sometimes it is helpful to separate the movement of water into a bulk flow due to hydrostatic pressure differences and an osmotic flow due to diffusion controlled by osmotic pressure differences.

The effect of an *osmotic pressure* difference is described by the equation

$$P_{\text{bop}} - P_{\text{top}} = P_{\text{cop}}$$

(i.e., blood osmotic pressure minus tissue osmotic pressure equals net capillary osmotic pressure).

The net flow through the capillary wall is determined by the difference between the net hydrostatic and the net osmotic pressure (Fig. 11-22). This situation is a special example of the Gibbs-Donnan equilibrium principle (see Chap. 10.1).

The net effective osmotic pressure of the blood plasma amounts to about 28 mm Hg. That of the tissue fluids, on the other hand, is only about 3 mm Hg. The net osmotic driving force for water is thus generally into the plasma to the extent of about 25 mm Hg.

The hydrostatic pressure in the capillaries (the "blood pressure") tends to force fluid out of the capillaries—this is the *filtration pressure.* In capillaries of the kidney this pressure is kept rather constant, but in most other capillary beds it varies a great deal. Hydrostatic pressure in the tissue spaces may tend to oppose the effective drive of capillary filtration pressure.

MOVEMENT OF FLUID THROUGH CAPILLARY WALLS. Whether fluid tends to move inward or outward in any particular capillary or set of capillaries depends on the balance of pressures and the permeability of the capillaries. Under conditions of inadequate oxygen supply, capillary walls tend to be leaky, and if the blood pressure is also high, there will be considerable tendency for outward movement of fluid.

When the blood is dilute and the hydrostatic pressure low, fluid tends to *return* to the capillaries. On the average, as much fluid returns to the circulation as leaves it. Along any particular set of capillaries, there is likely to be a gradient of hydrostatic pressure and of osmotic pressure that favors exit of fluid at the arterial end and return of fluid at the venous end.

This condition occurs if the net outward hydrostatic pressure at the arterial end sufficiently exceeds the net outward osmotic pressure difference. Water and small molecules and ions will then move out of the capillaries and into the tissue spaces. The blood protein molecules, too large to pass through the capillary walls, therefore increase in concentration as the blood moves toward the venous side of the capillaries. The total osmotic pressure of the blood on the venous side is thus somewhat greater than the osmotic pressure of the tissue fluids, which contain considerably less dissolved protein.

During blood flow, the hydrostatic pressure is dissipated and the pressure is

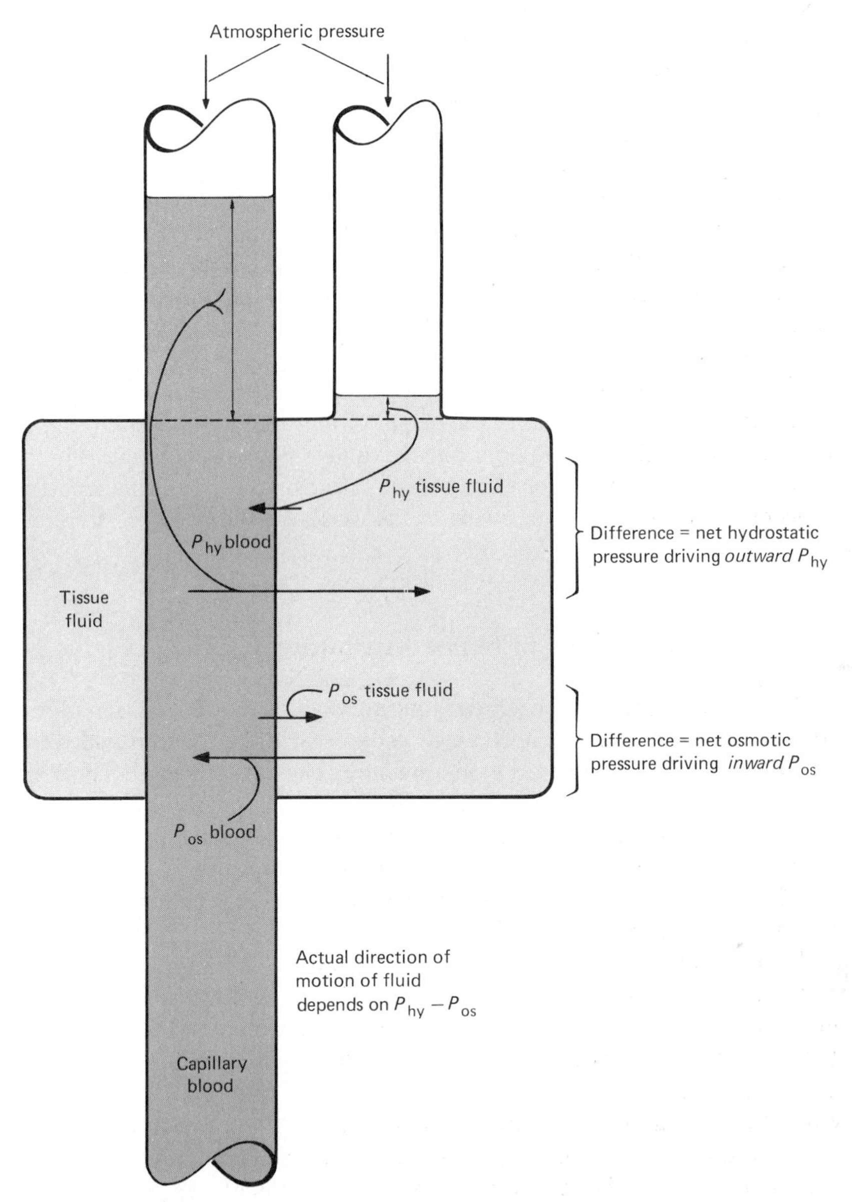

Fig. 11-22. Pressures governing fluid transfer through capillary wall. Relative magnitudes of pressures involved are shown by lengths of straight arrows. Ordinarily, the hydrostatic fluid pressure of the blood ($P_{\text{hp-blood}}$) exceeds that of the tissue fluids ($P_{\text{hp-tissue fluids}}$) and the osmotic pressure of the blood ($P_{\text{op-blood}}$) exceeds that of the tissue fluids ($P_{\text{hp-tissue fluids}}$).

lower at the venous end because of frictional losses of the energy available in the blood pressure. At the same time that the blood pressure becomes lower, the osmotic pressure increases. This situation occurs because the large molecules, together with the smaller ions that they attract, remain within the bloodstream when the water moves out of the capillaries. The hydrostatic pressure having decreased and the osmotic pressure having increased within the capillaries, water then tends to leave tissue spaces and return to the blood.

Occasionally the venous pressure becomes sufficiently great to oppose the return of fluids. Such obstruction may develop, for example, if the heart fails to pump the blood rapidly enough. Blood piles up in the veins and exerts a back pressure opposing the osmotic flow of water from the tissue spaces back into the capillaries. Quantities of fluid therefore accumulate, resulting in edema. Swelling of the ankles is a common example of this situation.

When we compare the vessels of the legs to those of the arms, we find a higher blood pressure in the legs because of the height of the column of blood from heart to leg level. Consequently, accumulation of tissue fluid is more likely to occur in the legs than in the arms. Such edema for example, may develop during pregnancy. It results from a decrease in venous return, owing to the weight of the body pressing on veins of the pelvis.

Reflex regulation of blood distribution

Both heart rate and arteriolar tone are regulated by chemical and pressure receptors. Receptors in both the aortic arch and the carotid arteries provide the brain with information about the arterial blood pressure. These receptors are nerve endings that discharge impulses according to the extent of stretch in the wall of the vessel in which they are imbedded. The receptors in the carotid are in the wall of a special, dilated region of the vessel at the point where the internal carotid leaves the common carotid. This *carotid sinus* has the particular function of monitoring the pressure of blood going to the brain.

If blood pressure is too low, blood will not be adequately supplied to the tissues, and the brain, in particular, may suffer from anoxia. If the pressure is too high, there is a possibility of injury to weak-walled vessels. The aortic and the carotid pressure receptors provide the information that allows the brain to adjust heart rate and arteriolar tone so as to maintain an appropriate blood pressure.

When the arterial wall is stretched by increasing blood pressure, there follows an increase in the frequency of impulses traveling in the afferent fibers of the IX and X nerves that innervate the stretch receptors in the walls of the carotid artery and aorta (see Fig. 11-23). The region of the medulla oblongata that these fibers enter controls the arteriolar tone and the frequency of heartbeat. These areas are therefore called the *vasomotor center* and the *cardioregulatory center*. If the arterial blood pressure rises too rapidly, the heart rate ought to be made to decrease, and arteriolar tone to diminish, in order to reduce the pressure. In each of the centers, two parts ensure that this will be done. The *cardioinhibitory* center of the cardioregulatory center is stimulated to increased activity and provokes an increased frequency of impulses in the vagal parasympathetic nerve fibers. Inhibition of the heart ensues. At the same time, the *cardioaccelerator center* of this cardioregulatory center is inhibited by the impulses from the pressure receptors. The frequency of impulses traveling out along the sympathetic fibers is reduced as a result, further helping to cut down heart rate. Meanwhile, the vasodilator center of the vasomotor center is stimulated to increase activity, while inhibition of the *vasoconstrictor center* further helps reduce the blood pressure. Very precise control of heart rate and of vasomotor tone is ensured by the reciprocal action of the opposing members of the pairs of centers.

When blood pressure declines, the pattern of activity is reversed. The lessened stretch in the aorta and carotid sinuses is followed by a decrease in impulse frequency in the afferent nerves. The cardioaccelerator center, no longer restrained, now discharges impulses at a higher rate, while the cardioinhibitory center is less active. Acceleration of the heart follows from the resulting combination of increased frequency of impulses in the sympathetic and decreased frequency in the parasympathetic fibers to the heart. The vasoconstrictor center shows an increase in activity, while the vasodilator center decreases its output. The resulting increase in heart rate and peripheral resistance tends to raise the blood pressure.

In addition to the arterial pressure receptors, there are also pressure receptors on the venous side of the circulation. For example, receptors in the right atrium signal pressure changes in that organ. An increase in pressure in this chamber implies that the blood is not moving rapidly enough out of the ventricles. Increased pressure in the right atrium is therefore logically followed by stimulation of the cardioaccelerator center (with inhibition of the cardioinhibitory center). Along with the cardioregulatory center, the vasomotor center is also affected in a manner appropriate to provide vasodilation if the blood pressure is too high and vasoconstriction if the blood pressure is too low.

Figure 11-23 shows the reflex pathway of impulses involved in vagal control of the heartbeat. When there is a rise in blood pressure in aorta or carotid (1), impulse frequency increases along the afferent fibers of nerves X and IX that innervate those receptors. In the medulla oblongata, where the fibers enter the central nervous system (2), these impulses cause, via interneurons, synaptic excitation of the neurons from which arise the vagus autonomic preganglionic fibers that contribute to the cardiac nerves. When they arrive at the heart, these impulses excite the postganglionic neurons that are buried in the substance of the heart at the nodes. The impulses in the postganglionic fibers diminish the excitability of the nodal cells and thus decrease the heartbeat frequency.

Chemoreceptors also help control blood distribution. A high CO_2 level of the blood implies that the blood is not being moved rapidly enough from the active tissues to the lungs. It is therefore appropriate that CO_2 and H^+ directly stimulate the vasoconstrictor center and also result in a more effective heartbeat.

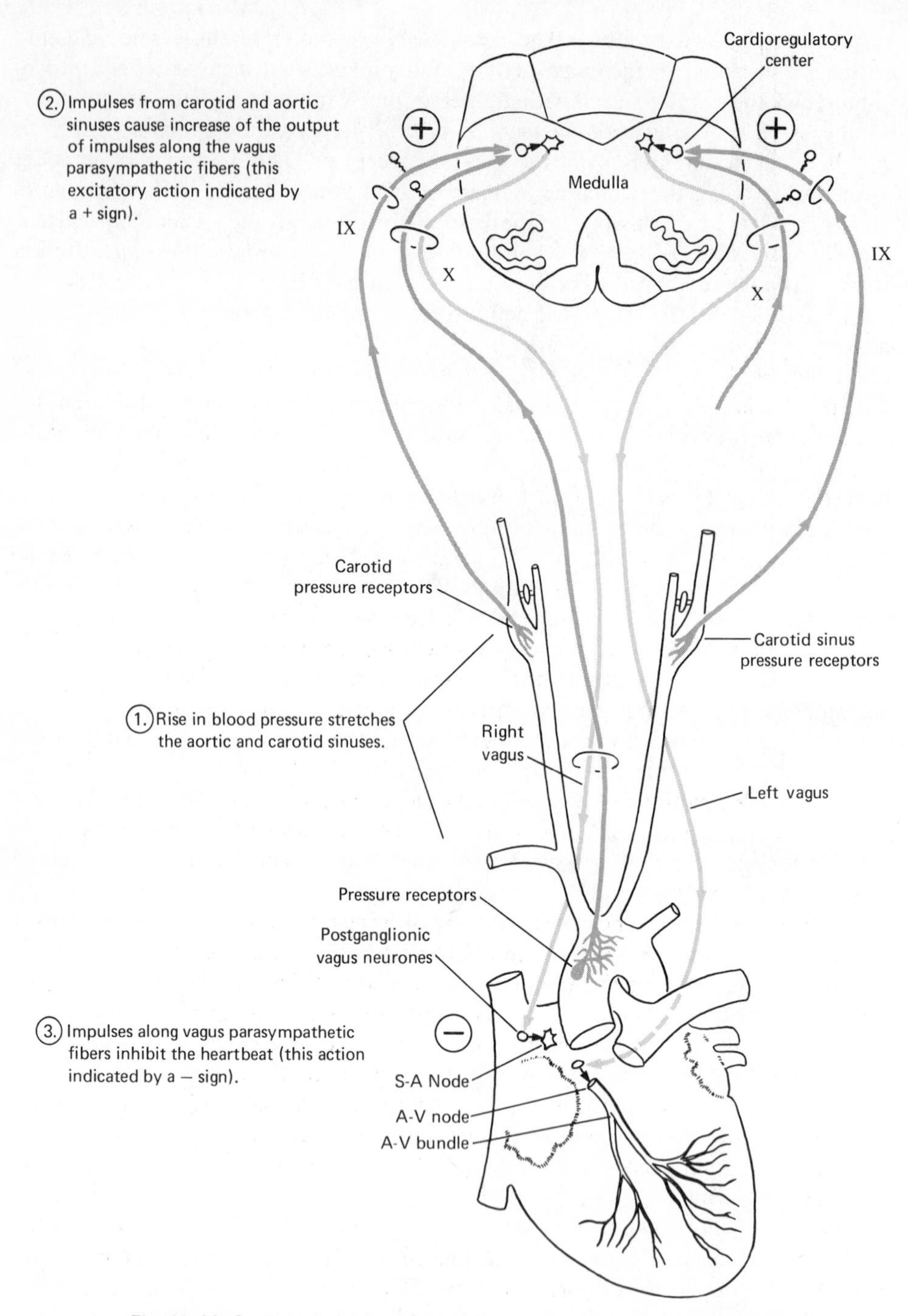

Fig. 11-23. Cardioregulatory mechanisms. Large arrows show the direction of nerve impulses.

Section 11: *Questions, Problems, and Projects*

1. Illustrate, by means of a pair of interconnected standpipes, the principle that blood flow is due to a pressure differential between different parts of the circulation.
   (a) In that highly simplified scheme of the circulation, draw an arrow to show the direction fluid would tend to flow.
   (b) Write an expression for the volume rate of flow $\dot{V}$ as a function of the pressure difference P and the resistance R.
   (c) Is the flow directly or inversely proportional to P?
   (d) Is the flow directly or inversely proportional to R?
2. What is the pressure in mm Hg of a column of water 1000 mm high? (Hg is 13 times as heavy as H_2O.)
3. The heart is about 1 m above the foot. If the blood pressure at the heart is 100 mm Hg, what is the blood pressure at the foot?
4. Write, in the sequence in which they become activated, the names of the nodes and bundles along which excitation proceeds in the mammalian heart.
5. Write, in the sequence showing the direction of blood flow, starting at the vena cavae and going to the lungs, the names of the chambers and the valves of the heart.
6. Do the same, starting at the lungs and going to the aorta.
7. What changes occur in the cardiac output when the stroke volume increases from 70 to 100 ml and the heart rate increases from 70 to 110 per minute?
8. On a sketch showing a (large) capillary and its arteriolar and venular ends, indicate at appropriate places the pressures shown in the chart.

	Pressure at Arteriolar End (mm Hg)	Pressure at Venular End (mm Hg)
Blood osmotic pressure (BOP)	28	32
Blood pressure (BP)	35	15
Tissue osmotic pressure (TOP)	4	4

   (a) In the diagram, draw arrow heads to show the direction in which each pressure factor tends to influence the movement of water.

(b) Indicate the sum of the forces tending to move water in or out as the case may be.

(c) Then state the net direction and force at each end of the vascular channel shown.

9. What is the fluid in which the fetus floats?
10. What is the function of the umbilical cord?
11. Describe the relation of the fetus and the mother's circulation.
12. The P_{O_2} of the maternal blood in the placental sinuses is 60 to 80 mm Hg and of the fetal blood 20 to 30 mm Hg. In which direction is the diffusion gradient for O_2?
13. Comment on the diffusion gradients between maternal and fetal bloods for nutrients and for excretory products.
14. Match properly:

    (a) Umbilical vein bypassing through liver — (1) Ductus venosus
    (b) Passage from right to left atrium — (2) Foramen ovale
    (c) Shortcut from pulmonary artery into aorta — (3) Ductus arteriosus

15. What happens to the ductus venosus after birth? Why does it become occluded?
16. How does the opening of the pulmonary circulation ensure the closing of the foramen ovale?
17. Explain how, although *in utero*, liver and lungs were bypassed, after birth blood flows through both these organs.

12

RESPIRATION AND GAS EXCHANGE

12.1 Mechanics and Control of Breathing

Most animal cells require a continual supply of oxygen. In addition, they require a means of getting rid of the surplus carbon dioxide that is generated when oxygen is used. *Respiration* includes the processes involved in using oxygen and producing carbon dioxide. It is convenient to consider these processes at three levels: *external respiration*, or breathing, which involves moving the air in and out of the lungs; *transportational respiration*, concerned with the means of carrying the gases between the lungs and the tissues; and *cellular* (*internal*) *respiration* or *metabolism*, the direct utilization of O_2 and production of CO_2 by the cells of the body.

In the vertebrates and in many other animals, the respiratory gases are carried between the atmosphere and the tissues by means of the blood. In order that the blood may acquire a load of oxygen and be relieved of its load of carbon dioxide, a large permeable area must be exposed to the environment. In fish, part of the body surface is pushed outward in innumerable featherlike projections called *gills*, while in air-breathing vertebrates, the surface is pushed inward to form *lungs*. In ourselves, the lungs consist of a multitude of tiny sacs that provide about 50 m^2 of moist surface for gas exchange between the air in the lungs and the blood in the capillaries.

The parts of the breathing apparatus

Venous blood from the body tissues comes to the lungs through the pulmonary arteries. At the terminations of the smaller divisions of the pulmonary arteries, the capillaries surround the lung air sacs, the *alveoli*, through the walls of which gas exchange occurs. CO_2 moves from the venous blood into the alveoli, and oxygen moves from the alveoli into the blood. Consequently, the blood entering the pulmonary veins is rich in O_2 and low in CO_2. In this condition it is forced toward the body tissues by the heart. Air enters the pulmonary system by way of the slit in the *glottis*, guarding the passage to the larynx, the "voice box" at the entrance of the *trachea* (Fig. 12-1). The voice depends on control of flow rate of air over the

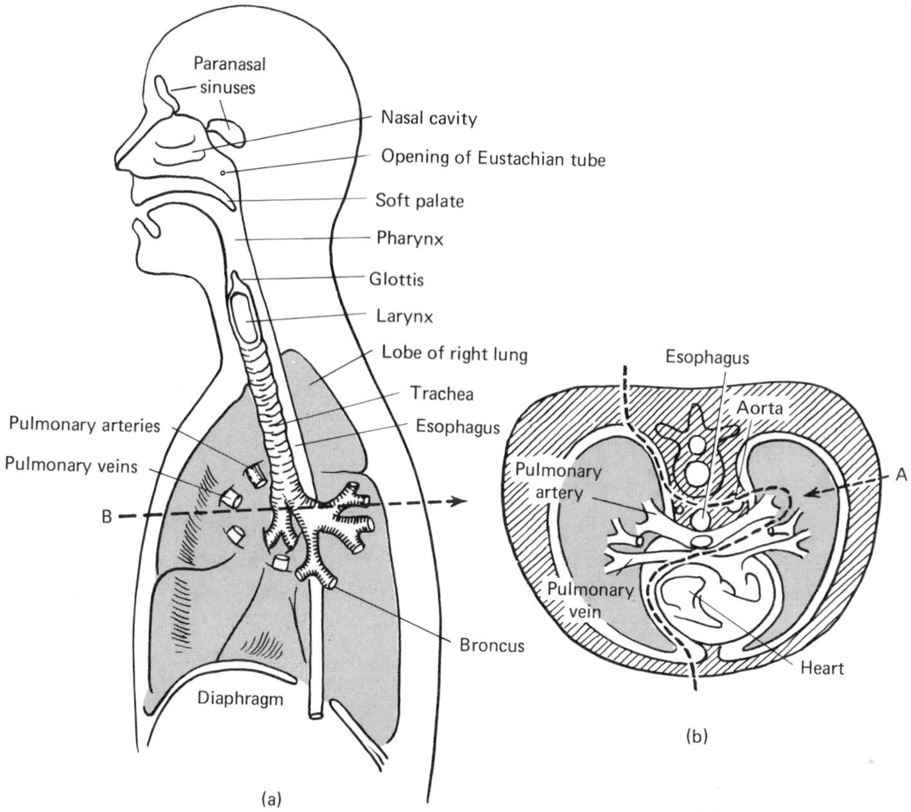

Fig. 12-1. The respiratory system. (a) Medial sagittal view of the thorax to show the lungs. (b) Generalized cross section of the thorax. Arrow *B* shows the approximate level of the section. Arrow *A* shows, to the left of the dashed line, the parts included in (a).

vocal cords, adjusted to appropriate tensions by muscles within the larynx. Breathing is suspended during swallowing, when the slit in the glottis closes, and the larynx moves upward, allowing the *epiglottis* to cover the glottal entrance as food slides into the esophagus. Ordinarily the trachea is kept open by incomplete cartilagenous rings, but, during swallowing, the food moving through the esophagus collapses the soft part of the rings. The trachea divides into two *bronchi*, one to each lung, formed of complete cartilaginous rings. These, in turn, divide into *secondary bronchi*, one for each lobe of each lung, and further division may continue as many as twenty times.

The intrapulmonic branches within the substance of the lungs are held firm by irregularly-shaped cartilaginous plates in their walls. These supports are gradually lost along the course of the passageways from bronchi to bronchioles. In the smaller divisions, the walls of the bronchioles are composed mainly of rings of smooth muscle fibers. When these fibers contract, they decrease the bronchiole diameter and slow the movement of air into and out of the lungs. In its upper reaches, the respiratory tract is lined by *ciliated epithelium* interspersed with mucus cells. Continual secretion of a watery or *serous* fluid along the respiratory tract ensures that the air will be saturated with water vapor as it enters the lungs, while a layer of sticky mucus prevents foreign particles from entering the depths of the lungs. The particles are removed by the cilia, whose whiplike motions move the mucus out toward the mouth.

In the deeper regions of the lungs, the respiratory lining is a simple squamous epithelium forming the alveoli, which extend in bunches of connected hollow spheres beyond the smallest bronchioles (Fig. 12-2). The alveoli are densely enveloped by the

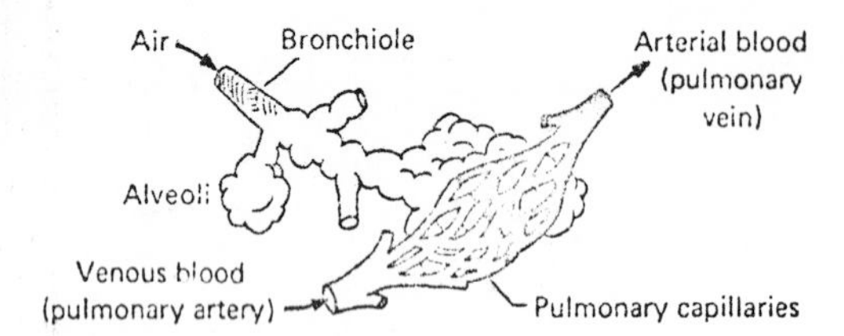

Fig. 12-2. An enlargement of a small portion of lung tissue. The pulmonary capillaries surround the alveoli.

pulmonary capillaries. The air in the depths of the lungs is separated from the blood by, at most, two thin cell layers, the alveolar epithelium and the capillary endothelium. The total distance through which gas molecules must travel between the air and blood is here usually no more than 1.5 μm, but during pulmonary disease the gas exchange may be slowed by thickening of the wall or by a heavy layer of exuded fluid.

Movement of air in the lungs

The thoracic cavity, within which the lungs are contained, is an airtight chamber of which the volume may be changed by movements of the diaphragm and ribs. The diaphragm is a thin, dome-shaped muscle forming a roof over the abdominal cavity and a floor for the thoracic cage. The fibers of this muscle originate along a line marked by the lower extent of the ribs and insert at the *central tendon* in the middle of the muscle. Ordinary *inspiration* occurs when air rushes in to occupy the extra volume made available for the lungs to expand into at the time when the floor of the thoracic cage is pulled downward during contraction of the diaphragm muscle fibers (Fig. 12-3). *Expiration* occurs when the diaphragm contraction ceases and the thoracic cage collapses to its resting volume, forcing air out of the lungs.

The lungs are entirely passive during breathing. They behave as would a thin-walled balloon, open to the atmosphere but otherwise enclosed in a partially evacuated space. Atmospheric pressure presses the lungs closely against the inside wall of the

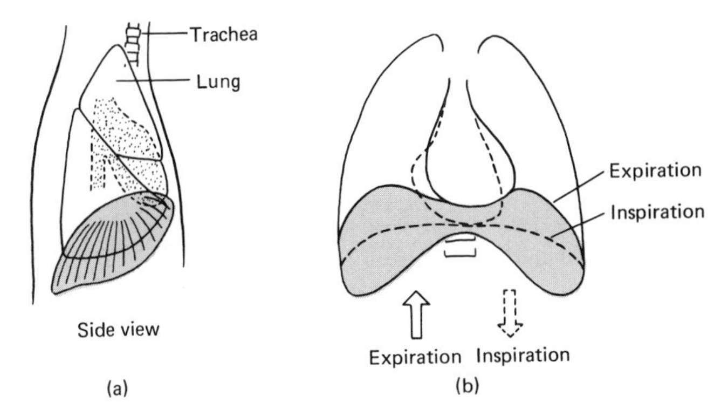

Fig. 12-3. Action of diaphragm in altering volume of thoracic cage. (a) The diaphragm muscle fibers are directed from the central tendon out to the lowermost limits of the ribs. (b) When the diaphragm muscle fibers shorten during inspiration, the dome shape of the diaphragm flattens. The heart also moves downward during the inspiration.

thorax even during rest, for the pressure in the *intrathoracic space* is below atmospheric. Between the outside of the lungs and this wall, a thin layer of fluid lubricates the surfaces, which are subject to continual movement upon one another during breathing.

If the thorax is opened to the outside, the lung will collapse. Breathing then becomes ineffective because the region outside the lung (inside of the thoracic cavity) remains at atmospheric pressure during respiratory movements. If the hole in the thorax is sealed to a pressure gauge, the following changes may be recorded during respiration: during *inspiration*, the pressure inside the thorax *decreases* to 752 mm Hg or about 8 mm Hg below normal atmospheric pressure as the thorax volume becomes larger; during *expiration*, the pressure inside the thorax *increases* to about 756 mm Hg or about 4 mm Hg below atmospheric pressure as the thorax volume becomes smaller. The changes in *intrathoracic pressure* accompanying changes in thoracic volume are responsible for the flow of air in and out of the lungs (Fig. 12-4).

The *intrapulmonic space* is continuous with the atmosphere. In the resting position of the thorax, this space is at atmospheric pressure. The walls of the lungs are elastic, like a balloon. Therefore they tend to pull away from the parietal pleura, thus tending to leave an intrathoracic space. This does not account, however, for the fact that the pressure in this space is always below atmospheric. It is necessary to postulate a continual withdrawal of fluid—and possibly, air—from this virtual space. It is this continual removal that keeps the pressure low in the *interpleural space.*

When inspiration begins, the pressure within the interpleural space decreases and remains at the lower level as long as the inspiration is sustained. In the intrapulmonic space, however, the pressure is atmospheric at the start of inspiration and at the time when the inspiration has reached its maximum. There is only a transient decrease in pressure during the transition of the thoracic cage from one position to another.

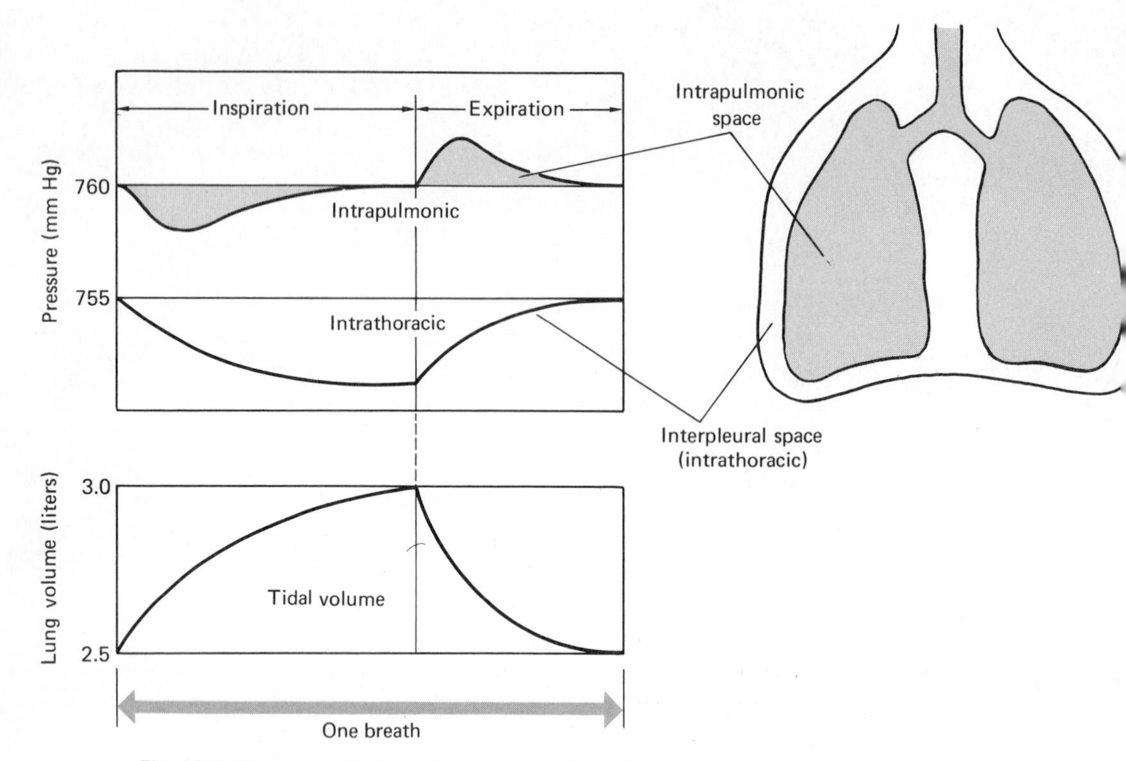

Fig. 12-4. Pressure and volume changes during breathing.

When the vigor of respiration is increased, the movements of the ribs make an increasingly important contribution to the volume changes of the thoracic cage. When the ribs move upward and outward, the volume of the cage is increased. Ribs 2 through 5 move outward and mainly increase the anterior-posterior distance of the thorax, while the movement of ribs 6 through 10 increases the transverse diameter as well.

The respiratory volumes

Changes in lung volume may be recorded if one breathes into a *spirometer* (Fig. 12-5). It is impossible to force out all the air from the lungs, simply because the thoracic cage cannot be made small enough. Consequently there is always a *residual volume* of air in the lungs at the end of an extreme expiration. The volume of rhythmic respirations is the *tidal volume*, occurring regularly like the tides. The maximum volume that can be moved *out* of the lungs following an extreme inspiration, or moved *into* the lungs following an extreme expiration, is the *vital capacity*. The vital capacity

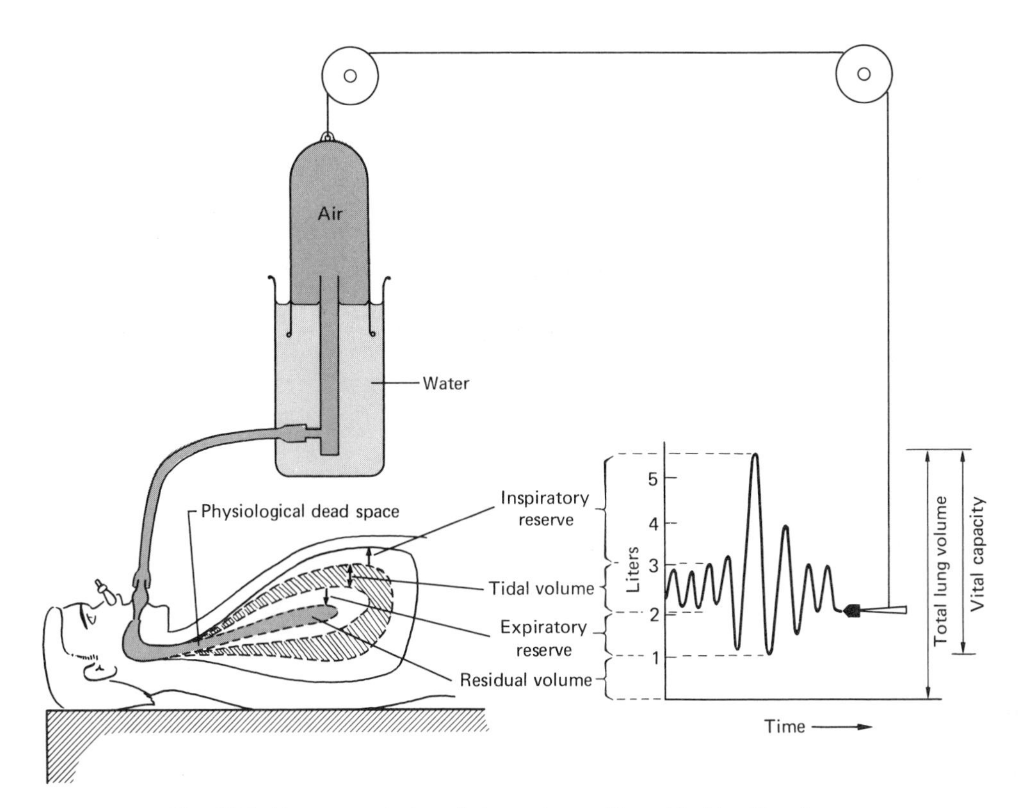

Fig. 12-5. Measurement of respiratory volumes. The subject breathes from a spirometer that provides a continuous record of respirations. The CO_2 produced by the subject may be absorbed so that normal breathing may continue. In an actual recording using the arrangement pictured, the record would then move upward, unless air is continually added to the spirometer to compensate for the O_2 used by the subject.

may be altered by the pooling of blood in the pulmonary circulation. If blood does not flow back to the left atrium as quickly as it should, the blood will reduce the volume available for air in the lungs. Exuded tissue fluid in the thorax will have the same effect.

Insufficient force of heartbeat during disease may produce a similar effect by allowing blood to back up in the pulmonary circuit. Vital capacity may be changed by exercise, which helps increase the size of the thoracic cage. The volume of air that can be added during an inspiratory effort in excess of the tidal inspiration is the *inspiratory reserve*; the volume that can be forced out in excess of tidal expiration is the *expiratory reserve*. The amount of reserve left during any respiration depends on the depth of the tidal respiration. During increased depth of breathing, the tidal volume increases at the expense of both reserves. The *inspiratory* and *expiratory reserves* are also changed under circumstances that alter the vital capacity.

Lung ventilation

By increasing the *depth* of breathing (i.e., the volume of tidal air), the rate at which O_2 is brought in and CO_2 is removed is increased. An increased *frequency* of breathing has a similar effect. Depth and frequency together determine the *ventilation* (i.e., the amount of air moved in and out of the lungs per unit time). When ventilation is increased, larger supplies of fresh air are made available more frequently to the respiratory epithelium. To calculate the ventilation during a certain time, we need only multiply the volume of air in each breath by the number of breaths during that time. It is convenient to take one minute as the standard time interval. Then ventilation ($\bar{V}$) equals volume (V) of air per minute (t) or,

$$\bar{V} = \frac{V}{t}$$

Volume of air moved per minute (V/t) equals the average volume per breath (v) times number of breaths per minute (n/t):

$$\frac{V}{t} = v \times \frac{n}{t}$$

Therefore

$$\bar{V} = v \times \frac{n}{t}$$

that is, ventilation equals the average tidal volume times the respiration frequency. During exercise, tidal volume and respiratory rate both increase, thus increasing the minute volume. The amount of new air exposed to the blood per minute therefore rises. Normal breathing is about 16 breaths per minute. At 0.5 liter per breath, the average ventilation is 8 liters per minute.

The movement of gases during respiration

The rhythmical contractions of the respiratory muscles ensure that the air will move in and out of the lungs and will thus be mixed with air in the depths of the lungs. The air deep in the alveoli of the lungs is never moved out all at once, even during forced expiration, but remains as the residual volume. As O_2 is taken into the blood, the amount of O_2 in the alveoli is reduced considerably below the concentration of O_2 in the outside air. As CO_2 is delivered from the blood, the amount of CO_2 in the alveoli is increased considerably above the concentration of CO_2 in the outside air. The fact that O_2 continues to be supplied to the blood moving through the alveolar capillaries and that CO_2 continues to be removed is due to the *diffusion* of the gases. By haphazard, undirected movements of the gas molecules, O_2 moves from the outside air toward the alveoli and CO_2 moves away—each gas toward its lowest concentration.

Diffusion of a gas depends on the continual motion of the gas molecules, which by bouncing about at random and ricocheting from each other tend to become uniformly distributed throughout their container. By means of the tidal respiration, that part of the lung volume nearest the outside is rhythmically replaced; the new volume then mixes by diffusion with the remaining gases deeper in the lungs. By changing the ventilation of the lungs, we keep the diffusion gradient high in spite of increased uptake of O_2 or increased release of CO_2 by the blood flowing through the pulmonary capillaries.

The description of diffusion gradients in gases

The movement of the respiratory gases proceeds through three completely different regions: air, blood, and tissue fluids. In air the gases are dispersed among one another; in blood they are dissolved in the water of the plasma and are also held by hemoglobin; in the tissue fluids they are dissolved in water, together with other substances in solution. The respiratory gases move between these regions in accordance with the concentration differences, but how may we compare concentrations in such different phases?

For most substances in solution, the concentration may be defined as weight of dissolved substance per unit volume of solution. For a gas it is often more convenient to speak of volume of gas present in a unit volume of solution. But the concentration of a gas in a mixture of gases cannot be directly compared with the concentration of a gas in a solution. The different situations may be compared by using the concept of the *gas pressure* in all instances.

In a mixture of gases, each constituent contributes to the total pressure its own pressure (called the *partial pressure*) in proportion to the relative amount of it present. For example, ordinary air is composed of about 20 percent O_2 and 80 percent N_2. The total pressure exerted by the air is equivalent to 760 mm Hg. If the O_2 is removed from a sealed container of air that is at atmospheric pressure, the pressure in the container will be reduced by 20 percent. The partial pressure of O_2 in ordinary air at ordinary pressure is evidently about 152 mm Hg.

Gas tensions

The gas that is present in a solution may be referred to in terms of the pressure of gas in an overlying atmosphere that would be necessary to provide the amount of gas that is present in the solution. This pressure is equal to the *tension* of the gas tending to escape from the solution.

CO_2 and O_2 will diffuse out of a sample of blood or tissue fluid taken from anywhere in the body and exposed to the air. But if the blood or tissue fluid is placed in a container together with the gases at appropriate partial pressures, the concentrations of O_2 and CO_2 in the solution will remain unchanged. This fact provides a method of comparing the concentrations of various gases all the way from the outside air to the tissues deep in the body. Samples of blood from various places in the circulatory system may be examined to determine CO_2 and O_2 content. In each instance, the pressure of each gas necessary to provide the observed concentrations can be determined. The pressures so defined are the *tensions* of the gases. Figure 12-6 shows

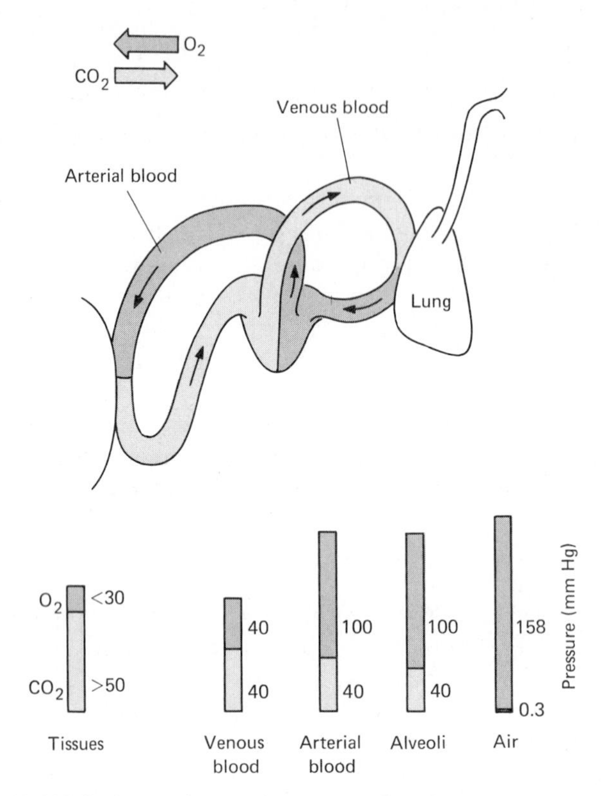

Fig. 12-6. Diffusion gradients and movement of respiratory gases. The lengths of the bars indicate the CO_2 and O_2 pressure in inspired air and in alveoli, and the CO_2 and O_2 tension in blood and tissues in mm Hg. Each gas moves toward the region of its lowest concentration (indicated by arrows at upper left).

the gas tensions for O_2 and CO_2 found in various places in the circulatory system, compared with the partial pressures of these same gases in the respiratory tract. O_2 tension is very low among the cells and may go down practically to zero in actively working tissues, which use the O_2 as fast as it can be delivered. CO_2 tension is high in active tissues producing CO_2 at a high rate. The O_2 tension of arterial blood is high because the blood has picked up O_2 from the alveoli. The CO_2 tension of venous blood is lower than that of the most active tissues because some of the blood has gone through capillaries in relatively inactive tissues where very little CO_2 is given off.

Neural control of breathing: The effector system

The cell bodies of motoneurons innervating the diaphragm are located in the gray matter of the cervical region of the spinal cord and send fibers into spinal roots $C_{3,4,5}$. As the *phrenic nerve*, these fibers arrive at the diaphragm, to which they carry trains of impulses during each inspiration.

As an inspiration builds up, the frequency of impulses in the phrenic fibers increases, and the number of active motoneurons (and hence motor units) also increases, until the inspiration reaches a maximum for the given situation. Then there is a sudden cessation of the phrenic activity, the diaphragm relaxes, and expiration ensues (Fig. 12-7).

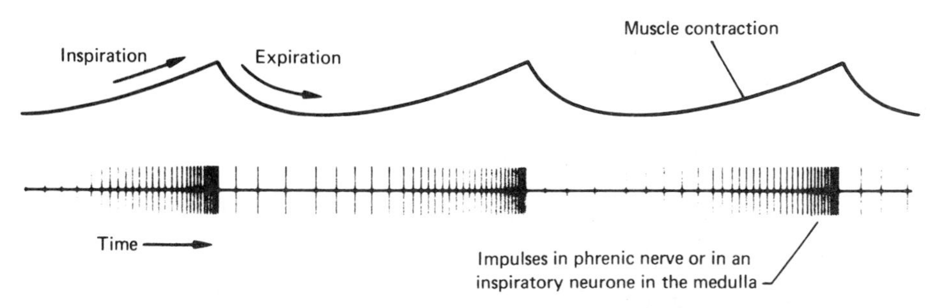

Fig. 12-7. Phrenic nerve impulses and diaphragm muscle contraction. Upper record represents contractions of a small part of diaphragm muscle. Lower record shows action potentials that might be taken from phrenic nerve fibers or from some neurons in the respiratory center.

During deepened respiration (*hyperpnea*) more phrenic-diaphragm motor units are activated, and the frequency of impulses in each active unit becomes greater. The *intercostal muscles* also come into action. Motor nerve fibers innervating the intercostal muscles arise from cell bodies in the gray matter of the thoracic region of the spinal cord and contribute fibers to spinal roots T_{1-12} (Fig. 12-8).

The external intercostal muscles elevate the ribs during inspiration, whereas the internal intercostal muscles lower the ribs during expiration.

Motoneurons to the diaphragm and intercostal muscles are driven by impulses coming down the spinal cord from the respiratory center in the medulla (Fig. 12-8). Like the sinoatrial node of the heart, the respiratory center seems to have a rhythm of its own, a rhythm that is present when all afferent nerves and all connections of the region with other parts of the nervous system are severed. It is this intrinsic rhythm that keep trains of impulses going to the diaphragm day in and day out in unending sequence. Like the basic heart rhythm, the fundamental rhythm of the respiratory center is modulated by a multitude of influences, including nerve impulses from

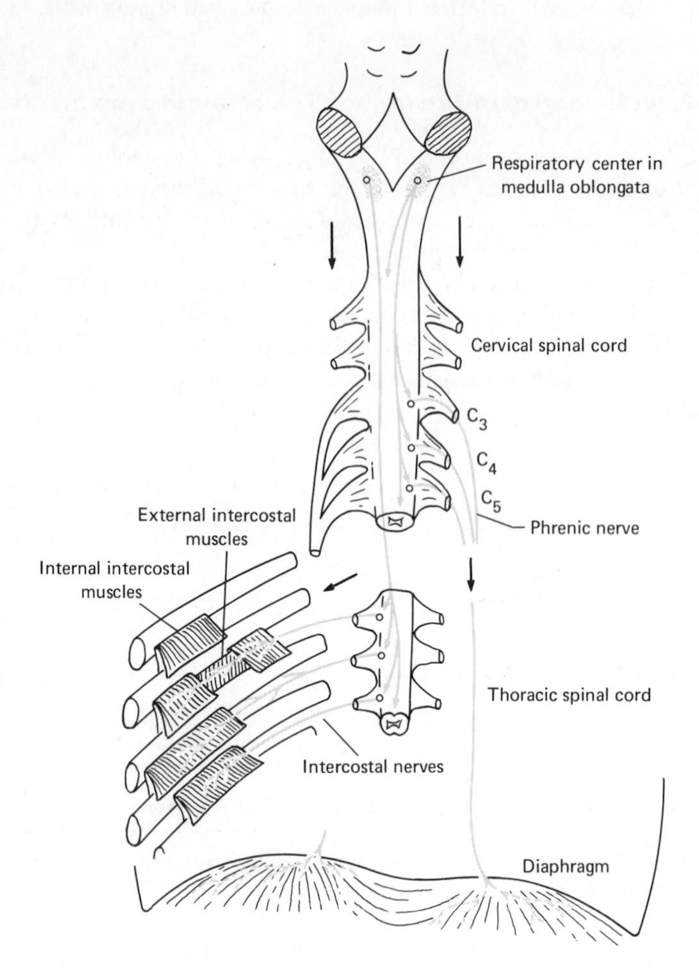

Fig. 12-8. Efferent control of respiration. Large arrows show direction of impulses traveling from respiratory centers of medulla to phrenic motoneurons of the cervical (C) region of spinal cord. In deep respiration, impulses also travel down to the intercostal motoneurons in the thoracic region of spinal cord.

chemoreceptors and pressoreceptors, plus the chemistry of the blood bathing the respiratory center.

The role of carbon dioxide

During strenuous activity the concentration of CO_2 diffusing from active muscles into the bloodstream increases greatly over the resting level and must be removed. It is therefore reasonable that this substance is a powerful stimulant to respiration. The CO_2 acts directly on the respiratory center, to which it is carried by the blood that arrives via the carotid artery after passing through the lungs. A higher than normal concentration of CO_2 in this blood indicates that the CO_2 is not being adequately forced from the lungs. The high CO_2 in the blood bathing the respiratory center stimulates discharge of impulses in fibers that travel from this center to the spinal motoneurons governing the diaphragm and intercostal muscles. The depth of inspiration is increased, more alveoli are opened, and the surface available for gas exchange is increased. Secondarily, the frequency of respiration is also increased.

Respiratory control

Ventilation is also modulated by several logically arranged controls that may be understood in terms of respiratory adjustment to the needs of the body. During inspiration the elastic tissues of the lungs are stretched more and more as the space made available to them is increased when the thoracic cage expands. Multitudes of tiny nerve endings buried in the lung tissue are distorted by these changes in tension and function as stretch receptors. The impulses that result from this stimulation travel into the respiratory center, where they have an inhibitory effect, decreasing the discharge of impulses to the spinal motoneurons (Fig. 12-9). Inspiration, therefore, terminates earlier than it would in the absence of the stretch receptor activity. The afferent fibers from these receptors travel in the vagus nerve.

The action of the pulmonary pressoreceptors is strikingly shown when the lung is suddenly inflated by an increase in air pressure applied into the trachea of an experimental animal. The diaphragm suddenly relaxes when this is done and it does not contract again until the CO_2 in the bloodstream has reached a level sufficient to overcome the influence of the maintained pressure stimulus.

Further control is exercised via the afferent impulses from *pressoreceptors* in the carotid sinus and the aortic arch. Even more important regulation, however, is furnished by information from *chemoreceptors*, which, located in the aortic arch and carotid bodies, detect changes in O_2 content of the blood. These receptors show an increased frequency of impulse discharge when the blood O_2 decreases below about half the normal concentration. The impulses increase respiratory center activity, and consequently the respirations become deeper and more frequent.

On the basis of the foregoing description, one may predict the respiratory response to low O_2 and high CO_2 after denervation of the chemoreceptors. Low O_2 serves as a respiratory stimulus in the unoperated animal, but after denervation of the chemoreceptors, the ventilation may actually decrease because the low O_2 in the blood depresses the respiratory center. The action of CO_2 is directly on the respiratory center and is not affected by the operation.

Respiratory and associated centers

The muscles of breathing are all striated skeletal muscles that are under voluntary control. The motoneurons innervating these muscles are the typical somatic motor type. The nerve fibers travel from the nerve cell bodies within the spinal cord,

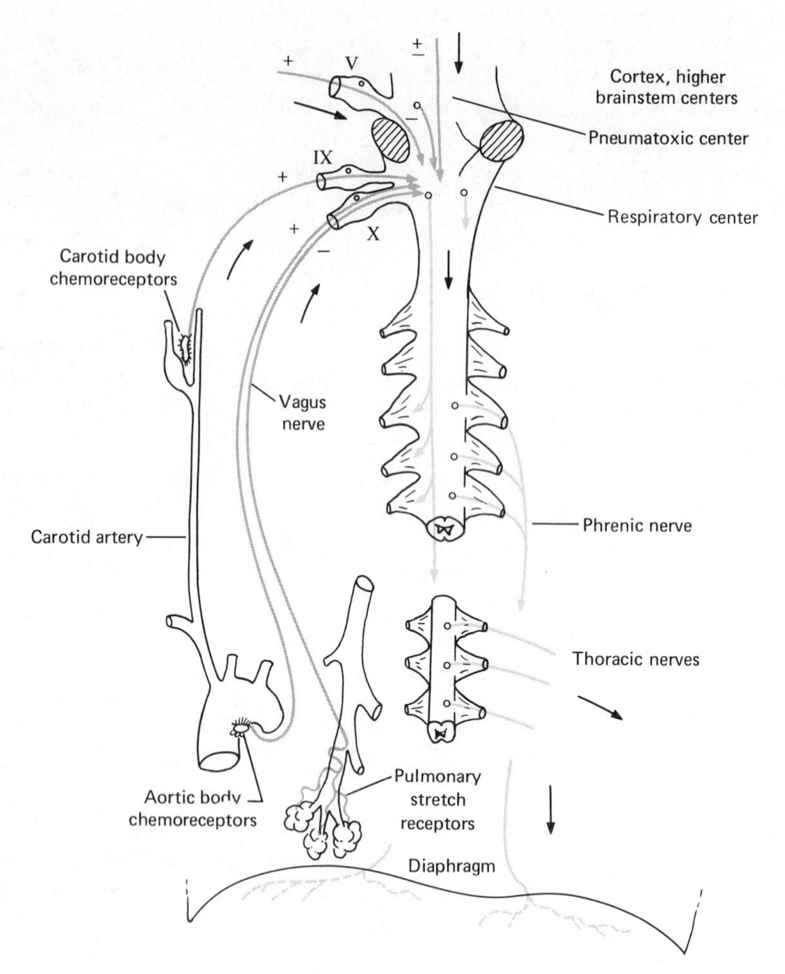

Fig. 12-9. Afferent influences on respiration. Only a few of the afferent influences on respiration are shown. + = stimulates increased respiratory movements. − = inhibits respiratory movements.

out to the muscle fibers, and there are no intervening peripheral synapses as there are in the autonomic nervous system. In spite of its *automatic* nature and its relation to the pulmonary viscera, breathing is clearly not an *autonomic* function. It is a *somatic* function.

Several centers or regions in the pons and medulla oblongata have been found to be directly concerned with breathing (Fig. 12-10). These centers can be located by noting the effects of electrical stimulation or of ablation (removal or destruction) of nervous tissue as well as by the recording of electrical activity in particular places after the afferent nerves have been severed.

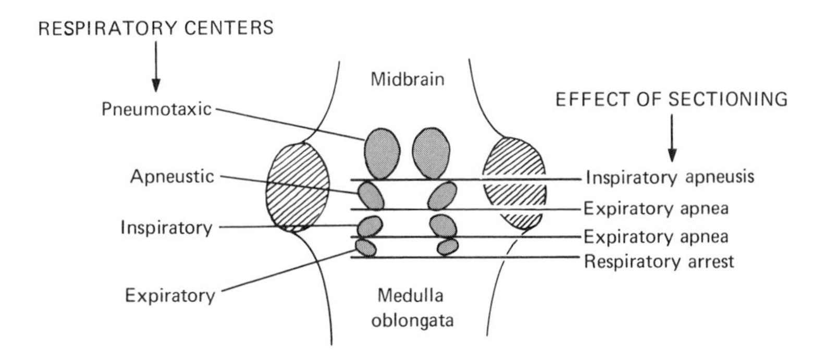

Fig. 12-10. Respiratory centers. Reticular centers in the pons and medulla oblongata that are concerned with breathing. Legend shows consequences of sequential transections after afferents have been cut.

The afferent nerves that provide an input to the respiratory center may be cut, without unduly interfering with breathing. The rhythm appears to be within the neurons of the respiratory center, and the afferent impulses from mechano- and chemoreceptors only provide a modulation of the intrinsic rhythm. The most important afferent fibers come from the stretch receptors of the lungs. When these fibers (in the vagus) are cut, breathing continues undisturbed even if the part of the brain above the pons is totally destroyed. When additional destruction includes the rostral one-third of the pons, the inspiratory phrase becomes exceedingly prolonged. There is, in fact, *inspiratory apneusis* punctuated by brief expirations. Evidently the region that was removed sustains the rhythmicity of the breathing in somewhat the same fashion as the vagi; that is, it terminates inspiration. Because that area or center is directed toward respiratory rhythm, it is called the *pneumotaxic center*. Now, if the destruction is extended to include the caudal two-thirds of the pons, respiration again becomes rhythmic. The region that was removed evidently was responsible for the apneusis, and it is called the *apneustic center*.

Electrical stimulation of the pneumotaxic center may increase the breathing rate and depth. Electrical stimulation of the apneustic center will cause a cessation of breathing at the inspiratory phase. The remaining rhythm of the respiratory center can be shown to have two aspects. In a *rostrodorsal* region, action potentials may be recorded during the *inspiratory* phase of breathing. Close by, *caudoventrally*, impulses are seen during *expiration*. The two regions are, respectively, the so-called *inspiratory* and *expiratory centers* of the medulla oblongata.

Considered as a reflex activity, breathing is not quite the same as a simple reflex that we usually think of as activating a particular set of muscles when a sensory receptor is stimulated. The respiratory motoneurons are already rhythmically discharging impulses, and the afferent input merely alters this frequency. In a control system sense, the afferent input provides negative or positive *bias* to the system.

In that sense, the respiratory center is rather like the sinoatrial node of the heart. Those modified muscle cells are also intrinsically active, but their firing frequency is modulated by impulses coming to them in vagus and sympathetic efferent fibers.

12.2 Transport of Gases by the Blood

The capacity of blood to hold O_2 and CO_2

The directions in which O_2 and CO_2 move between lungs and tissues depend solely on the diffusion that follows differences in concentration of these gases at different places (Fig. 12-11). In the alveoli, the mixing action of the movements of external respiration increases the concentration of O_2 and decreases the concentration of CO_2. The difference between each of these concentrations and the concentrations of these gases in the pulmonary capillaries forces diffusion of O_2 into the blood and of CO_2 into the alveoli. The circulation of the blood ensures that CO_2 will be picked up in the active tissues and will be made available in high concentration in the pulmonary capillaries. Similarly, the circulation of the blood ensures that O_2 will be picked up at the lungs, and made available in high concentrations at the tissues.

When the O_2 partial pressure is high (as in the lungs), O_2 is bound in temporary combination in the blood; when O_2 partial pressure is low (as in tissues), O_2 comes out of the blood. Similarly, when CO_2 partial pressure is low, as it is in the lungs, CO_2 comes out of the blood.

The ability of the blood to carry large amounts of O_2 and CO_2 lies in the red blood cells. Only very small amounts of these gases can be held dissolved in blood plasma lacking red cells. The amount of each gas *dissolved* depends strictly on the partial pressure driving each gas into solution. In any mixture of gases, each gas contributes to the total pressure in proportion to its amount as a fraction of the total amount of gas. Therefore each gas is said to provide a *partial pressure*, and the total pressure is the sum of all the partial pressures. If the pressure of O_2 is zero,

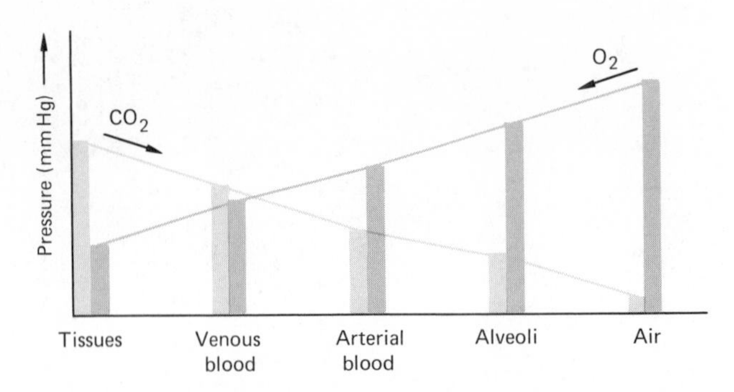

Fig. 12.11. Diffusion of each gas occurs "downhill" in the direction toward the lowest pressure of the gas (arrows). At each place, the *total* pressure is not changed, and is approximately equal to $pCO_2 + pO_2 + pN_2$.

there will be no O_2 in solution. If the partial pressure of O_2 is 40 mm Hg, there will be a definite small amount of O_2 in solution, and this amount will be doubled if the pressure is raised to 80 mm Hg. The relation between the partial pressure of oxygen in the overlying air and O_2 dissolved in a sample of *blood plasma* is plotted in Fig. 12-12. A straight line connects the points, and the amount of O_2 held in solution in 100 ml of plasma at any partial pressure of O_2 may be read from the graph. There is a direct proportionality between the pressure and the amount dissolved.

The oxygen-hemoglobin dissociation curve of blood

If red cells are added to the plasma, the amount of O_2 held by the blood at each O_2 pressure will be enormously greater. The relation between O_2 held and the O_2 pressure is no longer a simple proportionality; instead, an S-shaped curve results when the values are plotted (Fig. 12-12). If the pressure is dropped from 80 mm Hg down to 40 mm Hg, considerably less O_2 can be held, bound by the red cells in the blood at the lower pressure, and the amount of O_2 lost may be read directly from the graph. In the body, the partial pressure or the tension of O_2 at which the blood takes on O_2 (*loading tension* of O_2 in the blood) is about 100 mm Hg. In the tissues, the O_2 tension at which the O_2 leaves the blood (*unloading tension*) may be less than 40 mm Hg. Using these figures, one may read from the graph the amount of O_2 delivered to the tissues by each 100 ml of blood. The actual amount given up to the tissues is further increased because of the fact that the curve is shifted to the right when the temperature is raised or when the acidity is increased. Both changes occur in the active tissues.

Chemical combinations of CO_2 and O_2 in blood

When blood passes through the pulmonary capillaries surrounding the alveoli, it takes on O_2 and gives up CO_2. When this blood arrives at the tissues, the oxygen leaves and another load of CO_2 is taken on, the gases diffusing in directions

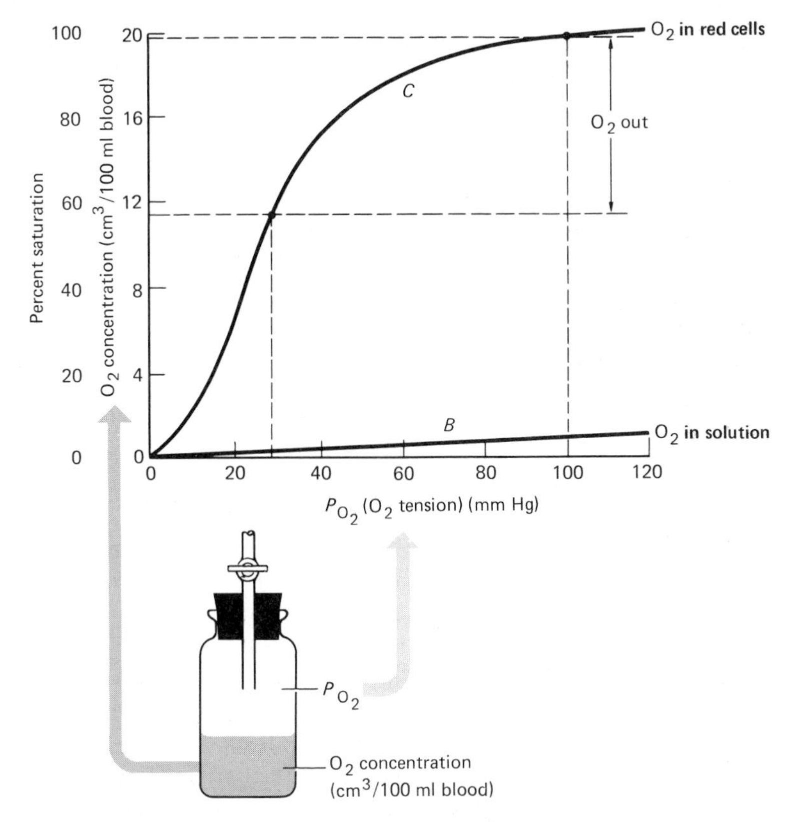

Fig. 12-12. The oxygen-hemoglobin dissociation curve. Curve C shows the amounts of O_2 held in whole blood (plasma plus red cells). This curve is called an oxygen-hemoglobin dissociation curve because it describes the extent to which hemoglobin is combined or associated with hemoglobin, according to the pressure driving the O_2 into combination. From vertical dashed lines you may estimate loss of O_2 when pO_2 drops from 100 to 28 mm Hg, for example. Gas is admitted into the bottle and goes into solution. In blood plasma or in an equivalent salt solution, the amount of gas dissolved is directly proportional to the pressure on the gas let into the bottle (curve *B*). The graph shows the amounts of O_2 in the solution (vertical axis) for any particular pressure in the bottle up to 120 mm Hg (horizontal axis).

determined by their concentration gradients. The circulation of the blood ensures that the concentration gradients will remain high, and diffusion is rapid enough to take care of the movement of the gases through the capillaries. However, as has already been pointed out, neither O_2 nor CO_2 is sufficiently soluble in the blood to account for the large volumes that are moved between the lungs and the tissues. Both gases are carried in temporary chemical combination in the blood.

When CO_2 dissolves in H_2O, the atoms undergo a rearrangement to form a new molecule, H_2CO_3, with the H_2O.

$$H_2O + CO_2 \rightleftharpoons H_2CO_3 \tag{1}$$

This chemical reaction is important in the transport of CO_2 by the blood because the H_2CO_3 is much more soluble than CO_2, and more H_2CO_3 than CO_2 can be carried by the blood. In cell-free plasma, however, this reaction occurs too slowly to take care of the large amounts of CO_2 produced by the tissues. The reaction is speeded up by a special enzyme in the red blood cells. This enzyme is *carbonic anhydrase* (CA). In its presence, the combination of CO_2 with water occurs more quickly inside the red cells than in the plasma. There are other combinations that the CO_2 makes with blood constituents, especially with hemoglobin, but the reaction described above is the most important.

By undergoing chemical combination, O_2 can be carried in greater quantity in the blood than if it were in simple solution in the plasma. The combination of O_2 is with *hemoglobin*, the red pigment inside the red blood cells:

$$HHb + O_2 \rightleftharpoons HHbO_2 \qquad (2)$$

The form of the oxygen dissociation curve of blood is due to this reaction occurring inside the red cells (Fig. 12-13).

The reversible combination of CO_2 in blood

How is the blood able to combine with or release the O_2 and CO_2 according to conditions in the alveoli or in the tissues? The ideas of the *reversible chemical reaction* and of *chemical equilibrium* are keys to understanding this problem. The horizontal arrows in Eqs. (1) and (2) above indicate that the reactions can go in either direction. That is, that they are *reversible*. In Eq. (1), if H_2O and CO_2 are present to begin with, some of the H_2O and CO_2 molecules will rearrange themselves to become H_2CO_3. If there is only H_2CO_3 at the start, some of the molecules will split apart as H_2O and CO_2. In either instance, a condition is soon reached when as many molecules are breaking up as are combining. This condition is called an *equilibrium*. If this condition of balance is upset by removal or addition of some of the constituents, then rearrangement of the atoms occurs in a direction to reestablish the equilibrium.

Carbonic acid (H_2CO_3) dissolved in water can split into two charged particles, ions of hydrogen and bicarbonate, and this, too, is a reversible reaction.

$$H_2CO_3 \rightleftharpoons H^+ + HCO_3^- \qquad (3)$$

Equations (1) and (3) may be combined, yielding

$$H_2O + CO_2 \rightleftharpoons H_2CO_3 \rightleftharpoons H^+ + HCO_3^-$$

These transformations are continually occurring in the blood (Figs. 12-13 and 12-14). The attainment of equilibrium is constantly frustrated by the ceaseless addition of CO_2 to the blood in the capillaries of the tissues and the removal of CO_2 at the lungs.

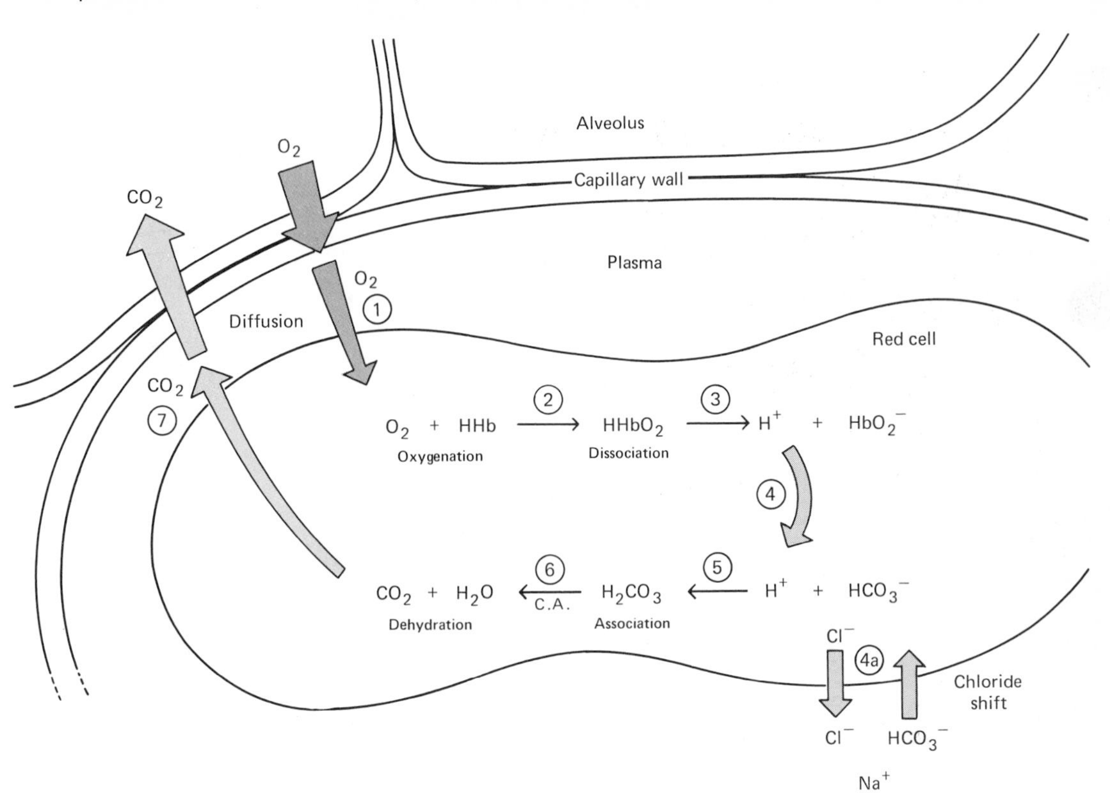

Fig. 12-13. O_2 and CO_2 transport at the alveoli. The means by which O_2 is taken on and CO_2 given up in the capillaries of the lungs.

1. Oxygen diffuses in the direction determined by the concentration gradient; that is, from alveolus into red cell.
2. Oxygen combines with the hemoglobin molecule that has previously lost its oxygen burden in the tissues.
3. Combined with oxygen, the hemoglobin more easily loses its H+ (becomes a stronger acid).
4. The H+ lost by the hemoglobin is made available to the bicarbonate. (4A) HCO_3^- diffuses into the cell in exchange for Cl^-.
5. H^+ and HCO_3^- combine to yield carbonate acid.
6. The carbonic acid formed breaks into water and carbon dioxide. The splitting goes rapidly, under the influence of carbonic anhydrase (C.A.).
7. The resulting carbon dioxide diffuses in the direction determined by its concentration gradient, that is, from red cell into alveolus.

In the one instance, the reaction is driven toward the right; in the other, toward the left. In the capillaries of active tissues, the CO_2 that diffuses into the blood combines with water to form H_2CO_3, which in turn splits into bicarbonate ion and hydrogen ion. Thus the CO_2 poured into the blood by the tissues is transformed into bicarbonate ion, and, in this form, the CO_2 is carried to the lungs.

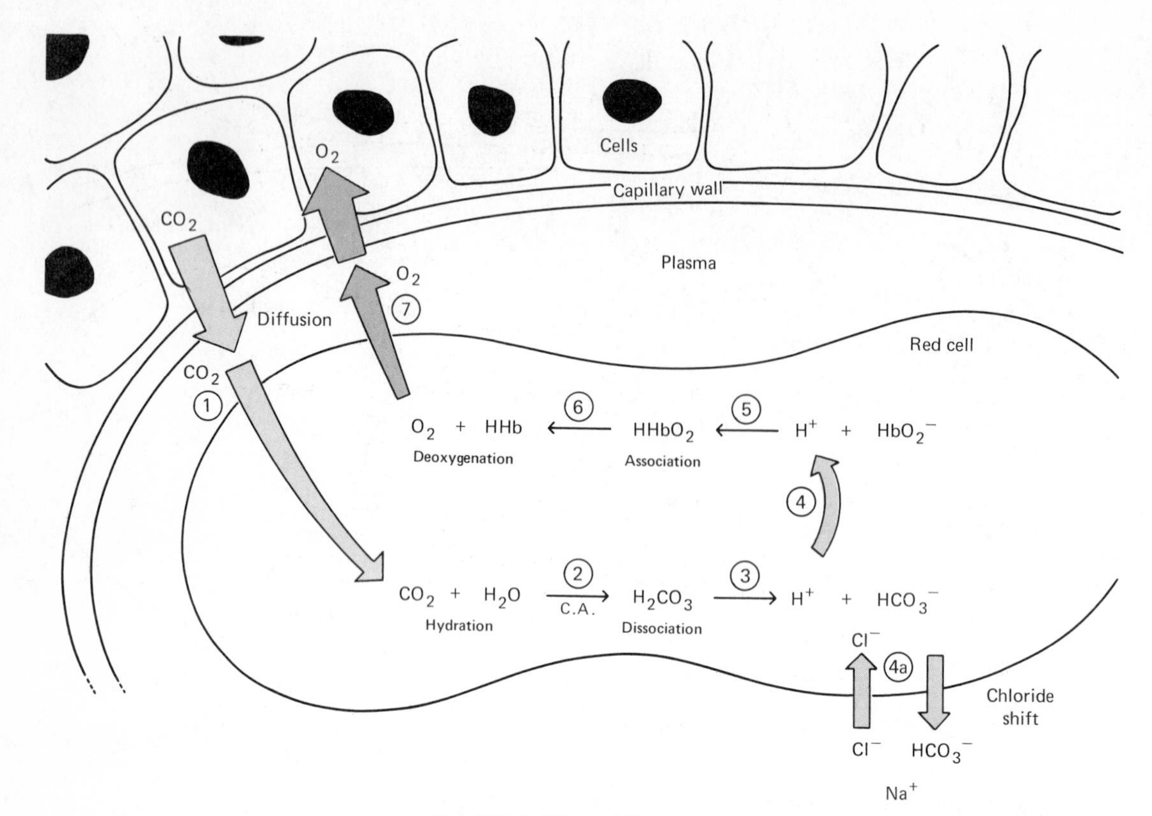

Fig. 12-14. CO_2 and O_2 transport at the tissues.

1. Carbon dioxide diffuses in the direction determined by its concentration gradient—from active tissues into red cells.
2. Carbonic anhydrase hastens the combination of the carbon dioxide with water to produce the carbonic acid.
3. Carbonic acid immediately dissociates into bicarbonate ion and hydrogen ion.
4. The H^+ lost by carbonic acid is made available to the oxygenated hemoglobin. (4A) The bicarbonate ion moves out of the red blood cell under the drive of its diffusion gradient. The electrical balance is not upset, since Cl^- moves into the cell in exchange for the HCO_3^-.
5. The hydrogen ion combines with the oxygenated hemoglobin.
6. The oxyhemoglobin loses its oxygen.
7. The oxygen diffuses in the direction determined by its concentration gradient—from red cell to the tissue.

CO_2 and O_2 diffusion occurs simultaneously. At the lungs, the diffusion of O_2 into the red cell forces toward the right the reaction making H+ available. Simultaneously, the diffusion of CO_2 out of the red cell uses up the H+ as the bicarbonate combines with H+. At the tissues, the large supply of CO_2 provides H+ ion. As O_2 diffuses out of the red cell, more O_2 is made available from the oxygenated hemoglobin still left which can soak up the H+.

As HCO_3^- increases in concentration inside the red cell, the ions diffuse through the red cell membrane and into the plasma. They cannot travel alone, however, because they carry an electric charge and are attracted to charges of opposite sign. Instead of moving along with + charged ions, they simply change places with Cl^- ions, which are always plentiful in the plasma. This exchange is called the "chloride shift." In doing so, they leave behind the H^+ ions, which, in conjunction with the Cl^- ions, would normally provide a strong acid—hydrochloric acid (HCl).

The HCl could be quite harmful to the cell were there not means of getting the H^+ out of the way. Fortunately, hemoglobin is a weak acid, and, like H_2CO_3, it has a considerable affinity for the H^+ ion. The hemoglobin, in other words, serves as a *buffer* against the added acid.

The reversible combination of O_2 in the blood

Hemoglobin becomes a slightly stronger acid, after it has combined with O_2. It then has less affinity for the H^+ ion, and dissociation proceeds as follows:

$$HHbO_2 \rightleftharpoons H^+ + HbO_2^- \qquad (4)$$

Equations (2) and (4) may be combined.

$$HHb + O_2 \rightleftharpoons HHbO_2 \rightleftharpoons H^+ + HbO_2^- \qquad (2), (4)$$

The oxyhemoglobin moves toward the capillaries of the tissues in the form shown on the right–hand side of the equation. The reaction goes toward the left as O_2 diffuses toward the active tissues. During this time the oxyhemoglobin soaks up the H^+ and provides a buffer for the H^+ that is made available in the red cell when bicarbonate ion diffuses out of the erythrocyte. Hemoglobin becomes a weaker acid as it loses its O_2, and for this reason more easily soaks up the H^+ ions. Conversely, hemoglobin to which the H^+ is attached gives up its O_2 more easily.

Thus hemoglobin, by combining with excess H^+ ions, helps the blood to carry CO_2. This same mechanism also helps deliver more O_2 to the tissues, for the O_2 is given up more easily by the oxyhemoglobin molecule to which the H^+ is attached. In brief, the means by which the blood carries CO_2 helps it to carry O_2, and the means by which the blood carries O_2 helps it to carry CO_2.

All these events may be combined as in Fig. 12-14. The heavy arrows show the direction the reactions proceed when the blood flows through the capillaries. In the lungs, all the reactions go in the direction opposite to that in the tissue. The CO_2 concentration gradient determines that CO_2 in the blood will move out into the alveoli. The loss of the CO_2 upsets the balance, which is partially restored by breakdown of H_2CO_3 to produce more CO_2. As the H_2CO_3 is used up, more bicarbonate and hydrogen ion can combine to form more H_2CO_3. Bicarbonate moves into the red cell from the plasma, whereas hydrogen ion is made available by the hemoglobin, which, having gained O_2, more easily parts with the H^+.

In the capillaries of both lungs and tissues, the slowest reaction is $H_2O + CO_2 \rightleftharpoons H_2CO_3$. In both places, the carbonic anhydrase hastens the attainment of an equilibrium that is determined by the affinities of the atoms for one another in a certain arrangement. However, the *direction* the reaction will go is determined by the concentration of the reacting materials.

Section 12: *Questions, Problems, and Projects*

1. About three-fourths of the molecules in a breath of expired air are N_2. What fraction of the total pressure in that air is due to N_2?
2. Write the following in the form of mathematical equations, letting P_a represent the partial pressure of gas a:
   (a) The total pressure P_T of a sample of gas is equal to the sum of the pressures of the individual gases, O_2, N_2, and X.
   (b) The relation between P_{O_2} and P_{N_2} when the amount of N_2 present is four times the amount of O_2 present.
   (c) The total pressure expressed as a function of P_{N_2} and P_x.
   (d) What is the magnitude of P_{N_2}/P_T when $P_x = 0$? ($P_T = P_{total}$.)
   (e) What is the magnitude of P_{N_2} if P_T is 760 mm Hg? 3 atm?
   (f) The partial pressure of a gas in a mixture bears the same relation to the total pressure as the number of moles of the gas bears to the total number of moles in the mixture.
   (g) If P_T is changed from 1 to 3 atm, what change occurs in the number of moles of gas per liter?
   (h) What is the percentage composition if $n_{O_2} = 120$ and $n_t = 600$?
   (i) What is the percentage of water vapor when $P_{H_2O} = 47$ mm Hg and $P_T =$ 760 mm Hg?
3. The concentration (c) of a gas in solution is directly proportional to its partial pressure p_a in contact with the solution.
   (a) Write the equation stating that relation, letting the proportionality constant be k_{sol}.
   (b) Show the dimensions of k when C_{O_2} is in milliliters O_2/100 ml H_2O and P is in mm Hg.
   (c) If $P_{O_2} = 142$ mm Hg and the numerical value of k_{O_2} is 0.003, what volume of O_2 is dissolved in 1 liter of the solution?
   (d) If $P_{CO_2} = 40$ mm H_2O, and $k_{CO_2} = 0.075$, what volume of CO_2 is dissolved per liter of solution?

(e) Write a simple equation to show the combination of hemoglobin with oxygen.

4. Label and draw the oxyhemoglobin dissociation curve for a *p*H of about 7.4. Let the blood be 60 percent saturated at 40 mm Hg.
   (a) The blood, when saturated, holds about 20 ml O_2 per 100 ml of blood. Indicate on the ordinate of the curve the volumes of O_2 held in the blood at the various levels of saturation.
   (b) Estimate from the curve the change in saturation and the O_2 lost from the blood in going from $P_{O_2} = 140$ mm Hg to $P_{O_2} = 40$ mm Hg if the blood *p*H is about 7.4.
   (c) Write the equation that shows the two successive chemical events that occur when CO_2 is dissolved in water.
   (d) Encircle in the equation the process affected by carbonic anhydrase.

5. If the tidal volume is 0.5 liter and the respiratory rate is 15 per minute, what is the ventilation rate in liters per minute?
   (a) Calculate the percentage increase when the breathing rate increases to 30 per minute and the tidal volume to 3 liters.
   (b) Describe whether the effect of each of the following on the respiratory center is excitatory or inhibitory: (a) stretching the lungs. (b) increase of CO_2 in the blood.
   (c) If the (H^+) is 10^{-6} moles liter, what is $1/(H^+)$?
   (d) What is log 1/(H) then?

6. Write the expression for the equilibrium constant for the equation showing the dissociation of carbonic acid.
   (a) Solve the equation for (H^+).
   (b) Take the reciprocal of each side of the equation.
   (c) Take logarithms of each side of the equation.
   (d) Now write the equation in terms of *p*H and *p*K.

13

PROCESSES OF DIGESTION

13.1 Structure and Function of the Digestive Tract

Digestion is the process that renders food materials soluble so that they may be absorbed into the bloodstream. Cutting and grinding action by the teeth and the squeezing action by the stomach and intestine mechanically break up the food. Thus large surface areas of the food are made available, and the chemical action to produce soluble particles can proceed swiftly. The chemical aspect of digestion involves the splitting of the food materials into molecules small enough to be absorbed through the walls of the digestive tract.

Origin and arrangement of the digestive tract

The cavity or *lumen* of the digestive tract is really outside the body, somewhat as the hole in a doughnut is outside the doughnut. The functional lining of the tract begins as a simple tube, the embryonic *gut*, the only derivative of the endoderm germ layer. During development, enormous evaginations or outpocketings growing out from this simple tube become the salivary glands, opening to the mouth, the pancreas, and the liver, together with the gall bladder opening into the small intestine. These glands, plus innumerable simple pockets of gland cells in the otherwise relatively flat epithelium of the tract, secrete watery solutions of mucus and digestive juices.

During embryonic development the surrounding mesoderm splits to form the *coelomic* (later the *peritoneal*) *cavity* and becomes the smooth muscle and blood vessels that surround the primary gut epithelium. The now complex tube remains attached to the middorsal body wall of the coelomic cavity through a thin, double sheet of squamous epithelium, the *mesentery*, which is continuous with the *parietal peritoneum* lining the body wall, and the *visceral peritoneum* covering the digestive tract. This layer is always moist, being lubricated by fluid in the small spaces between organs in the abdominal cavity. It is called a *serosa* in reference to the watery (serous) nature of the fluid and in contrast to the *mucosa*, the columnar epithelium inside the gut, where the more viscous mucus is secreted. During embryonic growth the digestive tract lengthens disproportionately compared to the midline from which it is suspended. Consequently, the mesentery fans out along with the lengthening gut.

The lengthy, meandering, sinuous coils of the small intestine are held together by their common mesentery attachment, which has a more-or-less whorled arrangement because of the circumstances of differential growth.

The stomach not only bends during its growth, it also twists and pulls the mesentery out into a long pocket that secondarily adheres to the mesentery of the small and large intestines (Fig. 13-1).

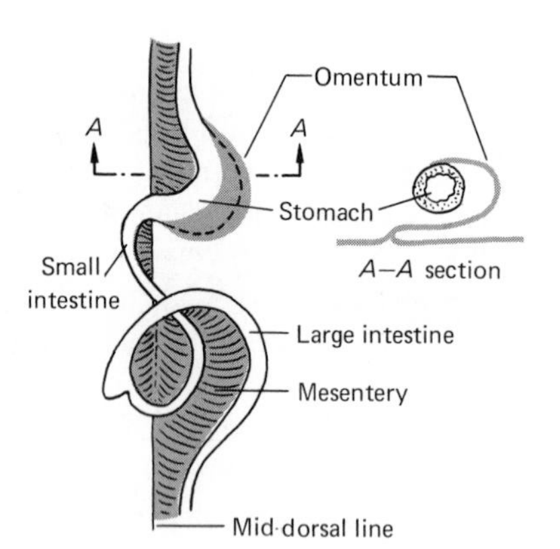

Fig. 13-1. Embryonic gut (about three months).

Smooth muscle constitutes most of the thickness of the gut wall. The smooth muscle fibers are small and spindle-shaped, with a single nucleus at the center. Longitudinal fibrils, showing no sign of the cross-banding characteristic of skeletal muscle, constitute the contractile apparatus of each cell. Continuity from cell to cell depends partly on a neural network.

The smooth muscle fibers are arranged in sheets. The simplest configuration exists in the small intestine, where the fibers of the inner sheet just beneath the mucosa are circumferentially arranged. Outside is the sheet of longitudinally disposed fibers. Between the muscle layers is an extensive layer of nerve cells that form, with their axons and highly branching dendrites, an extensive network, the *myenteric plexus*, which provides a rich innervation to the muscle cells. To this layer come particularly

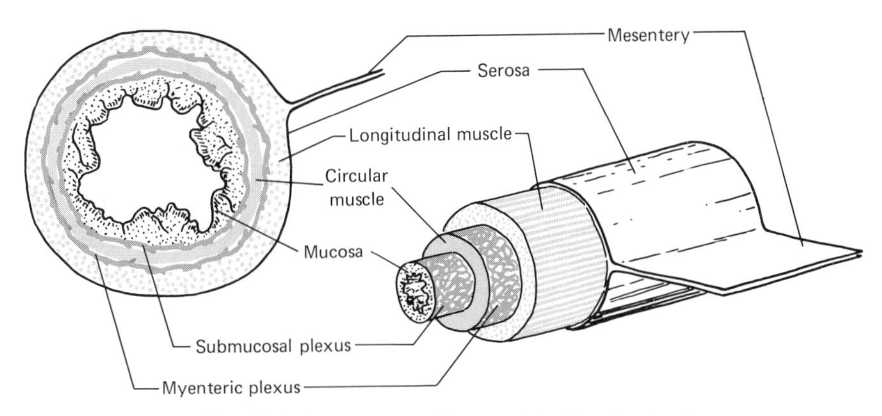

Fig. 13-2. Arrangement of layers of the digestive tract.

preganglionic parasympathetic fibers. The *submucosal plexus* of nerves is in the submucosal layer between the mucosa and the muscle layer of the gut (Fig. 13-2).

At each end of the small intestine, heavy cuffs of smooth muscle are organized into *sphincters*.

Swallowing

Digestion begins in the mouth when the food is mechanically broken down by the grinding action of the teeth. When food in the mouth has been sufficiently mixed with saliva, it will slide readily and is pushed by the tongue to the back of the mouth, where it begins its descent into the esophagus (Fig. 13-3). The start of the swallowing process is under voluntary control, but the subsequent trip is an autonomic reflex. During swallowing, muscles surrounding the back of the pharynx relax, and the tongue presses the food backward into the space made available. Respiration is inhibited, the *epiglottis* bends backward and is pressed close on the *glottis* as the larynx moves upward. The esophagus passes down through the neck and through the thoracic cavity dorsal to the larynx. The opening into the larynx from the pharynx is ventrally located in such a way that food sliding over it may get lodged in this passageway, except that the glottis closes and the epiglottis covers the entrance. This covering is produced by the upward raising of the larynx during swallowing (see Chap. 12.1). The food slides over the epiglottis into the esophagus and is propelled toward the stomach by *peristaltic* action, in which a wave of constriction preceded by relaxation moves along the esophagus. When the peristaltic wave arrives at the stomach, the smooth muscle between esophagus and stomach (called the cardiac sphincter) relaxes, and the food is forced into the stomach. Peristalsis moves material toward the stomach, even against the force of gravity.

Structure and function of the stomach

The esophagus passes out of the thoracic cavity through the central tendon of the diaphragm and opens into the *cardia*, so called because it is the part of

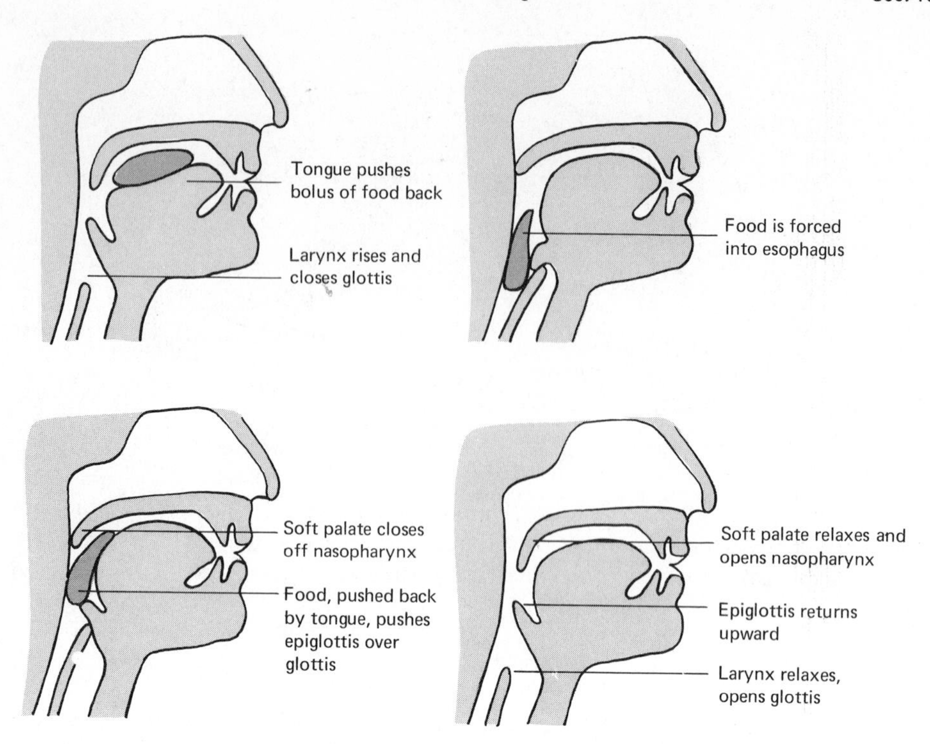

Fig. 13-3. Deglutition, or swallowing.

the stomach close to the heart (cardiac) (Fig. 13-4). The entrance from the esophagus into the *cardia* is guarded by the *cardiac sphincter*. This is a ring of smooth muscle that ordinarily remains constricted except when forced open by peristaltic waves moving food and water down the esophagus. It also opens when food is regurgitated during *reverse peristalsis* when the peristaltic waves of the stomach move in the wrong direction.

Normally food passed through the cardiac sphincter into the *cardia* of the stomach, moves on to the *fundus* and is subjected to the churning action of the smooth muscle contractions of the stomach (Fig. 13-5). At the same time, the digestive juices are secreted by the glands of the *muçosa* (as the inner lining of the stomach wall is called because of the ample quantities of mucus it produces). From the fundic region come mainly digestive enzyme and mucus. Special cells of the mucosal glands in the *body* of the stomach secrete the hydrochloric acid that provides a suitable *p*H for the enzyme to function. The *antrum* of the stomach is the slightly dilated region just before the *pylorus*. The stomach contents in this region may be less acid due to backflow of relatively alkaline material from the small intestine. The *pyloric sphincter* guards the entry from stomach into intestine.

Digestion continues in the stomach for 1 to 4 hours, depending on the type of food. Churning movements help to provide ever new surfaces of food to the enzyme and hasten the digestive process. Occasional peristaltic waves moving toward the

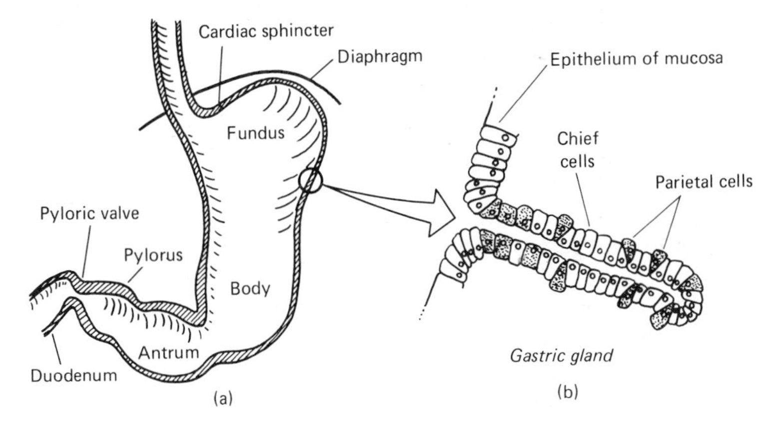

Fig. 13-4. The stomach. The shape of the stomach varies considerably, according to its fullness. The fundus is just under the diaphragm. (b) shows an enlargement of a gastric gland of the mucosa.

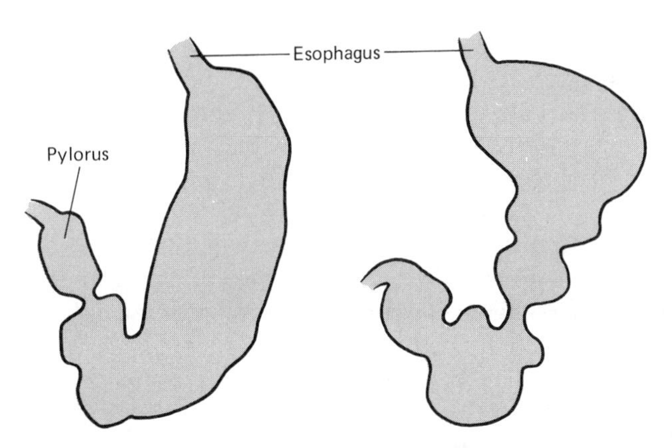

Fig. 13-5. Outlines of x-ray pictures of stomach showing usual and extreme peristaltic waves.

pylorus may deepen and practically squeeze the stomach in two. The force of strong waves of contraction will open the pyloric sphincter and propel food into the small intestine.

Digestion in the intestine

Most digestion, and almost all absorption of the products of digestion, occurs in the *small intestine*. This long tube is small only in diameter (2–3 cm). In much convoluted fashion it extends from the pylorus of the stomach for 6–7 meters to the *ileocecal* junction at the beginning of the large intestine. To it are attached, via their ducts, the *pancreas* and the *liver*, two of the most important glands of the digestive tract. The upper part of the small intestine, together with these glands, is the only really essential part of the digestive system. Digestion of all the major foodstuffs—carbohydrates, fats, and proteins—occurs within the narrow lumen of the small intestine. The *p*H of its contents is on the alkaline side of neutrality, appropriate for the most favorable action of the many enzymes found there.

Movements of the intestine

The entire mass of partially digested food and digestive juices moved from the stomach into the small intestine is called the *chyme*. The chyme is stirred and mixed by the contraction of smooth muscle in the small intestine, so that new opportunities are presented for enzyme and substrate to combine. The small components produced by hydrolysis during digestion are then absorbed through the small intestinal wall. The movements of the intestine serve several purposes. They churn and mix the food and move it along in the tract. In addition, the massaging action of the contractions assists in the flow of blood and especially of lymph through the vessels of the intestine.

The chyme is mixed by segmentation movements of the intestine. The food mass is squeezed by rings of contraction that occur in the circumferentially arranged smooth muscle. These rings are equally spaced, each only 2–5 cm from the next. Successive contractions squeeze between previous contraction rings, mixing the food back and forth.

Progression of the food along the gut is due to peristaltic waves that sweep along as a ring of contraction preceded by an area of relaxation. The relaxation is perhaps really a contraction of longitudinal muscle, which, in shortening would tend to enlarge the diameter of the intestine.

Absorption of the products of digestion takes place almost entirely through the wall of the small intestine. Many of the substances pass through the mucosal wall merely by diffusing in the direction determined by their concentration gradients. The intestine is also able to move certain dissolved materials through its wall into the blood, even when the concentration of the substance is lower in the intestine than it is in the blood. Metabolic energy is provided by the cells to carry out this transport. Smooth muscle contractions assist absorption by continually stirring the chyme over the intestinal surface. Absorption is further aided by the large surface area made available by the *villi*, which are tiny fingerlike processes projecting from the mucosa into the lumen of the intestine. There are about 5 million villi, and they provide about 10 m^2 of surface area. The villi exhibit a pumping action, as they alternately extend into and retract from the chyme. Within each villus is a capillary loop and a lymphatic terminal, called a lacteal. The lacteal receives its name from the milky appearance the droplets of absorbed fat or *chyle* give to the lymphatic vessels. After a meal, the lymphatic vessels of the mesentery appear milky from the content of chyle absorbed by the lacteals.

The functions of the liver

The amino acids from protein, the monosaccharides from carbohydrates, and some of the products of fat digestion are absorbed into the capillaries. They are then carried to the liver via the hepatic portal vein. This vein supplies three–quarters of the blood that flows to the liver. The hepatic artery, delivering only one-quarter of the blood flow, supplies the oxygen required by the liver cells. The liver cells have a rich blood supply and thus have easy access to the dissolved food materials. From the amino acids, the liver manufactures blood proteins. From the glucose, the liver synthesizes glycogen as an energy store to be drawn on by the body when required. Phagocytic cells line the labyrinthine passageways of the sinusoids through which the blood passes and remove undesirable particulate matter from the blood. Aged red blood cells are consumed in the liver, where the hemoglobin is converted into another pigment and excreted via the gall bladder. In the gall bladder are also stored the bile salts that are manufactured by liver cells.

The large intestine

By the time the food material has passed through the great length of the small intestine, all the digested substances have been absorbed. Undigested debris and water remain. The contents of the *ileum*, the lowermost segment of the small intestine, are forced by peristaltic action through the *ileocolic valve* and into the ascending colon. The ascending colon is located on the right side of the body and is the first part of the large intestine. The small is joined to the large intestine perpendicularly, as the stem to the top of a T. One end of the top corresponds to the blind-ended cecum. The other continues as the colon. Small in man and in carnivores, the cecum is, however, very large in herbivorous animals. The appendix (*vermiform* to distinguish it from other projections of the large intestine) arises at the junction and is attached on the underside of the cecum. It probably serves the function of lymphatic nodules. Movement of this material is then upward in the *ascending colon*, across toward the left in the *transverse colon*, downward on the left in the *descending colon*, and finally into the rectum, from which the material is evacuated from the body through the anus. In the course of this trip, a considerable amount of the water is absorbed, along with some food substances made available by the action of bacteria.

The large intestine secretes no digestive enzymes. There are glands in the lining of the large intestine, but they secrete mucus that protects the intestinal lining against mechanical injury and against injury due to irritating substances such as acids.

A considerable part of the intestinal contents consists of several species of bacteria serving two very useful functions: first, the bacteria produce enzymes that may digest some of the material not previously broken down. The products made available can be absorbed, along with water, through the mucosal wall and into the bloodstream. Second, some of the bacteria synthesize vitamin K, important in blood clotting, while other useful bacteria synthesize some of the B vitamins from simple compounds. The bacteria of the large intestine are helpful and apparently do no harm as long as they remain in the intestine and do not get into the bloodstream. It is doubtful whether the complaints accompanying constipation can be attributed to them. The discomfort of constipation can be attributed simply to the pressure resulting from the large amount of material crowded into the large intestine, rather than to any poisonous substances.

Aside from the small contribution the resident bacteria make to the available food, the large intestine serves as a temporary storage organ for debris remaining after the digestive enzymes have done their work. The process of getting rid of these materials is called *elimination.* Most of the materials eliminated have never participated in the body metabolism. They are merely left over in the digestive tract. The bile pigments are exceptions, for they are derived from hemoglobin of dead red blood cells. Other materials (e.g., heavy metals) are also excreted with the bile and from the mucosa of the digestive tract. These materials are eliminated with the feces, but they may be considered true products of excretion, for they originally were removed from the bloodstream.

13.2 Chemical Aspects of Digestion

Specific enzymes secreted into the digestive tract, act on each of the three basic food materials—carbohydrates, fats, and proteins. Although these food materials differ, the chemical mechanism of digestion is essentially the same in each instance. The large molecules are made into smaller ones by insertion of water molecules at appropriate places. This process is called *hydrolysis*, or "splitting by means of water." The hydrolytic enzymes of the digestive tract take no direct part in making energy available to cells. The enzymes catalyzing reactions that yield energy are within cells. The hydrolytic enzymes merely break large molecules into small ones.

Composition of saliva

The food is more easily chewed and swallowed when it is moistened by saliva. There is always a small amount of salivary secretion, but salivation is especially strong in response to food. Food placed in the mouth may stimulate flow of as much as 20 ml of saliva per minute. The average total daily production is only 1 to 1.5 liters, most of which is usually swallowed and reabsorbed in the digestive tract. Saliva is a solution produced by the salivary glands from the blood, and almost all the substances in the blood can be found in the saliva, although usually at lower concentrations. Into the saliva from the blood come not only salts but also small amounts of amino acids and even blood proteins. In some individuals the blood type can be determined from antibodies in the saliva. Sex hormones from the blood also pass into the saliva in sufficient amounts to permit a pregnancy test in the case of the female. NaCl, a major constituent of saliva, may be present at only one-seventh of its concentration in blood. It is important nevertheless, for salivary digestive enzyme requires Cl^- ion to activate its hydrolytic function. In addition to an activator, the enzyme requires a suitable balance of acid and base if it is to work effectively. This balance is provided by bicarbonate secreted in the saliva.

Salivary amylase

The digestive enzyme of the mouth hydrolyzes *starch.* A starch molecule is a long, chainlike structure of several hundred sugar molecules. Animal starch, stored in liver or muscle, is called *glycogen.* A starch-splitting enzyme is called *amylase*, from *amylum*, another name for starch, and *ase*, a suffix referring to enzyme. Many enzymes are named in this sensible way, according to the name of the substance on which they act. Because there are other amylases, the one in the mouth is called *salivary amylase. Ptyalin* is an older name for this enzyme.

If a starch were to be left in water solution for a long enough time, it would be hydrolyzed into its constituent sugars, even in the absence of a hydrolytic enzyme. The enzyme only helps to make the reaction proceed more rapidly. Salivary amylase speeds the reaction shown below.

Starch or glycogen $(C_6H_{10}O_5)_n$
↓ ↓ ←amylase acts mainly here
Dextrins $(C_6H_{10}O_5)_{6-8}$
↓ ↓ ←amylase has only little effect here
Monosaccharides $C_6H_{12}O_6$

The main effect of the salivary amylase is to hydrolyze the starch into smaller molecules, each of which still consists of a combination of several single sugar molecules (see Fig. 13-6).

Salivary amylase is generally found only in animals like ourselves that consume a mixed diet. Cats and dogs, carnivores that subsist mainly on meat, do not have the enzyme, nor do herbivores, such as sheep and cattle, that feed mainly on grass. The effectiveness of the enzyme (the *enzyme activity*) also varies in humans according to the diet. The salivary amylase activity may be ten times as great in persons on a fruit-and-vegetable diet as in persons on a meat diet.

Digestion in the stomach: An acid environment

The food arriving in the stomach encounters an environment quite different from that in the mouth. Like the saliva, the gastric secretions contain Na^+, K^+, and other common ions of the blood, but, although the saliva is nearly neutral in

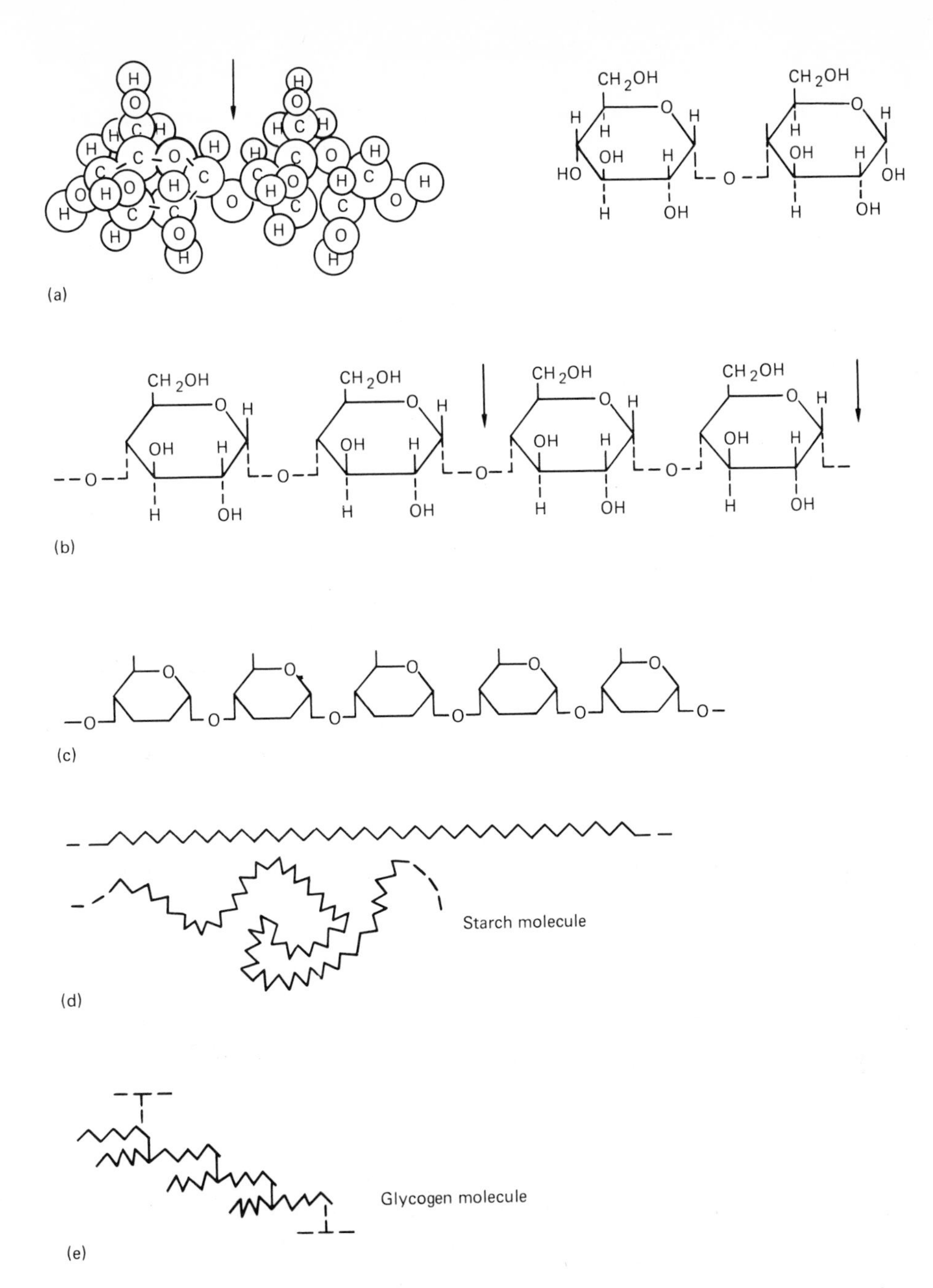

Fig. 13-6. Hydrolysis of starch. (a) A three-dimensional representation of maltose—consisting of two glucose molecules. The same molecule is shown above in the usual structural formula. (b) The double units of 1 are obtained by hydrolysis of long chains shown in 2. (c) Is a simplified version of (b) and includes more units. (d) The structure is further simplified to show a large number of glucose units hooked together in a long chain to form starch. (e) Similar to (d), except the glucose units are attached sidewise to form a branching structure, as in glycogen.

hydrogen ion activity, the stomach contents are extremely acid. When swallowed food mixes with the food already in the stomach, the action of the salivary amylase is soon suspended due to the high acidity, which is inimical to the action of the salivary amylase.

When the stomach contents are not acid, digestion cannot properly occur. Special cells secrete a solution of hydrochloric acid isosmotic to the blood. The concentration of HCl as it is secreted is about 0.165 mole/liter. Such a solution has a *p*H about 0.85. The stomach contents are never actually as acid as this, however, since the secretion is always mixed with food and mucus, which tend to soak up some of the acid and buffer the solution to a less acid level of about $p\text{H} = 2$.

The stomach glands are tiny pockets in the mucosa, lined with secretory cells (see Fig. 13-4). The largest number of these, the *chief cells*, produce the digestive enzyme. Other cells, fewer in number and distinguished by staining properties as well as by location at the periphery of each gland, are called *parietal cells*. They secrete HCl into tiny intracellular canals that open to the lumen of the stomach.

Most organs pour acid metabolites into the blood, but the venous blood leaving the stomach after a meal is, in contrast, slightly more alkaline than the blood arriving at the stomach. Correspondingly, the urine (which reflects the properties of the blood) is slightly more alkaline after a meal. Finally, there is an increase in the rate at which CO_2 is breathed out after a meal has been consumed. These and other observations suggest that the production of HCl in the parietal cell occurs according to a sequence of reactions much like that in the red blood cell during the taking on of CO_2, as the following scheme shows (see Fig. 13-7).

1. H^+ is produced from ionization of carbonic acid. The reaction has already been described in relation to the carrying of CO_2 by the blood (Chap. 10.2).

$$H_2O + CO_2 \overset{\text{carbonic anhydrase}}{\rightleftharpoons} H_2CO_3 \rightleftharpoons HCO_3^- + H^+$$

2. The HCO_3^- diffuses out of the cell into the blood, exchanging for Cl^-.
3. H^+ and Cl^- then diffuse together into the canals of the parietal cell. The HCO_3^- getting into the blood is expired as CO_2 after undergoing the following reaction.

$$HCO_3^- + H^+ \rightleftharpoons H_2CO_3 \rightleftharpoons H_2O + CO_2$$

4. The Na^+ ion that was left in the blood after the departure of Cl^- into the cell may be excreted by the kidney.

Gastric enzymes

The most important enzyme of the stomach is a *gastric* proteinase known as *pepsin*, which is secreted by the *chief* cells. Pepsin catalyzes the breakdown of

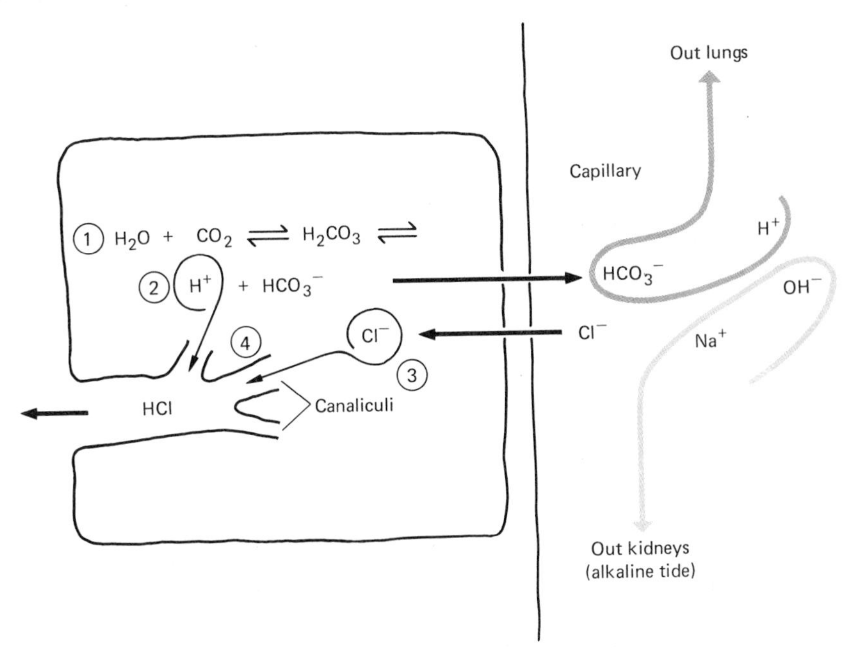

Fig. 13-7. An interpretation of HCl secretion. In the parietal cell of the gastric gland, hydrogen ion is provided from the ionization of carbonic acid. The bicarbonate ion exchanges for chloride ion from the bloodstream. The resulting HCl gets into the stomach lumen by way of the intracellular canals of the parietal cell.

protein and, like all enzymes, is itself a protein. The shape of the surface of the pepsin protein enzyme is such that a particular protein structure fits on it. When the protein is stretched out on the enzyme surface, it is readily broken at a place where a water molecule can be inserted. Pepsin ordinarily carries the digestion of the protein molecules only part way to *polypeptides*, which are combinations of several amino acids. The polypeptides resulting from the action of pepsin are called *proteoses* and *peptones*.

Pepsin is able to catalyze the hydrolysis of protein only when the *p*H is very low. *p*H influences enzyme activity by affecting the structure of the protein. The amino acids of which protein is composed can combine with acid or base, whichever is available. As an example, the effects of adding acid and base are shown for an amino acid in Fig. 13-8. When H^+ ion is added, the net *positive* charge of the molecule is increased. On the other hand, when OH^- is added, the net *negative* charge of the molecule is increased. The reactive groups of the constituent amino acids are also present in the protein molecule undergoing digestion. When the charge of the enzyme is altered, the attraction that the surface will have for the substrate molecules is changed, and the enzyme activity may be altered or lost completely.

The secretory cells are made of protein and might, under certain conditions, themselves be digested by the pepsin. Mucus, continually secreted from the gastric mucosa, helps protect the stomach wall. In addition, the pepsin is secreted in an

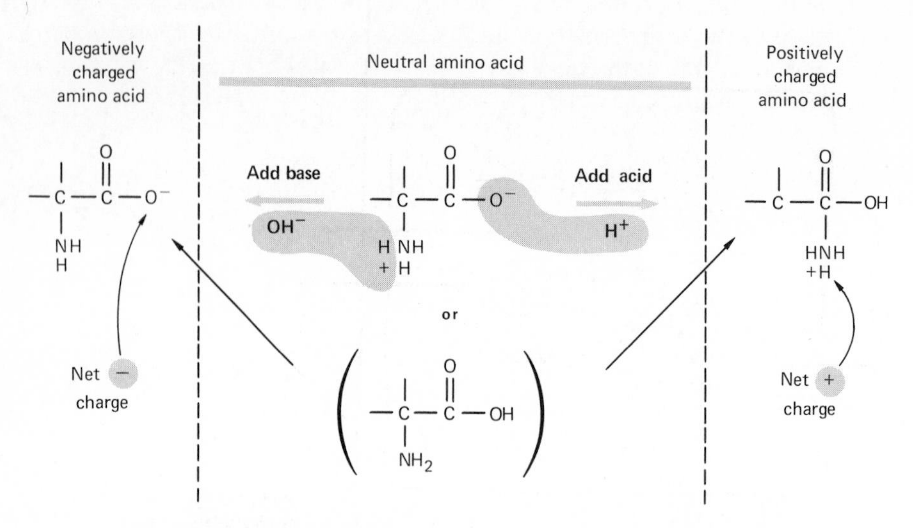

Fig. 13-8. Effect of acid and base on amino acid structure. When acid is added to amino acids or to a protein, the H+ is driven onto the amino acid in the manner shown, leaving the molecule with a net positive charge. When a base is added, the OH− combines with H+ from the amino acid, and the molecule is left with a net negative charge. In solution, the amino acid molecule in the middle may simultaneously carry a negative charge (on the carboxyl end) at a particular *p*H, or it may have no charge at either end (below). In either instance, when the *p*H is decreased by adding acid to the solution, the molecule may develop a net + charge, while if base is added, a net − charge will result.

inactive form, *pepsinogen*, which is transformed into active *pepsin* only under the influence of HCl.

$$\text{Pepsinogen} \xrightarrow{\text{HCl}} \text{Pepsin}$$

It is important that the pepsin be made available only when it can be soaked up in the food. Then it will not be so likely to digest the stomach wall. Nerve impulses and hormone secretions ensure that pepsin and acid are secreted at appropriate times.

Amylase and proteinase in the small intestine

The digestion of carbohydrates, which was suspended in the stomach, proceeds again in the small intestine. The enzyme responsible is similar to ptyalin but is called *pancreatic amylase* in recognition of its origin. Its action appears to be the same as that of salivary amylase.

Protein digestion in the small intestine is also due to a secretion from the pancreas. As in the stomach, the enzyme is secreted as an inactive precursor, *trypsinogen* in this instance. The transformation into trypsin is catalyzed by *enterokinase* (1), a secretion from the wall of the small intestine. Molecules of *trypsin* (2) serve the same purpose, and the conversion of trypsinogen to trypsin occurs very rapidly:

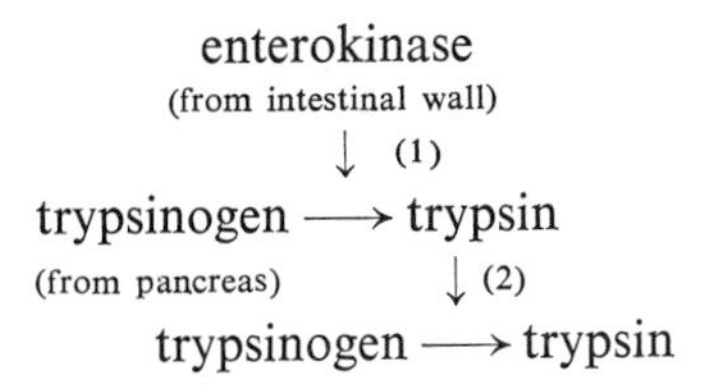

Trypsin splits proteins and polypeptides at the peptide link, where water is inserted. In this sense its action, although occurring in an alkaline medium in which the protein is negatively charged, is similar to pepsin. However, trypsin attacks portions of the molecule that differ structurally from portions affected by pepsin. As a result of trypsin action, the polypeptides produced differ slightly from those resulting from pepsin action:

$$\text{proteins} \xrightarrow{\text{trypsin}} \text{polypeptides} \xrightarrow{\text{trypsin}} \text{smaller polypeptides}$$

Enzymes from the intestinal mucosa

The final stages of carbohydrate and protein digestion are brought about by enzymes secreted from a multitude of small glands lining the wall of the small intestine.

Several of these enzymes act on the polypeptide fragments remaining after peptic and tryptic digestion. These *polypeptidases* (aminopeptidase, carboxypeptidase, and so forth) provoke the final hydrolysis into separate amino acids that can be absorbed into the bloodstream.

Another group of enzymes from the mucosa of the small intestine acts on the disaccharides that remain after the amylases have done their work. Maltose, sucrose (beet sugar), and lactose (milk sugar) are disaccharides that differ according to the kinds of monosaccharides of which they are composed. The specific *disaccharases* of the intestinal juices are able enzymatically to hydrolyze each disaccharide into its constituent monosaccharides.

Digestion of fats

The hydrolysis of fats is made rather difficult by the insolubility of fat in water. The interior of a globule of fat is inaccessible to react with any suitable enzymes that may be in the surrounding water. However, the smaller the globules can be made, the relatively larger is the surface of fat particles that can be exposed to the solution. It is appropriate, therefore, that an *emulsifying* agent is poured into the small intestine to help keep the lipids as tiny globules that do not coalesce. The emulsifying agent, *bile salts,* is secreted by the liver and stored in the gall bladder, which squeezes its contents into the duodenum at suitable times.

Following this mechanical breakdown, a lipase from the pancreas is able to catalyze the hydrolysis of some of the lipid molecules to their constituent parts.

$$\text{lipid} \xrightarrow{\text{pancreatic lipase}} \text{fatty acids} + \text{glycerol}$$

13.3 Neural and Endocrine Control of Digestion

The digestive enzymes do not flow uniformly nor continually but are adjusted in rate according to the kinds and amount of food in the digestive tract. The primary stimulus is food in the mouth, which is followed by reflex secretion of saliva and of gastrointestinal juices. This reflex involves the parasympathetic nervous system. Food materials in the gastrointestinal tract stimulate the gut wall to secrete hormones that get into the circulation and induce the secretion of gastrointestinal digestive juices. There are specific hormones for each group of digestive enzymes.

Unconditioned reflex control of salivation

The quantity and composition of salivary secretion are controlled by the nervous system and depend on reflex action. For example, the *volume* of saliva may be five times as great for a bit of bread as it is for an equal quantity of milk, and the Cl^- *concentration* may also be five times as great. The latency for secretion is different for different substances and is different at different temperatures, being greatest at the temperature of the mouth and less for materials that are colder or warmer. That is, cold or warm materials cause rapid salivation.

Saliva is reflexly secreted when food is placed in the mouth (Fig. 13-9). The receptors in this reflex are the *taste cells* in the *taste buds* of the surface of the tongue and pharynx. The taste buds on the anterior one-third of the tongue are innervated

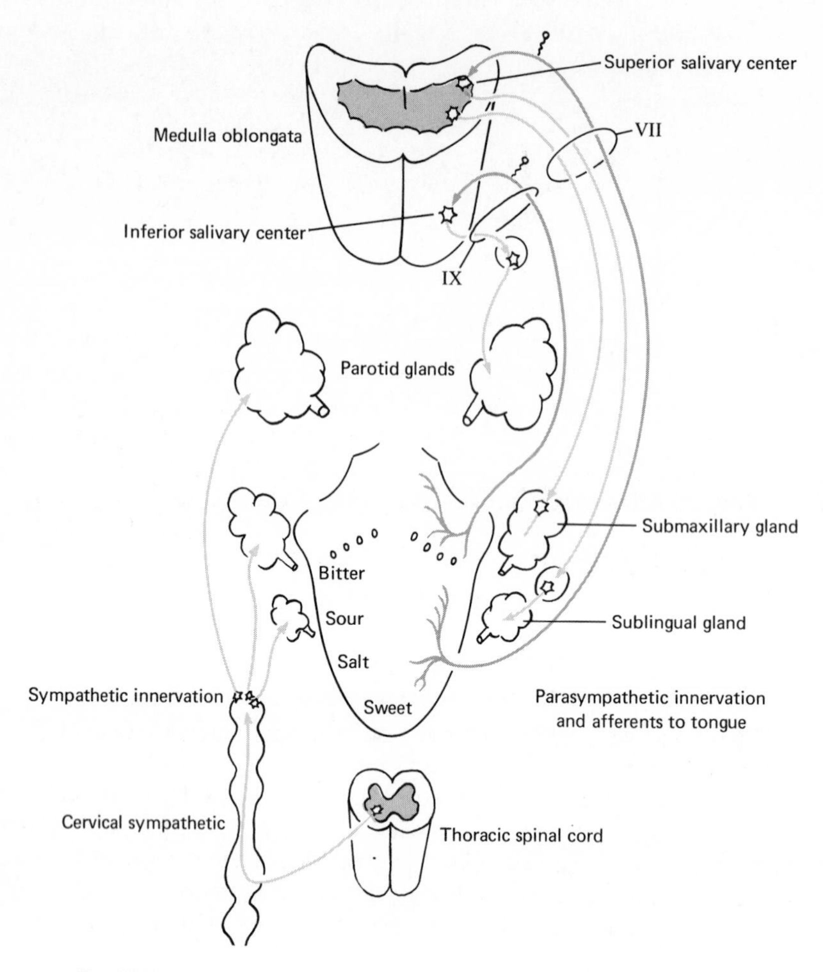

Fig. 13-9. Reflex regulation of salivation. The direction of impulse traffic from taste receptors into the salivary nuclei of the medulla is shown. Impulses in parasympathetic nerves result in an abundant flow of saliva. Secretion is also influenced by sympathetic impulses originating in the thoracic region of the spinal cord.

mainly by the afferent fibers of nerve VII, whereas those in the posterior two-thirds are supplied principally by afferent fibers of nerve IX. Fibers from the taste buds enter the *salivary center* in the medulla oblongata of the brain stem. The efferent limb of the reflex consists of autonomic nerve fibers. Preganglionic parasympathetic fibers of the VII and XI cranial nerves originate in the salivary center and travel out to ganglia situated near the salivary glands. The parotid, the largest salivary gland, is stimulated by efferent impulses in nerve IX, governed mainly by stimulation of afferent fibers of nerve IX. The sublingual and submaxillary glands, of which the secretions are more mucoid and more viscid than those of the parotid, are activated by efferents in cranial nerve VII. Various parts of the tongue differ in their relative sensitivities to the four basic tastes, and the appropriate quality and quantity of saliva are secreted in response to stimulation of these receptors. On arrival at the salivary glands, impulses from the cervical sympathetic chain provoke very little salivation.

Conditioned reflex salivation

Secretion of saliva often does not require the presence of food in the mouth. An odor, a picture, or a word associated with food will often start secretion. This kind of reflex secretion is not built in. It develops only after the nonfood stimulus has been presented several times in association with food. At first, the sight of food is followed by the presence of food in the mouth, which stimulates salivation. Finally, the sight of food is sufficient. Reflex salivation to stimulation other than of the taste receptors depends on the *condition* that the otherwise indifferent stimulus be presented in association with the food. The response aroused in this indirect way is described as a *conditioned reflex*, in contrast to the *unconditioned reflex* evoked by direct stimulation of the taste receptors.

During occasions of stress there is a reduction of digestive processes. Parasympathetic impulses to the salivary glands become less frequent, and the sympathetic innervation takes over, resulting in sparse, sticky saliva high in mucus. This is the condition of dry mouth that occurs in moments of fear or anger. This reaction can also be conditioned, for the drying up of the saliva may occur merely from a word or picture that recalls a stressful situation.

Stomach contractions

As in other viscera, there are afferent nerve fibers innervating the stomach. The nerve endings are stimulated by the pulling and stretching in the stomach wall during the smooth muscle contractions. These contractions occur almost continually in the stomach. When the blood sugar is low, the motility of the stomach increases. The contractions help tell us we are hungry, and, in extreme cases, give us hunger pangs. When the stomach is distended by food, another pattern of impulses, informing us of this fullness, provides a sensation of satiety rather than of hunger. Similar sensations can often be produced by drinking a large quantity of water. Contractions of the stomach may be visualized by recording the changing pressures in a balloon that is swallowed at the end of a long tube, and is then inflated when it arrives in the stomach. Hunger pangs show up as increased pressure in the balloon.

Nervous regulation of gastric digestion

Like the salivary glands, the stomach glands are regulated by nerve impulses. The most important nerve distributed to all the gastric and intestinal digestive glands is the *vagus parasympathetic*. Impulses travel from the origin of this nerve in the medulla to the nerve plexus beneath the mucosa. Here the postganglionic neurons of the submucosal plexus are stimulated to deliver more impulses to the cells of the gastric glands where their fibers end. Under the impact of this bombardment, the chief cells pour forth their pepsinogen. When the frequency of impulses to the chief cells decreases, the secretion is reduced.

The frequency of impulses in the vagus nerve efferent fibers depends on the extent to which the vagal centers in the medulla are themselves excited or inhibited by nerve impulses from various places. The most effective way to increase the frequency of nerve impulses to the chief cells is to place food in the mouth. The afferent nerves VII and IX from the taste cells then carry impulses into the medulla and there stimulate the vagus motor centers to greater activity. The resulting secretion is the *cephalic phase* of digestion. As with salivation, conditioned gastric reflexes are easily established, and the sight or mention of food becomes a stimulus for secretion. Because of this reflex control, there are always gastric secretions awaiting the food that arrives in the stomach. If food is unappealing or if conditions surrounding food intake are disturbing, the stomach may not be appropriately prepared for the food it receives. There is good physiological reason, therefore, for food to be not only nourishing but also palatable.

Secretions of acid and pepsin may occur in situations not followed by food and having often no obvious relation to food. When HCl and pepsin appear in the absence of food to dilute and neutralize the secretions, ulceration of the stomach wall may result. Changes in gastric blood flow, and in the effectiveness of the alkaline-mucous secretions (both under nervous control), also affect the propensity to develop gastrointestinal ulcers.

Not only do the nerves affect gastric secretion, but they also control the contractions in the muscular wall. Between the layers of gastric smooth muscle is the *myenteric plexus*. By way of this plexus, vagus impulses regulate stomach motility. Regular waves of contraction begin near the middle of the stomach and move toward the pylorus. As they do so, they pinch off quantities of food, and the wall of the antrum is stretched as this food is pressed onward.

Hormonal control of gastric digestion

Mechanical pressure of food against the stomach wall initiates endocrine control, the *gastric phase* of digestion. Stretch of the stomach wall causes cells of the pyloric antrum to release a substance that stimulates the gastric secretory cells, particularly the parietal cells that produce HCl. This material, the hormone *gastrin*, diffuses into the circulation and is distributed throughout the body (see Fig. 13-10). It has no particular effect until it has returned to the stomach and to the target organs, the parietal cells. Of the many other substances that stimulate secretion, one of the most potent is *histamine*, which resembles gastrin very closely in its action. Some foods stimulate secretion directly by their presence in the stomach (e.g., spinach, which has a high histamine content).

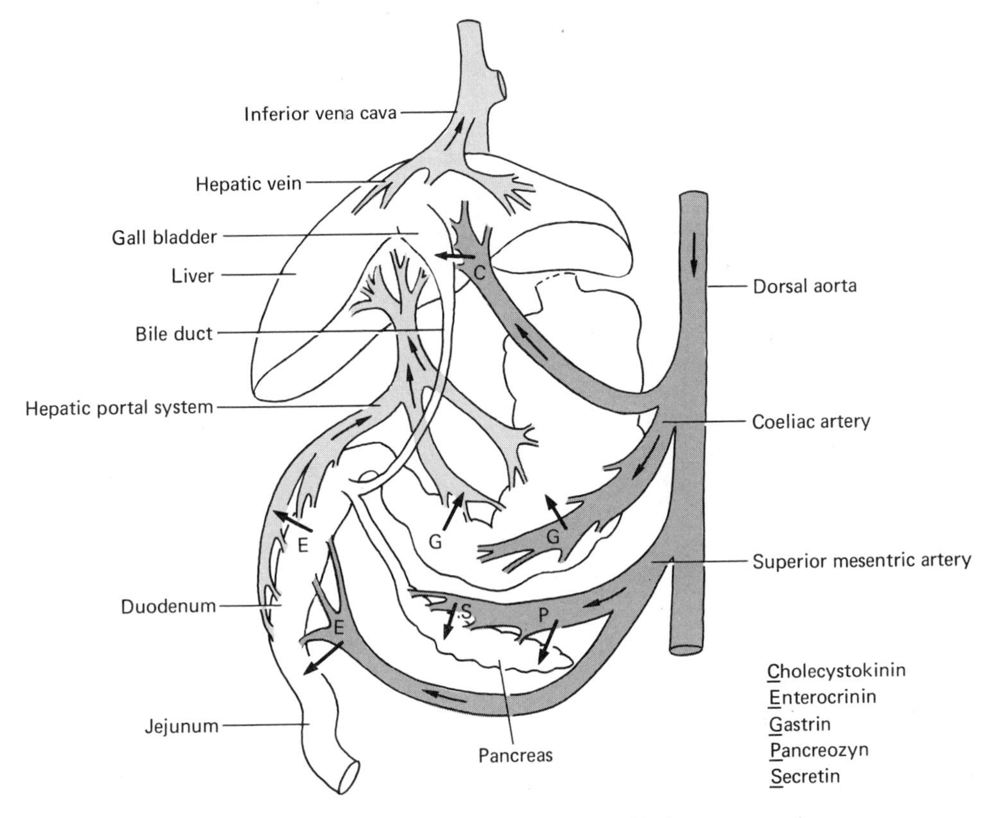

Fig. 13-10. Endocrine control of digestive secretions. The hormones are shown moving away from their sources into the hepatic portal veins, and arriving at their target organs by way of the arterial branches from the dorsal aorta.

Nervous and endocrine control of digestion in the small intestine

Secretory activity of pancreas and small intestine and contraction by the gall bladder are controlled by nervous and hormonal means. Even before food has been placed in the mouth, impulses in the vagus parasympathetic nerve fibers stimulate secretion from the pancreas as well as from the intestinal mucosa. This *psychic secretion* is followed by the *hormonal phase* of secretion when food arrives in the intestine. The hormonal phase is triggered by the acid and by the partially digested protein molecules that enter the duodenum from the pylorus. Two hormones from the intestinal mucosa enter the bloodstream and stimulate pancreatic secretion (Fig. 13-10). One, *secretin*, increases the flow of fluid and salts that provide the alkaline environment so appropriate for the enzymes of the small intestine. A *p*H of about 7.8 is attained by the secretion of bicarbonate from the pancreas.

The other hormone, *pancreozymin*, chiefly stimulates the flow of pancreatic enzymes, trypsin (as trypsinogen) and amylase. Gall bladder contraction is stimulated by *cholecystokinin*, another hormone from the intestinal mucosa. The glands of the intestinal wall secrete their enzymes when stimulated by a hormone released from other cells of the intestinal mucosa. This arrangement is comparable to the hormonal mechanism of the stomach. In the intestine, *enterocrinin* is the hormone that, being released into the blood, returns via the circulation to the secretory cells. Here it stimulates the output of enzymes that carry out the last stage of the hydrolysis of foods, prior to absorption. Table 13-1 summarizes some features of gastrointestinal hormones and enzymes.

Table 13-1 Some gastrointestinal secretions

Source	Hormone	Target	Enzyme or Other Secretion	Action on
Gastric mucosa	Gastrin	Gastric mucosa	(HCl) Pepsin	Proteins
Intestinal mucosa	Secretin Pancreozymin	Pancreas	($NaHCO_3$) Pancreatic amylase	Starches
			Trypsin	Proteins
			Pancreatic lipase	Fats
"	Enterocrinin	Intestinal mucosa	Dipeptidases	Dipeptides
"	Cholecystokinin	Gall bladder	Bile salts	Fats

Section 13: *Questions, Problems, and Projects*

1. Draw the structural formula of a simple amino acid, such as alanine. Beside it draw another similar molecule—for example, glycine.
   (a) Indicate which is the amino and which is the carboxylic acid group on the molecules.
   (b) Show how a water molecule might be taken out between the molecules. Draw the dipeptide thus formed and indicate the location of the peptide bond.
   (c) Illustrate hydrolysis by adding a molecule of water at the peptide bond and showing the resulting products.
2. Draw a molecule of glycerol and place beside it in suitable position three molecules of fatty acid.

(a) Encircle the appropriate atoms in the molecules, to show how water may be taken out and allow a new larger molecule to be formed.
(b) Identify the class to which the new molecule belongs.
(c) What is the name applied to the chemical process involving replacement of the water molecule?
(d) Give a name to an enzyme that might catalyze this splitting process.

3. Draw a molecule of glucose as a ring compound.
   (a) Identify the class of compound you have just drawn.
   (b) Draw an identical molecule beside it, and show how water may be taken out between the molecules to form a larger molecule.
   (c) Identify the molecule thus formed.
   (d) Name an enzyme that would catalyze the splitting of the new molecule by the restoration of water.
4. Considering that secretion and other activities of the digestive tract are essentially suspended during such stress as exercise, what part of the autonomic nervous system would you predict activates the salivary glands and pancreas?
   (a) What is the optimum pH for the functioning of salivary and pancreatic amylase? Of pepsin? Of trypsin?
   (b) What is the approximate pH of the mouth? Of the stomach? Of the small intestine?
   (c) What is the function of gastrin?
   (d) Why is gastrin called a hormone?
   (e) What would happen to the secretion of pepsin by the stomach if enterogastrone were injected into the bloodstream?
5. Where in the digestive tract would you find the enzymes pepsin and trypsin?
   (a) Where would you find their precursors?
   (b) What must be the substrate of a proteinase?
   (c) What two amino acid residues seem to form the active centers of both trypsin and chymotrypsin, as well as of many other enzymes?
   (d) Why do not the digestive enzymes, secreted into the small intestine, digest the cells that secrete them?
   (e) What happens to transform an enzyme like trypsin from an inactive to an active form?

14

METABOLISM

14.1 Heat Production and Control

The source of energy for living things

All things in the universe have a tendency to deteriorate and become dispersed in an unorganized fashion. On a grand scale, the substance of the sun is transformed into light energy that is dissipated in all directions. A very tiny fraction of this energy reaches our earth, and a small part of that fraction provides the fuel for life. The most important chemical reaction for life is *photosynthesis*, "putting together by means of light," which may be described as follows:

Carbon dioxide + water + energy from sunlight
becomes
sugar + oxygen

or

$$6CO_2 + 6H_2O + E_{\text{sunlight}} \longrightarrow C_6H_{12}O_6 + 6O_2 \quad (1)$$

The usual result of mixing CO_2 and H_2O is merely to produce carbonic acid. A reaction that occurs thus, without any "outside help," is said to be a "spontaneous reaction." In photosynthesis the reaction does not occur spontaneously, but work must be done to rearrange the atoms in quite another way. The energy to force the atoms into the arrangement that they have in a sugar molecule comes from the photons of light that are absorbed by *chlorophyll*, the characteristic green pigment of plants. With the help of a number of enzymes, the energy of the photons is used to disrupt the forces holding together the atoms of CO_2 and H_2O, so that rearrangement into molecules of sugar may take place.

Use of stored energy by cells

The energy of sunlight stored in sugar molecules during photosynthesis is subsequently used by the plant to manufacture other substances. These products, as well as sugar, are used by animals both as fuel to provide energy and as raw materials from which to construct their own protoplasm. The plants also provide the atmospheric oxygen that animals use in consuming the energy sources. Most of the oxygen is formed by plants in the oceans.

The energy stored in the sugar molecule is released when sugar is burned.

Sugar + oxygen
becomes
carbon dioxide + water + energy

or

$$C_6H_{12}O_2 + 6O_2 \longrightarrow 6H_2O + 6CO_2 + E \tag{2}$$

This chemical change, in which the molecules of sugar are broken apart and combined with oxygen, is called *oxidation.* It is essentially the reverse of the photosynthetic reaction, with an important exception: the energy released is not in a form that could be used in place of the energy of sunlight to resynthesize the carbon dioxide and water.

As a result of oxidation, or burning, the atoms become rearranged in a form less capable of further alteration. When sugar, for example, is burned in a flame, the rearrangement occurs in one cataclysmic step, producing CO_2 and H_2O with a violent release of energy that appears immediately as heat. In living cells, the rearrangement of atoms and the release of the energy take place in many small steps, regulated by enzymes arranged in orderly fashion in the cells. These steps involve attachment of the food molecules to other molecules in the cell, thus producing chemical structures that fulfill a variety of functions. They may replace worn-out parts in an existing protoplasmic framework, make possible a change in length of muscle substance, provide charged particles in cell membrane, or synthesize secretion products. Whatever the changes, the food molecules themselves take part in the rearrangements and are broken gradually into pieces that become progressively less capable of undergoing further energy-yielding alterations. The end products are the same as if the food had been burned in a flame, although the intermediate steps are much different. A molecule of sugar yields the same volumes of CO_2 and H_2O, and the same amount of heat is produced in both instances.

Heat and temperature

All the chemical reactions involving energy transformations in the body are collectively described as *metabolism.* When food materials are oxidized, the reactions produce heat. It is possible to use the heat production to determine the rate of metabolism and thus to measure the "speed of living." To see how this is so, it will be useful to understand what *heat* is and how it is measured.

The heat of an object arises from the collision of the atoms or molecules of which the object is composed. In general, there is a direct relation between the heat of an object and the rapidity of motion of its particles. The faster the atoms or molecules move, the more frequently they collide. The more often they bounce against one another, the farther they push apart. For this reason, most materials expand when heat is applied to them. This fact provides a convenient way of noting when heat is added to or removed from any material; that is, it provides a means of making a *temperature scale.*

In an ordinary *thermometer*, the volume change of mercury in a narrow tube is noted. The Celsius (centigrade) *scale* is used in scientific work. The length of the mercury column at the freezing point of water is marked and called 0°; the boiling point of water is marked and called 100°; and the distance between is divided into 100 equal parts, or degrees. Heat is measured in calories. One *calorie* is the amount of heat required to raise the temperature of one gram of water one degree centigrade (specifically between 14.5 and 15.5°). Because the calorie is a very small quantity of heat, use is frequently made of the *kilocalorie* (kcal), defined as 1000 calories. Heat produced by the body is ordinarily measured in kilocalories.

Direct versus indirect calorimetry

If a person's heat production is to be measured *directly*, he must be placed inside a well-insulated box, a *calorimeter*, surrounded by water to absorb his heat output. The increase in the temperature of the water provides a measure of the number of calories of heat produced.

The heat production may be more conveniently measured *indirectly*. The amount of heat produced by the body when a certain mixture of food materials is consumed has been found to be approximately the same as that produced if the same mixture is burned in a flame. From principles of chemistry, we could have predicted such a result. For any chemical reaction, if we know the starting materials and the products, we can predict the overall heat production (it does not matter what the intermediate changes may have been). This is a very convenient rule, for it allows us to predict the heat production by the body when certain foodstuffs are being consumed, in circumstances when it is inconvenient to make a direct measurement of heat production. For a particular combination of food materials, a particular volume of O_2 will be used for oxidation, and a characteristic amount of heat will be produced.

Chemical reactions and heat production

Only three general classes of foodstuffs are burned to supply energy for the chemical reactions of the body: *carbohydrates*, *proteins* and *fats*. For each of these classes of compounds, a definite amount of heat is produced from the breakdown of a standard amount of the substance.

When carbohydrate is burned:

$$\text{sugar} + \text{oxygen} \longrightarrow \text{carbon dioxide} + \text{water} + \text{heat}$$

Each compound in the equation has a particular structure, and we may write as follows:

$$[C_6H_{12}O_6] + 6[O_2] \longrightarrow 6[CO_2] + 6[H_2O] + \Delta H \text{ calories} \tag{3}$$

Here each symbol representing a molecule is bracketed (ΔH means "an amount of heat given off"). The molecules on the left of the arrow are the starting materials; those on the right are the products that result when the starting materials break up and recombine during the chemical reaction. Note that in the equation the number of molecules has been adjusted so that the same total number of *atoms* is present at the end as at the beginning of the reaction. In an actual reaction there will be millions upon millions of molecules, but they will always react in the proportions shown in this equation.

Although we cannot ordinarily see molecules—still less directly count such large numbers of them—we can easily mix known substances in any desired proportions of their molecules. This is possible because weights of atoms relative to one another are known. It is necessary, therefore, only to make the mixtures according to the relative weights of the molecules that are constructed of these atoms. Oxygen is the standard for comparison, and the oxygen atom has been assigned a weight of 16 atomic units. Oxygen gas exists ordinarily as a molecule of two atoms, O_2, having a molecular weight of 32 atomic units. The hydrogen atom is only about one–sixteenth as heavy as oxygen; therefore it has an atomic weight of about 1. The water molecule, H_2O, has a molecular weight of 18 atomic units. The oxygen gas molecule and the water molecule have weight ratios of 32 : 18, and equal numbers of each of these molecules will always have the same weight ratios; conversely—and this is of special importance—weights of these substances in this ratio will always contain equal numbers of molecules. We know then, for example, that 32 g of oxygen gas contain the same number of molecules as 18 g of water. The weight of a substance in grams numerically equal to the molecular weight is called the *gram-molecular weight* (see Chap. 3.1).

Heat production during burning of food

The oxidation of sugar shown in Eq. (3) produces about 673 kcal of heat for each gram-molecular weight of sugar burned in the calorimeter; the same overall reaction occurs in the body. With this information and the rules noted above, we can calculate the heat corresponding to each gram of sugar and to each liter of oxygen used in the oxidation of sugar. In the equation, 1 gram-molecular weight of sugar

($C_6H_{12}O_6$) is burned. This amounts to 180 g, the atomic weight of carbon being 12, of oxygen 16, of hydrogen 1.

$$\text{Heat/g} = 673 \text{ kcal}/180 \text{ g} = \underline{3.7 \text{ kcal/g sugar}}$$

When starch, rather than sugar, is burned, 4.1 kcal/g are produced. Equation (3) also tells us that the oxidation of 1 gram-molecular weight of sugar requires 6 gram-molecular weights of oxygen. One gram-molecular weight of any gas occupies about 22.4 liters under conditions of standard temperature and pressure. With this additional information, we discover that the *volume* of oxygen required for this oxidation is 134.4 liters; then

$$\text{kcal heat/liter of } O_2 = 673 \text{ kcal}/134.4 = \underline{5 \text{ kcal/liter } O_2 \text{ burning sugar}}$$

This procedure may be carried out with each of the foodstuffs.

Fats vary in their composition, but the figures below provide average values that have been estimated for each of the pertinent quantities.

$$\begin{aligned} \text{fat} + O_2 &\longrightarrow CO_2 + H_2O + \text{heat} \qquad (4)\\ 1 \text{ g fat} + 2 \text{ liters } O_2 &\longrightarrow 1.43 \text{ liters } CO_2 + \underline{9.5 \text{ kcal heat/g fat}}\\ \text{fat} + 1 \text{ liter } O_2 &\longrightarrow \underline{4.7 \text{ kcal heat/liter } O_2 \text{ burning fat}} \end{aligned}$$

Proteins vary even more than fats in their composition, but average values are

$$\begin{aligned} \text{protein} + O_2 &\longrightarrow CO_2 + H_2O + (N) + \text{heat} \qquad (5)\\ 1 \text{ g protein} + 0.97 \text{ liter } O_2 &\longrightarrow 0.78 \text{ liter } CO_2 + 0.16 \text{ g N}\\ &\qquad + \underline{4.3 \text{ kcal heat/g protein}}\\ \text{protein} + 1 \text{ liter } O_2 &\longrightarrow \underline{4.5 \text{ kcal heat/liter } O_2 \text{ burning protein}} \end{aligned}$$

Note that the products of combustion are the same for fats, carbohydrates and proteins, with the addition of a nitrogenous product in the case of the latter.

From the preceding considerations, you can see that if it is known what food materials are being used by the body during a particular time, then the amount of heat produced by the body during that interval can be estimated from the amount of oxygen used to oxidize these substances.

The calculation of body-heat production

Food that is absorbed is utilized, or it is stored as glycogen in liver and muscles, as fat in various places, and (to a limited extent) as protein. At any particular time, all three of these materials may be drawn upon to provide the fuel requirements of the body chemistry. To calculate the heat production of the body, we must know in what amounts these materials are consumed. The total heat production may then be calculated from the sum of the contributions of each food material. This information cannot be obtained directly but must be inferred from the oxidation products that are produced. All the foods use O_2 and produce CO_2 during their *oxidation*, but protein is unique in yielding, in addition, a nitrogenous product that appears in the urine. About 16 percent of a protein is nitrogen. Therefore a determination of urinary nitrogen provides an immediate measure of the amount of protein that was used in the interval during which the collected sample of urine was being formed:

$$0.16 \times (\text{g protein burned}) = \text{g N in urine}$$

$$\text{g protein} = \frac{\text{g N}}{0.16}$$

The oxygen consumption and the carbon dioxide production are also measured for the same interval of time as that during which the urine sample was being formed. The amount of O_2 and CO_2 associated with the calculated amount of protein may then be estimated.

It will be noted that the ratio

$$\frac{CO_2 \text{ produced}}{O_2 \text{ consumed}} \text{ (called the Respiratory Quotient, or RQ)}$$

is a characteristic value and different for each food consumed [Eqs. (3), (4), (5)]; for carbohydrate, RQ = 1, for fat, RQ = 0.7, for protein, RQ = 0.8.

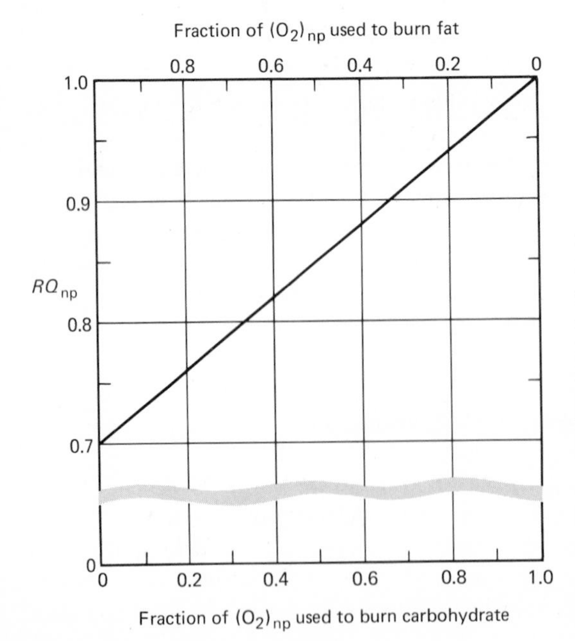

Fig. 14-1. Nonprotein respiratory quotient. After O_2 and CO_2 due to protein have been subtracted from the volumes of gas involved in the measured RQ, the remaining O_2 and CO_2 are due to the metabolism of a mixture of carbohydrate and fat. If the entire volume of O_2 in question was used to burn carbohydrate, the RQ_{np} would be 1; if all was used to burn fat, the RQ_{np} would be 0.7; if 50 percent for carbohydrate, the RQ_{np} would be halfway between 1 and 0.7, and so forth. This curve shows this relationship graphically, and it may be used to calculate the volume of O_2 used for burning carbohydrate and fat when the RQ_{np} is known.

When CO_2 and O_2 attributable to protein are subtracted from the directly measured volumes of CO_2 produced and O_2 used, the remainder gives an RQ that is due to a mixture of carbohydrates and fat. If this nonprotein respiratory quotient (RQ_{np}) is 0.7, all the O_2 was used to *burn fat*. If it is 1.0, all the O_2 was used to burn carbohydrate. If it is somewhere in between, there will be a proportionate amount of each (Fig. 14-1).

A specific example of indirect calorimetry

Suppose that during a certain interval 25 liters CO_2 were produced and 30 liters O_2 were consumed by a subject. Then

$$RQ = \frac{CO_2}{O_2} = \frac{25}{30} = 0.83$$

The total volumes of CO_2 and O_2 must be divided properly among carbohydrates, fats, and proteins. Assume that during the time the respiratory gases were measured, 1.6 g (N) was found in the urine. Then [from Eq. (5) above], 7.8 liters CO_2 were produced, and 9.7 liters O_2 were used in the consumption of protein during that time. 1.6 g (N) corresponds to 10 g protein, which would generate *43 kcal heat due to burning of protein*. The volume of gases due to protein consumption may be subtracted from the original totals of gases.

$$\frac{CO_{2_t} - CO_{2_p}}{O_{2_t} - O_{2_p}} = \frac{25 - 7.8}{30 - 9.7} = \frac{17.2 \text{ liters } CO_2}{20.3 \text{ liters } O_2} = 0.84 = RQ_{np}$$

The nonprotein RQ is 0.84—that is, the RQ due to combustion of carbohydrate plus lipid, and excluding protein.

How can the gas volumes that determine the nonprotein respiratory quotient, RQ_{np}, be allocated between carbohydrate and fat? If $RQ_{np} = 0.7$, all the O_2 used would provide heat from burning of fat. If $RQ_{np} = 1.0$, all the O_2 used would provide heat from burning of carbohydrate. Evidently if half the O_2 was used to burn fat, and half to burn carbohydrate, the RQ would be halfway between 0.7 and 1.0. The relationship is shown in the graph (Fig. 14-1).

The fraction of the total O_2 used to burn carbohydrate may be read from the graph, or it may be calculated. Approximately 0.5 of the O_2 was used for carbohydrate combustion. The "nonprotein" O_2 is 20 liters, of which 10 liters were used for carbohydrate.

Since the $RQ_{carb} = 1$, the volume of CO_2 produced during burning of the carbohydrate is also 10 liters. Therefore

$$RQ_{fat} = \frac{CO_{2np} - CO_{2carb}}{O_{2np} - O_{2carb}} = \frac{17 - 10}{20 - 10} = \frac{7}{10} = 0.7 = RQ_{fat}$$

In the combustion of carbohydrate,

$$10 \text{ liters } O_2 \text{ at } 5 \text{ kcal/liter } O_2 = \underline{50 \text{ kcal due to burning of carbohydrate}}$$

In the combustion of fat,

$$10 \text{ liters } O_2 \text{ at } 4.5 \text{ kcal/liter } O_2 = \underline{47 \text{ kcal due to burning of fat}}$$

The total heat production in this instance is *140 kcal.*

Significance of the metabolic rate

If the heat production of an individual is known, either from direct or indirect measurement, the overall rate of chemical action in that individual may be compared directly with the rate in another person of exactly the same dimensions. But we expect a large person (who has more protoplasm) to generate more heat than a small one (who has less). We must, therefore, adjust the actual heat production for a whole person to take into account his size, if we wish to make a comparison of rates of metabolism in individuals of different sizes. We are really interested in the heat production of comparable units of metabolizing tissue. A person's metabolism is usually expressed in terms of relative surface area on the assumption that the rate of heat production is proportional to the rate of heat loss, which, in turn, must be proportional to the relative surface area.

Why is the heat production proportional to the surface area rather than to the volume (or weight)? The effect of surface area can be realized in a simple model. Consider two pieces of metal of equal volume (and weight), one a cube, the other a sheet (Fig. 14-2).

If the cube is $2 \times 2 \times 2 = 8 \text{ cm}^3$ and the sheet is $\frac{1}{2} \times 2 \times 8 = 8 \text{ cm}^3$, the volumes are equal, but the areas are, respectively, 24 cm^2 and 42 cm^2. Heat loss must be directly proportional to the surface area, since the opportunities for loss by radiation and by collision with air molecules depend directly on the surface available. The ratio of the surfaces for these equal volumes is $\frac{42}{24} = 1\frac{3}{4}$; that is, heat loss would be $1\frac{3}{4} \times$ as fast for the thin plate as for the cube. It follows that if the two pieces are to be kept at the same temperature, say $37\frac{1}{2}$°C, the flame must evidently provide more heat in case B if the amount of heat supplied is to keep up with the rate of heat loss.

Similarly, in two people of equal weight, if one is rotund, the other lanky, the lanky one must produce heat at a faster rate if he is to keep up with the heat loss.

Besides body weight, the surface area, sex, age, nutritive state, and degree of general body activity affect the rate of heat production.

The basal metabolic rate of an individual is commonly expressed as a percent deviation from normal. Thus a BMR of +15 percent means that the heat production is 15 percent higher than the normal average rate found in tables of data assembled

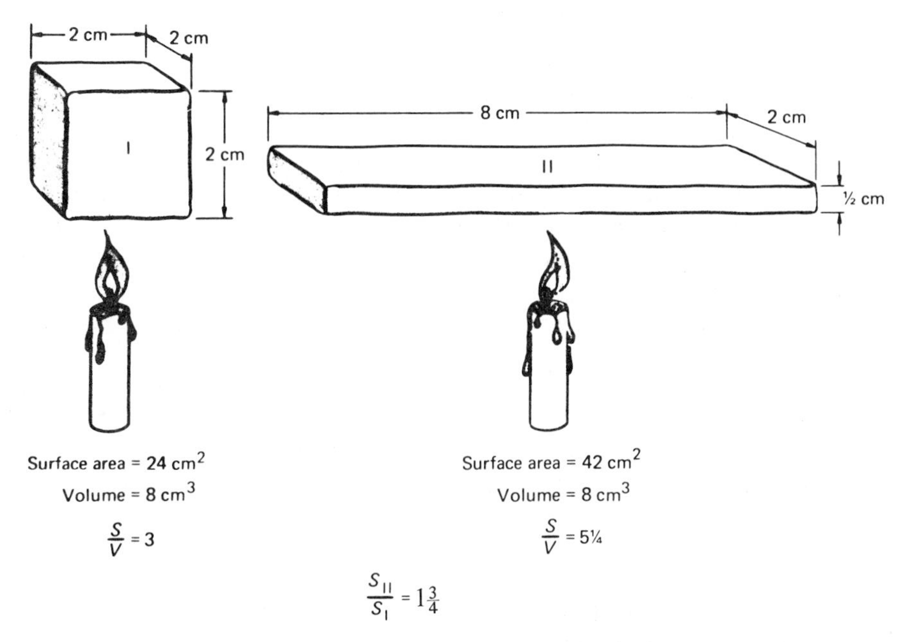

Fig. 14-2. Surface area relevant to heat dissipation.

from hundreds of measurements of apparently normal persons of the same age and sex. The BMR has been used mainly to check thyroid function. A BMR more than +15 percent above normal suggests a hyperthyroid condition, whereas one more than 15 percent below normal suggests a thyroid deficiency.

Regulation of metabolism

In all cells there is a continual exchange of constituents of the protoplasm with the food materials in the tissue fluids. The constancy of body structure is purchased at the price of unceasing effort expended to keep things the same. The energy involved in this effort is the basal metabolism. In growing children, some of the energy in the consumed food is not dissipated as heat but is stored in body structure as molecules of food material rearranged into protoplasm. The high rate of synthesis during growth requires considerably more energy than is necessary simply for maintenance of an already completed structure. Thus the metabolic rate of children is higher than that of adults under comparable conditions. In persons gaining weight, oxidizable material is stored and not used, and the energy in food eaten exceeds the metabolic heat produced. In persons losing weight, only part of the heat produced is obtained from the food eaten and there is a net decrease in stored oxidizable material.

The fate of energy contained in foods

The energy in the food molecules can be temporarily stored but eventually will appear as heat. It cannot be lost; it can only be changed into different forms. When the energy exists in the forces holding atoms together, it can be used by the body. The bonds break explosively. Part of the energy drives the atoms into a new arrangement, but most of it merely increases the random motion of the particles; that is, it appears as heat. Energy as heat cannot be used directly by the body metabolism. Because of the limited way it can be used—only to raise the temperature—heat is called the lowest form of energy. The heat produced by the body cells helps provide the constant temperature necessary for the proper functioning of the metabolic system.

Under basal conditions these packets of energy are used only as necessary to allow minimum functions to be carried out: to provide energy for the heartbeat and certain other muscle contractions, for essential glandular secretions, and the functioning of the nervous system. Energy cannot disappear, but it appears as heat. This is the source of the heat production under basal conditions.

More heat is generated during activity because energy for muscle contraction is provided by the breakdown of a large number of energy-bearing compounds. As in measurement of basal metabolism, the heat production accompanying this activity may be calculated indirectly from the estimated amount of food oxidized and will be equal to the directly measured extra heat produced by the muscles. However, only part of the energy represented by this extra heat production can appear as useful work done by the muscles. The efficiency is perhaps 30 percent or less.

Energy expenditure and body weight

The growing individual requires extra food for his higher metabolism. When he continues his youthful eating habits into adulthood, the extra food (no longer required for growth) may be stored as a reserve in the form of fat. The extra weight thus gained may be lost if metabolism is increased through exercise, or it may be lost by reducing food intake. In either case, the body reserves will be consumed to supply some of the energy required by the body.

The following example will illustrate the effectiveness of each procedure: assume that an obese individual restricts his daily intake to 1200 kcal, together with the essential vitamins, minerals, and so forth, and that he has a daily requirement of 2800 kcal (1800 for basal metabolic needs, 1000 for light daily activity). The extra 1600 kcal that he requires must come from his own stores of surplus body fat. Fat supplies 9.3 kcal/g.

$1600/9.3 = 172$ g of body fat will be consumed per day to provide the necessary energy above that available in the food. In 100 days the person would have lost 17.2 kg of fat by eating less food.

The extra fat might be oxidized by increasing the daily activity instead of by restricting food intake. Walking increases the metabolic rate 2 to 3 times above the basal level. The basal requirement per hour is 1800/24 = 75 kcal/hr. Assume an increase of 150 kcal/hr in walking. Then 150/9.3 = 16 g extra body fat is oxidized during 1 hour of walking. To lose 17.2 kg, about 3 years would be required (17,200/16 = 1075 days), assuming that the food intake remained only sufficient for basal requirements plus light daily activity, not including the one hour of walking. The best combination for weight reduction is, obviously, restricted food intake plus exercise.

Temperature regulation

The temperature of the body at any moment is the result of a particular balance between heat production and dissipation. The central core of the body is kept at a rather constant temperature of about 37.5°C. The temperature at the body surface and in the extremities varies considerably, according to the need of the body to conserve or dissipate heat to maintain the core temperature constant. To maintain this constancy, the heat gain by the body must equal the heat loss. Heat is gained from the chemical reactions of metabolism, particularly from muscle work, and by uptake from the surroundings if the external temperature is very high.

Heat is lost from the body by radiation and conduction of heat and is aided by convection or movement of air currents, bringing new air to the body surface (Fig. 14-3 and Table 14-1).

Table 14-1 Temperature regulation

Means of Heat Distribution
1. Bring heat to body surface.
(a) Dilation of blood vessels.
(b) Production of sweat, which evaporates (0.54 Kcal/g H_2O).
2. Get rid of heat at the surface.
(a) Radiation (IR)
(b) Conduction (direct contact):
to solid objects
to moving air (by convection currents).
For Heat *Conservation*, Reverse the Above

Heat is brought from the core to the body surface during vasodilation of surface blood vessels. Blood brought from central regions is here separated from the air only by the thickness of the skin. Heat dissipates easily through this thin layer. Considerable loss occurs with the help of the sweat glands, for the evaporation of water in sweat uses up 0.54 kcal heat for each gram of water changed from liquid into vapor.

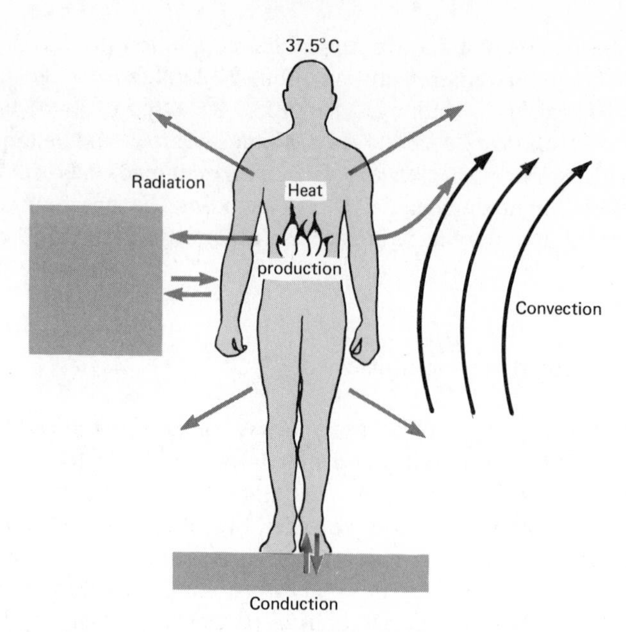

Fig. 14-3. Body heat balance. Heat production symbolized by internal fires. Arrows directed outward indicate heat dissipation. If heat production and dissipation are equal, body temperature remains constant. Heat radiated from the body surface may heat a nearby object, it may be directly conducted to an object in contact with the body, or it may be lost to currents of air (convection).

The heat for this transformation comes from the skin; thus heat is removed from the body during sweating. Air currents aid this process by bringing to the body surfaces fresh air unsaturated with water vapor.

The means for regulating the production and dissipation of heat are controlled mainly by the nervous system. During exposure to cold, the body heat production is increased, especially by tetanic contraction of skeletal muscle. These contractions may be organized into movements or may be random, as in shivering. In the cold, sweat glands stop working; vasoconstriction in the extremities shunts the blood into the central core of the body, and the hair stands on end. This latter reaction is useful in animals with a good coat of fur, for it increases the layer of air serving as an insulation at the surface. In ourselves, it is little more than "goose-flesh," a useless adaptation.

There is a temperature regulating center in the hypothalamus. An anterior part protects against heat; a posterior part protects against cold. Stimulating the anterior part causes peripheral vasodilation and sweating, and hence loss of heat. Stimulation of the posterior part results in shivering, vasoconstriction of peripheral vessels, decreased sweating, and conservation of heat. The hypothalamic temperature-regulating center, like the respiratory center in the medulla, is influenced by the properties of the blood, by impulses from sense organs, and the higher centers in the nervous system. The anterior part goes into action when the blood temperature rises and when impulses arrive from the warmth receptors of the skin.

The ultimate source of the heat dissipated by the body is the food supply, which may be stored in varying amounts as carbohydrates, fats, and proteins. Under conditions in which a basal metabolic test is given, 100 to 150 g glycogen is stored: one-tenth of it is in the liver, the rest in the muscles. There is always a supply of glucose available in the blood to supply tissues like the brain that do not store an energy source.

14.2 Cellular Metabolism

The meaning of "cellular work and chemical energy"

We have described the overall reactions involving the breakdown of the food substances to CO_2, H_2O, and nitrogenous products, and we have shown that a certain amount of heat is associated with these overall chemical reactions. But the body is not a heat engine. The chemical reactions important for doing the work of the cells of the body are those reactions *in between* the beginning and the end, as shown in the equations below. What are these reactions *in between*, and in what way are they important?

$$C_6H_{12}O_6 + 6O_2 \xrightarrow[\text{What happens here?}]{\text{- - - - - - - -}} 6CO_2 + 6H_2O + 673 \text{ kcal heat}$$

$$\text{fat} + O_2 \xrightarrow[\text{and here?}]{\text{- - - - - - - -}} CO_2 + H_2O + \Delta H \text{ (heat)}$$

$$\text{protein} + O_2 \xrightarrow[\text{and here?}]{\text{- - - - - - - -}} CO_2 + H_2O + (N) + \Delta H \text{ (heat)}$$

The dashed lines indicate sequences of reactions that make energy available for the accomplishment of physiological work.

The "physiological work" is essentially the synthesis, in the cells, of many kinds of molecules that are relevant for muscle contraction, nerve impulses, secretions, and transport of water and solutes through cell membranes. To say that the chemical reactions make energy available for physiological work is merely to say that there exist conditions favorable to particular sequences of changing arrangements of molecules.

Work implies productive effort. For example, energy is said to be consumed in doing work to lift the ball from position II to position I, as shown in the diagram (Fig. 14-4). Actually, when a muscle contracts, there is not so much a *consumption* of energy as a *transformation* in the cells of chemical substances able to do work, into molecules that are incapable of doing that work under the prevailing conditions.

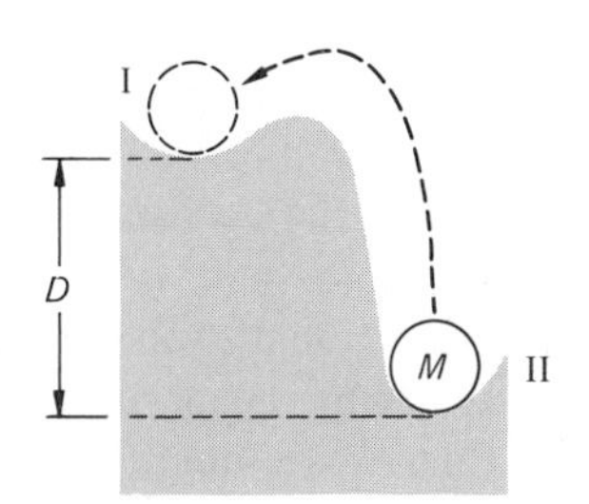

Fig. 14-4. Work and energy. Work is done in moving the mass against gravity, from position II to position I. The mass in position I has more energy (determined by distance *D*) than it has in position II.

Our muscles do not break the universal law that energy can be neither created nor destroyed, and that the energy of the universe remains constant. Like other events in the universe, the physiological processes occurring in our bodies involve shifting relationships of parts, in particular, the rearrangement of atoms into combinations that cannot be used further by the body to do work.

"Consumption of energy" in the body, means making use of molecules that are capable of doing work. The energy involved in holding together atoms of these molecules is changed in form. The energy exists in the work done by the cells, in the forces holding the atoms together in the new arrangement, in the changes in vigor of movement of the molecules (we observe these changes as heat), and in vibrations that we may see as electromagnetic radiations, such as light. When the body uses energy, the energy is *lost* only insofar the body's capability to use it is concerned. The energy is not destroyed; it appears in different forms. When our muscles contract, they may do useful external work, as in lifting a ball, but, in addition, our temperature rises, we give off a brighter glow of invisible infrared radiation, and the food we consume becomes carbon dioxide and water.

The source of chemical energy: Molecular stability and the direction of chemical reactions

The elemental relevant forces that drive the biological machinery reside in the forces holding atoms to one another. These bonds may be described in terms of electron orbital arrangements. According to this picture, for example, a bond exists between two carbon atoms when the bonding electrons move in a pathway around the nuclei of both atoms.

Among the infinity of combinations that may be entered into by atoms, some combinations are more stable—that is, the bonds are less easily broken than in others. When the atoms of a molecule are held together very tightly, a great deal of work must be done to pull them apart. From the standpoint of being a source of energy to *do* work, such a molecule is very poor indeed. Work may be done when a relatively labile combination of atoms falls into a more stable configuration. Thus *it follows that in a molecule having high bonding energy* the atoms are held together tenaciously, but such a molecule has low energy for getting chemical work done so far as a cell is concerned. It is the probability of falling into a more stable configuration—from a loose connection to a tighter connection—that makes a molecule have "high energy" so far as driving chemical reactions are concerned.

Energy may be represented in terms of work. In Fig. 14-4 work must be done to move the ball up from position II to position I. We may say that the energy in the ball, because of the position of the ball, is greater in position I than in position II. The difference is directly proportional to the distance that the ball was moved against the force of gravity. It is greater by the amount of work that has to be done to lift the ball to the new position. Energy is a quality of a system that changes when work is done on or by the system.

Another form of energy is *heat*, a quality reflecting motion of the particles of which an object is composed. A similar diagram might represent by *D* not distance but an increment of heat added to the object.

In general, for the various kinds of energy, we can speak of a lower or a higher energy state, corresponding to positions II and I in the diagram.

Chemical energy resides in the capability of a particular arrangement of atoms to undergo a rearrangement into another form. Consider the familar example:

$$C_6H_{12}O_6 \longrightarrow 6CO_2 + 6H_2O + \text{Energy}$$

Some arrangements of atoms are more preferred—that is, more stable—than others, as illustrated in Fig. 14-5, which is another representation of the preceding equation. In this diagram the arrangement of C,H,O as $C_6H_{12}O_6$ is shown to be in a higher energy state than the arrangement as CO_2 and H_2O. In effect, the atoms in $C_6H_{12}O_6$ are held to one another with less tenacity than the atoms of CO_2 and H_2O.

The figure also shows that work must be done (energy must be added) to bring the $C_6H_{12}O_6$ molecules into the *activated state* (at A), in which they could undergo a rearrangement to CO_2 and H_2O. This extra energy may be provided as heat; in this case, the increased violence of molecular motion will momentarily put such a stress on the molecules that they are literally pulled apart and then may fall together in various ways. Those combinations that are most stable, and that require the greatest amount of work to disrupt, tend to accumulate. These molecules are in the lower trough of the figures. The work required to separate all the atoms of a molecule of $C_6H_{12}O_6$ from one another can be measured, and it is less than the work that must be done to separate the atoms of the six molecules of CO_2 and the six molecules of H_2O. The difference is a measure of how much easier it is to rearrange the atoms of $C_6H_{12}O_6$ than CO_2 and H_2O. The combination of atoms that is easier to rearrange and more likely to be rearranged is said to be at a higher energy level.

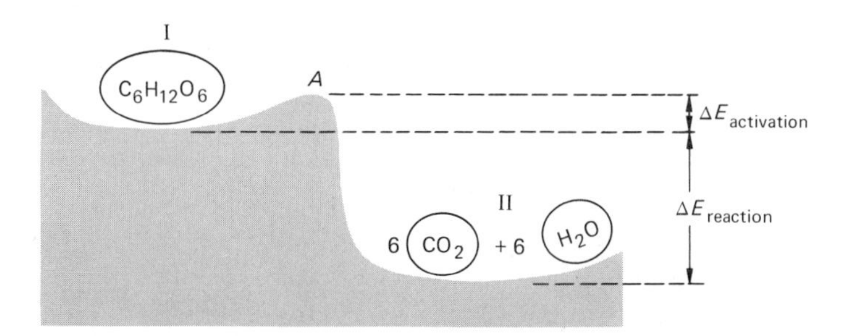

Fig. 14-5. System I is at a higher energy level than system II. $F_{activation}$ is energy required to bring system I to condition (at A) in which it can fall to the lower energy level of system II. $F_{reaction}$ is net energy made available by transition from I to II.

The energy difference between these two levels is called the "free energy." As the figure shows, and as should be clear from the discussion, a reaction that results in a decrease in free energy (and an increase in the stability of the system) is more likely to occur than the opposite. That is the most obvious reason that the reaction

$$C_6H_{12}O_6 + O_2 \rightleftharpoons 6CO_2 + 6H_2O$$

goes to the right during metabolism. Another reason is that a product of the reaction, CO_2, is continually being removed from the scene. As the figure shows, it takes a good deal of energy to kick $CO_2 + H_2O$ up to point *A*. From point *A*, the excited atoms might as likely fall into the combination $C_6H_{12}O_6$ as back into the combination of CO_2 and H_2O. But again, since the latter combination is more stable, that form tends to accumulate.

When $C_6H_{12}O_6$ and O_2 are the only molecular species around, nothing more might be expected to happen than what we have described. When other kinds of molecules are present, as in a cell, other rearrangements are possible.

Energy-rich phosphate bonds: Coupled reactions and common intermediates

Consider the compound, *adenosine triphosphate* (ATP) (Fig. 14-6). The terminal phosphate is easily hydrolyzed by the insertion of a water molecule, as shown. We may call the conditions of the mixture of ATP and H_2O on the left side of this equation State I; the condition of the mixture on the right, State II of the system. In certain circumstances, this reaction will occur with a production of energy as heat equivalent to about 8 kcal. The arrangement in State II is at a lower energy level than the arrangement in State I.

Consider the splitting of another compound, glucose-6-phosphate (Fig. 14-7). In going from State I on the left to State II on the right, there is a decrease in free energy of 4 kcal. Considered by itself, each of these reactions has a tendency

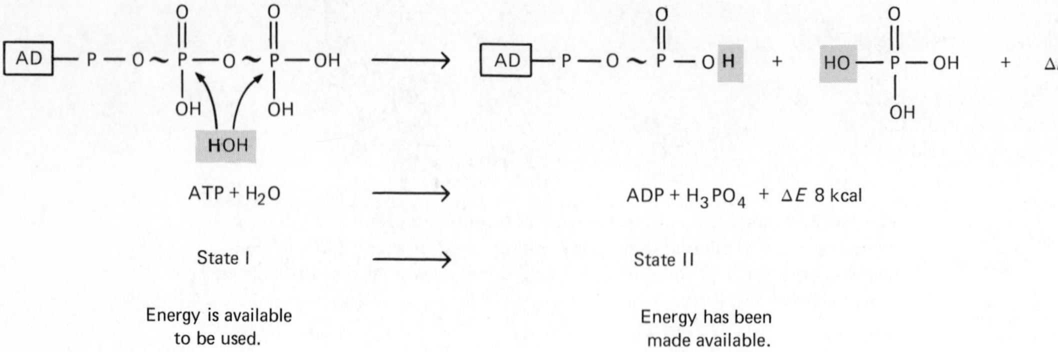

AD — P — O ~ P — O ~ P — OH (with =O and —OH on each P; HOH) ⟶ AD — P — O ~ P — OH + HO — P — OH + ΔE

$ATP + H_2O \longrightarrow ADP + H_3PO_4 + \Delta E$ 8 kcal

State I ⟶ State II

Energy is available to be used. Energy has been made available.

Fig. 14-6. Energy involved in hydrolysis of ATP.

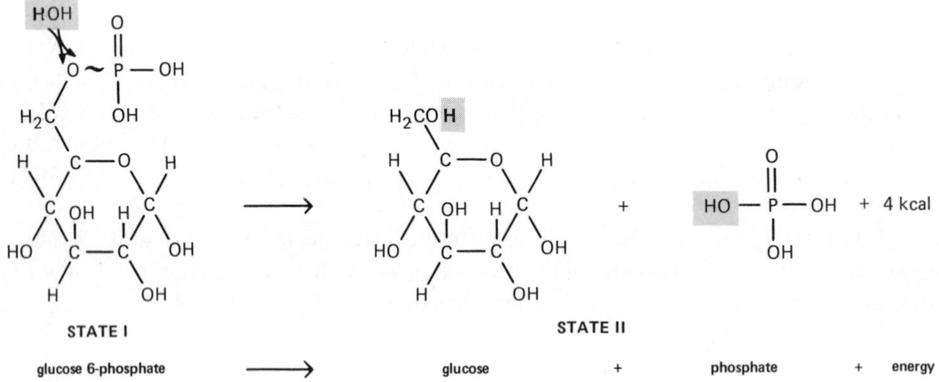

HOH + glucose 6-phosphate ⟶ glucose + HO — P — OH + 4 kcal

STATE I STATE II

glucose 6-phosphate ⟶ glucose + phosphate + energy

Fig. 14-7. Hydrolysis of glucose 6-phosphate makes energy available.

to go toward the right. But, note that in terms of the energy made available, the tendency for the splitting of ATP is more pronounced than the tendency for the splitting of glucose-6-phosphate. Put the two reactions together and the former reaction drives the latter reaction toward the *left* (Fig. 14-8). You can see that as ~Ⓟ tends to pile up from the first reaction, it forces the second reaction back toward the left. The product of the first reaction is used by the second reaction. The ~Ⓟ here is a common intermediate in the sense that it is *common to both reactions.* Look at it another way: consider that the ~Ⓟ is not allowed to get loose but is transferred from the ATP to the glucose (Fig. 14-9). The example in Fig. 14-9 illustrates the way ATP is used in cellular processes to provide the chemical energy

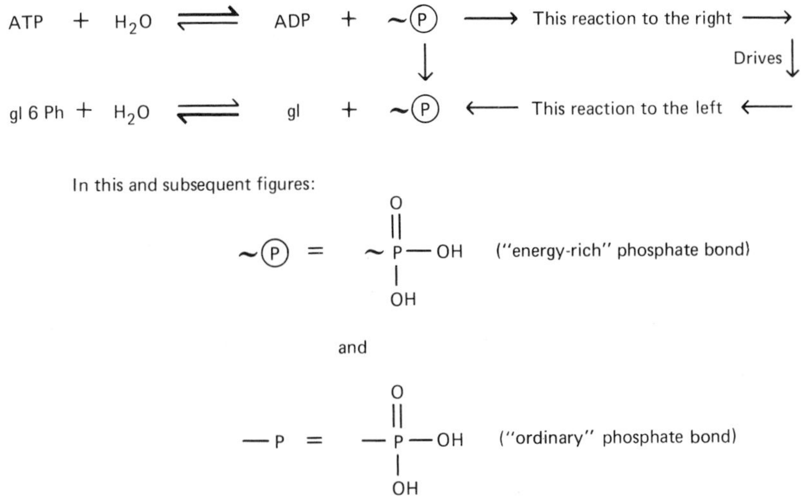

$$ATP + H_2O \rightleftharpoons ADP + \sim\text{(P)} \longrightarrow \text{This reaction to the right} \longrightarrow$$

$$\downarrow \qquad \text{Drives} \downarrow$$

$$gl\ 6\ Ph + H_2O \rightleftharpoons gl + \sim\text{(P)} \longleftarrow \text{This reaction to the left} \longleftarrow$$

In this and subsequent figures:

$$\sim\text{(P)} = \sim\overset{\overset{\large O}{\|}}{\underset{\underset{\large OH}{|}}{P}}—OH \quad (\text{"energy-rich" phosphate bond})$$

and

$$—P = —\overset{\overset{\large O}{\|}}{\underset{\underset{\large OH}{|}}{P}}—OH \quad (\text{"ordinary" phosphate bond})$$

Fig. 14-8. Reactions coupled by common intermediates. ~(P) is the intermediate in common in this instance.

to drive or force certain reactions to occur. The direction the reactions will occur depends on their spontaneous tendency—that is, the direction for the formation of the most stable products—but it will also depend on the relative concentrations and on whether the products of the reactions are used up by another reaction. Taking advantage of all these factors, the cellular machinery can carry out a great many chemical transformations that, each taken separately, would seem unlikely to occur.

The role of enzymes: Lowering the energy of activation and helping reactions to go

Left to themselves, however, all these substances mixed together in a test tube would be very sluggish in their transformations. Refer back to the diagram that illustrates the energy difference between two states of a system (Fig. 14-4). Work is done on the system—that is, energy must be added if the system is to go from State II to I. Therefore this is an unlikely direction, but as the preceding example illustrates, chemical energy can be added to make the reaction go. The change from I to II results in an overall decrease in free energy and also requires a source of energy to get up to the first bump before it can fall to the lower energy state.

To look at it another way, we can say that the molecules must be transformed into an excited or activated state before they are able to go in the direction circumstances dictate. Heat might be added to the test tube in order to increase the probability of reaction, but body temperature is kept relatively constant and is not high enough to provide the necessary activation energy. Suppose that the molecules could

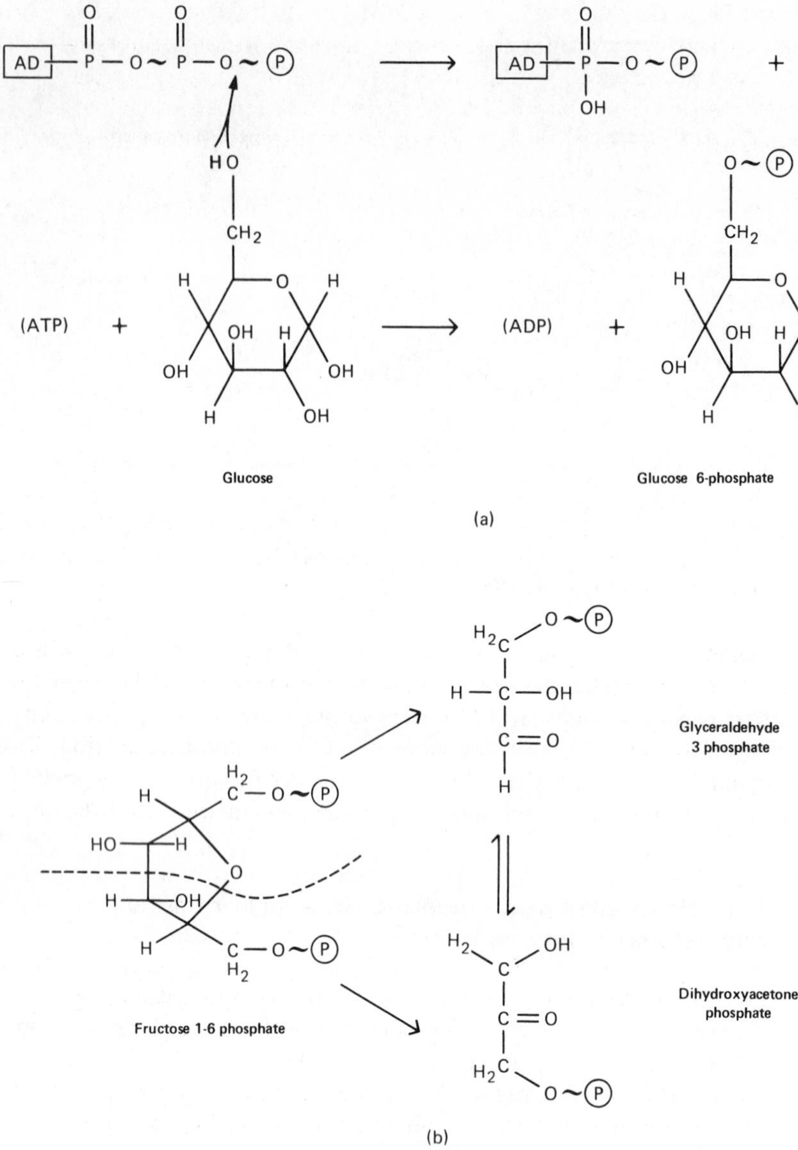

AD — P — O ~ P — O ~ (P) ⟶ AD — P — O ~ (P) + OH

HO CH$_2$ O ~ (P) CH$_2$

(ATP) + H O H OH H OH OH H OH ⟶ (ADP) + H O H OH H OH OH H OH

Glucose

Glucose 6-phosphate

(a)

H$_2$C O ~ (P)

H — C — OH

C = O

H

Glyceraldehyde 3 phosphate

H H$_2$ C — O ~ (P)

HO — H

O

H — OH

H C — O ~ (P) H$_2$

Fructose 1-6 phosphate

H$_2$ C OH

C = O

H$_2$C O ~ (P)

Dihydroxyacetone phosphate

(b)

Fig. 14-9. (a) ATP energy used to phosphorylate glucose. (b) Hexose split into trioses.

be individually spread out and held in a vulnerable position. Then the susceptible parts of different molecules would be more likely to react. This is essentially what happens when the appropriate enzyme is present.

An *enzyme* helps the reaction to occur easily. For the reaction described above, the particular enzyme is a protein with a surface shape to which the glucose molecule conforms. Stresses are imposed on the glucose molecule and a swap occurs—a particular H on the glucose is exchanged for the phosphate group given up by the ATP. In order to show the similarity to hydrolysis, the H in question is in boldface type (Fig. 14-9).

In relation to another enzyme, a second phosphate may be attached and the molecule slightly rearranged to form fructose 1–6, diphosphate.

Each of these changes involves only small changes in free energy. The changes may go in the opposite direction in any particular case if conditions are changed one way or the other. That is, the reaction is reversible.

Oxidation of glucose to provide energy

When we say that glucose is a source of energy in the body, we really mean that conditions exist that allow the breakdown and rearrangment of the atoms into a form suitable for use in specific activities. The production of "energy-rich" phosphate bonds is one of the consequences of these rearrangements. The major energy reservoir in the glucose molecule lies in bonding forces holding the hydrogen atoms to the carbon backbone skeleton of the glucose. An enzyme that forces the removal of the H atoms is called a *dehydrogenase*.

The removal of hydrogen is called *oxidation*, and the glucose molecule is said to be *oxidized* when hydrogen is removed. The molecule that receives the H is said to be *reduced*. The glucose or other molecules from which the H atoms are torn are the *substrate* molecules. The enzyme *dehydrogenase* lowers the *activation energy* required to bring about the oxidation of the substrate molecules. There are many different enzymes, but they work together each with one of two or three available hydrogen acceptors called *coenzymes*. When glucose is metabolized in cells, some of the energy of the bonds between atoms of the sugar molecule appears in high-energy phosphate compounds that can be used more directly by the cellular machinery. The steps of this process involve first phosphorylation of the glucose, as shown in Figs. 14-7 to 14-9. Atoms of the glucose molecule then become rearranged in a slightly different form as *fructose phosphate*. In this form, the molecule can add another energy-rich phosphate (again with the help of ATP). Having an energy-rich phosphate at each end, the molecule, *fructose 1-6-phosphate*, can now fall into two portions, two slightly different triose (3-carbon) sugars that are interconvertible into one another, and each of which bears a ~Ⓟ [Fig. 14-9(b)]. The first step is a phosphorylation of the glyceraldehyde 3-phosphate, at which time two hydrogen atoms are consumed by the NAD (Fig. 14-10). In this step in the metabolic processes, not only is ATP stored up but, in addition, a molecule of coenzyme is reduced that will subsequently be oxidized

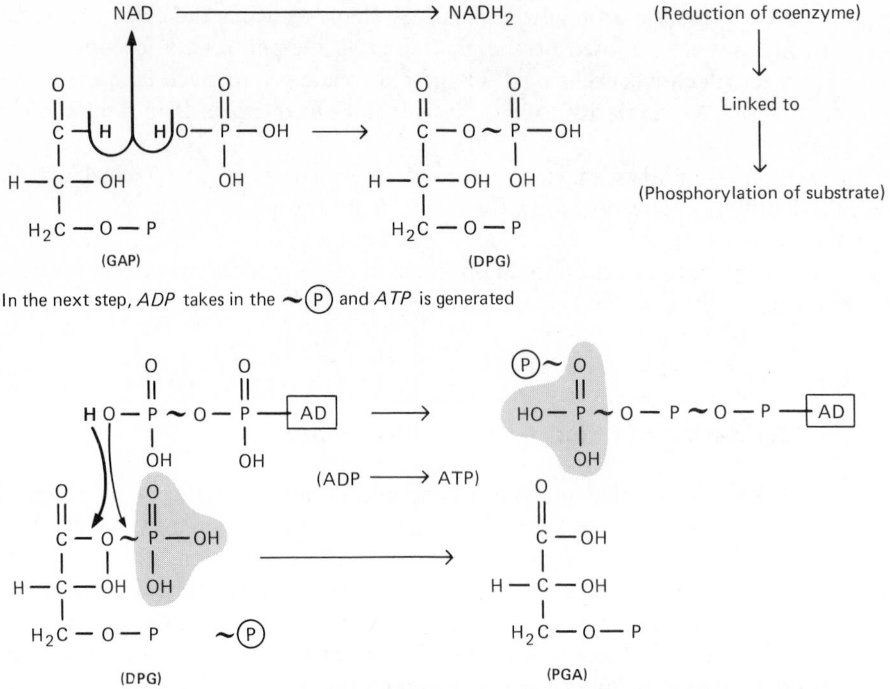

NAD → $NADH_2$ (Reduction of coenzyme) ↓ Linked to ↓ (Phosphorylation of substrate)

(GAP) → (DPG)

In the next step, *ADP* takes in the ~Ⓟ and *ATP* is generated

(ADP → ATP)

(DPG) → (PGA)

Fig. 14-10. Reduction of coenzyme linked to phosphorylation of substrate during glycolysis. GAP ≡ glyceraldehyde phosphate; DPG ≡ diphosphoglyceraldehyde; PGA ≡ Phosphoglyceric acid.

in another sequence, where it will give up its hydrogen to oxygen and generate more ATP in the process.

The next step the substrate takes toward destruction involves a rearrangement in the presence of enzymes, then a removal of water, to produce another high-energy phosphate compound (Fig. 14-11).

By the same process as before, ADP takes on the ~Ⓟ and ATP is generated. What remains is *pyruvic acid*, which may now go to *lactic acid* by taking on 2H from $NADH_2$, *or* the pyruvic acid may enter the oxidation pathway to destruction (Fig. 14-12). Each step requires the appropriate enzyme, named in most instances according to the reaction that it catalyzes.

Oxidation of other substrates than glucose

Up until now the discussion has concerned only the way a cell can begin to utilize carbohydrate for its energy requirements. The energy made available by the rearrangement of the food or substrate molecules into more stable configuration is used to construct other more generally useful molecules, particularly by donating H to the coenzyme NAD, and generating ATP. The whole process depends on the

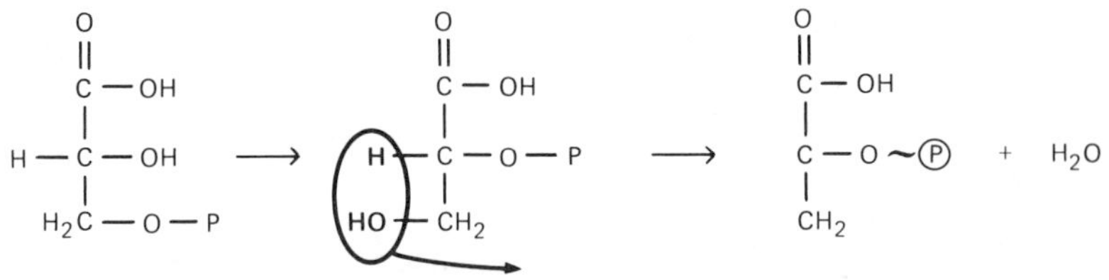

```
     O                    O                       O
     ||                   ||                      ||
     C — OH               C — OH                  C — OH
     |                    |                       |
H — C — OH   ⟶      H — C — O — P     ⟶        C — O ~ (P)   +   H2O
     |                    |                       |
H2C — O — P         HO — CH2                     CH2
```

3 Phosphoglyceric acid — 3PGA; 2 Phosphoglyceric acid — 2PGA; Phosphoenolpyruvic acid — PEPA

Fig. 14-11. A step in glycolysis—rearrangement of molecule preparatory to ATP generation.

Lactic acid

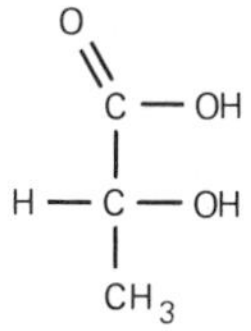

```
     O
      \\
       C — OH
       |
  H — C — OH
       |
      CH3
```

Fig. 14-12. Products of anaerobic glycolysis.

Pyruvic acid

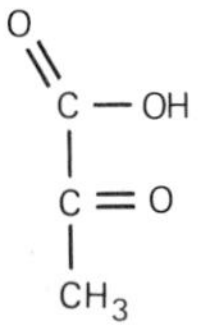

```
 O
  \\
   C — OH
   |
   C = O
   |
   CH3
```

Urea

```
NH2
   \
    C = O
   /
NH2
```

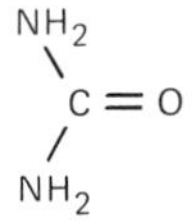

Ammonia NH_4OH

Fig. 14-13. Products of protein metabolism.

presence of particular enzymes, which are complex protein molecules manufactured by the cell in a process involving the nucleic acids, described elsewhere. In order to synthesize these proteins, the cell must have available to it the proper mixture of amino acids, best provided in a diversified diet. The NAD is a form of niacin, vitamin B_2, which is therefore an essential vitamin.

Amino acids in the diet may not only be linked together to form protein; they may also be burned in the "metabolic flame," as are carbohydrates and lipids. This process involves splitting off the amino group, which may then be excreted as urea and ammonia (Fig. 14-13). Many lipids are incorporated into cellular structures, but they, too, may be split to provide energy for the synthesis of ATP.

Lipids are generally a combination of glycerol and fatty acids, and both constituents can be chewed up in the metabolic mill. The structure of glycerol is much like the structure of the 3-carbon compounds that arise during glycolysis, and as you would expect, this molecule therefore has no trouble entering the scheme. The fatty-acid portion is handled by progressive removal of two-carbon fragments that are poured into the mill.

H and CO_2 removal in the tricarboxylic acid cycle

A common meeting ground for carbohydrates, lipids, and proteins is a sequence of reactions variously called the *Krebs cycle*, the *citric acid cycle*, the *succinic acid cycle*, or the *tricarboxylic acid cycle*, after the name of its discoverer and of some of the molecules that participate in the series of reactions. In this cycle, two-carbon fragments of carbohydrate and lipid are incorporated. As they are added at some points, CO_2 and H are pulled off at other points (Fig. 14-14).

The breakdown of carbohydrate to pyruvic acid can proceed as long as the pyruvic acid can be removed by the circulation or can be transformed into lactic acid. That sequence of changes is called *anaerobic glycolysis* because it can proceed in the absence of O_2. However, it undoubtedly proceeds in the presence of O_2 as well, in normal cells. The citric acid cycle is considered to be *aerobic*; that is, it requires the presence of O_2. This requirement involves the necessity to remove from the scene the reduced coenzyme, $NADH_2$, which is generated as the substrate is consumed. In glycolysis, the $NADH_2$ may donate its H to pyruvic acid, which thereby becomes lactic acid. The anaerobic functioning of glycolysis occurs at the price of accumulating lactic acid. The tricarboxylic acid cycle does not have that alternative. The $NADH_2$ must be oxidized to NAD if the cycle is to continue.

The tricarboxylic acid cycle is not a sequence of reactions with a beginning and an end; rather, it is a sequence that turns back upon itself (Fig. 14-14).

It may help in getting acquainted with this scheme to note first the carbon skeletons of the participating molecules (Fig. 14-15). It is convenient to begin at the top, where acquisition of a two-carbon acetyl portion contributed by acetyl coenzyme A, transforms a four–carbon organic acid into a six-carbon organic acid. Two hydrogens are nibbled off by NAD, and then two CO_2 molecules are produced, leaving the skeleton again with four carbons. At this point, about halfway around the cycle, coenzyme A comes in again, making an energy-rich bond that can be used to generate ATP. Then two successive dehydrogenations, with an intervening addition of water, leave oxaloacetic acid, which is ready to take on an acetyl group, and the cycle continues around again. To what purpose? The important feature of the cycle is that it is a way in which the participating molecules can be stripped of their hydrogens—they are oxidized, as the dehydrogenase coenzymes are reduced (Fig. 14-15). The NAD and another coenzyme, *flavine-adenine-dinucleotide* (FAD), can acquire the hydrogens of the substrate because with these hydrogen atoms attached, the

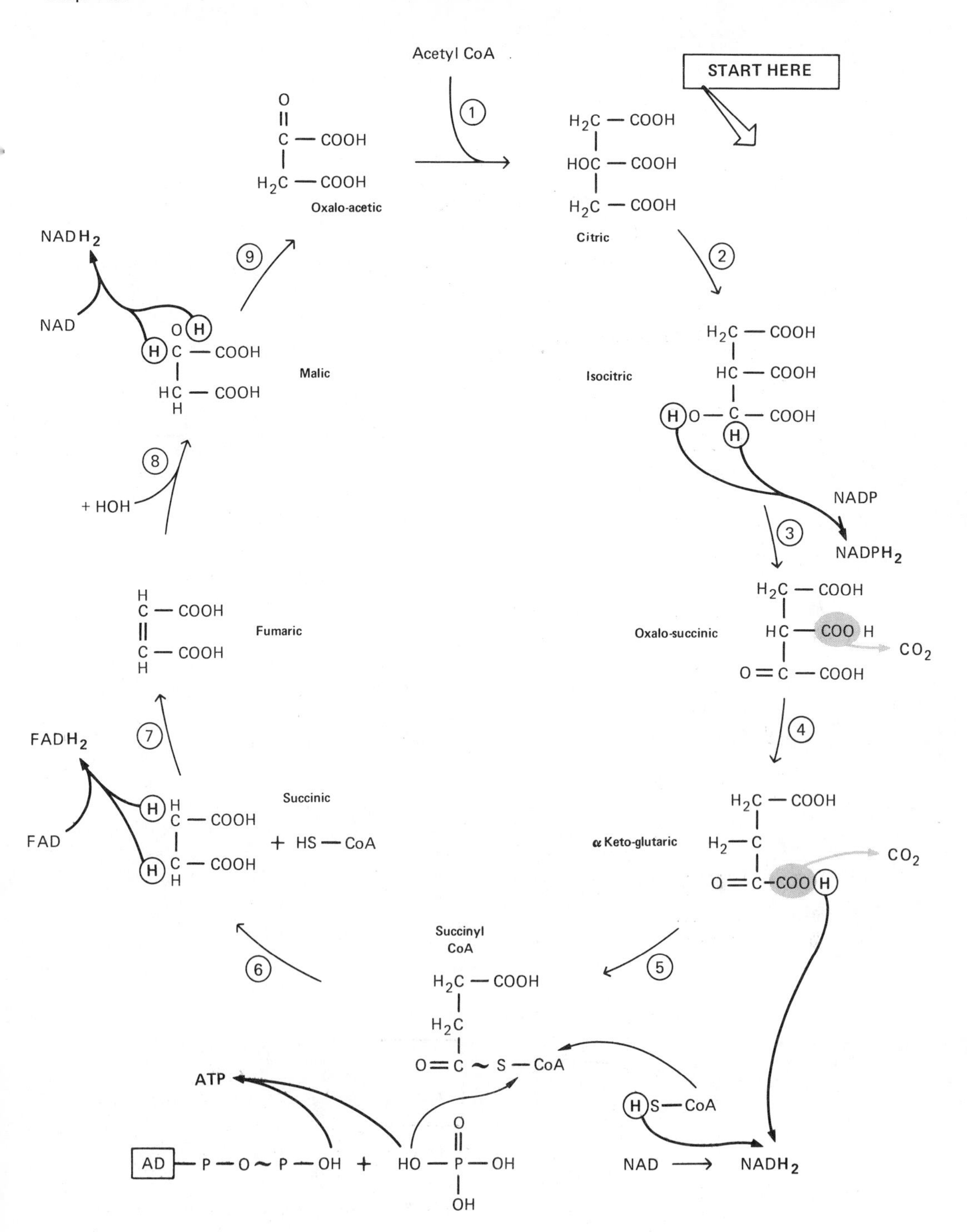

Fig. 14-14. The tricarboxylic acid.

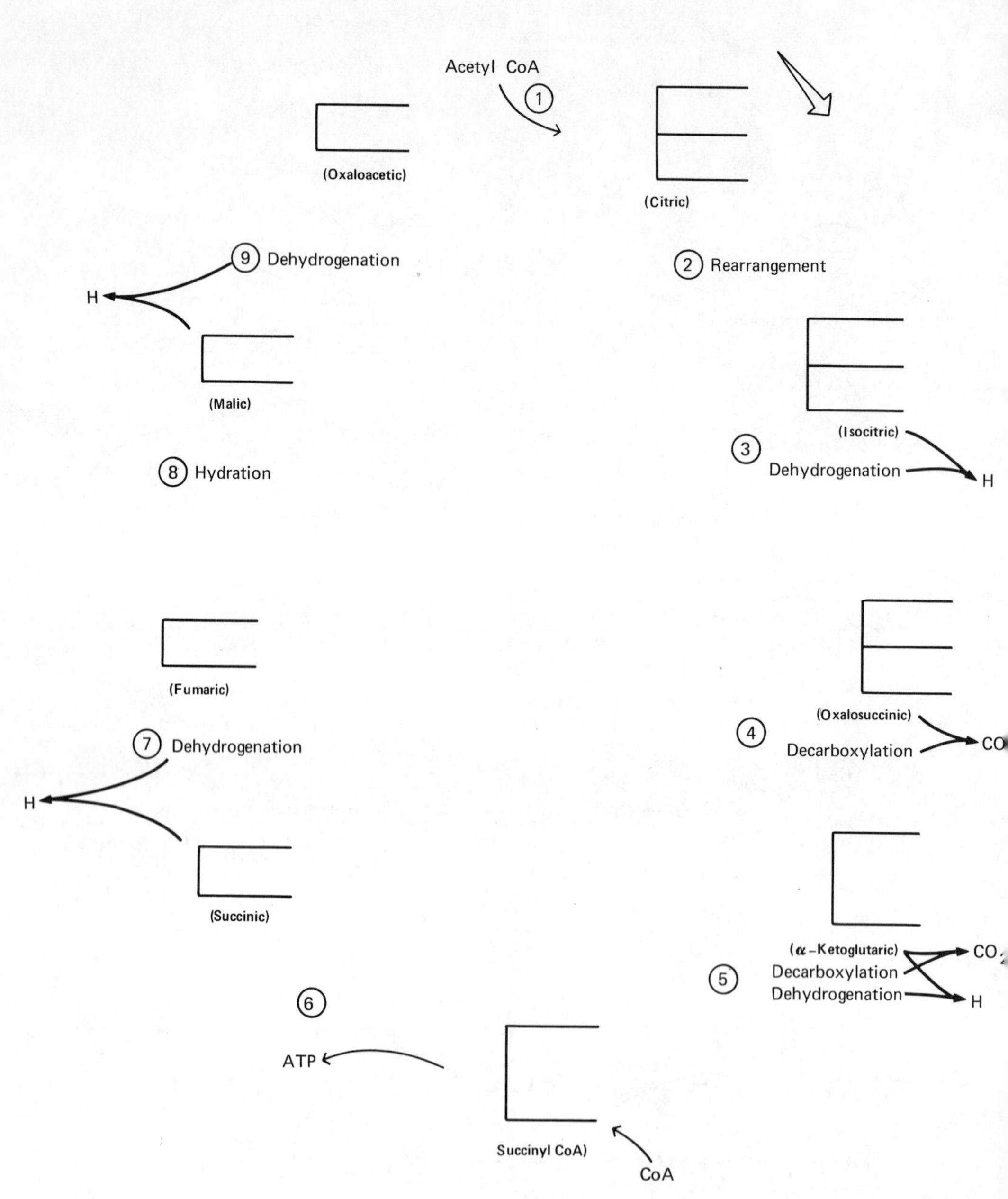

Fig. 14-15. Tricarboxylic acid cycle—simplified scheme.

coenzymes are at a lower energy level than the substrate molecules from which they have robbed the H atoms. In addition, the resulting $NADH_2$ and $FADH_2$ are immediately shunted into another metabolic sequence. The loss of CO_2 can be considered a useful procedure to transform the substrate molecules into forms that can be stripped of their hydrogen.

The reduced coenzymes will now provide the means to generate more ATP molecules. This step occurs in relation to the combination of these hydrogens with oxygen brought to the cells by the blood.

The reaction between the reduced coenzymes and the oxygen is an indirect one. It involves the separation of electrons from the hydrogens. The electrons are conducted along a special metabolic "ladder" and delivered to the O_2. Since we have been considering a *pair* of hydrogens, we will consider a *pair* of electrons.

$$2e + \tfrac{1}{2}O_2 \longrightarrow \tfrac{1}{2}O_2^{=} \quad \text{and} \quad 2H^+ + \tfrac{1}{2}O_2^{=} \longrightarrow H_2O$$

This is the fate of the hydrogens of the substrate—to combine with oxygen and form water.

Between the reduced coenzyme $NADH_2$ and $FADH_2$, which carry the hydrogen, and the final production of water, there is a sequence of steps during which the electrons creep up to pounce on the O_2, and ATP is synthesized from ADP. In this sequence, iron Fe^{3+} atoms are sequentially reduced and oxidized by the electrons until these electrons arrive at the O_2. The scheme shows the result, if not the mechanism, of these events (Fig. 14-16).

The electrons are drained off in the following reaction:

$$2Fe^{++} - 2e^- \longrightarrow 2Fe^{3+}$$

The Fe^{++} exists in relation to molecules very much like hemoglobin in structure. The organic-iron complexes are pigments, similar to hemoglobin, in structure and color. They are called *cytochromes* and are distinguished as *a*, *b*, *c*, c_2, etc., depending on their absorption spectra (Fig. 14-17). They form a graded series according to their relative affinity for electrons, a series that is not the same as the alphabetical sequence by which they are named. In Fig. 14-16 each level shows the change that occurs in the iron as it receives an electron and then gives it up to an Fe^{3+} that has a greater affinity for the electron. Each Fe^{++} or Fe^{3+} is part of a different cytochrome molecule. Finally, the electrons are delivered to the oxygen, which, now that it is in an activated state, can combine with substrate protons to yield water.

Coenzymes and vitamins

One striking feature of the evolutionary process is that the loss of metabolic capabilities is the price paid for the development of specialization of function. We are wholly dependent on what we consider to be simple and more primitive organisms to supply many substances that our own bodies cannot make.

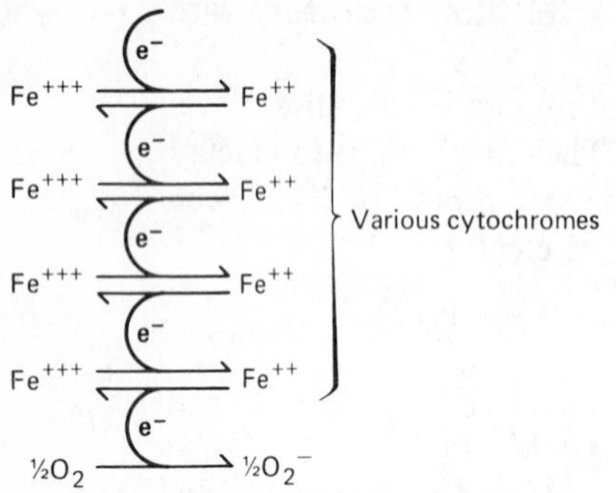

Fig. 14-16. Reduction-oxidation steps carrying electrons from substrate to oxygen.

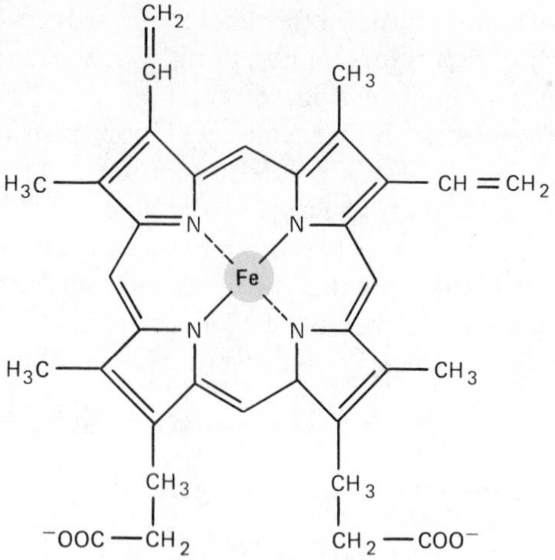

Fig. 14-17. General structure of heme prosthetic group in cytochromes.

Some primitive organisms can synthesize from CO_2 and N_2, together with the mineral salts, all the complex molecules required for their life. The cellular biochemical picture that we have described is only the merest sketch of the complex and intricate story that has been worked out. All of it, with modifications here and there, is found throughout the living world that we know. As animals, we have been completely dependent on plants for a supply of basic food materials. Theoretically, there is no reason why we might not eventually be able to synthesize all our nutritional requirements by test-tube chemical methods from the elements, but our cellular biochemical machinery is incapable of that feat.

The three foodstuffs—proteins, carbohydrates, and lipids—constitute the basic triad of food required in the diet to supply energy for the cellular machinery, as well as to supply the materials necessary for the structure of cells. In the cellular structure there is a continual dynamic flux of the building blocks. The biochemical sequences responsible for the synthetic processes have been worked out in many instances.

Besides the basic triad, certain vitamins and minerals are also essential constituents in the diet. A *vitamin* is any organic substance that cannot be synthesized by the body out of the three fundamental food classes but must be added to the diet. The substances are often characterized by dramatic effects associated with a deficiency of supply in the diet. Their effects are usually very general in the body but are most easily seen when they affect the skin and the nervous system.

Let us return to the schema of cellular metabolism in order to describe the role of some vitamins, especially in relation to certain coenzymes. Each step in the metabolic sequence is catalyzed by an enzyme. A reversible combination of substrate molecule with the enzyme accounts for the action of the enzyme. In many instances, the part with which the substrate molecule undergoes a combination is a small molecule that is only momentarily attached to the larger protein part of the enzyme. This loosely associated portion of the enzyme is called a *coenzyme.* Many vitamins are coenzymes. Figure 14-18 shows the participation of some coenzymes and their vitamin sources or equivalents in the metabolic sequences. The formulas of the coenzymes are shown later in Figs. 14-19 to 14-26.

Energy of the substrate food materials is captured in the process of stripping hydrogen from those molecules and passing the H to a coenzyme. Several molecules are capable of accepting the hydrogens. All of them are vitamins. They cannot be synthesized by the body's metabolic machinery, but they are made by certain plants and microorganisms. Primary sources are plants, particularly whole grains and fruits, yeasts, and bacteria. Many vitamins are synthesized by these organisms in amounts beyond their immediate metabolic requirements. Animals that eat the plants provide us with a secondary source of some of the vitamins when these substances are stored, especially in the organ meats—liver, kidney, and heart.

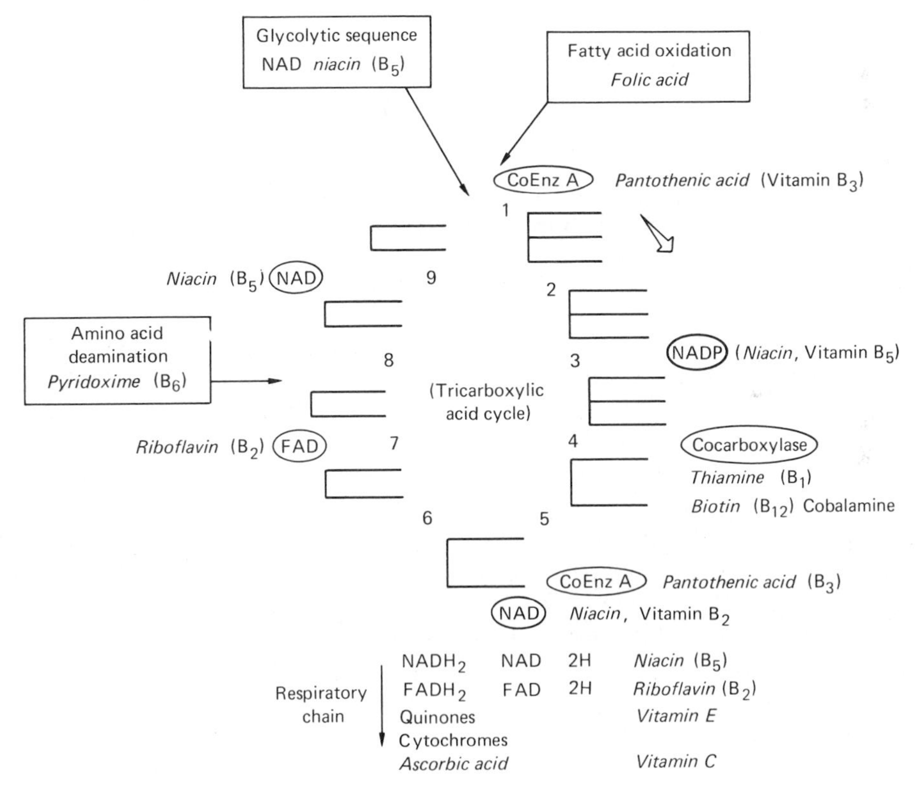

Fig. 14-18. Vitamins and cellular respiration.

Refining of grains removes a large fraction of the vitamins, leaving only the starch. The refined food will keep better than the same food unrefined, but, just as the refining process makes the food unpalatable to organisms that cause spoilage, so also refining may make the food less nutritious for human consumers. Vitamin content of food is also decreased by overcooking (prolonged intense heat will oxidize many of the vitamins) and by washing out materials and discarding the cooking water.

A substance that is a vitamin for one animal is not necessarily a vitamin for another. For example, *ascorbic acid* is a vitamin for man and for the guinea pig, but not for most other mammals. Although the ascorbic acid is an essential constituent in the metabolic processes of all animals, most animals are able to synthesize the molecule from other sources. There may also be considerable differences in the requirements of various individuals for the vitamin, depending on extent of storage, as well as individual differences in the rate of utilization of the enzymes. The vitamins are widely distributed in our foods. Culinary action in accordance with these facts, coupled with a *varied* diet and a moderate intake, would probably eliminate many health problems arising from dietary deficiencies.

The metabolic schema that we have described applies with modifications to all organisms and, with varying emphases, to all the cells of our body. Therefore it is reasonable that vitamin deficiencies have generally widespread effects. Restriction of growth in the young child follows an inadequate vitamin supply. Severe shortage affects dividing cells, capillary permeability, and nerve function very rapidly.

Review of some metabolic sequences with special reference to the role of coenzymes

The first energy-yielding step in glycolysis occurs when glyceric acid phosphate takes on a second phosphate that is an energy-rich linkage. This linkage is generated in connection with reduction of the molecule, and the two hydrogens that are stripped off are accepted by *nicotinamide adenine dinucleotide* (NAD) (Fig. 14-19). This coenzyme arises from *niacin, vitamin B_5*. This same coenzyme is used at two places in the Krebs tricarboxylic acid cycle, where it accepts hydrogens from α ketoglutaric acid and from malic acid. *NADP, nicotinamide-adenine-dinucleotide phosphate* is another form of niacin. It is a hydrogen acceptor at an earlier step of the tricarboxylic acid cycle, and it also participates in H transfer in other places not discussed here. The mechanism of action is shown in the figure. The $NADH_2$ can pass its hydrogen on to FAD.

Flavin-adenine dinucleotide (FAD), derived from *vitamin B_2, riboflavin* (Fig. 14-20), receives hydrogens at another stage in the Krebs cycle. It may also receive hydrogens from NAD and pass them on directly to oxygen. NAD and FAD are both coenzymes of dehydrogenases.

Inflammation of the skin (dermatitis), nervous disorders, and pellagra are associated with niacin deficiency, which may be avoided by consuming milk, wheat germ,

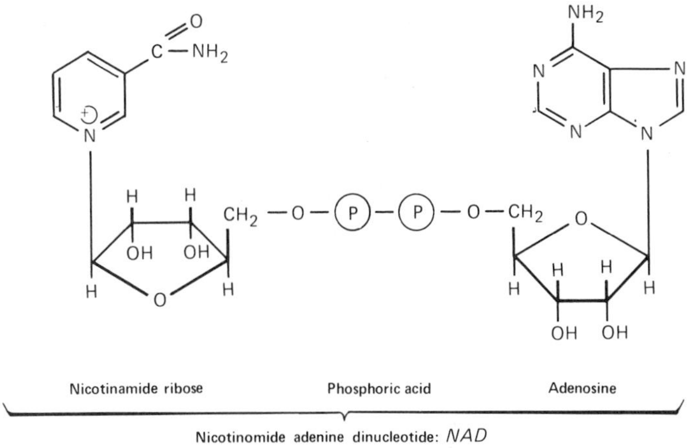

Nicotinamide ribose Phosphoric acid Adenosine

Nicotinomide adenine dinucleotide: *NAD*

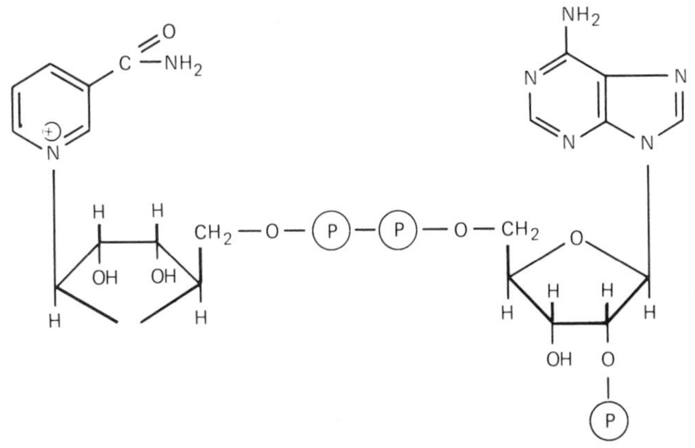

Nicotinamide adenine dinucleotode phosphate: *NADP*

Fig. 14-19. Coenzymes NAD and NADP.

liver and yeast, all of which contain good supplies of niacin. A deficiency of riboflavin provokes dermatitis, plus inflammation of vasculature of the eye. Riboflavin is stored in kidney, liver, and lean meat.

Coenzyme A, derived from *pantothenic acid, vitamin B_3* (Fig. 14-21), carries both carbohydrate and lipid into the Krebs cycle as 2-carbon (acetyl) fragments. In addition, it participates in the generation of ATP halfway around the cycle, where α ketoglutaric acid becomes succinic acid.

Pantothenic acid is found in rice, bran, liver, and kidney and is essential in all animals studied. A deficiency of this vitamin leads to graying of the fur in rats, but

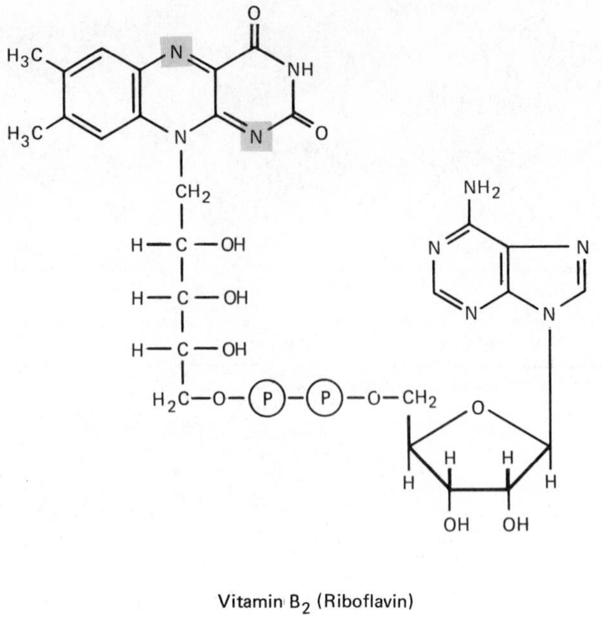

Vitamin B_2 (Riboflavin)

Fig. 14-20. Flavin adenine dinucleotide—vitamin B_2 (riboflavin).

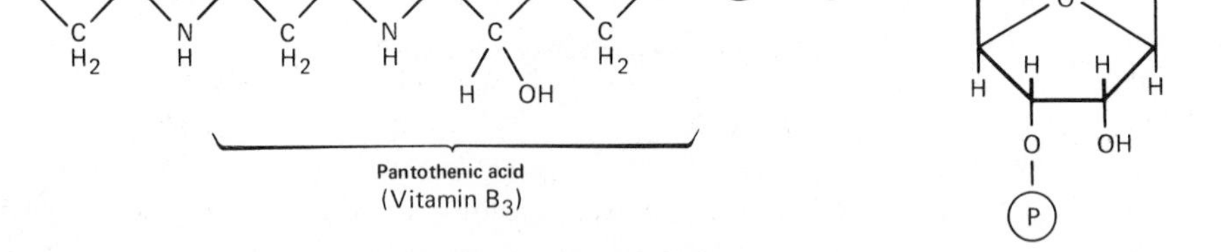

Pantothenic acid
(Vitamin B_3)

Fig. 14-21. Coenzyme A.

specific human requirements have not been ascertained, perhaps because of its nearly ubiquitous occurrence.

Folic acid (*pteroylglutamine*) (Fig. 14-22) is another B vitamin, which appears to be involved in fatty-acid oxidation, in the transfer of single carbon fragments of the substrate molecules.

Several vitamins are associated with decarboxylation, the splitting of CO_2 from

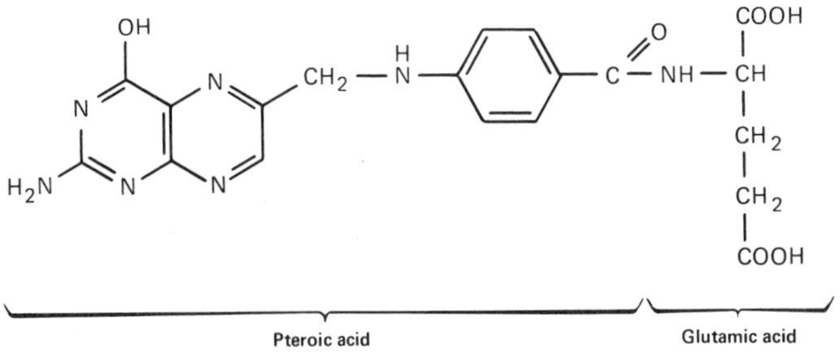

Fig. 14-22. Pteroylglutamic acid or folic acid.

the substrate molecules. *Vitamin* B_1, *thiamine*, is the best known; *biotin* and *cobalamine* (*vitamin* B_{12}) are also involved in this important process (Fig. 14-23).

Thiamine is known as the *antineuritic* vitamin. It is especially important in the metabolism of nerve. A deficiency is characterized by inflammation of peripheral nerve and paralysis. The effect is dramatically reversed in less than an hour when the vitamin is administered by injection.

Cobalamine is a unique vitamin in having a metal atom—cobalt—as part of its structure. This vitamin is an *antianemic* factor. When there is a deficiency, red cells do not adequately differentiate.

Pyridoxine (*vitamin* B_6) (Fig. 14-24) is also concerned with decarboxylation, but, in addition, it is concerned with deamination of amino acids. It plays a role, then, in the entry of amino acids into the tricarboxylic acid cycle.

Ascorbic acid (*vitamin C*) (Fig. 14-25), the antiscurvy vitamin, probably functions in the respiratory chain. It easily yields an H^+ and provides two electrons to cytochrome oxidase. Ascorbic acid deficiency is accompanied by a host of problems involving the mucous membrane, capillary walls, bones, and joints. General debility, connective tissue deterioration, slow healing of wounds—all these symptoms show the very general nature of the effects of its absence.

All the vitamins mentioned so far are water-soluble. The vitamins remaining to be discussed are lipid-soluble. Only *tocopherol*, *vitamin E*, has been given a tentative place in the general metabolic scheme. It is a quinone and can undergo reversible reduction and oxidation. Its affinity for hydrogen places it between FAD and cytochrome A. Tocopherol is known as an antisterility vitamin because it prevents sterility —at least in rats and mice a deficiency of the vitamin interferes with reproduction.

There are other vitamins for which rather special effects have been observed, but which have not been assigned a place in the general metabolic scheme. The role of *vitamin A* in vision has been previously described. It is also important in epithelial surfaces generally. Dry cracked skin is characteristic of a deficiency of this vitamin. *Vitamin D* is involved in calcium absorption, and its function in bone growth has been discussed already. *Vitamin K* has a role in blood clotting, which has also been discussed.

There are other substances generally required for good nutrition. Of these,

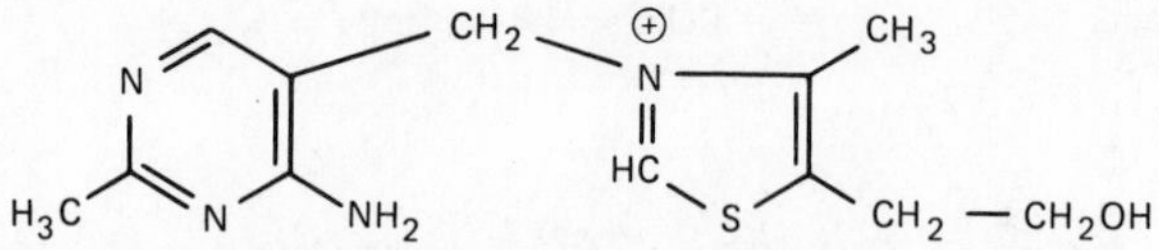

CH_2 ⊕ CH_3
N N
‖
HC
H_3C N NH_2 S $CH_2 - CH_2OH$

Thiamine
(Vitamin B_1)
(a)

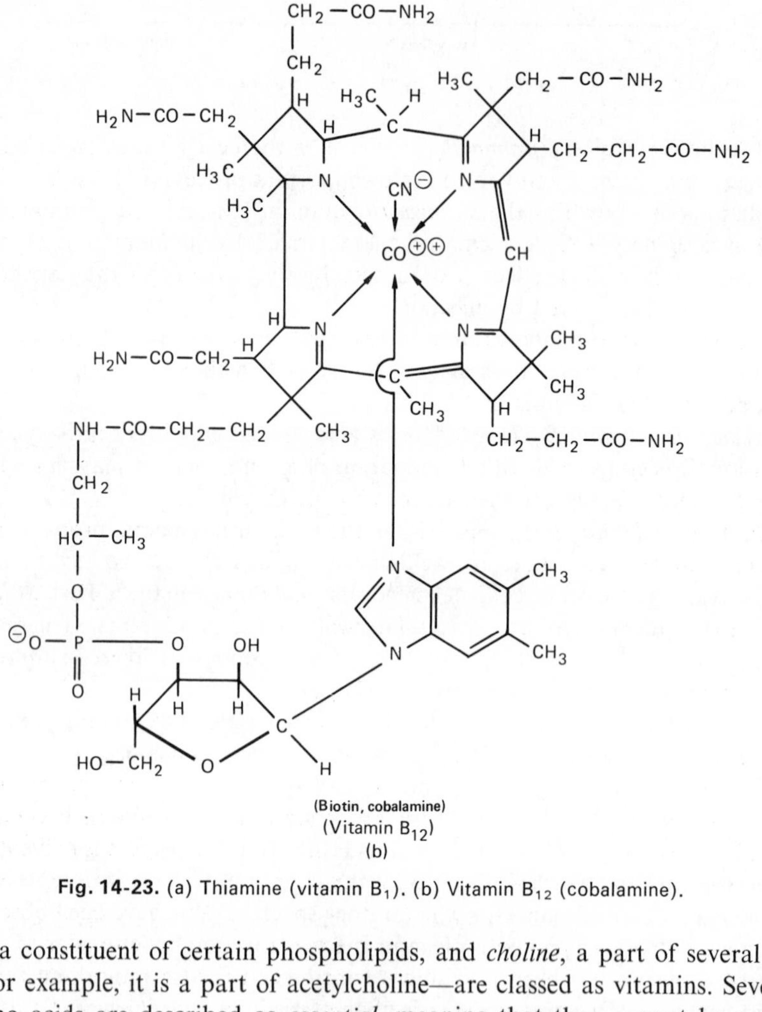

$CH_2-CO-NH_2$
CH_2
H_3C $CH_2 - CO - NH_2$
H H_3C H
$H_2N-CO-CH_2$ C
H H
$-CH_2-CH_2-CO-NH_2$
H_3C N $CN^{\ominus}$ N
H_3C
$Co^{\oplus\oplus}$ CH
H N N CH_3
H
$H_2N-CO-CH_2-$ C
CH_3
CH_3 H
$NH-CO-CH_2-CH_2$ CH_3 $CH_2-CH_2-CO-NH_2$
CH_2
$HC-CH_3$
O
$^{\ominus}O-P$ O OH N CH_3
‖
O H H H N CH_3
C
$HO-CH_2$ O H

(Biotin, cobalamine)
(Vitamin B_{12})
(b)

Fig. 14-23. (a) Thiamine (vitamin B_1). (b) Vitamin B_{12} (cobalamine).

inositol, a constituent of certain phospholipids, and *choline*, a part of several molecules—for example, it is a part of acetylcholine—are classed as vitamins. Several of the amino acids are described as *essential*, meaning that they cannot be made by conversion from other amino acids. In that loose sense, they may be called vitamins. At least *isoleucine*, *leucine*, *lysine*, *methionine*, *phenylalanine*, *threonine*, *tryptophan*, and *valine* (see Chap. 3.2) must be considered essential in the human. For other

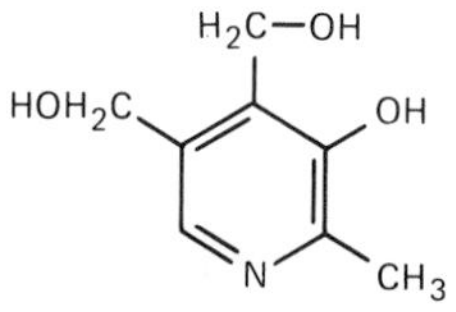

Pyridoxine
(Vitamin B_6)

Fig. 14-24. Pyridoxine (vitamin B_6).

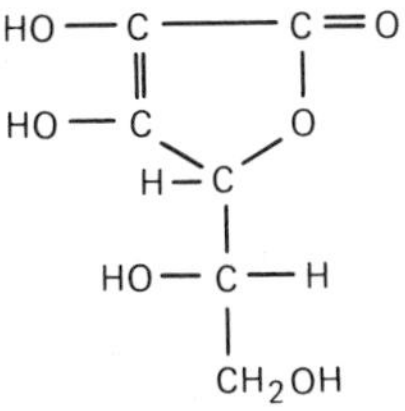

Ascorbic acid
(Vitamin C)

Fig. 14-25. Ascorbic acid (vitamin C).

animals, other amino acids are essential, and some of those mentioned may not be. Leucine, isoleucine, serine all have a branched carbon chain backbone, and perhaps the metabolic machinery does not conveniently make branched chains. In threonine, the branch is an oxygen rather than a carbon atom. Phenylalanine and tryptophan are unsaturated ring compounds that are evidently also inconvenient for the cellular machinery to synthesize. These compounds can, however, be used as the starting points for synthesizing a variety of other ring compounds. Lysine has a terminal amino (NH_2) group in its backbone and is therefore a *basic* amino acid. Methionine contains sulfur and is the source of material from which molecules with this atom are synthesized. S—S bonds are of great importance in determining the tertiary structure of protein.

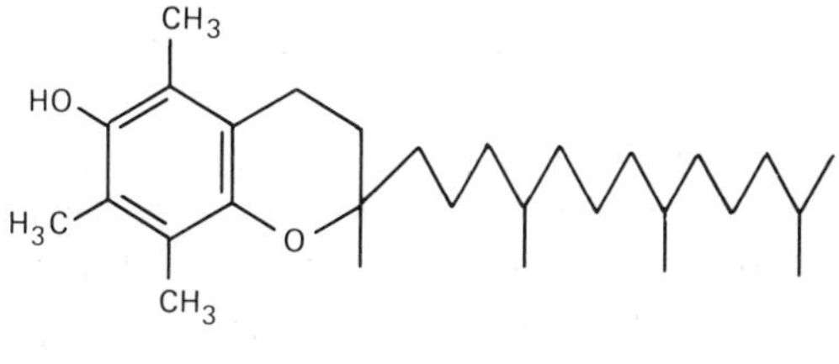

α-Tocopherol
(Vitamin E)

Fig. 14-26. α-Tocopherol (vitamin E).

In order to complete the tabulation of substances essential to our life, we must mention various minerals. Iodine is part of thyroxine, the hormone of the thyroid gland. Iron (Fe) is an essential part of the hemoglobin and the cytochrome. Zinc (Zn) and copper (Cu) apparently also make complexes with enzymes. Fluorine (F) has a special usefulness in bone formation and possibly in other cellular processes. Magnesium (Mg) and calcium (Ca) help hold cell membranes together. Sodium (Na), potassium (K), chlorine (Cl) have been discussed previously. And it is worth repeating that water is an essential nutritional substance.

Section 14: *Questions, Problems, and Projects*

1. What is energy?
2. Name four functions of the body that require energy.
3. What is metabolism?
4. What is the relation between metabolic rate and heat production?
5. Why is the heat production increased during exercise?
6. What effect does the sympathetic nervous system have on metabolic rate?
7. What is the effect of increasing temperature on the metabolic rate? Conversely, what is the effect on heat output of increasing the rate of metabolism?
8. Heat production increases after any meal, but protein increases metabolism substantially more than does fat or carbohydrates. What is this effect called?
9. What is the specific dynamic action of foods?
10. How must the following factors be controlled to provide a basal condition: exercise, mental activity, air temperature, meals, body temperature?
11. What is the primary factor determining differences in basal metabolic rate from one individual to the next? Therefore such measurement is most useful for testing the functioning of what organ?
12. What is a calorimeter? How would it be used for measuring human heat production?
13. When 1 g carbohydrate is metabolized, 4.1 kcal of heat is produced. When 1 liter O_2 is used for the oxidation of carbohydrate, 5.05 cal of heat is produced. How much carbohydrate is burned when 1 liter of O_2 is used?
14. One g of fat yields 9.3 kcal, but 1 liter O_2 used to burn fat produces 4.7 cal. How much fat is burned when 1 liter O_2 is used for the burning? The corresponding figures for protein are 4.1 and 4.6 kcal. Make the corresponding calculation for protein.
15. Calculate the volume of a cube 2 cm on a side. Compare with the volume of a cube 2.5 cm on a side. What is the ratio of the volumes? Calculate the surface area of each cube. What is the ratio of the surface areas of the cubes? What is the surface-to-volume ratio (S/V) in each instance?
16. Heat is dissipated according to the surface area available. Considering that a constant temperature is maintained, do you think that the steady-state heat production of different individuals is proportional to the volumes or the surface areas?

17. There are 686,000 cal/mole of glucose. What weight of glucose does that figure represent? Thirty-eight moles of ATP are produced in the reaction ADP + Ⓟ ⟶ ATP during the oxidation of the mole of glucose. At 7000 cal/mole, how much energy (of the terminal ~Ⓟ bond) is represented in that amount of ATP? Where is the rest of the energy? Calculate the efficiency of the energy conversion.
18. How do fats and amino acids enter into the metabolic pool that generates energy–rich phosphate?
19. Write the equation showing the origin of adenosine triphosphate.
20. What serves as a reserve supply of phosphate bond energy in many cells?
21. What is the consequence of accomplishment of work faster than oxidative metabolism can supply the energy?
22. What is oxygen debt?
23. What is a kilocalorie (kcal)?
24. Before it gets to the site of utilization for cellular processes, most of the energy in the foodstuffs has been lost. What has happened to it?
25. If a person remains essentially immobile for one hour, and during that time uses stored food as a source of energy, what has happened to the energy contained in the food? If the cellular processes for which energy has been used are proceeding in exactly the same fashion at the end of the hour as before, how could you determine how much energy was contained in the food used?
26. If the protein in the diet comes exclusively from vegetables, nearly 50 percent more is required than if the protein comes from animal sources. Explain.
27. What is an unsaturated fat?
28. Explain why strenuous exercise requires more energy than is needed during rest.
29. If a basal metabolic level requires about 80 kcal/hr, what would be the daily rate of heat production?
30. Vitamin A is important for epithelial structure generally, and it has a special role in the eye. What is that role?
31. A vitamin deficiency will generally have multiple effects on an individual. Why?
32. Pantothenic acid is utilized in the synthesis of coenzyme A. What purpose does coenzyme A serve?
33. If you have a severe injury, would you avoid vitamin C? Why?
34. Would you favor excluding vitamin E from the diet? Why?
35. What general function does vitamin D serve? Vitamin K?
36. What general functions are served by Na, K, and other ions of the extracellular fluid?
37. What is the consequence of tissue destruction and/or infection on the body temperature?
38. What does it mean to set the hypothalamic thermostat higher?

39. If the hypothalamic thermostat is set higher than normal, would you predict that the skin would feel cold or warm when the rising body temperature was still below the temperature determined by the hypothalamic thermostat?
40. If the body temperature is 39°C and rising (the normal being 37.5°C), would you expect the mechanism for heat conservation or heat dissipation to be overly active? If the condition is due to a high setting of the hypothalamic thermostat, would you expect sensations of warmth or cold?
41. Does a chill mean that you are decreasing or increasing your temperature?
42. When the patient's temperature begins to fall during the course of a disease, which is more dominant: (a) the heat-dissipating or (b) the heat-conserving process? As his temperature falls, are his sensations of warmth or of cold?
43. Suggest some positive effects of a fever. What is the adverse effect of excessive fever?
44. What is the effect on the metabolic rate of decrease in temperature?
45. What would you expect to be the effect on temperature-regulating mechanisms when cool blood is circulated through the hypothalamus?
46. Glycolysis involves the breakdown of glucose to pyruvic acid. Does this process require oxygen?
47. What is this compound:

$$CH_3-\overset{\overset{\displaystyle O}{\|}}{C}-COOH?$$

48. In what part of the metabolic scheme is most of the energy of foodstuffs made available?
49. What is a tricarboxylic acid?
50. Carbon dioxide is split off in the course of the tricarboxylic acid cycle. What other important event occurs during this cycle?
51. Where is the energy that is contained in the glucose molecule?
52. The energy in the glucose molecule passes in small steps finally to what molecules?
53. When hydrogen is stripped from the substrate molecules, where does it go?
54. Oxidative metabolism provides 18 times as much energy as does the glycolytic cycle. Where does this energy appear?
55. Explain: in glycolysis only enough energy is released to produce two molecules of ATP per molecule of glucose metabolized, but 34 molecules of ATP are produced during the stage of oxidation of water.
56. If 28 liters of O_2 are consumed during 2 hours, what is the heat production, assuming 4.825 kcal/liter of O_2? If the surface area is 2.3 m^2, what is the metabolic rate? If the normal rate for a person of the same age, weight, and height is 42 kcal/m^2/hr, what is the metabolic rate expressed as a percent different from normal?

57. What tissue makes the largest contribution to the heat production of the body?
58. What is heat?
59. Distinguish between radiation and conduction as methods of heat loss.
60. How do convection currents help heat loss by conduction?
61. The hand placed in hot water becomes red; the hand placed in cold water is blanched. Explain why.
62. Evaporation of 1 g H_2O requires about 0.5 kcal. Where does this heat come from if the evaporation is at the body surface? How much evaporation would be required to remove the average heat of basal metabolism in a 70-kg man whose surface area is 2 m^2 and whose heat production is 50 kcal/m^2/hr. Under what circumstances could the heat be removed by this means?
63. How does sweating assist in heat dissipation from the body?
64. Explain how moving air helps keep you cool.
65. Why is wet clothing an inadequate heat insulator?
66. Describe the effect on body temperature of changing the relation between heat production and heat loss.
67. What is the effect of strenuous exercise on body temperature? Explain.
68. What specific changes occur in the skin to raise the body temperature?
69. How does vasoconstriction in the skin vessels reduce heat dissipation?
70. How does sympathetic stimulation help increase the body temperature?
71. In adaptation to cold, production of thyroid hormone increases. Explain.
72. Of what benefit is pilo-erection in heat conservation?
73. Several body changes may help reduce body temperature. Explain the mechanisms: vasodilation of skin vessels, decreased sympathetic stimulation, decreased metabolism, increased sweating and panting.
74. Panting involves an increase in respiratory rate. Why then does it not result in alkalosis?
75. The hypothalamus is a thermostat. Explain.

15

EXCRETION AND OSMOTIC CONTROL

15.1 The Kidney

The excretory system

In the chemistry of the body, the products of some reactions serve as starting materials for others, but certain molecules, especially terminal products, may accumulate in excess and disturb the smooth flow of chemical reactions if they are not removed. These materials are called *excretory* substances, and they are produced in different amounts by all cells.

Excretion refers to the process of getting rid of excretory substances. Excretion may occur through the salivary glands, through the skin (especially the glands of the skin), through the lungs, and through the wall of the digestive tract. Out of the lungs pass CO_2 and other products of metabolism (such as acetone, in cases of diabetes) or substances absorbed from food (e.g., garlic). The salivary glands excrete various salts, particularly NaCl. The sweat glands produce not only water but also salts, amino acids, vitamins, and other materials.

The kidneys are the main excretory organs. They are particularly important in ridding the body of the products of incomplete metabolism of protein. Normally carbohydrates and fats are completely burned to carbon dioxide and water. Proteins, on the other hand, are not completely oxidized, for a residue of nitrogenous product remains in the form of *urea*. In addition to excreting urea, the kidneys get rid of surplus salts, such as potassium chloride; they also help maintain a constant *p*H and a constant osmotic pressure of the blood.

Water is taken into the body, along with, or separately from, food and the air we breathe. Water is also produced during the oxidation of foods. Water is lost from the body in the breath, in feces, in urine, and from the external body surface, both as droplets of perspiration and by evaporation. The kidneys help to control the loss of water so that in spite of fluctuations in amounts of water lost or gained through these avenues, the amount of water in the blood and tissues remains constant. When various salts are added to or taken from the body fluids, great changes in osmotic pressure of blood and of tissue fluids may be threatened. The kidneys are the instruments that control the concentration of the salts, and hence the osmotic pressure of the blood and tissue fluids.

Substances like the cellulose walls of plant cells cannot be broken down by our digestive enzymes and are eliminated via the bowel without having been involved in metabolic processes. During elimination, excretory products, such as bile pigments from the liver, may also be removed.

General structure of the kidneys

Unlike most of the abdominal viscera, the kidneys are covered by the peritoneum only on the side that faces into the abdominal cavity. They are therefore *retroperitoneal* in location (Fig. 15-1).

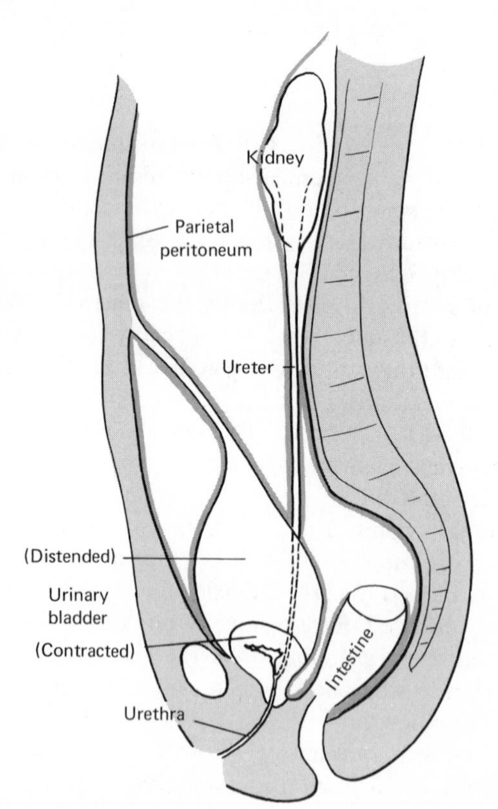

Fig. 15-1. Sagittal section through the body to show relations of urinary system.

The formation of urine begins with the filtration of blood through capillaries arising from the renal arteries. Together the two kidneys receive about one-quarter of the total body blood volume every minute. This enormous blood flow is provided by the renal arteries that branch off one from each side of the dorsal aorta. The smaller subdivisions of each renal artery control the passage of blood to the million or so functional units of each kidney—the nephrons—where purification of the blood takes place. Blood comes to each nephron via an *afferent arteriole* that terminates in a small knot of capillaries, the *glomerulus*. The million glomeruli provide an enormous surface for filtration of the blood. The nephron is a tiny convoluted tube. Its inverted blind end, *Bowman's capsule*, surrounds the glomerulus (Fig. 15-2).

As is true of capillaries generally, some of the water and small molecules dissolved in the blood entering a glomerulus are filtered through the glomerular capillary walls. The remainder leaves the glomerulus by way of the *efferent arteriole*, from which it

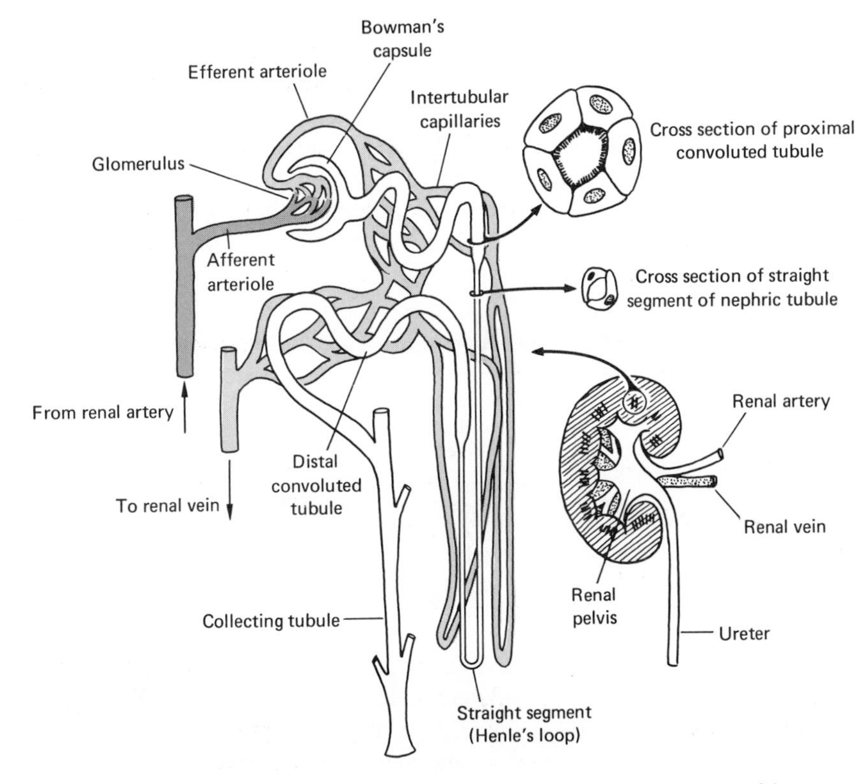

Fig. 15-2. Anatomy of the nephron. The glomeruli are in the cortical layer of the kidney, while the straight segments of the tubules loop down into the medullary part. Note thickness of cells in the convoluted versus the straight segments. Note also long loops of capillaries accompanying the straight segment. This arrangement presumably speeds up absorption of water from the straight segment.

gains access to another set of capillaries surrounding the convoluted nephric tubules. By movement of substances between the nephron and the capillaries of the tubules, the composition of the blood filtrate moving along in the nephron is changed. From the nephrons, the fluid moves to *collecting tubules* and into the *ureter* leading to the *urinary bladder*, where the urine is temporarily stored.

The means by which the nephron produces urine has been studied directly in some vertebrates. In frogs, for example, individual living kidney units may be easily seen under a microscope, and samples of fluid taken by means of tiny pipets may be analyzed by micromethods. Micropunctures have also been made in mammals, but direct study is usually impracticable, and most knowledge of kidney function in animals like ourselves has been obtained by analyses of the concentration of various substances in urine and in blood.

Composition of urine

In the face of large fluctuations in solute and water intake, the need exists to keep constant the osmotic pressure, *p*H, and relative concentrations of the various constituents in the blood, in order that the body cells may be bathed in a medium that allows optimal functioning of the cells. The kidney has a large responsibility in maintaining this constant condition. If the composition of the food fluctuates widely, so, too, may the output in the urine show wide variations, reflecting the need to excrete undesirable materials. Table 15-1 shows the composition of a relatively concentrated sample of urine in comparison with blood plasma. The osmotic pressure

Table 15-1 Composition of urine and blood plasma

	Concentration in Plasma (mM/1)	Concentration in Urine (mM/1)
Urea	7	750
Na^+	147	260
Cl^-	111	250
K^+	4	77

of the urine in this instance is about five times that of the blood. In another situation, the materials in the urine may be diluted to about half that in the blood, and the relative proportions of the various solutes may also differ greatly.

The urea is an organic product of the metabolism of proteins (see Chaps. 3.2 and 14.2). The inorganic salts reflect the excess of salt present in the food, but any particular ion of these salts may just as well have come from the food or from the cells or body fluids, for there is a continual interchange of the various body constituents with the fluids that are taken into the body. It is essential that particular concentrations be maintained within cells and in the blood, the lymph, and the cerebrospinal fluid.

Within certain limits, it is the kidney that determines the amounts of various small ions and molecules that will be retained in these fluids. If the intake of water is relatively excessive, the urine will be dilute and large in volume. If the intake of inorganic salts is low, or if salts are lost in perspiration, then there will be only low concentrations of these salts in the urine.

Filtration of blood through the kidneys

When blood flows through the capillaries of the kidney, a fraction of it is forced out of the glomeruli and into Bowman's capsule of the nephric tubules, while the rest travels on into the capillary network that surrounds the tubules. About 80 percent of the blood continues on, in the blood vessels, and only about one-fifth of the blood plasma coming to the glomeruli enters the nephron.

As the filtered plasma moves onward within the tubules, most of it moves out through the tubule walls, into the interstitial spaces, and back into the capillaries that surround the tubules. Therefore the volume of blood leaving the kidney in the renal veins, which receive the blood from those capillaries, is only slightly less than the volume flow arriving via the renal arteries. The small amount of blood plasma that does not return to the circulation is very important, however. It contributes to the volume of urine, formed at about 1 ml/min under average conditions.

Renal clearance: A special instance of the dilution principle

By the action of the kidneys, a substance present at low concentration in the blood plasma may appear at a high concentration in a small volume of urine.

Let us ask the question, "What would be the volume of the urine sample if it were diluted until the concentration of the material in the urine were the same as its concentration in the blood plasma?" This is, of course, the converse of the question we really want to answer: "Considering the concentration of the material in the blood plasma, what volume of blood would contain the amount of the substance that is found in the urine sample?" Or since the material was, in effect, removed from or *cleared* out of the blood, "What is the virtual volume of blood plasma that was cleared of the material during the time the urine sample was formed?"

We can answer this question if we can place in the bloodstream a substance that will be filtered by the glomerulus and not be otherwise acted on by the kidney, but will remain in the glomerular filtrate. *Inulin*,* a plant polysaccharide having a molecular weight of about 5000, is such a substance. If inulin is present in the bloodstream, it will appear in the glomerular filtrate at the same concentration as it is in the blood plasma, but it will be more concentrated in the urine. We may then apply the *dilution principle* to discover what volume of plasma would have yielded the

* *Inulin* is not to be confused with *insulin*, with which it has no connection whatever.

amount of the inulin found in the urine. The inulin cannot be taken by mouth if it is to serve as a clearance marker, for it would be hydrolyzed into its constituent sugar molecules by the digestive enzymes. There is no amylase in the bloodstream, and therefore the molecules remain intact if injected into the circulation.

It is usual to make measurements in terms of the volume of blood that must be filtered to provide the amount of the substance found in a sample of urine produced in one minute. Of course, it is inconvenient to collect urine at one-minute intervals, except by the use of a catheter. Instead, the volume per minute—that is, the *minute volume*—may be calculated from the volume of a sample of urine passed after a known period of time has elapsed since a previous collection.

The volume of blood that is cleared of a substance during one minute is called the *clearance* for the substance. In the instance of inulin, that volume is also the glomerular filtration rate, or the *minute volume* of flow of plasma through the glomeruli. Let us translate this discussion into a simple formula, applying the *dilution principle* (Fig. 15-3).

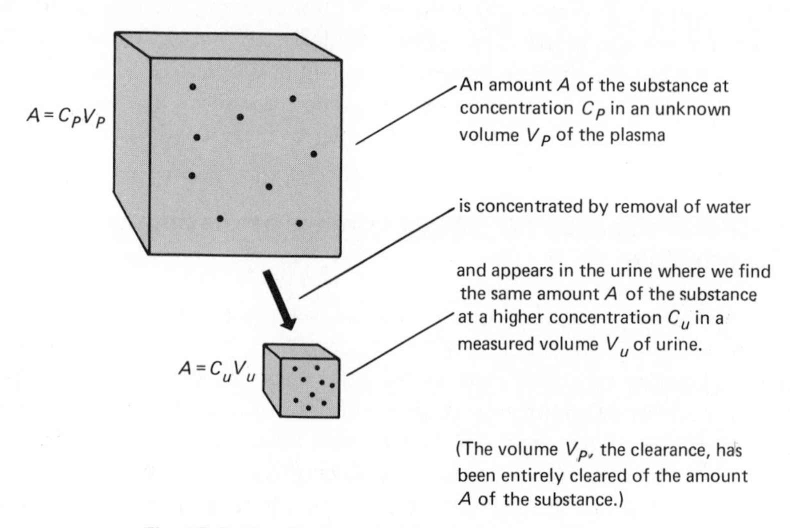

Fig. 15-3. The dilution principle applied to renal clearance.

We consider that the amount of a substance in a certain large volume of plasma is the same as the *amount* in a particular small volume of urine. The *concentration* of the substance in the large volume is lower than the concentration in the small volume. In fact, the volumes and the concentration are inversely related because, by definition, $C = a/V$, and $a = CV = C_2V_2$, and

$$\frac{V_1}{V_2} = \frac{C_2}{C_1}$$

In this instance, C_1 equals the original concentration of the material in the blood plasma, V_1 equals the volume of blood plasma that was cleared of the substance during a particular time, and V_2 equals the volume of urine flowing during that same time, and in which the material is found at concentration C_2. We want to know V_1. We know that

$$V_1 = \frac{V_2C_2}{C_1}$$

or writing it another way,

$$V_p \frac{V_uC_u}{C_p}$$

where the subscripts indicate urine (u) or plasma (p). V_p equals the renal clearance for the substance in question.

Glomerular filtration: Inulin clearance

In a particular instance, we might make the following measurements for inulin.

$$\text{Concentration in plasma} = C_p = 10 \text{ mg/100 ml}$$

(i.e., 10 mg inulin per 100 ml of plasma)

$$\text{Rate of urine flow} = V_u = 15 \text{ ml/10 min}$$

$$\text{Concentration in urine} = C_u = 800 \text{ mg/100 ml}$$

Substituting these values into the formula, we find:

$$V_p = \frac{800 \text{ mg/100 ml} \times 15 \text{ ml/10 min}}{10 \text{ mg/100 ml}} 120 \text{ ml/min}$$

According to this calculation, the renal clearance for inulin is 120 ml/min, and we know, therefore, that the glomerular filtration rate is 120 ml/min., and 120 ml of blood filtrate passing into the nephron every minute amounts to 172.8 liters per day. Thus the entire blood volume is filtered about 34 times per day.

Renal blood flow

Inasmuch as only one-fifth of the renal plasma flow passes through the glomeruli and into the renal tubule, the inulin that is cleared out of the blood is not actually all removed from the blood during a single pass of the blood through the renal circulation. It *is* removed from that volume that *is* filtered. There are, however, some substances that are removed entirely from the blood that flows through the kidney. A substance of that sort must not only be filtered through the glomerulus, for that removes only one-fifth of the material, but it must also move from the capillaries into the tubules. It does so by a process of active secretion from the peritubular capillaries into the tubules.

Such a substance, which is entirely cleared from the plasma during one passage of the blood through the kidney, gives us a means to determine the renal blood flow by the clearance method. The example in Fig. 15-4 illustrates this.

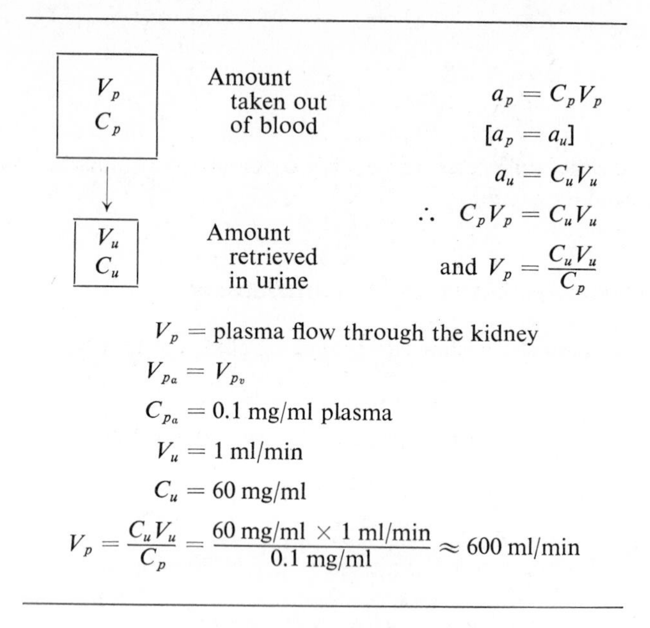

Fig. 15-4. Calculation of renal clearance.

The volume of plasma flow through the kidney is 600 ml/min according to this example. We can find the total blood flow if we know the fraction that the plasma is of the total. This information comes from the *hematocrit*, which is the fraction of cells in a sample of blood. Since $V_{\text{total}} = V_{\text{cells}} + V_{\text{plasma}}$, we may write, by definition,

$$\text{Hct} = \frac{V_{\text{cells}}}{V_{\text{total}}}$$

Then

$$\text{Hct} = \frac{V_{\text{total}} - V_{\text{plasma}}}{V_{\text{total}}}$$

and

$$V_t \, \text{Hct} = V_t - V_p$$

This may be rewritten

$$V_p = V_t - \text{Hct}\, V_T$$
$$V_p = V_t(1 - \text{Hct})$$

and

$$V_t = \frac{V_p}{1 - \text{Hct}}$$

If the hematocrit is 0.45, then

$$V_t = \frac{600}{1 - 0.45} = \frac{600}{0.55} = 1090 \text{ ml}$$

that is, more than 1 liter of blood flows through the kidneys every minute.

Penicillin is an example of a substance that is filtered through the glomeruli and secreted from peritubular capillaries into the renal tubule. The rapidity with which it is removed from the circulation is, of course, the reason this antibiotic must be used in large quantities and frequent doses. Certain dyes are more convenient for studying the blood flow, since they can be more easily measured. Actually, it is not strictly correct to say that these substances are entirely removed from the renal circulation during one pass through the kidney. First of all, some of the material becomes bound to plasma protein, and, second, the blood that comes in contact only with non-nephric structures in the kidney does not become unburdened of its load. When these factors are taken into consideration, the method provides an accurate measure of the "effective" renal blood flow.

Return of fluid to the capillaries from the nephric tubules

Inasmuch as the glomerular fluid filtered in one day is 34 times the entire blood volume of the body, although the daily urine flow is only about 1 to 2 liters, it follows that about 99 percent of the filtered fluid is returned to the bloodstream.

Water that you drink will be absorbed from the digestion tract and distributed throughout the fluid compartments of the body in less than an hour. If a large excess of water is suddenly imbibed, there will certainly be no stimulus to conserve the water, yet only a small fraction of the plasma filtered at each moment is excreted during that time. Specifically, the maximum rate of urine flow is only about one-fifth of the rate of glomerular filtration (Fig. 15-5). That is, during any particular interval of time, say one hour, 80 percent of the water in the plasma that filters into the glomerulus is returned to the blood, even when there is already too much water in the body.

What accounts for the large volume of *obligatory reabsorbtion* of glomerular filtrate? As in any capillary system, there is, in the renal capillary network, a decrease in hydrostatic blood pressure and an increase in colloid osmotic pressure as the blood flows toward the venous end of the capillaries (see Chap. 11-3). This condition might be expected to favor a return of fluid from the interstitial spaces back into the blood. However, the pressures can be experimentally equalized and movement of water *still* occurs from tubule into capillaries. The movement stops when metabolic energy is not available. The prevailing view is that the metabolic machinery pumps Na^+ ions out of the proximal tubules and the Na^+ ions drag along the water mole-

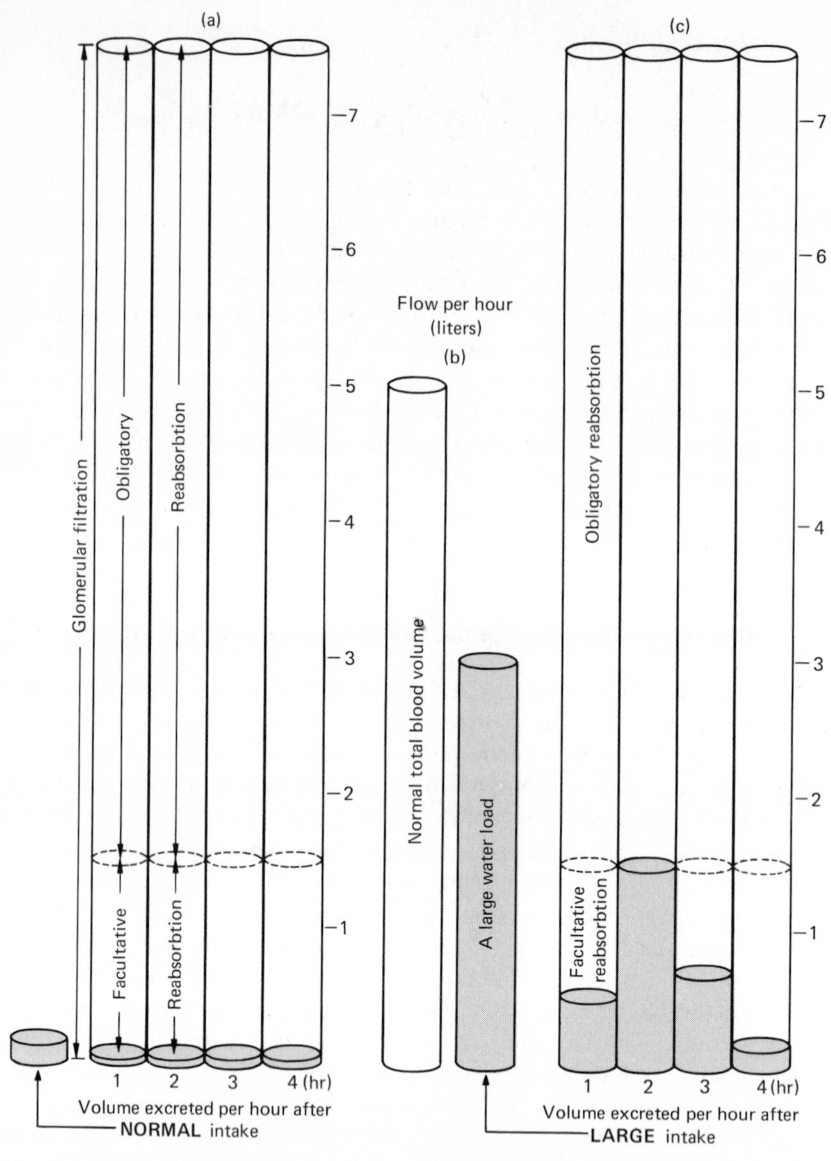

Fig. 15-5. Volumes of water movement involved in kidney function. (a) Total volume of plasma filtered through glomeruli per hour—$7\frac{1}{2}$ liters. Of this, about 4, 5, or 6 liters returns to circulation by obligatory reabsorption. Most of the rest, about $1\frac{1}{2}$ liters, returns by facultative reabsorption and a small volume, about 0.1 liter, is excreted per hour. (b) Total normal blood volume is less than hourly glomerular filtration rate. The large volume of water that has been swallowed is excreted as shown in (c). Volumes of water excreted per hour, after subject drinks large water load shown in (b). Obligatory transfer unchanged, facultative transfer is decreased.

cules. We may say that the water molecules are osmotically obligated to move in the same direction along with the Na^+ ions.

Inasmuch as Na^+ is the principal cation of the blood plasma, the Na^+ pump serves a very useful purpose in preventing the loss of an essential blood constituent.

Glucose transport: Tubular maximum for reabsorption

Glucose is another substance that is actively pumped out of the tubules and back into the circulation. Like any other small molecule, its transport into glomerular filtrate is directly proportional to its concentration in the blood plasma (Fig. 15-6). Note, however, that the proportionality breaks down at concentrations

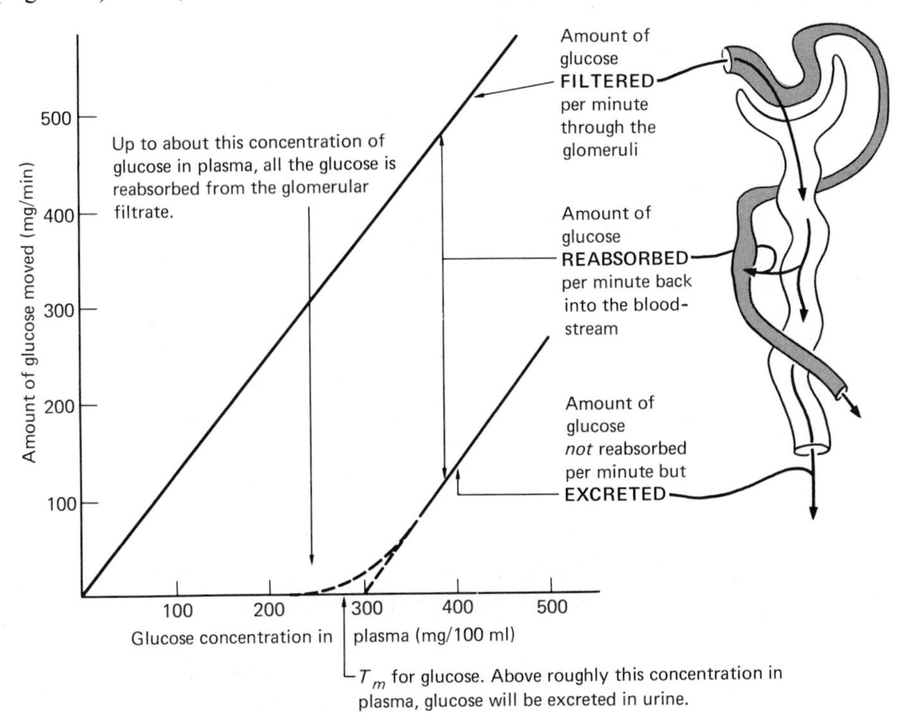

Fig. 15-6. Glucose transport by the kidney. When the concentration of glucose in the blood is above a particular level, as might happen in the case of insulin insufficiency, the kidney tubules cannot return all the glucose back to the bloodstream, and the glucose appears in the urine. The graph shows the amount of glucose moved (ordinate) as a function of the concentration of glucose in the bloodstream. Note that when the blood glucose concentration is 100 mg/100 ml, then in one minute the amount of glucose filtered is the amount in 125 ml of plasma because the filtration rate (calculated from inulin clearance) is 125 ml/min. That volume will of course contain 125 mg glucose, all of which will be reabsorbed. If the glucose concentration in the plasma is as high as 300 mg/100 ml (abscissa), then a small amount may appear in the urine, because the tubules cannot transport the glucose back into the circulation at a faster rate than about 375 mg/min (ordinate).

somewhat above the normal level, around 0.01 mole/liter (200 mg/100 ml). Above this level, the glucose begins to appear in the urine in an amount directly proportional to the concentration of glucose in the plasma. Under these conditions a clearance value for glucose can be measured, although the clearance for glucose is ordinarily zero of course, since no glucose normally appears in the urine.

The following data might be obtained for glucose in an abnormal situation.

Concentration in plasma = C_p = 300 mg/100 ml
Rate of urine flow = V_u = 15 ml/10 min
Concentration in urine = C_u = 9000 mg/100 ml

The clearance for glucose in this instance is

$$V_p = \frac{9000 \text{ mg}/100 \text{ ml} \times 15 \text{ ml}/10 \text{ min}}{300 \text{ mg}/100 \text{ ml}} = 45 \text{ ml/min}$$

The blood glucose level in the example is more than twice normal. Such a condition might be present in *diabetes mellitus*. In this disease there is an inadequate secretion of the hormone insulin, secreted by special cells in the pancreas. Therefore the blood sugar is inadequately utilized and inadequately stored, and hence the blood sugar rises. The excess above a threshold level is not returned to the blood but appears in the urine. There is nothing wrong with the kidney, however; its resorption capacity has simply been exceeded. The tubules have a maximum capacity to return sugar to the circulation; above this critical concentration, the surplus appears in the urine. There is a critical concentration for each of the other substances that the tubules are capable of returning to the circulation.

By measuring the clearance of a particular substance, one may determine whether the substance returns, like glucose, back to the circulation; whether, like some other materials, it is added to the nephron contents by moving through the wall from the capillaries surrounding the nephric tubule, as well as moving through the glomerulus; or whether, like inulin, it moves only through the glomerulus and not through the tubule wall.

Figure 15-7 compares the renal clearance values for three substances and relates the clearance to the filtration, secretion, and reabsorption rates.

Facultative reabsorption of water

A remarkable feature of the kidney is the rather constant rate of glomerular filtration. Filtration rate can be constant only if the blood pressure in the afferent arteriole remains unchanged in spite of the variations in systemic blood pressure that normally occur, for if the renal blood pressure increased along with the systemic pressure, the filtration would correspondingly increase. Very high pressures such as may be reached in strenuous exercise may override the control system, but ordinarily

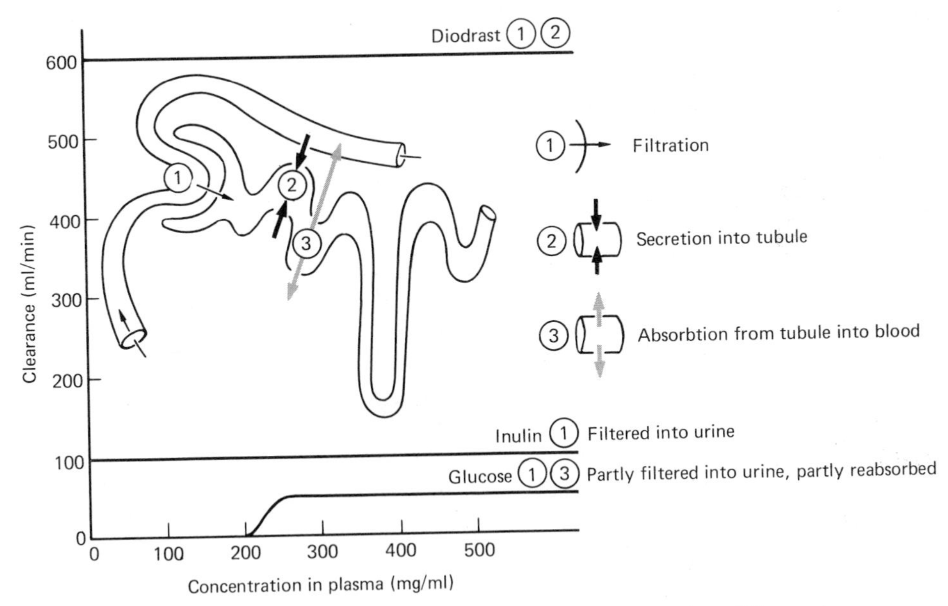

Fig. 15-7. Renal clearance. Ordinate = clearance volume in milliliters of plasma cleared of the substance per minute. Abscissa = concentration of substance in milligrams/milliliter.

1. Clearance volume is the plasma volume that is filtered through glomerulus. Only the inulin in this volume determines the renal clearance for inulin.
2. Clearance volume is apparent plasma—volume from which the material is removed when the glomerular filtrate moves along in the peritubular capillaries. Penicillin and diodrast are present in the glomerular filtrate *and* are actively secreted into the tubules from the plasma flowing in the peritubular capillaries
3. Clearance volume is reduced, because material is returned back to the circulation. Glucose clearance is zero at less than "threshold" concentration —above that level, the clearance for glucose rises quickly to a constant fraction of the inulin clearance, because the amount that is returned to the circulation is a constant amount.

the tone of the afferent arterioles is adjusted to control the blood flow and the pressure in the glomeruli.

Water balance is not, therefore, determined by changes in filtration rate, but by regulation along the nephric tubules.

The osmotic concentration of the urine may vary considerably, although the osmotic pressure of the systemic blood remains remarkably constant. This variability cannot be due to changes in obligatory reabsorption of water, for the osmotic pressure of the renal tubular fluid remains unchanged as the fluid moves along the proximal convoluted part of the renal tubule. Both water and solute move back to the blood through tubule wall in equal proportions, but selective absorption of water occurs in the more distal regions of the nephron.

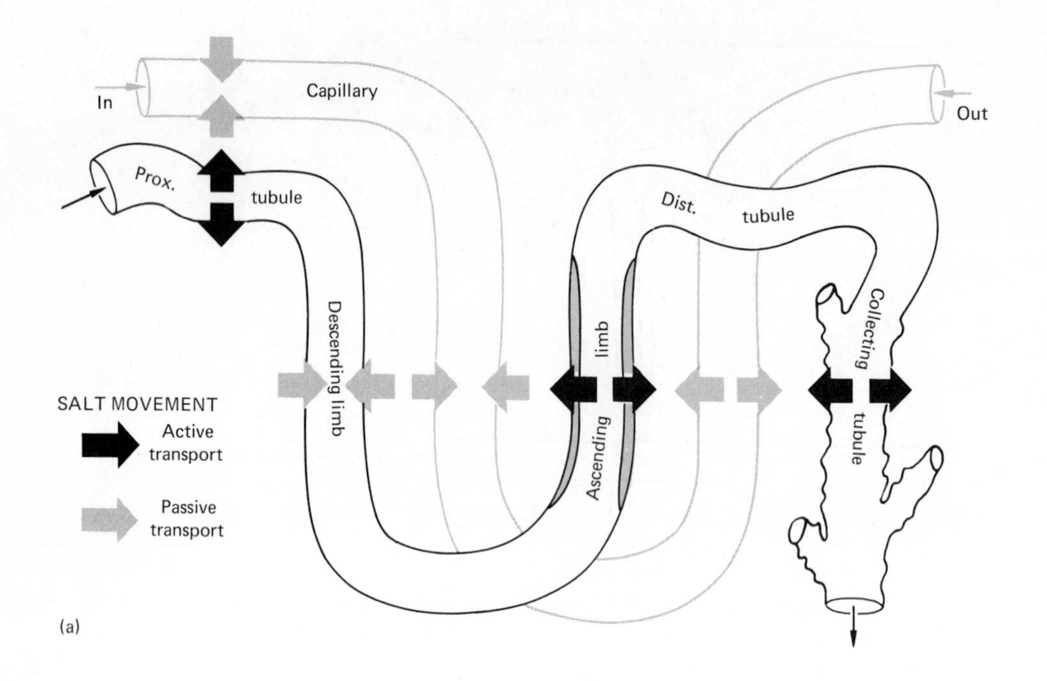
Capillary
In
Out
Prox.
tubule
Dist.
tubule
Descending limb
Ascending limb
Collecting tubule
SALT MOVEMENT
Active transport
Passive transport
(a)

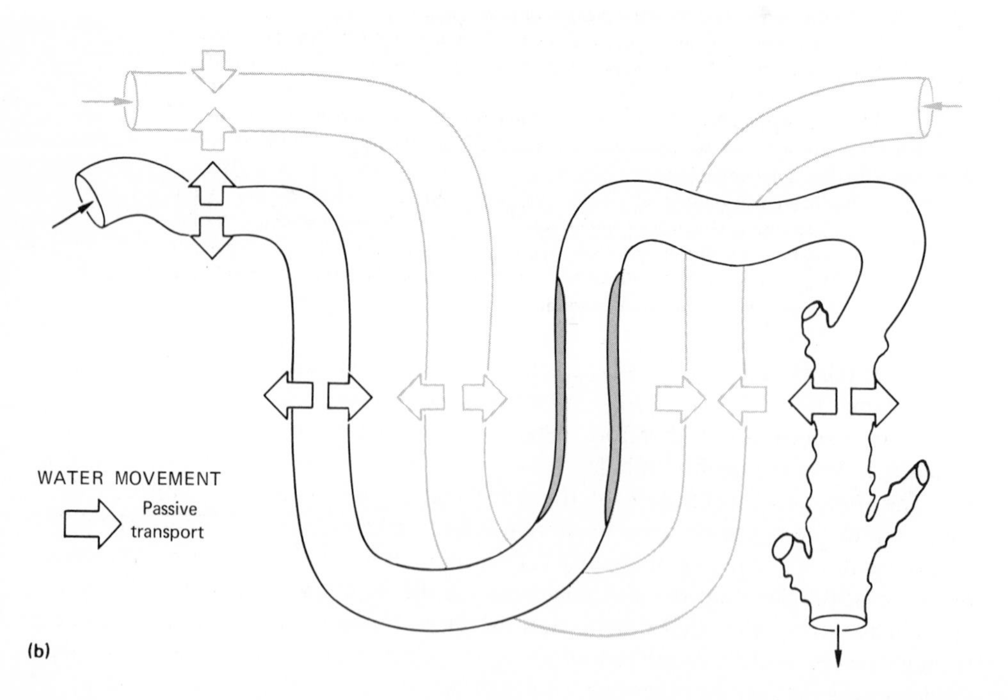
WATER MOVEMENT
Passive transport
(b)

The *facultative reabsorption* of water, concentrating or diluting the solution, apparently occurs mainly in the *collecting tubules* of the nephron. These structures are particularly long in the desert rat, which must conserve all the water it possibly can. Another clue to the mechanism of facultative reabsorption of water is provided in the arrangement of the thin segment (loop of Henle) and the collecting tubules in parallel arrangement alongside the long loops of blood vessels (Fig. 15-8).

Renal counter-current system for water reabsorption control

The arrangement of the capillary and nephric loops constitutes a *counter-current multiplying system.* The meaning of this can be made clear by referring to Fig. 15-9. In the loop, the direction of the current flow in one limb is counter—that is, in the opposite direction—to the other. Consider that the ascending limb of the loop is actively pumping Na^+ out of the tubule but is relatively impermeable to water. The concentration of Na^+ increases in the interstitial spaces, and the concentrated solution will diffuse into the descending limb of the loop. The following picture will give you an idea of how a high concentration of salt solution might develop at the turn of the loop, compared with the inflow and outflow ends of the loop. Although the wall of the outflow side is impermeable to H_2O, the wall of the inflow side of the loop *is* permeable to water and to the solute, and the solution in this part of the tubule, therefore, easily comes into equilibrium with the interstitial fluids (Fig. 15-9).

Let us imagine a beginning to the process when the condition of water impermeability begins to exist in the outflow limb. Just before this moment, there is uniform concentration throughout. That condition is upset as the Na^+ pump ejects, say, 100 Na^+ ions into the interstitial space, and these ions become equally distributed between incoming solution and interstitial fluid. Now the fluid going round the turn of the loop has a higher concentration of Na^+ to present to the pump. Another 100 Na^+ ions ejected by the Na^+ pump will become evenly distributed between the interstitial space and the in-current side of the loop. As the in-current fluid flows around the turn and becomes the out-current, the pump ejects another 100 ions from the higher concentration that is presented to it. The process is repeated as more fluid moves around the turn in the loop. Finally, there exists a considerable concentration gradient between the turn and the in- and out-current ends of the loop, although the gradient between the loop fluid and the interstitial fluid remains small and essentially unchanged at all levels along the loop. The osmotic pressure of the tubular fluid becomes

---

Fig. 15-8. (a) Salt movement. *Heavy arrows* show region of *active transport* of Na^+. *Thin arrows* show passive movement of Na^+. (b) Water movement. Arrows show direction of movement of water. Water and salts move out of proximal tubule into the tissue spaces (obligatory water loss). Descending limb of loop of Henle is freely permeable to water and salts, ascending limb is impermeable to water. Therefore, tissue fluids gain salt at the medullary region of the kidney.

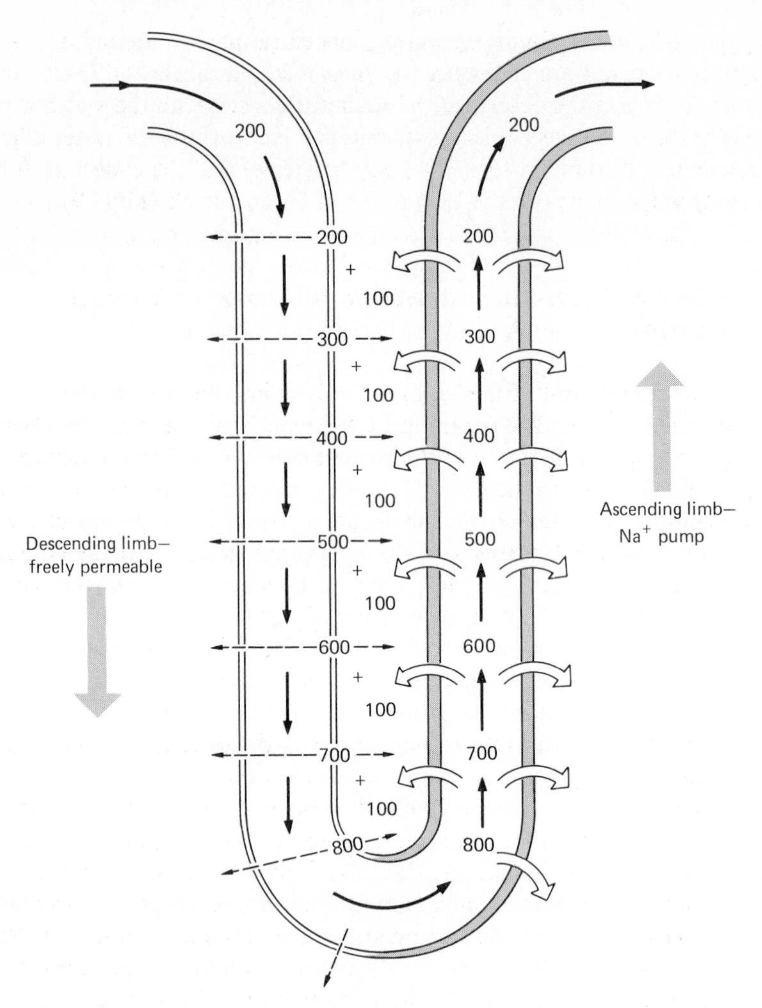

Fig. 15-9. Arrows show direction of flow of renal tubular fluid. ↓ means toward the turn in the loop; → indicates around the turn of the loop, and ↑ is away from the turn of the loop. The epithelium of the outflow limb is relatively impermeable to H_2O, but pumps out Na^+ into the interstitial spaces. Numbers indicate concentration of Na^+. See text for explanation.

progressively greater as the fluid moves toward the turn in the loop, while the outflow becomes progressively lower in osmotic pressure as it flows away from the turn.

Now that we have seen how a concentration gradient along the loop may be established, let us make use of the gradient to do renal work. Alongside the loop runs a collecting tubule that is exceedingly permeable to water. More accurately, the permeability of the wall of the collecting tubule to water is greatly increased when antidiuretic hormone (ADH) is present. The renal fluid that enters the collecting tubule is approximately isotonic to blood (or it may even be hypotonic—that is, more dilute than blood). But as it flows alongside the nephric loop, the osmotic gradient becomes increasingly great as the fluid in the collecting tube approaches the turn of the loop of the nephron. Thus the fluid leaving along the collecting tubule loses more and more of its water, and the urine becomes a concentrated solution compared to the plasma filtrate that was its beginning.

The capillary loops that run parallel to the nephric loops and collecting tubules are subject to the same influences as the interstitial fluid. Blood flow in this region is essential to carry away, into the general circulation, the water and/or the salts that may be poured into the interstitial fluid by the nephron. By removing these materials, the circulation maintains appropriate concentration gradients. If the blood flow or the salt concentration changes, new steady-state gradients will be established.

In order to realize the significance of the capillary loop, consider what would be the consequence of a straight rather than a looped pathway for the blood. The concentration gradient along the nephric tubule could not be sustained because blood of normal osmotic concentration would be flowing nearby at all points along the tubule. Normally, the gradient along the tubule can be maintained because the blood in the capillary loop has the same kind of concentration gradient as is present in the nephron loop: Blood at the turn of the loop has a higher salt concentration than the blood at the incurrent and outflow ends of the capillary.

Endocrine control of water balance: Antidiuretic hormone

By the time the glomerular filtrate has reached the collecting tubule, 80 percent of the fluid has normally been returned to the circulation, and most of the remaining volume will move into the interstitial spaces because of the high osmotic pressure of fluid in these spaces compared to the relatively dilute fluid in the collecting tubules. The movement of the water depends on the concentration difference between these two components, plus the permeability of the tubular wall to water. Water permeability of the tubules is quite variable and remarkably well adjusted to the requirements of the body to conserve or get rid of water. This remarkable homeostatic system involves *osmoreceptors* in the hypothalamus that control secretion of a hormone from the posterior lobe of the pituitary gland. The one-fifth of the water transfer that is adjustable or *facultative* depends on the action of a pituitary hormone. This hormone increases the permeability of the collecting tubules and allows water to return from the ducts back into the capillaries. The posterior pituitary lobe that secretes the hormone is a neurosecretory organ. It consists of nerve endings whose cell bodies are in the hypothalamus of the brain. These nerve endings are distended with secretory products that, at appropriate times, are poured into the capillaries with which this region is so richly supplied. Endocrine secretions from the posterior pituitary seem to have three main effects: a vasopressor material causes a rise in blood pressure, probably by acting on the arterioles; oxytocin stimulates contraction of muscle of the pregnant uterus; and *antidiuretic hormone* reduces diuresis, or free flow of urine, by increasing the return of fluid from the nephric tubules into the circulation.

By increasing or decreasing the amount of water returned to the blood, the kidney acts as a regulator of the blood osmotic pressure, which must be kept at a constant level if cellular functions are to be carried out properly. This constancy is maintained in the face of large quantities of water or of salt poured into the digestive tract at various times. During the summer we lose large quantities of water in perspiration. The osmotic pressure of the blood increases slightly as a result of this loss. The increase in osmotic pressure stimulates the neurosecretory cells of the hypothalamus and posterior pituitary to secrete more antidiuretic hormone into the circulation. Arriving at the kidney, this hormone hastens the return of water from the glomerular filtrate to the blood. Urine flow is reduced, and the osmotic pressure of the blood is decreased to the normal level. In colder weather, when heat must be conserved, evaporation of sweat is reduced, and ingested water dilutes the blood slightly. The lowered blood osmotic pressure provides a lessened stimulus to the neurosecretory cells, and the output of antidiuretic hormone declines. The return of glomerular filtrate to the blood is slowed down, and the surplus water is passed out in the urine. When the posterior pituitary neurosecretory system is not functioning, as may happen in diabetes insipidus, urine flow may rise to 30 liters per day. This condition is, of course, accompanied by great thirst, since much drinking is necessary to restore the lost fluid.

Section 15: *Questions, Problems, and Projects*

1. Draw a diagram of a kidney tubule and label the following structures: afferent arteriole, efferent arteriole, glomerulus, Bowman's capsule, loop of Henle, proximal convoluted tubule, distal convoluted tubule, collecting tubule.
   (a) Show, by an arrow, where filtration occurs (label *F*).
   (b) Show, by an arrow, where reabsorption of fluid first occurs (label *R*).
   (c) In the figure, show, by arrow heads on the appropriate lines, the direction of active transport of Na^+ (label *A*).
   (d) What direction does water ordinarily go between capillaries and tubules?
2. Indicate whether the quantity in question is more, less, or the same compared to the second. (Diabetic refers to diabetes mellitus.)
   (a) The ratio of the concentration of glucose in glomerular fluid to that in the plasma in normal compared to diabetic individuals.
   (b) Glucose concentration in glomerular filtrate compared to concentration in tubular fluid in normal or diabetic individual.
   (c) Tubular maximum for transport of glucose in normal compared to diabetic individual.

(d) Glucose concentration in bloodstream of normal compared to diabetic individual.

(e) Concentration of glucose in urine of normal compared to diabetic individual.

(f) The volume rate of flow of glomerular filtrate in a normal individual compared to the flow in a diabetic individual.

(g) What is meant by the T_m of the kidney?

3. (a) In the terms of the T_m for glucose, why does glucose appear in the urine of the individual having diabetes mellitus?

(b) The osmotic pressure of the blood in the capillaries is kept high by the active transport of Na out of the tubules. On the basis of that information, how do you explain reabsorption of water from the glomerular filtrate?

(c) If the osmotic pressure of the blood is raised higher than normal, what would you expect to happen to urine flow?

(d) What hormone increases reabsorption of water from the collecting tubules?

(e) Write the equation for the production of bicarbonate ion from carbon dioxide in the bloodstream.

(f) What happens to the concentration of dissolved CO_2 when the pH of the glomerular filtrate falls?

(g) What is the role of carbonic anhydrase in the tubular cells?

(h) Write the equation describing the combination of ammonia with water.

4. What is meant by *renal clearance*? Draw a simple diagram to illustrate the concept of clearance. Indicate on the diagram, with appropriate symbols, volumes of fluid and amounts of substance to be cleared. Write an equation that describes the concentrations of the substance in the plasma in terms of those quantities. Do the same for the same substance in the urine. Derive the equation for the clearance V_p.

(a) Calculate the renal clearance for inulin when the concentration of the inulin in the plasma is 10 mg/100 ml, the rate of urine flow is 15 ml/10 min, and the concentration of the inulin in the urine is 800 mg/100 ml.

(b) From the preceding calculations of inulin clearance, determine the volume of glomerular fluid filtered during 24 hours. If the volume of urine voided is 2 liters per day, what proportion of the glomerular filtrate is reabsorbed?

(c) What is the normal concentration of glucose in the urine, and therefore what is the normal renal clearance for glucose?

(d) If a substance is filtered through the glomeruli and also secreted along the renal tubules, from capillaries into tubules, then all the blood flowing through the kidney is cleared of that substance. From these considerations, calculate the renal blood plasma flow if the concentration of such a substance is present at 2 mg/ml in the blood plasma, at 800 mg/ml in the urine, and the urine flow is 15 ml/10 min.

16

THE SURVIVAL OF THE SPECIES

16.1 The Reproductive System

Sexual reproduction in general

Living things are self-perpetuating. Biologically we exist to reproduce, and all our physiological mechanisms are generally adjusted to help us survive as individuals and to reproduce our species. The fact that some individuals do not reproduce does not in any way contradict this idea.

Some primitive organisms reproduce simply by dividing in two when a certain size is reached, or new individuals may develop from pieces broken off accidentally. In such organisms, the offspring are identical to the parent. In most animals and plants, reproduction occurs sexually by the combination of contributions from each of two individuals of different sex during *fertilization*. Offspring produced in this way develop from the combination of two sex cells, one from each parent, and have some features characteristic of each parent. The two cells, called *gametes*, are rather similar in that the nucleus of each carries a set of hereditary factors, but they differ in other ways. One cell, the egg (*ovum*), carries a great deal of cytoplasm, is relatively immobile, and is produced by the female. The other cell, the *sperm*, has very little cytoplasm, is rather motile, and the individual that produces it is called male. The sex cells are produced in special organs, the *gonads*—known as *ovaries* in the female, *testes* in the male. In some of the lower animals, differences in the gonads and the sex cells are the only certain means of distingushing male from female.

After fertilization of the egg by the sperm, food stored in the cytoplasm provides an energy source for development of the fertilized egg (*zygote*), until the new individual is capable of securing his own food. In many aquatic animals and plants, the combination of egg and sperm occurs in the surrounding watery medium. In land animals, the eggs are not delivered to the outside until after the male, by means of a special *copulatory organ*, has fertilized the eggs by depositing the sperm inside the female genital tract. In these instances of internal fertilization, relatively few eggs are produced, but the chances of fertilization occurring are much greater than in the case of external fertilization.

In mammals, such as ourselves, the zygote remains in a special, watery environment within the *uterus* in the body of the female, where it develops as it absorbs food from the circulatory system of the mother. Thus, after its own meager food supply is exhausted, it grows as a sort of parasite in the body of one parent.

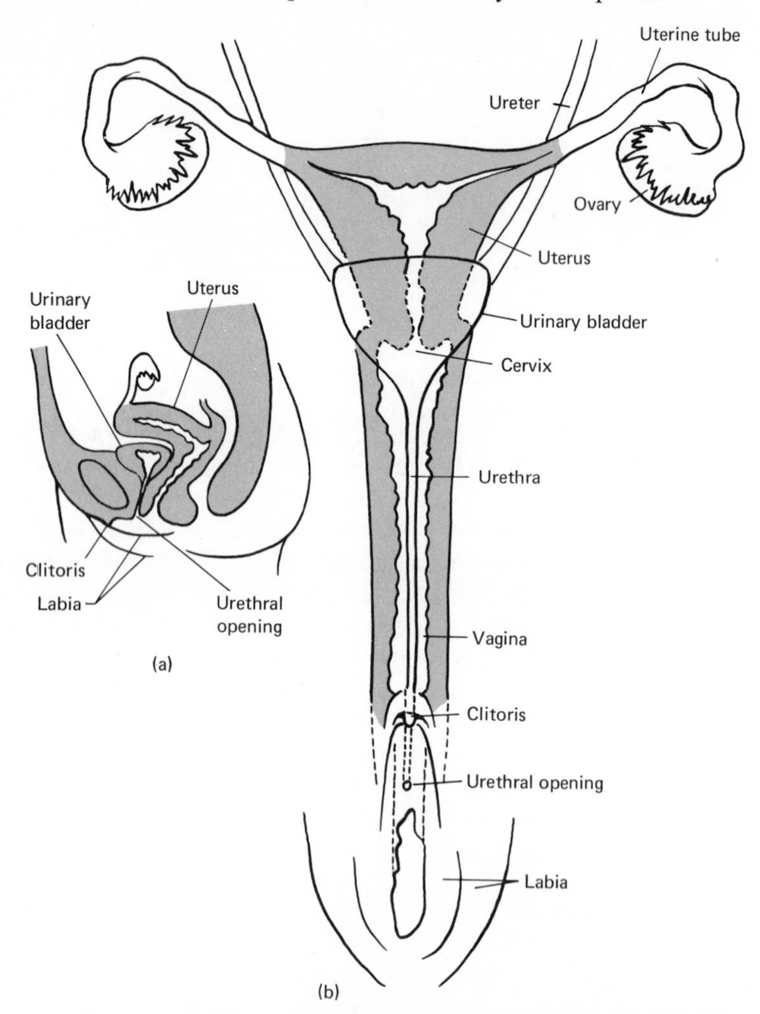

Fig. 16-1. Female reproductive system. (a) Medial sagittal section through female pelvis, showing reproductive organs. (b) Diagrammatic frontal view of reproductive organs. Uterus and vagina are shown in section; lower end of vagina is shown in transparency behind external sex organs.

Sexual differences in humans

The *primary sexual differences* between the male and female are in the *gonads*. The female *ovary* produces eggs; the male *testis* produces sperm. Structures directly related to the functioning of the primary sex organs are called *accessory sex organs*. In the female, the *uterine tube* (*oviduct*), *uterus*, *vagina*, as well as the *external genitals*, are accessory sex organs (see Fig. 16-1). In the male, the *sperm duct* and associated secretory glands, as well as the external genitals, are accessory sex organs (see Fig. 16-2). The distribution of hair, the proportions of the skeleton, the development of the breasts, certain behavioral qualities that we associate with maleness or femaleness—these are examples of *secondary sex characteristics*. Although most individuals can be placed clearly in one category or the other, the secondary characteristics are by no means always reliable criteria of sex, and even confusion in the anatomy of the accessory organs occurs.

The production of sex cells is heralded by *adolescence*, a time of transition from sexual immaturity to sexual maturity. At this time the primary and accessory organs become fully functional, and the secondary sex characteristics develop. These changes occur through the action of hormones secreted from the anterior pituitary as well as from the gonads themselves.

Sexual development in the male

During adolescence in the male there is increased cell division in the *seminiferous tubules* that make up the substance of the testis. In many adult mammals, the production of sperm depends on the amount of light and dark during the changes of the seasons. Cell division in the testis of man is nearly continuous, however. Cells resulting from these divisions become the sperm, and older sperm are moved away along the sperm tubules by the pressure of more recently matured sperm. At the same time, the seminal *vesicles*, the *prostate*, and the *bulbourethral* glands become active. The secretions of these glands dilute the suspension of sperm during ejaculation and provide a suitable hydrogen ion concentration along the urethral canal, which normally is relatively acid from the presence of urine. The sperm, together with the secretions from the accessory glands, constitute the white, viscid fluid called *semen* or *seminal fluid*. Emission of semen is not under voluntary control but involves contraction of smooth muscle associated with these glands. In the absence of specific sexual stimulation, it may occur at more or less regular intervals every few weeks, as nocturnal emissions associated with dreams.

Sexual development in the female

In the female mammal (including the human), each egg or ovum develops within a tiny fluid-filled sphere, or *follicle*, just under the surface of the ovary.

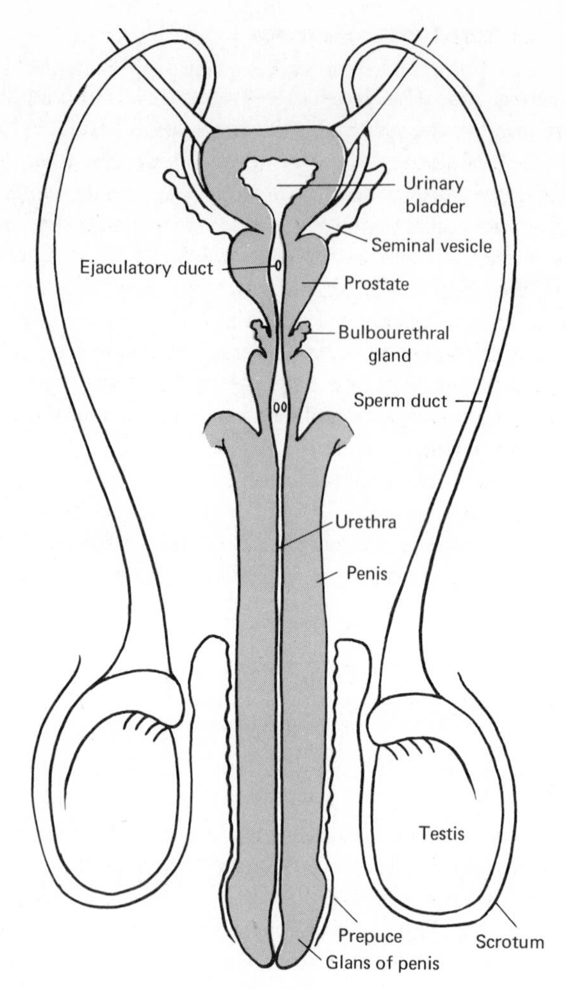

Fig. 16-2. Male reproductive system. Diagrammatic frontal view of male reproductive organs. For medial section of male pelvis see Fig. 16-5.

As the egg grows, the follicle enlarges and pushes out as a small elevation on the surface of the ovary.

All the eggs to be produced during the lifetime of the individual are already present in the ovaries of the girl child. With the onset of adolesence, the eggs begin to mature one at a time. Approximately one lunar month (about 28 days) passes between each *ovulation* as the release of an ovum is called. The expelled egg travels down the uterine tube toward the uterus. For somewhat more than a week prior to each *ovulation*, cell division occurs in the lining of the uterus (*endometrium*). The resulting stratified epithelium and numerous tiny convoluted glands, with their rich blood supply, build up the thickness of the wall. About 2 weeks after ovulation, the thickened wall of the uterus breaks down, and remnants of tissue, together with blood flowing from the broken blood vessels, leave the uterus by way of the cervix, pass into the vagina, and out of the body. This process, *menstruation*, usually occupies less than a week. The beginning of menstruation marks the menstrual period in the human female. The moment of ovulation is less easily determined but is usually about 14 days after onset of mensturation. Some women experience a slight pain in the abdomen at the moment of ovulation. Electrical potential changes can be measured in connection with the event, and a change in basal body temperature can be noted during its occurrence (Fig. 16-3).

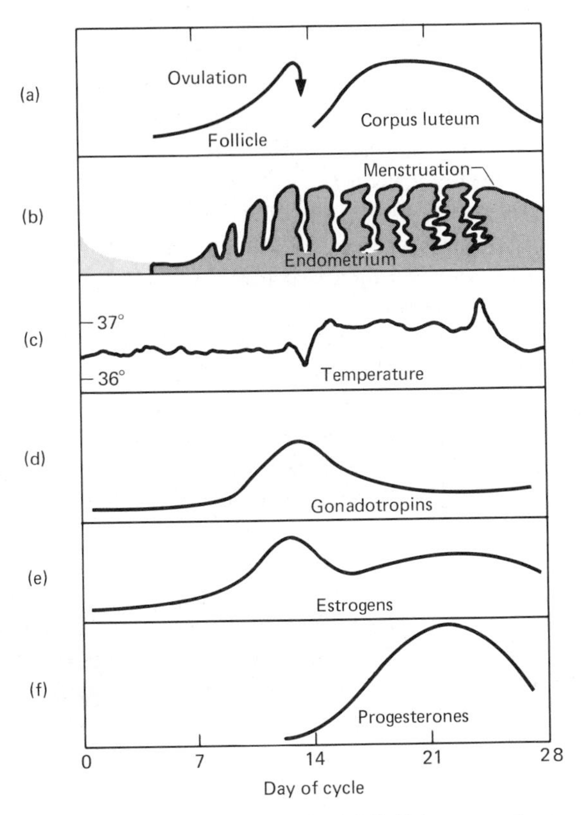

Fig. 16-3. Events of the menstrual cycle. (a) Follicle grows, then regresses after ovulation, at about the fourteenth day of the menstrual cycle. Corpus luteum grows subsequent to ovulation, then regresses when fertilization does not occur. (b) After menstruation, endometrium and its glands grow and develop under influence of estrogen from follicles and of progesterone from the corpus luteum. (c) Body temperature during the menstrual cycle drops then rises at ovulation. A plateau is then maintained until just before menstruation. (d) Gonadotropin secretions from anterior pituitary rise to a peak at ovulation. (e) Estrogen follows the gonadotropin secretion but is maintained thereafter. (f) Progesterone levels increase as the corpus luteum develops.

The union of egg and sperm

Sperm are made available to fertilize the egg during sexual union, *coitus*, when the erect penis (phallus) is thrust into the vagina. During each ejaculation about 3 ml of seminal fluid containing about 100 million sperm per ml are ejected against the cervix, whence they move into the uterus and up into the uterine tubes. Fertilization generally occurs as a sperm encounters an egg moving down this tube.

The ovum moves down the tubes by means of coordinated action of cilia lining the central regions of the folded wall of the tube, while the sperm are moved upward in more peripheral channels, perhaps assisted by cilia beating in the opposite direction. Sperm also move by lashing their tails, but their main motive power probably comes from peristaltic movements of uterus and Fallopian tubes. The sperm may ascend from the cervix all the way up the tubes in less than an hour. The ovum requires about 2 days for its trip down to the uterus. It must be fertilized within less than a day after ovulation if it is to continue development. When one sperm has penetrated the egg, entry of other sperm is prevented, perhaps through the formation of a *fertilization membrane*, as has been demonstrated to occur in some animals. Nevertheless, if the concentration of sperm in the semen is below about 60 million per ml, fertilization may not occur, probably because the low sperm density reduces the chances of a sperm encountering the egg. In addition, the sperm secrete an enzyme that assists the entry of a sperm through the membrane and cell layer surrounding the *unfertilized egg*. A decrease in the number of sperm diminishes the amount of this enzyme. If the sperm or the egg has been released more than a day or two before their encounter, fertilization will not be successful, or, if successful, the fertilized ovum cannot survive. If the egg is not successfully fertilized, it may reach the uterus, where it disintegrates. If fertilization has been successful, the fertilized ovum, called a *zygote*, undergoes cell division and development as it moves down the uterine tube. Arriving in the uterus, it becomes implanted in the soft endometrium, where it develops for approximately the next 9 months. If *implantation* is successful, the expected next menstrual period does not take place. However, a delayed or skipped period does not *necessarily* mean pregnancy. The nervous and hormonal controls of these events are so complex that changes in timing are to be expected.

Hormonal regulation of female sex cycle

Menstruation, ovulation, and associated monthly body changes constitute the female sex cycle. This cycle is controlled primarily by the anterior pituitary gland, which, in turn, is affected by the nervous system via the hypothalamus (see Fig. 16-4). Hormonal changes in relation to other events of the cycle are shown in Figs. 16-3 and 16-4.

In mammals other than man, the nervous system control may be rather specific. For example, in the rabbit, ovulation is stimulated by coitus, as well as by almost any

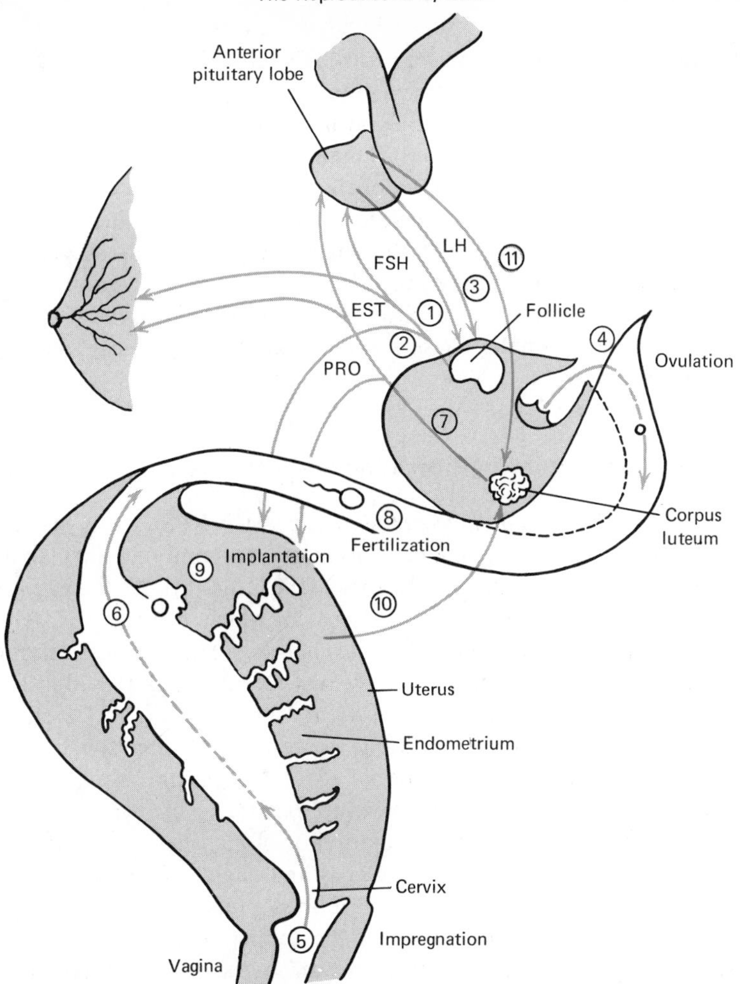

Fig. 16-4. Endocrine control of female reproductive system. (1) FSH from anterior pituitary lobe (AP) stimulates growth of follicles in ovary. (2), (3) The follicles secrete estrogen (EST), which stimulates the differentiation of endometrium and mammary glands. As the level of estrogen builds up in the bloodstream, it tends to inhibit the production of FSH while production of LH by AP increases. (4) LH, together with FSH, triggers ovulation.

(5), (6) If impregnation has occurred, the sperm start their ascent into the uterus and up the oviduct. (7) After ovulation, the remaining scar tissue of the follicle differentiates, under the influence of LH, into the corpus luteum. Progesterone stimulates further development of the uterine endometrium as well as of the mammary glands, and acts back on the pituitary to inhibit production of gonadotropins.

(8), (9) The egg, descending the oviduct, will be fertilized. Continuing down the oviduct, it will implant in the endometrium, and the implantation will be maintained by the progesterone. (10) As development of the embryo proceeds, the placenta becomes itself an endocrine gland, secreting choriogonadotropic hormone (CG) that helps sustain the corpus luteum and ensures continued progesterone production. (11) Luteotropic hormone (LtH) has similar action and also triggers the production of milk from mammary glands after parturition.

If pregnancy does not ensue, stages 8-11 will not occur. In the absence of choriogonadotropin, pituitary LH is insufficient to sustain the corpus luteum, progesterone declines, the endometrium deteriorates, and menstruation ensues.

other strong excitation. The mechanism is through activation of the hypothalamus by means of afferent nerve impulses carried into the central nervous system from the reproductive organs during coitus. The hypothalamus, in turn, causes the pituitary to release hormones that induce ovulation. In many mammals the sexual cycle is an annual affair and is initiated by the influence of changes in the length of daylight. Presumably the susceptibility of the human female cycle to nervous influences arises from alteration in the secretory rhythm of the anterior pituitary cells. A period may be missed, and pseudo-pregnancy may even develop in a very suggestible individual.

The secretions from the anterior pituitary gland influencing the sexual cycle are quite similar in male and female. These *gonadotropins* have, of course, different target organs in each sex.

Follicle stimulating hormone and estrogen

In the female, the *follicle-stimulating hormone* (FSH) acts on the ovarian follicular cells, stimulating enlargement of follicles, secretion of follicular fluid, and growth of the egg contained within the follicle. (Fig. 16-4 (1)). As the egg grows, the surrounding cells of the follicle secrete *estrogen* (2), which, in turn, is largely responsible for the development of the secondary sex characteristics in the maturing female and for the onset of events associated with sexual receptivity in the adult female.

Estrogen means to give rise to *estrus*, or mating behavior. By its effect on the nervous system, estrogen encourages female acceptance of attention by a prepared male. In human beings, a primary hormonal effect on behavior is made less obvious by social conditioning. Many women, however, experience changes in emotional responsiveness corresponding to time of ovulation or of menstruation. A more obvious effect of estrogen is on the lining of the uterus and vagina. Estrogen stimulates the growth of the uterine endometrium, which is thus prepared for the arrival of the fertilized egg. It also favors proliferation of the vaginal epithelium and presumably makes coitus easier by toughening the vaginal wall. In the adolescent female, estrogen is responsible for differential skeletal growth, particularly the relatively wider pelvis compared to that of the male. The wide pelvis provides a sufficiently large passageway for the birth of a child. The development of the breasts is stimulated by estrogen. Size of breast is not a necessary indication of estrogen level, however—measurements of certain movie actresses notwithstanding. A large breast may have much fatty tissue; a small one may have a larger proportion of functional tissue.

Luteinizing hormone and progesterone

The increase in concentration of estrogen in the bloodstream of the female inhibits the secretion of FSH from the pituitary, while it stimulates an increase in the production of *luteinizing hormone* (LH) (3) from the anterior pituitary lobe. LH acting together with FSH precipitates ovulation, which generally occurs in only one follicle at a time in the human. The pressure of fluid inside the follicle rises, perhaps via secretion and the contraction of follicular smooth muscle. When the follicular fluid pressure is sufficiently great, the follicle bursts and the egg is ejected into the funnel of the uterine tube (4). Following ovulation, LH transforms the follicular cells into a knot of yellow tissue, the *corpus luteum* (7). The *corpus luteum* becomes a new endocrine organ, which secretes a hormone, *progesterone*. Progesterone has a special role in preparing the uterus for the pregnancy that might be expected to follow the coitus encouraged by the estrogen.

Progesterone means before gestation (pregnancy). This hormone provides the final preparation of the endometrium for the implantation of the ovum and helps the endometrium to remain intact. Maintenance of the corpora lutea depends on a continuing supply of appropriate gonadotropic hormone. The pituitary's production of gonadotropin is adequate to sustain functional corpora lutea for only a few days. In the absence of adequate support from gonadotropin, the corpora lutea stop secreting progesterone. The endometrium cannot remain intact in the absence of adequate amounts of progesterone in the circulation. As progesterone declines, the uterine lining deteriorates, and menstruation ensues.

If the egg has been fertilized, (8, 9), the cells resulting from the subsequent divisions will be partly those of the embryo itself and partly of membranes that will surround the embryo. The developing tissues will attach to the endometrium and form, together with the maternal tissue, the *placenta* (10). This structure, which permits the developing embryo to receive nourishment from the mother, also functions as an endocrine gland. It secretes *chorionic gonadotropins*, which, like pituitary gonadotropins, stimulate continued life in the corpus luteum, and, hence production of progesterone and maintenance of the intact condition of the endometrium. In addition, it secretes estrogen and progesterone and thus, directly as well as indirectly, contributes to the hormonal milieu necessary for its own survival.

If no ovum is available in the Fallopian tube when coitus occurs, there will be no conception. Suppression of ovulation is, therefore, a very effective method of contraception. Ovulation normally occurs as a result of the concerted action of FSH and LH, present at appropriate concentrations at a critical time. Taking advantage of the feedback principle of hormone action, we realize that if we maintain estrogens and/or progesterones at suitable levels in the bloodstream, the adequate production of FSH and/or LH will be prevented, and ovulation will not occur.

Fortunately, for contraceptive convenience, synthetic estrogen and progesterone are manufactured that, if taken orally at regular intervals, will inhibit ovulation. The empirical combination that seems to function best has 10 to 20 times as much progesterone as estrogen and probably acts, therefore, mainly on the LH production while also keeping a slight rein on the FSH.

A steady suppression of pituitary production of FSH and LH is not the usual circumstance for the female body. We should not be surprised, then, if slow, long–term metabolic changes, whether advantageous or disadvantageous, might occur when the condition is continued for some time.

Hormonal regulation of male sex cycle

In the male, FSH stimulates spermatogenesis in the epithelium of the seminiferous tubules of the testes (Fig. 16-5). LH in the male is known as *interstitial cell-stimulating hormone* (ICSH). The interstitial cells of the testes are located among the connective tissue cells surrounding the seminiferous tubules of that organ. Under

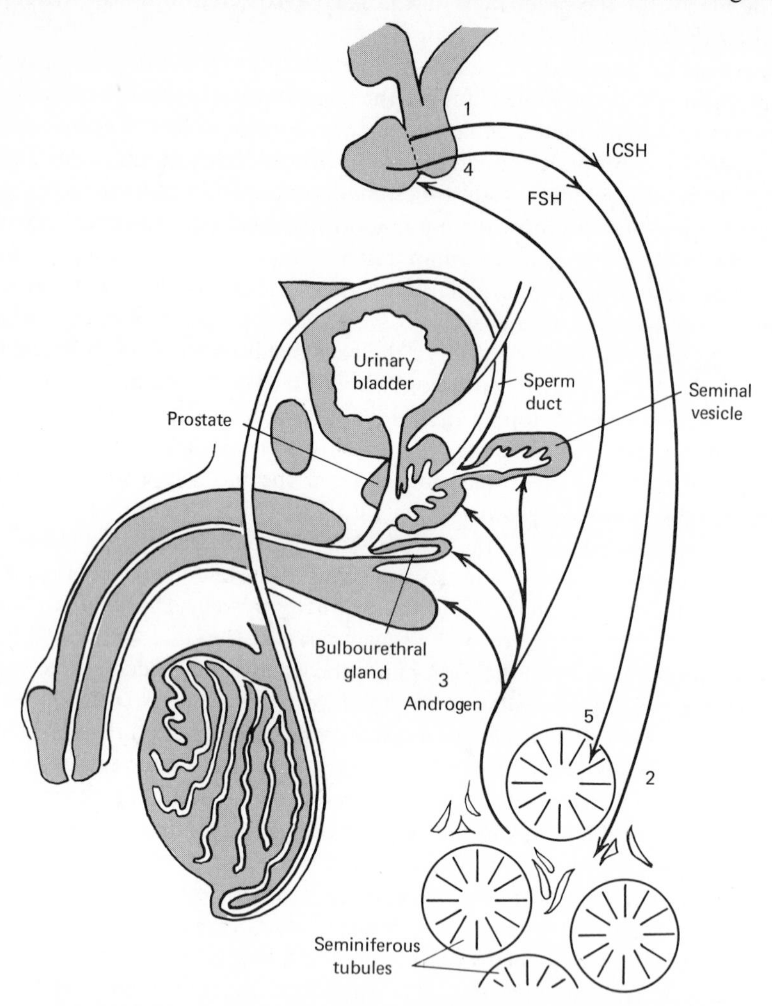

Fig. 16-5. Endocrine control of male reproductive system. (1) Anterior pituitary secretes interstitial cell stimulating hormone (ICSH), corresponding to LH of the female. (2) ICSH stimulates production of androgens by interstitial cells of testes. (3) Androgens (e.g., testosterone) stimulate growth and development of primary and secondary sex characteristics. (4) Anterior pituitary secretes follicle stimulating hormone (FSH). (5) FSH acts on seminiferous tubules of testes to favor maturation of sperm.

the impact of ICSH, they secrete *androgens*, mainly *testosterone*, responsible for the development and maintenance of the secondary sexual characteristics in the male.

Parturition

For reasons not too well understood, pregnancy is terminated at about 9 months in the human. The production of choriogonadotropic hormone presumably declines to a critical level in the aging placenta, which can, therefore, no longer adequately ensure its own survival. During pregnancy, estrogen, together with progesterone from the corpus luteum, as well as growth hormone from the anterior lobe of the pituitary, have stimulated the development of the uterine musculature. Another hormone, *oxytocin*, is secreted from the posterior lobe of the pituitary at the time when pregnancy is about to terminate. Oxytocin stimulates powerful rhythmical contractions of the uterus and thereby assists in expelling the fetus through the dilated cervix and vagina.

During pregnancy, the mammary glands become prepared for milk production through the action of estrogen and progesterone. After the child is born, lactation or milk production is triggered by the hormone *prolactin*, secreted by the anterior pituitary lobe. Since the actual flow of milk is favored by oxytocin, the mother will be prepared to suckle her offspring immediately. Oxytocin is also released as a reflex response to suckling.

Pregnancy and parturition (childbirth) mobilize the full physical and emotional strength of the mother. Among other animals generally, and among physically active people, the birth is accomplished with the mother's full assistance. In modern society parturition is generally associated with hospitalization and anesthetization. In terms of the protection from disease offered, this situation is often desirable and necessary. In another sense, however, it reflects a lack of physical and emotional preparedness by the sophisticated woman for one of the most important events of her life. Like her mate, she has used the conveniences of modern life to diminish the effort to develop her muscles, her mind, and her emotions. In doing so, she is no more derelict than her mate, however, for the sedentary life of the modern male is equally deleterious for his health. In time we will, hopefully, arrange to live according to physiological rules that will give us good health and allow us to enjoy our lives fully, including our sexuality and our reproduction.

16.2 Heredity

Meiosis: A modified mitosis

During the fertilization of an egg, the haploid nucleus of the sperm joins with that of the egg to form the diploid nucleus of the zygote. If a doubling of nuclear material occurred in this fashion in each successive generation, and was maintained during mitosis, the cells would soon become astronomically large. Reduction of chromosome number from diploid to haploid is brought about during a special modification of mitosis called meiosis (Fig. 16-6). Meiosis takes place during *gametogenesis,* the processes involved in the production of the eggs (*oögenesis*) and of sperm (*spermatogenesis*). In the ovary, the *oögonia* (diploid cells from which eggs arise) divide mitotically until they are ready to undergo meiosis. At this time they are called *primary oöcytes.* In the testis, *spermatogonia* about to undergo meiosis are called *primary spermatocytes.* Like mitosis, *meiosis* is a series of movements of the chromosomes that results finally in the splitting off of material that was already generated prior to the visible appearance of separation. Meiosis can be conveniently seen as two major stages: the *first meiotic division,* during which the visible number of chromosomes in each cell is halved but each separately discernible chromosome is actually double, and the *second meiotic division,* during which the double chromosomes are split apart and the resulting daughter chromosomes are allocated to separate cells. These are true haploid cells, which have half the somatic mumber of chromosomes.

First meiotic division

Although the chromosomes behave independently during mitosis, each chromosome sticks to its homologous mate during prophase of meiosis. Prophase of meiosis is prolonged, compared to prophase of mitosis, and is divided into several substages, named according to the appearance of the chromosomes.

The filamentous chromosomes (*leptonema*), each consisting of two chromatids, pair at corresponding points with their homologous mates (*zygonema*). The paired chromosomes then become shorter and thicker as their coils tighten up (*pachynema*). While thus paired, each chromosome is seen to be double. Each chromosome is now revealed as composed of two chromatids lying alongside one another, and separate, except at the centromere in each chromosome. The double-appearing chromosomes then begin to move apart (*diplonema*). Separation of the chromosomes occurs at the centromeres and at the ends of the chromosomes. When an X (*chiasma*) arrangement in chromosomes is seen, the inference is made that prior to the time when the chromosomes become visible as such, exchange of parts has taken place between chromatids of homologous chromosomes. The pieces retain their new attachments, and separation of the chromosomes continues, the chromatids undergoing rotation toward the arrangement shown in Fig. 16-7.

By this time, the chromosomes have become arranged at the *metaphase* equatorial plate. A spindle is formed, and, during *anaphase*, the members of each pair of separating chromosomes move apart, one toward each pole of the dividing cell. The chromosomes thus distribute themselves randomly into the two daughter cells, but the chromosomes are not the same as they were when the meiosis began in the original cell. The example in Fig. 16-7 shows that one chromatid in each of the original chromosomes has lost a part and has gained in exchange a corresponding part of a chromatid from the other homologous chromosome. The original nuclear material has been divided between two cells, and the *telophase* nuclei that result are haploid, since the pair of chromatids in each chromosome is held to the other by a single centromere.

The first meiotic division is now complete. In the testis, the resulting cells, still diploid, are *secondary spermatocytes*. Those in the ovary are *secondary oöcytes* (see Fig. 16-6).

Equational division of meiosis: Spermatogenesis

The secondary gametocytes undergo a final division before they become functional gametes. This final division, called the second meiotic or equational division, is a mitotic division, with the slight difference that the chromatids are already present in *prophase*. The chromatids separate completely as the centromeres divide. The new chromosomes move away from the *metaphase* plate—migrate toward the centrioles—during *anaphase*, and the *telophase* nuclei reorganize while the two daugh-

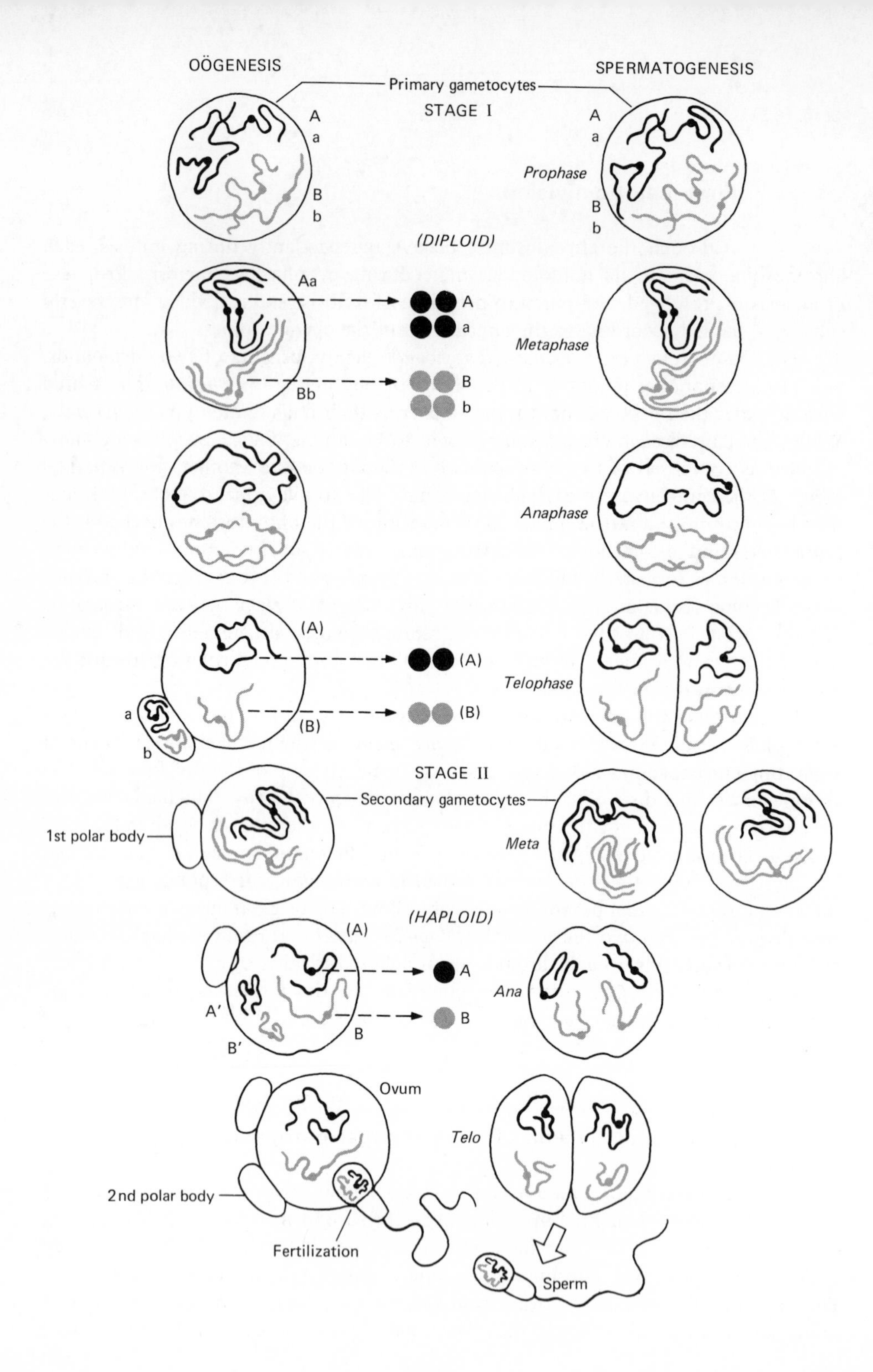
OÖGENESIS
SPERMATOGENESIS
Primary gametocytes
STAGE I
A
a
B
b
Prophase
(DIPLOID)
Aa
Bb
A
a
B
b
Metaphase
Anaphase
(A)
(B)
a
b
Telophase
STAGE II
Secondary gametocytes
1st polar body
Meta
(HAPLOID)
(A)
A
A'
B
B'
Ana
Ovum
Telo
2nd polar body
Fertilization
Sperm

ter cells are pinching off from one another. The four cells that result from division of each primary spermatocyte are morphologically similar. Each cell undergoes a further change, called *maturation*, during which the sperm tail and other apparatus are developed and a fully functional sperm is produced.

Meiosis in ovogenesis

With regard to nuclear events, oögenesis proceeds in the same general manner as spermatogenesis. However, whereas four sperm arise from each spermatogonium, only one egg is produced from each oögonium. The division of the cytoplasm is unequal during the first meiotic division of oögenesis and one set of chromosomes is left with virtually no cytoplasm. This *first polar body* sits on the outside of the human egg, while the egg, a secondary oöcyte, travels down the uterine tube. In the human, only one egg is produced during each reproductive cycle. The unequal cytoplasmic division ensures that adequate food material will be available to sustain this egg during its early development as it moves toward its implantation site in the uterus.

The second meiotic division in the egg is, like the first, an unequal one. This division does not occur in the human until after the sperm has penetrated the egg during the trip down the uterine tube. In other animals it may happen at some other stage of the meiotic cycle. The *second polar body*, like the first, is extruded from the egg but during the second meiotic division. The first polar body may occasionally undergo the second meiotic or equational division, but both polar bodies are fated for deterioration.

The combination of chromosomes at fertilization

In *fertilization*, the diploid number of chromosomes is restored by the union of the haploid egg nucleus with the haploid sperm nucleus. The chromosomes, now double structures, line up independently at the equatorial plate and undergo mitotic splitting. The zygote is on its way to becoming an embryo.

Each member of a pair of homologous chromosomes in a diploid cell can be considered to have come, through successive mitotic divisions, from one or the other

---

Fig. 16-6. Sequence of changes during cell division producing sex cells (meiosis). At left, stages in oogenesis. At right, stages in spermatogenesis. Behavior of two pairs of chromosomes is shown:

1. Black. Solid and dashed are homologous (matched) chromosomes of one pair.
2. Gray. Solid and dashed are homologous (matched) chromosomes of another pair.

Stage I. Homologous chromosomes (each one is double) pair up at metaphase plate, then move apart (each still double) at anaphase, into separate daughter cells (telophase). *Stage II.* The double chromosomes move apart from one another into separate cells, ovum, or sperm. Each primary oöcyte yields only one ovum—the extra chromosomes are discarded in the polar bodies—but each primary spermatocyte yields four sperm.

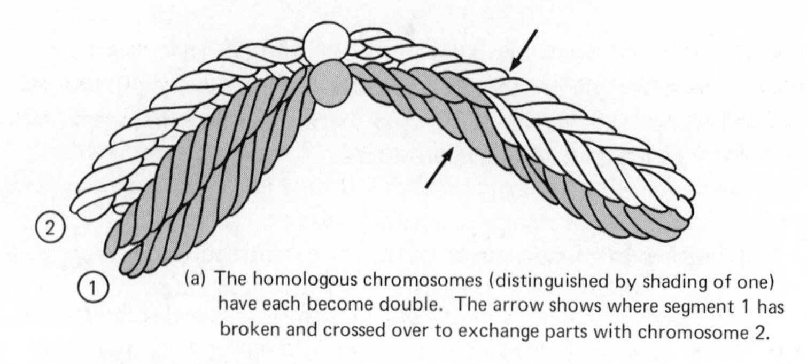

(a) The homologous chromosomes (distinguished by shading of one) have each become double. The arrow shows where segment 1 has broken and crossed over to exchange parts with chromosome 2.

(b) As a result of the crossing over, the chromosomes form a chiasma (x) as they separate.

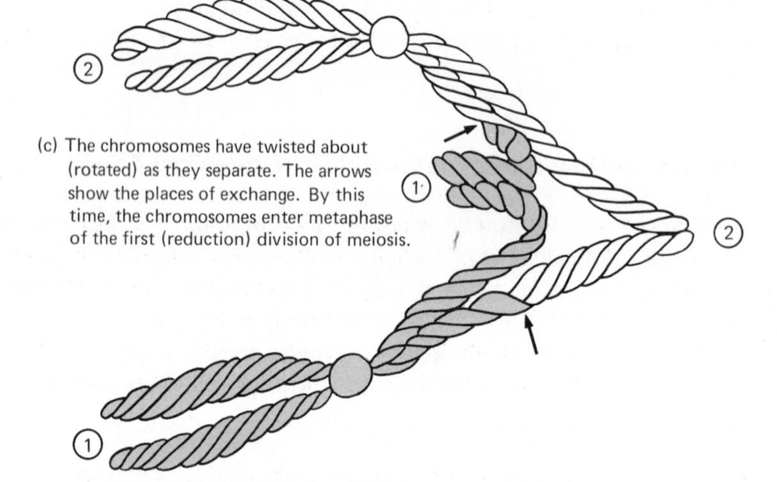

(c) The chromosomes have twisted about (rotated) as they separate. The arrows show the places of exchange. By this time, the chromosomes enter metaphase of the first (reduction) division of meiosis.

Fig. 16-7. Crossed-over appearance of chromosomes during meiosis. (a) The homologous chromosomes (distinguished by shading of one) have each become double. The arrow shows where segment I has broken and crossed over to exchange parts with chromosome 2. (b) As a result of the crossing over, the chromosomes form a chiasma (x) as they separate. (c) The chromosomes have twisted about (rotated) as they separate. The arrows show the places of exchange. By this time, the chromosomes enter metaphase of the first (reduction) division of meiosis.

parent of the individual of which the cell is a part. In gametogenesis, it is entirely a matter of chance which particular chromosomes will move along with certain other ones into the gametes during the equational division, although each gamete must receive one of each different kind of chromosome. Two sets of chromosomes thus randomly assembled are combined during fertilization. The chromosomes in the egg can theoretically be traced back in their origin to the parents of the female bearing the egg; those in the sperm, to the parents of the male contributing the sperm. The zygote is, in this sense, the product of its grandparents' chromosomes.

The chromosomes as bearers of the hereditary characteristics

The *result* of various combinations of chromosomes can best be observed in an adult after cell division and tissue differentiation have provided adequate physiological function, or complete anatomical form. A comparison of certain specific characteristics in the individual with similar characteristics in the individual's parents and offspring allows inferences to be made concerning the distribution of the chromosomes from which the individual developed. Some specific characteristics can be shown to result from a direct biochemical influence, of a very small part of a chromosome corresponding to a particular part of a DNA molecule. In the lower animals and in plants, in which genetic experiments can be carried out more precisely than in man, hundreds of these tiny regions, each responsible for a particular characteristic or for several characteristics, have been precisely localized on the chromosomes. These hereditary factors are the *genes*.

It is a long way from all the anatomical and physiological features that define an adult, back to the primary action of genes of the chromosomes. The occurrence of characteristics in families does, however, follow the same rules that the distribution of chromosomes follows. The linked characteristics can be shown to break up and to associate with another combination in a way that follows the rules of crossing over in the chromosomes. In this way, the arrangement of genes along the chromosomes can be inferred from the recombination of genetic factors in an individual, in his progeny, and in his progenitors. The series of genes in both members of each pair of homologous chromosomes controls the development of the same kind of hereditary qualities. For example, a particular pair of genes (one on each homologous chromosome) might control chemical events leading to the synthesis of eye pigment. If both controlled the production of a brown pigment (melanin) in the cells of the iris of the eye, the individual would have brown eyes. If the genes that would be concerned with synthesis were both incapable of stimulating brown pigment synthesis, the eyes would be of a different color.

Distribution of hereditary factors: Dominant and recessive characteristics

By careful selection, plant and animal breeders are able to obtain individuals that breed true and therefore carry only one kind of genetic factor for a particular characteristic. Such individuals are said to be *homozygous* for that genetic factor. They have the same kind of gene for that characteristic in both homologous chromosomes. If, in a human pedigree, all members have brown eyes, it may be inferred that the members of the family are homozygous for the gene determining brown eyes. If a member of such a family marries a blue-eyed person, the children will all be brown-eyed. The gene determining brown eyes is, therefore, said to be *dominant* over the gene for blue eyes, which in turn is recessive to the gene for brown eyes. In this combination, each child receives from the brown-eyed parent a gene for brown eyes (B) and from the blue-eyed parent, a gene for blue eyes (b). A dominant factor is generally designated by its initial capitalized letter, while its corresponding recessive is designated by the same letter in lower case. The child's *genotype* (or gene combination) with respect to eye color in the present instance is, therefore, designated Bb. This is a *heterozygous genotype*, for it consists of two genes that individually would determine different expressions of a physical feature (*phenotypes*). The *homozygous genotypes* of the parents would be BB for brown eyes and bb for blue.

♀ ○ $Bb \times Bb$	Parents		
Parental genetic factors	B	b	
B	BB $\frac{1}{4}$	Bb $\frac{1}{4}$	Genotypes of offspring
b	Bb $\frac{1}{4}$	bb $\frac{1}{4}$	

Fig. 16-8. Random assortment of genetic factors in cross of heterozygotes. Random assortment of genetic factors into offspring of parents heterozygous for the factor.

A particularly interesting result follows the mating of two heterozygous individuals (Fig. 16-8). Each parent will produce gametes of which one-half will contain the gene B determining brown eye pigment, and one-half will contain b, determining blue eye pigment. The possible zygotic combinations that are produced as a result of random union of gametes are $\frac{1}{4}$ BB, $\frac{1}{2}$ Bb, $\frac{1}{4}$ bb. The gene for brown eye color being dominant over that for blue, the phenotypic ratio will be $\frac{3}{4}$ brown, $\frac{1}{4}$ blue. This is an average result when a large number of children or a great number of crosses are considered. In any one small family, the ratio may be very different.

Combinations of hereditary factors and combinations of chromosomes

Because the development of specific adult characteristics can be referred back to specific places on the chromosomes, we can illustrate the behavior of hereditary factors by means of a diagram similar to those used to show the allocation of chromosomes to sperm and eggs (Fig. 16-9). In the human there are 23 pairs of homologous chromosomes. A gene for blue or brown eye color may be assumed to be carried on each member of one particular pair of chromosomes. The figure represents the behavior of such a pair of chromosomes bearing dominant and recessive genes. The shaded and unshaded areas on the homologous chromosomes in the diagram represent the location of genes considered to affect one and the same genetic characteristic (e.g., brown or blue eye color). The shaded area represents the locus of a dominant gene; the unshaded area, the locus of a recessive gene. Such genes located at corresponding loci on homologous chromosomes are called *alleles*. If the gene at the shaded locus is dominant to its allele at the unshaded locus, then, in this cross, all

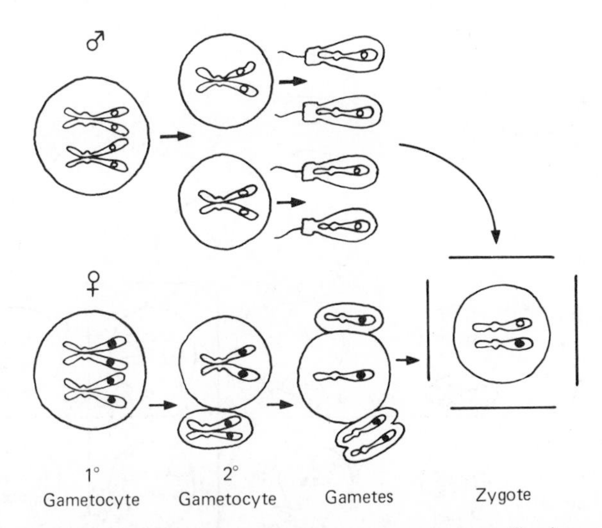

Fig. 16-9. Random assortment of genes on chromosomes. Random assortment of genetic factors visualized in terms of distribution of chromosomes carrying the genes. Shaded and unshaded bands represent allelic factors. Cross of two homozygotes, one recessive, one dominant. In a simpler fashion, this cross may be represented as:

BB × bb genotypes of parents
↘ ↙
Bb genotypes of children

the offspring will be heterozygous in genotype, but in phenotype all will resemble the parent that is homozygous for the dominant gene. When both parents are heterozygous (Fig. 16-10), half the sperm will carry the dominant gene, and the rest will carry the recessive gene. There is, therefore, an equal probability that the egg will be fertilized by a *sperm* carrying the dominant gene (probability equals $\frac{1}{2}$) or the recessive gene (probability, $\frac{1}{2}$). The probability is the same that the available *egg* will carry the dominant gene (probability equals $\frac{1}{2}$) or that it will carry the recessive gene (probability, $\frac{1}{2}$). The proportions of zygotes bearing the several gene combinations may be calculated from the product of the probabilities with which the genes occur in the gametes.

Most genetic crosses are more complicated than this *monohybrid* (or single-factor cross), which involves only one pair of genetic factors (one pair of alleles). The next most complicated situation involves two different pairs of genes or factors. If these different genes are located close together on the same chromosome, then these *linked* factors will be distributed to the offspring in the same manner as the single factors just described (Fig. 16-11). If the factor pairs are on different chromosomes, they will be assorted independently in all possible combinations (Fig. 16-12).

When the factor pairs are on one and the same pair of chromosomes, *but are sufficiently far apart*, they tend to become separated from one another by crossing over during meiosis. Most of the offspring will be similar to one or the other parent

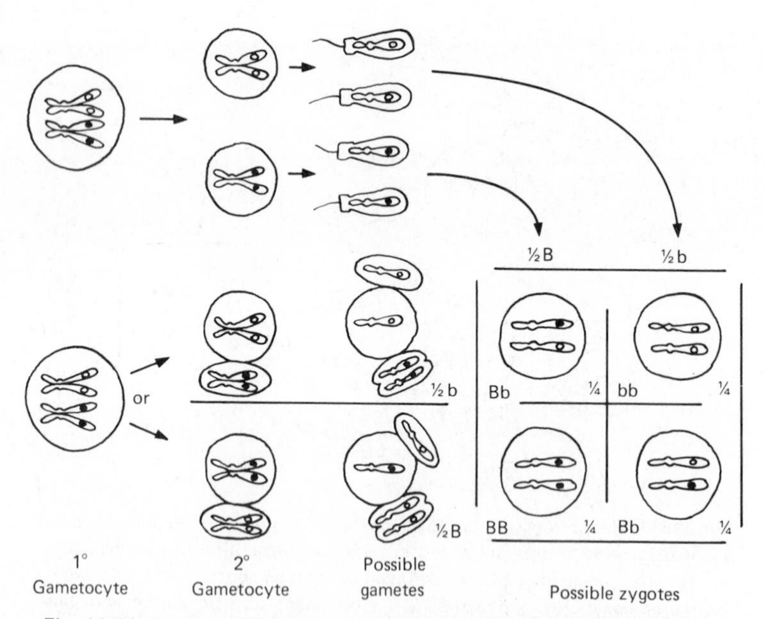

Fig. 16-10. Random assortment of genes in chromosomes in cross of heterozygotes. As in previous diagram: Segregation of genetic factors in gametes of parents and union of these genetic factors in the zygotes. Cross of heterozygotes. (Compare with Fig. 16-9.)

in both genetic characteristics, while a smaller fraction will have qualities of both parents as far as the linked genes are concerned.

♀ ○ *Ba/bA* × *Ba/bA*	Parental genotypes		
Gametes	*Ba*	*bA*	
Ba	*Ba/Ba*	*Ba/bA*	Offspring genotypes
bA	*bA/Ba*	*bA/bA*	

Fig. 16-11. Assortment of genes in offspring if a gene is always carried along with another on the same chromosome. Assortment of two pairs of alleles close together on one pair of chromosomes. The slanted line separates the two chromosomes on which the alleles are located. The different alleles are imagined to be completely linked.

♀ *Bb Aa*	○ *Bb Aa*			
	$BA\ \frac{1}{4}$	$Ba\ \frac{1}{4}$	$bA\ \frac{1}{4}$	$ba\ \frac{1}{4}$
$BA\ \frac{1}{4}$	*BBAA*	*BBAa*	*BbAA*	*BbAa*
$Ba\ \frac{1}{4}$	*BBAa*	*BBaa*	*BbAa*	*Bbaa*
$bA\ \frac{1}{4}$	*BbAA*	*BbAa*	*bbAA*	*bbAa*
$ba\ \frac{1}{4}$	*BbAa*	*Bbaa*	*bbAa*	*bbaa*

Fig. 16-12. Assortment of genes in offspring when two pairs of alleles are on different chromosomes. Independent assortment of two pairs of alleles present on different pairs of chromosomes. Identical genotypes in the offspring are indicated by arrows joining them.

Changes in heredity

A large range of hereditary possibilities is made available during meiosis, but the hereditary pattern becomes fixed at fertilization. Generally speaking, the genes are stable from one generation to the next. However, any gene may be altered and changed under the impact of suitable irradiation (x rays, cosmic rays) or certain chemical substances. For example, the gene for brown eyes may lose its ability to control the synthesis of iris pigment. Such a change could be attributed to a localized alteration in the structure of the deoxyribose nucleic acid-protein complex at the particular region of the chromosome where that gene is located. This general type of gene change (or *mutation*) occurs with considerable regularity—probably in all genes at a low rate, once in several million gametes. Thus, in a single ejaculation of semen, there may be several hundred sperm that carry at least one mutated gene.

In an adequately functioning animal, the physiological processes are all well-balanced and appropriate for adjustment to the environment. When a mutation occurs, one or more of these processes may be interfered with, or changed, so as to function differently. The interdependence of the parts in an animal is such that a disturbance at one place often has ramifications elsewhere. Most mutations, therefore, are probably harmful, although advantageous ones would be less likely to be detected. Occasionally a mutation may result in an alteration that will help an animal to survive more easily in a hostile environment, but such mutations are rare. A more–or–less random change in any part of an already efficient organism is unlikely to improve the functioning of the various physiological processes. It is more likely to upset the balance among them and to diminish the chances of survival.

Section 16: *Questions, Problems, and Projects*

1. The ovum is fertilizable only during 8 to 24 hours and moves only a quarter of the way down the Fallopian tube during that time. Where, therefore, must it be fertilized?
2. Why does only one sperm of the millions available enter the egg?
3. What is the chromosome number of the fertilized ovum after the male and female pronuclei have combined?
4. Prior to vascularization of the placenta, how does the developing embryo receive its nutrition?
5. The corpus luteum remains active for the first 3 to 4 months of pregnancy. Why is it important that this activity be maintained during this period?
6. What is chorionic gonadotropin?
7. List effects of testosterone at puberty.
8. Identify: ovary, ovum, Fallopian tube, uterus.
9. Where is the germinal epithelium of the ovary? Contrast with testes.
10. What effect does the gonadotropin hormone have on the primary ovarian follicles? What is the source of the gonadotropic hormones at puberty?

11. After a follicle has ovulated, the other follicles regress. Why?
12. Ordinarily, how many ova are ovulated each month per female?
13. What is the chromosome number of each ovum?
14. Where does the first polar body come from? The second polar body?
15. What is the female homolog of the male penis?
16. How are the sperm aided in their ascent up the Fallopian tube, since their own propulsive powers are rather meager?
17. What follicles are stimulated by the follicle-stimulating hormone?
18. Lutein refers to the yellow color of the corpus luteum that replaces the ruptured follicle. What is the function of luteinizing hormone in the female? In the male?
19. The corpus luteum secretes progesterone if properly stimulated. What adenohypophyseal hormone provides this stimulus?
20. When does secretion of follicle-stimulating hormone begin? What is its source? What is its effect on the female?
21. Which hormone ripens the follicle to the point of rupture (ovulation)?
22. What is the origin of the corpus luteum, and what hormone stimulates its appearance?
23. Luteotropic hormone is directed toward what target organ and produces what result?
24. When the corpus luteum regresses after about 2 weeks, what secretion does the adenohypophysis begin again to release?
25. Where are estrogens and progesterone mainly elaborated?
26. List the effects of estrogens on the female body.
27. What is the endometrium?
28. Progesterone makes final preparations for the fertilized ovum. In terms of uterus and mammary glands, what is the nature of this action?
29. Describe the hormonal interactions responsible for the monthly female sexual cycle.
30. What is menopause? What is the endocrine basis for it?
31. What is the stimulus to transform follicle cells into corpus luteum?
32. What effect do estrogens have on the endometrium? What additional influence does progesterone have on the same site?
33. With fertilization and implantation, corpus luteum life is prolonged; otherwise it ceases to function. Explain.
34. What happens to the hormone levels when the corpus luteum ceases to function?
35. What is menstruation? What causes it?
36. When, during the female sexual cycle, does ovulation occur?
37. Considering the viability of sperm and ovum, what is the time, in relation to menstruation, during which fertilization may occur?
38. Where is the genetic material in a sperm?

39. How does a sperm move? How fast, under its own power?
40. What are the seminiferous tubules?
41. Arrange these terms in proper developmental sequence and define them: spermatogonium, primary spermatocyte, secondary spermatocyte, spermatid, spermatozoa.
42. What are interstitial cells?
43. What is the chromosome number of the spermatids (and of the spermatozoa) in the human?
44. What is the chromosome number of the primary spermatocyte and of somatic cells?
45. Show by means of a simple diagram the possible kinds of sperm and of eggs, so far as the X, Y chromosomes are concerned. Show, also, the possibilities of combinations to produce male or female offspring.
46. The epididymis is a system of tubules continuous with the seminiferous tubules. What are stored in these tubules?
47. Identify: vas deferens, seminal vesicles, prostate gland, urethra, glans.
48. How does the parasympathetic innervation influence the erectile tissue?
49. What purpose do the bulbo-urethral glands serve?
50. What part of the autonomic nervous system is concerned with the peristalsis involved in ejaculation?
51. What is the composition and average volume of semen?
52. What special function does the prostate fluid serve?
53. How many sperm penetrate and fertilize the ovum? Why then is a male sterile if the number of sperm goes below about 150×10^6 per ejaculate?
54. At what age does follicle-stimulating and luteinizing hormones begin to be secreted? What organ secretes these hormones?
55. What is the effect of follicle-stimulating hormones on the germinal epithelium of the testis?
56. What is the effect of luteinizing hormone in the male?

INDEX

Italicized numbers refer to pages with figures.